Brassinosteroids and Abiotic Stress

Brassinosteroids and Abiotic Stress

Recent Developments on Physiological and Molecular Aspects

Edited by
Anket Sharma
Zhejiang A&F University, Hangzhou, China

Renu Bhardwaj
Chandigarh University, Mohali, Punjab, India

Deepshikha Thakur
Amity University, Noida, Uttar Pradesh, India

Bingsong Zheng
Zhejiang A&F University, Hangzhou, China

Durgesh Kumar Tripathi
Amity University, Noida, Uttar Pradesh, India

WILEY

This edition first published 2026

Registered Office(s)
John Wiley & Sons, Inc., 111 River Street, Hoboken, NJ 07030, USA
John Wiley & Sons Ltd, New Era House, 8 Oldlands Way, Bognor Regis, West Sussex, PO22 9NQ, UK

For details of our global editorial offices, customer services, and more information about Wiley products visit us at www.wiley.com.

The manufacturer's authorized representative according to the EU General Product Safety Regulation is Wiley-VCH GmbH, Boschstr. 12, 69469 Weinheim, Germany, e-mail: Product_Safety@wiley.com.

Wiley also publishes its books in a variety of electronic formats and by print-on-demand. Some content that appears in standard print versions of this book may not be available in other formats.

Library of Congress Cataloging-in-Publication Data:

Names: Sharma, Anket, editor. | Tripathi, Durgesh Kumar, editor.
Title: Brassinosteroids and abiotic stress : recent developments on physiological and molecular aspects / edited by Anket Sharma, Durgesh Kumar Tripathi.
Description: 1. | Hoboken, NJ : John Wiley & Sons Ltd., 2025. | Includes bibliographical references and index.
Identifiers: LCCN 2025029778 | ISBN 9781119883449 (hardback) | ISBN 9781119883456 (adobe pdf) | ISBN 9781119883463 (epub) | ISBN 9781119883944
Subjects: LCSH: Brassinosteroids: | Plants—Effect of stress on
Classification: LCC QK898.B85
LC record available at https://lccn.loc.gov/2025029778

Cover Design: Wiley
Cover Image: © George "TheHouseKeeper" Mateo/Getty Images

Set in STIXTwoText 9.5/12.5 by Lumina Datamatics

Contents

List of contributors *vii*

1 **Roles of Brassinosteroids in Plant Growth and Development: A Revisit** *1*
Vimal Pandey, Amar Kumar Muska, Raju Mallick, Neha Behar, Anjali Srivastava, Arya Srivastava, Ritu Badgujar, Sahil Mehta, and Om Prakash Narayan

2 **Brassinosteroid-mediated Regulation of Plant Biology: An Omics Approach** *35*
Indu Sharma, Ashish Sharma, Ajay Sharma, Vikas Sharma, Pooja Sharma, Sonika Sharma, Rasha S. El-Serafy, Renu Bhardwaj, and Ashutosh Sharma

3 **Regulation of Plant Primary Metabolism by Brassinosteroids Under Abiotic Stress** *63*
Slimane Mokrani and El Hafid Nabti

4 **Regulation of Plant Secondary Metabolism by Brassinosteroids Under Abiotic Stress** *89*
Dhriti Sharma, Bharat Kapoor, and Dhriti Kapoor

5 **Brassinosteroids Improve Resilience Against Nutritional Stress: Dissecting the Responses in Crops** *119*
Allan Klynger da Silva Lobato, Roberto Rivelino do Nascimento Barbosa, Leidy Alves dos Santos, Lorene Bianca Araújo Tadaiesky, and Ynglety Cascaes Pereira

6 **Role of Brassinosteroids on Postharvest Physiology of Plants** *139*
Naveed Mushtaq, Muhammad Ahsan Altaf, and Zhiwei Wang

7 **Brassinosteroid-mediated Physiological and Molecular Mechanisms in Plants Under Heavy Metal Stress** *159*
Deepak Kumar, Mohd Ali, Raman Tikoria, Parkirti, Roohi Sharma, Vikram, Rajesh Kumari Manhas, Harish Changotra, and Puja Ohri

8 **Brassinosteroid-mediated Physiological and Molecular Mechanisms in Plants Under Heat Stress** *187*
Neetu Jagota, Swapnil Singh, Ravneet Kaur, Harleen Kaur, and Ashish Sharma

9 **Brassinosteroids Biology in Plants Under Salt Stress: Physiological and Molecular Aspects** *207*
Muhammad Ahsan Altaf and Riaz Ahmad

10 **Brassinosteroids Biology in Plants Under Pesticide Stress: Physiological and Molecular Aspects** *223*
Palak Bakshi, Pooja Sharma, Sukhmeen Kaur Kohli, Dhriti Kapoor, Nahida Rehman Mir, Nitika Kapoor, and Renu Bhardwaj

11 **Brassinosteroids Biology in Plants Under Drought Stress: Physiological and Molecular Aspects** *245*
Muhammad Ahsan Altaf, Ravinder Kumar, Milan Kumar Lal, Rahul Kumar Tiwari, Hafiza Muniba Din Muhammad, and Riaz Ahmad

12 **Illuminating Insight: Exploring Brassinosteroid Biology in Plants Amidst High Light and UV Exposure** *257*
Sukhmeen Kaur Kohli, Harsimran Kaur, Pooja Sharma, Priyanka Sharma, Manu Sharma, Shalini Dhiman, Nikita Kapoor, and Renu Bhardwaj

13 **Brassinosteroids: Fascinating Roles on Stomatal Mechanism, Physiological, and Anatomical Responses** *291*
Maria Antonia Machado Barbosa and Allan Klynger da Silva Lobato

14 **Regulation of Photosynthesis by Brassinosteroids Under Optimal and Suboptimal Conditions** *309*
Rohit Chhabra, Neetu Jagota, Kompal Gaur, Tavisha Singh, and Ashish Sharma

15 **Brassinosteroids and Its Crosstalk with Other Plant Hormones Under Stressful Environments** *335*
Mohammad Faizan, Ghafoor Unnisa, Aurosha Fayaz, Ira Khan, S. Maqbool Ahmed, and Shamsul Hayat

16 **Role of Brassinosteroids on Antioxidants Regulation Under Abiotic Stresses** *353*
Mohammad Faizan, Anjuman Hussain, Shamsul Hayat, Nayab Ishrat, Aurosha Fayaz, and S. Maqbool Ahmed

17 **Implications of Brassinosteroids Application: A Review** *367*
Saurabh Awasthi, Abhay Chand, Om Prakash Narayan, and Sahil Mehta

Index *383*

List of Contributors

Riaz Ahmad
Department of Horticulture
The University of Agriculture
Dera Ismail Khan, Pakistan

S. Maqbool Ahmed
Department of Botany
School of Sciences
Maulana Azad National Urdu University
Hyderabad, India

Mohd Ali
Department of Zoology
Guru Nanak Dev University
Amritsar, Punjab, India

Muhammad Altaf
Key Laboratory for Quality Regulation of Tropical Horticultural Crops of Hainan Province
School of Breeding and Multiplication (Sanya Institute of Breeding and Multiplication)
Center of Nanfan and High-Efficiency Tropical Agriculture
Hainan University
Sanya, China

Key Laboratory for Quality Regulation of Tropical Horticultural Crops of Hainan Province
School of Tropical Agriculture and Forestry
Hainan University
Danzhou, China

Hainan Yazhou Bay Seed Laboratory
Sanya, China

Saurabh Awasthi
Department of Botany
University of Delhi
New Delhi, India

Ritu Badgujar
Department of Botany
Hansraj College
University of Delhi
New Delhi, India

Palak Bakshi
Department of Botany
School of Biosciences
Baba Ghulam Shah Badshah University
Guru Nanak Dev University
Rajouri, Jammu and Kashmir, India

Maria Antonia Machado Barbosa
Grupo de Estudos em Fitotecnia
Universidade do Estado de Minas Gerais
Ituiutaba, Minas Gerais, Brazil

Roberto Rivelino do Nascimento Barbosa
Núcleo de Pesquisa Vegetal Básica e Aplicada
Universidade Federal Rural da Amazônia
Paragominas, Pará, Brazil

Neha Behar
Department of Biotechnology
D.L.S. P.G. College
Bilaspur, Chhattisgarh, India

Renu Bhardwaj
Plant Stress Biology Lab
Department of Botanical and Environmental Sciences
Guru Nanak Dev University
Amritsar, Punjab, India

University Centre for Research & Development
Chandigarh University
Mohali, Punjab, India

Department of Botany
School of Biosciences
Baba Ghulam Shah Badshah University
Guru Nanak Dev University
Rajouri, Jammu and Kashmir, India

Abhay Chand
Department of Life Sciences
Hansraj College
University of Delhi
New Delhi, India

Harish Changotra
Department of Biochemistry and Molecular Biology
Guru Nanak Dev University
Amritsar, Punjab, India

Rohit Chhabra
Department of Botany and Environment Studies
DAV University
Jalandhar, Punjab, India

Allan Klynger da Silva Lobato
Núcleo de Pesquisa Vegetal Básica e Aplicada
Universidade Federal Rural da Amazônia
Paragominas, Pará, Brazil

Deepak Kumar
Department of Zoology
Guru Nanak Dev University
Amritsar, Punjab, India

Shalini Dhiman
Plant Stress Biology Lab
Department of Botanical and Environmental Sciences
Guru Nanak Dev University
Amritsar, Punjab, India

Leidy Alves dos Santos
Núcleo de Pesquisa Vegetal Básica e Aplicada
Universidade Federal Rural da Amazônia
Paragominas, Pará, Brazil

Rasha S. El-Serafy
Horticulture Department
Faculty of Agriculture
Tanta University
Tanta, Egypt

Mohammad Faizan
Department of Botany
School of Sciences
Maulana Azad National Urdu University
Hyderabad, India

Aurosha Fayaz
Department of Botany
School of Sciences
Maulana Azad National Urdu University
Hyderabad, India

Kompal Gaur
Department of Floriculture and Landscaping
Punjab Agricultural University
Punjab, India

Shamsul Hayat
Department of Botany
Aligarh Muslim University
Aligarh, India

Anjuman Hussain
Department of Botany
Aligarh Muslim University
Aligarh, India

Nayab Ishrat
Department of Biotechnology
Era University
Lucknow, India

Neetu Jagota
Department of Botany and Environment Studies
DAV University
Jalandhar, Punjab, India

Bharat Kapoor
Department of Tourism and Hospitality
Guru Nanak Dev University
Amritsar, Punjab, India

Dhriti Kapoor
Department of Bioengineering and Biosciences
Lovely Professional University
Jalandhar, Punjab, India

Nitika Kapoor
Department of Botany
School of Biosciences
Baba Ghulam Shah Badshah University
Guru Nanak Dev University
Rajouri, Jammu and Kashmir, India

Plant Stress Biology Lab
Department of Botanical and Environmental Sciences
Guru Nanak Dev University
Amritsar, Punjab, India

Harleen Kaur
Department of Botany and Environment Studies
DAV University
Jalandhar, Punjab, India

Harsimran Kaur
Post Graduate Department of Botany
Khalsa College Amritsar
Punjab, India

Ravneet Kaur
Department of Botany and Environment Studies
DAV University
Jalandhar, Punjab, India

Ira Khan
Department of Botany
School of Sciences
Maulana Azad National Urdu University
Hyderabad, India

Sukhmeen Kaur Kohli
Department of Earth and Climate Sciences (ECS) and Biology
Indian Institute of Science Education and Research (IISER) Tirupati
Tirupati, Andhra Pradesh, India

Milan Kumar Lal
Division of Crop Physiology and Biochemistry
ICAR-National Rice Research Institute
Cuttack, Odisha, India

Raju Mallick
School of Comparative Indic Studies and Tribal Sciences
Kalinga Institute of Social Sciences
Bhubaneswar, Odisha, India

Rajesh Kumari Manhas
Department of Microbiology
Guru Nanak Dev University
Amritsar, Punjab, India

Sahil Mehta
Department of Botany
Hansraj College
University of Delhi
New Delhi, India

Slimane Mokrani
University Mustapha Stambouli,
Department of Agronomy
Laboratory of Research on Biological
Systems and Geomantic
Mascara, Algeria

Laboratoire de Maitrise des Energies
Renouvelables
Faculté des Sciences de la Nature et de la
Vie
Bejaia, Algeria

Hafiza Muniba Din Muhammad
Department of Horticulture
Bahauddin Zakariya University
Multan, Pakistan

Naveed Mushtaq
Key Laboratory for Quality Regulation of
Tropical Horticultural Crops of Hainan
Province
School of Breeding and Multiplication
(Sanya Institute of Breeding and
Multiplication)
Center of Nanfan and High-Efficiency
Tropical Agriculture
Hainan University
Sanya, China

Key Laboratory for Quality Regulation of
Tropical Horticultural Crops of Hainan
Province
School of Tropical Agriculture and Forestry
Hainan University
Danzhou, China

Hainan Yazhou Bay Seed Laboratory
Sanya, China

Amar Kumar Muska
School of Comparative Indic Studies and
Tribal Sciences
Kalinga Institute of Social Sciences
Bhubaneswar, Odisha, India

El Hafid Nabti
Laboratoire de Maitrise des Energies
Renouvelables
Faculté des Sciences de la Nature et de la
Vie
University of Mustapha Stambouli
Mascara, Bejaia, Algeria

Om Prakash Narayan
Division of Chemical Biology and
Medicinal Chemistry
University of Texas at Austin
Austin, TX, USA

University of Florida
Gainesville, FL, USA

Puja Ohri
Department of Zoology
Guru Nanak Dev University
Amritsar, Punjab, India

Vimal Pandey
School of Comparative Indic Studies and
Tribal Sciences
Kalinga Institute of Social Sciences
Bhubaneswar, Odisha, India

Parkirti
Department of Zoology
Guru Nanak Dev University
Amritsar, Punjab, India

Ynglety Cascaes Pereira
Núcleo de Pesquisa Vegetal Básica e
Aplicada
Universidade Federal Rural da
Amazônia
Paragominas, Pará, Brazil

Ravinder Kumar
ICAR-Indian Agricultural Research Institute
New Delhi, India

Nahida Rehman Mir
Department of Botany
School of Biosciences
Baba Ghulam Shah Badshah University
Guru Nanak Dev University
Rajouri, Jammu and Kashmir, India

Ajay Sharma
Department of Chemistry, Career Point University
Tikker-Kharwarian, Hamirpur, Himachal Pradesh, IndiaIndia

Ashish Sharma
Department of Botany and Environment Studies
DAV University
Jalandhar, Punjab, India

Ashutosh Sharma
Faculty of Agricultural Sciences
DAV University
Jalandhar, Punjab, India

Dhriti Sharma
Department of Bioengineering and Biosciences
Lovely Professional University
Jalandhar, Punjab, India

Sidharth Govt. College Nadaun
Hamirpur, Himachal Pradesh, India

Indu Sharma
Department of Life Sciences
University Institute of Sciences
Sant Baba Bhag Singh University
Jalandhar, Punjab, India

Manu Sharma
Plant Stress Biology Lab
Department of Botanical and Environmental Sciences
Guru Nanak Dev University
Amritsar, Punjab, India

Pooja Sharma
Department of Microbiology
DAV University
Sarmastpur
Jalandhar, Punjab, India

Department of Botanical and Environmental Sciences
Guru Nanak Dev University
Amritsar, Punjab, India

Department of Biotechnology
DAV College
Amritsar, Punjab, India

Roohi Sharma
Department of Zoology
Guru Nanak Dev University
Amritsar, Punjab, India

Sonika Sharma
Department of Molecular Biology and Genetic Engineering
Jalandhar, Punjab, India

Vikas Sharma
Faculty of Agricultural Sciences
DAV University
Lovely Professional University
Phagwara, Kapurthala, Punjab, India

Swapnil Singh
Department of Botany and Environment Studies
DAV University
Jalandhar, Punjab, India

Tavisha Singh
Department of Botany and Environment Studies
DAV University
Jalandhar, Punjab, India

Anjali Srivastava
Department of Botany
Hansraj College
University of Delhi
New Delhi, India

Arya Srivastava
Department of Botany
Hansraj College
University of Delhi
Delhi, India

Lorene Bianca Araújo Tadaiesky
Núcleo de Pesquisa Vegetal Básica e Aplicada
Universidade Federal Rural da Amazônia
Paragominas, Pará, Brazil

Raman Tikoria
Department of Zoology
Guru Nanak Dev University
Amritsar, Punjab, India

Rahul Kumar Tiwari
Division of Crop Protection, Indian Institute of Sugarcane Research
Raibareli Road, P.O. Dilkusha, Lucknow, India

Gafoor Unnisa
Department of Botany
School of Sciences
Maulana Azad National Urdu University
Hyderabad, India

Vikram
Department of Zoology
Guru Nanak Dev University
Amritsar, Punjab, India

Zhiwei Wang
Key Laboratory for Quality Regulation of Tropical Horticultural Crops of Hainan Province
School of Breeding and Multiplication (Sanya Institute of Breeding and Multiplication)
Center of Nanfan and High-Efficiency Tropical Agriculture
Hainan University
Sanya, China

Key Laboratory for Quality Regulation of Tropical Horticultural Crops of Hainan Province
School of Tropical Agriculture and Forestry
Hainan University
Danzhou, China

Hainan Yazhou Bay Seed Laboratory
Sanya, China

1

Roles of Brassinosteroids in Plant Growth and Development: A Revisit

Vimal Pandey[1], *Amar Kumar Muska*[1], *Raju Mallick*[1], *Neha Behar*[2], *Anjali Srivastava*[3], *Arya Srivastava*[3], *Ritu Badgujar*[3], *Sahil Mehta*[3], *and Om Prakash Narayan*[4*]

[1]*School of Comparative Indic Studies and Tribal Sciences, Kalinga Institute of Social Sciences, Bhubaneswar, Odisha, India*
[2]*Department of Biotechnology, D.L.S. P.G. College, Bilaspur, Chhattisgarh, India*
[3]*Department of Botany, Hansraj College, University of Delhi, Delhi, India*
[4]*Division of Chemical Biology and Medicinal Chemistry, University of Texas at Austin, Texas, USA*

*Corresponding author: om.tufts@gmail.com

1.1 Introduction

In recent years, the world has witnessed an alarming increase in the challenges faced by plants in their struggle to thrive and flourish, ultimately affecting agricultural productivity (Shahzad et al. 2021; Sharma et al. 2021). The escalating impact of abiotic and biotic stressors has cast a shadow over global crop production, inciting growing concerns within the agricultural and environmental communities (Sharma et al. 2020). Due to their sessile nature, plants are unable to avoid these indications from their surroundings. Plants have evolved a complex, multidimensional defense system in response to the increasing stresses of environmental stressors (dos Santos et al. 2022). This system is a complex web of physiological and molecular alterations that are coordinated by a wide range of phytohormones to strengthen plants against the growing assault of adversity (EL Sabagh et al. 2022). To address these concerns and unlock new avenues for enhancing plant resilience, researchers have identified a new group of phytohormones known as Brassinosteroids (BRs), a class of polyhydroxylated steroidal phytohormones (Peres et al. 2019; Zhang et al. 2023). BRs, whose extraordinary qualities have generated great curiosity and enthusiasm among scientists, agriculturalists, and environmentalists alike, are at the vanguard of this intricate molecular symphony. Therefore, systematic studies on BRs are pivotal to understanding the effect of concurrent abiotic and biotic stress conditions on crop productivity.

With groundbreaking work by Mitchell and others, the discovery of BRs about 50 years ago marked a turning point in the area of plant biology and established an extensive investigation of this extraordinary plant hormone (Mitchell et al. 1970). These organic extracts from *Brassica napus* pollen have the amazing capacity to encourage stem elongation and

Brassinosteroids and Abiotic Stress: Recent Developments on Physiological and Molecular Aspects,
First Edition. Edited by Anket Sharma, et al.
Published 2026 by John Wiley & Sons Ltd.

cell division in plants. The discovery that Brassinolide (BL) is the substance that gives rise to similar physiological effects as brassins created was a major turning point in the research on BRs (Grove et al. 1979). Shortly after, the first formulations of BL and related compounds, identified in several plant species with similar functions, were published (Thompson et al. 1981; Sakakibara and Mori 1982). Furthermore, structurally and functionally analogous to BL are coined as the BRs family. BRs – deficient mutants in *A. thaliana* mark its significance in the plant kingdom and are acknowledged as the sixth class of plant hormones. Throughout the BRs discovery, more than 65 compounds have been detected or isolated from more than 100 plant species, from algae (*Chlorella vulgaris, Cystoseira myrica*, and *Hydrodictyon reticulatum*), bryophytes such as liverworts (*Marchantia polymorpha*), and moss (*Physcomitrella patens*), lycophytes (*Selaginella moellendorffii* and *Selaginella uncinata*), pteridophytes (*Equisetum arvense*), gymnosperms to angiosperms. These findings demonstrate the compounds' widespread distribution throughout the plant kingdom (Bajguz 2011; Yokota et al. 2017). The most naturally occurring and abundant BRs are BL, Castasterone (CS), 6-deoxocastasterone (6-deoxoCS), 28-norcastasterone (28-norCS), and Teasterone (TE) and Typhasterol (TY) in plants (Bajguz and Tretyn 2003) (Figure 1.1).

However, their concentrations drastically vary within the tissues as well as from species to species (Bajguz and Tretyn 2003). Simultaneously, following the detection and isolation of BRs from different plants, investigations were carried out to confirm their potential advantageous impacts on crops and to ascertain their hormonal roles offering a strong foundation for comprehending their effects on plant growth and development (Cohen and Meudt 1983; Sasse 1985), including BL's function in pollen tube germination and growth.

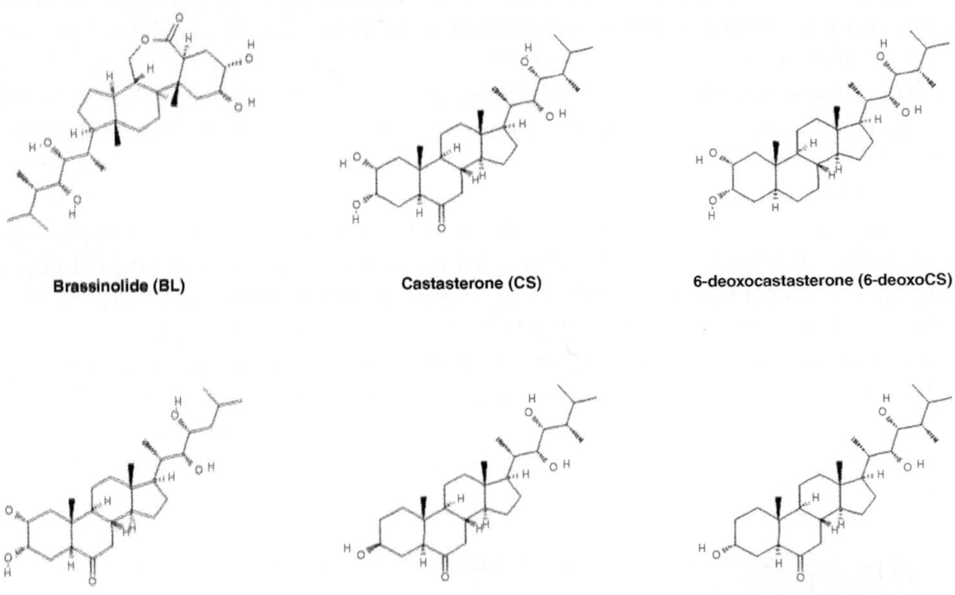

Figure 1.1 Naturally occurring brassinosteroids in plants. The chemical structures were adapted from PubChem on 12 December 2023.

The action of BRs was soon the subject of molecular analyses, and further investigations into the molecular mechanisms of BRs' actions were made possible by the identification of BR-deficient mutants, BR-signaling mutants, and BR biosynthesis inhibitors (Li and Chory 1997; Nomura et al. 1997; Asami et al. 2000). Therefore, by presenting current articles within the context of "BR research related to plant development" over the past 27 years, we aim to emphasize the significance of BRs. These publications showcase the most recent research on BR perception, signaling, development, and stress responses (Figure 1.2).

BRs are steroid hormones that have several functions in the growth and development of plants, including cell division, cell elongation, expansion, stem cell maintenance, vascular development, and floral transition (Fàbregas and Caño-Delgado 2014; Singh and Savaldi-Goldstein 2015; Lv et al. 2018; Wang et al. 2020). Additionally, BRs are involved in a variety of processes such as seed and pollen germination, pollen tube growth and development (Vogler et al. 2014), shoot and root growth (Hacham et al. 2011; Planas-Riverola et al. 2019; Bajguz et al. 2020), hypocotyl elongation (Oh et al. 2014), stomata development and patterning (Kim et al. 2012; Khan et al. 2013), xylem formation and differentiation (Saito et al. 2018; Lee et al. 2019a), tracheary elements differentiation (Yamamoto et al. 2001), plant reproduction, photomorphogenesis (Ahammed et al. 2020; Nolan et al. 2020), and senescence (Yin et al. 2002). BRs also contribute to overgrowth in hydroponically grown plants (Arteca and Arteca 2001), manage leaf cell proliferation (Nakaya et al. 2002), initiate the cell cycle, as well as control the cell cycle's progression (Zadvornova et al. 2005; González-García et al. 2011). The protective role of BRs in mitigating the detrimental effects of both abiotic and biotic stresses underscores their significance in plant resilience.

To mitigate the negative effects of biotic stresses like pathogen attacks and abiotic stresses like cold, drought, heat, salinity, and exposure to organic pollutants such as herbicides and heavy metals, BRs play a protective role by altering important physiological and biochemical processes, including antioxidant defense, nutrient uptake, osmotic adjustment,

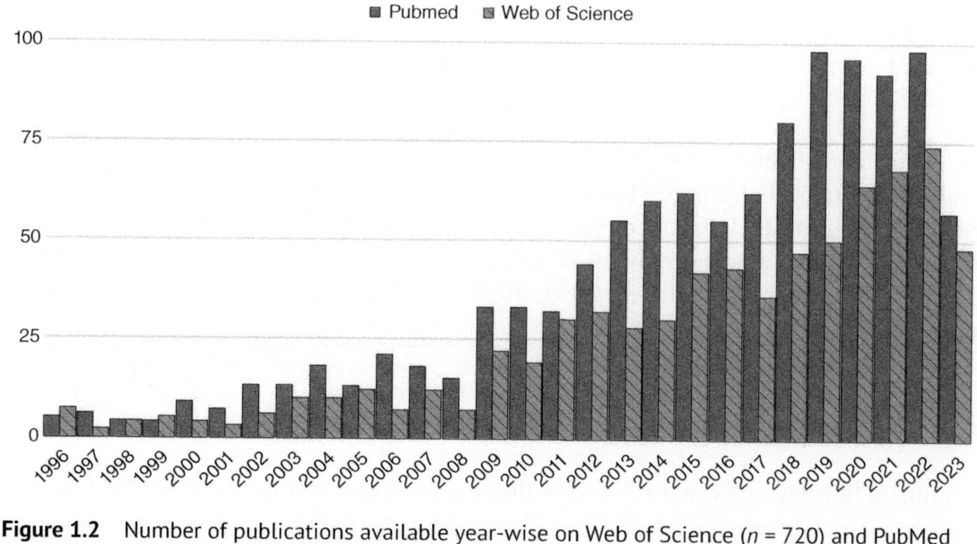

Figure 1.2 Number of publications available year-wise on Web of Science (n = 720) and PubMed (n = 1007) databases with the topic "Roles of Brassinosteroids in Plant Growth and Development."

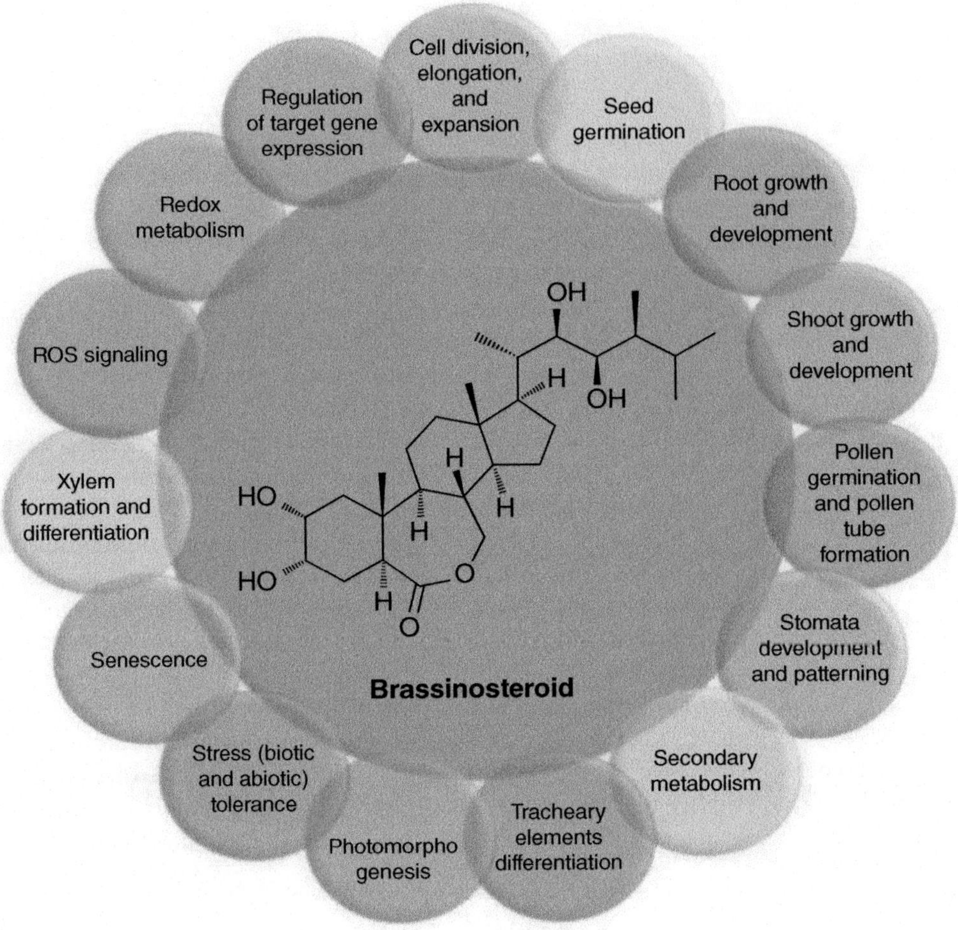

Figure 1.3 Major roles of brassinosteroids in plant growth and development, and stress migration.

and stomatal regulation (Nolan et al. 2020; Djemal et al. 2023; Gruszka et al. 2023; Wang et al. 2023; Zhang et al. 2023) (Figure 1.3). The potential role of BRs is not only limited to plant growth and development or stress combating, but many scientists reported that it also boosts crop growth and yield, including broad beans (Helmy et al. 1997), groundnuts (Vardhini and Rao 1998), maize (Hu et al. 2017a), and rice (Krishnan et al. 1999; Castorina and Consonni 2020).

1.2 Brassinosteroids in Plant Growth and Development

Prominent bioactive BRs that have been thoroughly researched for their effects on plant growth and metabolism include BL, 24-epibrassinolide (24-EBL), and 28-homobrassinolide (28-HBL). These compounds are associated with various physiological responses, such as stem elongation, pollen tube growth, induction of ethylene (ET) biosynthesis, activation of the proton pump, xylem differentiation, protein synthesis, and enhanced photosynthesis (Fariduddin et al. 2009; Zhiponova et al. 2013; Alyemeni and Al-Quwaiz 2016). According to a report, BRs also contribute to nodulation, nitrogen fixation, and the enhancement of metabolite content, as well as being involved in cell enlargement through effects on gene expression and enzyme activities (Ali et al. 2008; Wei and Li 2016; Perez-Borroto et al. 2022). Advancements in biochemistry, proteomics, and molecular genetics have unveiled BR-mediated signaling pathways regulating gene expression and plant development (Zhu et al. 2013; Nolan et al. 2020; Kour et al. 2021). The binding of BR to the cell surface receptor kinase BRI1 activates downstream kinases like CDG1 and BSK1, leading to the inactivation of the GSK3-like kinase BIN2 by BSU1 (Kim et al. 2009; Wang et al. 2016; Song et al. 2023). This inactivation results in the dephosphorylation of transcription factors BZR1 and BZR2, which migrate to the nucleus, regulating gene activation or repression upon binding to target gene promoters (Zhu et al. 2013; Kim et al. 2017). BRs have a significant impact on stomatal conductance, leaf bending, organ boundaries, cell elongation and proliferation, and other aspects of plant architecture (Bertolino et al. 2019; Castorina and Consonni 2020; Nolan et al. 2020; Manghwar et al. 2022). According to Gudesblat et al. (2012) and Bell et al. (2012), the accumulation of BRs inactivates BIN2, which decreases stomatal production, while the accumulation of BZR1 causes organ fusion phenotypes by blocking the expression of CUP-SHAPED COTYLEDON (CUC).

1.2.1 Seed Germination and Seedling Growth

Important factors for a healthy crop's production are seed germination and seedling growth. Stress is one of the minor external stimuli that can easily harm seeds during germination because they are so delicate (Pandey and Pati 2016, 2017). Seed dormancy is also one important process that keeps seeds dormant to wait for ideal growth conditions in the environment before they germinate. Gibberellic acid (GA) and abscisic acid (ABA) are plant hormones well known for controlling dormancy and germination. Brassinosteroid (BR) hormones are becoming more and more fascinating due to newly available information. By deactivating the signaling pathways linked to ABA and activating downstream GA genes, these hormones promote seed germination. A key component of the BR signaling cascade is the receptor kinase BRI1 (BRASSINOSTEROID INSENSITIVE 1). The extracellular domain of BRI1 detects the BR signal and creates a complex by attaching itself to BAK1 (BRI1-ASSOCIATED RECEPTOR KINASE 1). The stimulation of BKI1 results in the dissociation of BRI1/BKI1 compounds, phosphorylation, separation from the plasma membrane, and binding of the inhibitor to the 14-3-3 protein (Li et al. 2021b). Kim and colleagues (2019) demonstrated that the delayed germination of seeds in the BR-insensitive mutant, *bri1-5*, was significantly improved through the use of fluridone, an inhibitor of ABA production (Kim et al. 2019). BR improves the rate of germination and ultimately boosts seedling growth under several stressful conditions (Özdemir et al. 2004; Yusuf et al. 2012; Abd_Allah et al. 2018; Hasan et al. 2020; Gupta and Seth 2023). Mobilization of the seed reserve, its transport, and reuse are some of the key physiological and biochemical processes in seed germination. In rice, glutelin is the main storage protein. Cysteine proteinase (REP-1) expression is induced by BR and GA during seed germination (Xiong et al. 2021). In the endosperm, REP-1 breaks down the storage protein glutelin to release amino acids that are necessary for the embryo to grow. The scutellum subsequently carries the amino acids produced in the endosperm to the embryo for germination and post-germination growth. For common beans to germinate, it has been demonstrated that BRs and GAs are involved in the production of cysteine proteinases. To provide enough nutrition for the germination and growth phase, BRs and GAs control the synthesis of cysteine proteinases, which break down stored proteins in the cotyledons of bean plants (Yamauchi 2007). Furthermore, the GRMZM2G091919 gene model of maize encodes glycosyl transferase, which is essential for preserving cell homeostasis (Hu et al. 2017a). Glycosyl transferase has been demonstrated to control BR activity. *Arabidopsis* transgenic plants exhibit BR-deficient dwarf morphologies with ectopic overexpression of UDP-glycosyltransferase 73C5 (UGT73C5); exogenous administration of 24-EBL reversed this dwarfism to wild-type morphology (Poppenberger et al. 2005). The essential genes that regulate seed germination, G-protein α-subunit 1 (GPA1) and G-protein coupled receptor 1 (GCR1), are under the control of BR. GPA1 controls early germination and seedling growth through interaction with Pirin1 (Lapik and Kaufman 2003; Chakraborty et al. 2015). To regulate hypocotyl elongation during seedling growth, FERONIA (a receptor-like kinase necessary for normal pollen tube development and cell elongation) participates in BR and ET signaling, suggesting an antagonistic relationship between BR and ET (Deslauriers and Larsen 2010).

1.2.2 Root and Shoot Growth

Application of BR at very low concentrations greatly promotes plant growth, while higher concentrations have negative early-stage effects on growth and other growth-related parameters (Choudhary et al. 2012). Plants exhibit prompt responses to BR application, resulting in enhanced growth of both shoots and roots (Yamagami et al. 2017). Numerous prior studies have indicated that elevated BR levels during plant growth result from the elongation and expansion of cells (Catterou et al. 2001; Caño-Delgado et al. 2004; Zhiponova et al. 2013). Mutants lacking BR function show markedly smaller cells as well as fewer leaf epidermal and mesophyll cells, while plants that overexpress the BR receptor BRI1 show significantly larger epidermal and mesophyll cells (Hong et al. 2002; Caño-Delgado et al. 2004). A crucial initiator of the BR signal pathway, Brz-insensitive-long hypocotyl 4 (BIL4) interacts with the BR receptor BRI1 and dynamically regulates cell elongation: when compared to wild-type plants, BIL4-OX mutants displayed larger and taller cells, whereas BIL4-RNAi mutants showed a decrease in these characteristics (Caño-Delgado et al. 2004; Sharma and Bhardwaj 2007). BRs communicate via the receptor kinase BRI1 located in the plasma membrane. Yin et al. (2002) revealed that the BES1 gene, a semidominant suppressor of the BRI1 gene, displays constitutive BR response characteristics such as bent and long petioles, curling leaves, and accelerated senescence. Plant architecture and patterning are determined by the spatiotemporal regulation of the development of organ primordia and organ boundaries from the specific plant stem cell of the shoot apical meristem (SAM). The CUC family of organ boundary genes is repressed by the BR-activated transcription factor BZR1 (Gendron et al. 2012). Organ fusion phenotypes and repression of CUC gene expression are caused by the activation of BR signaling. Because they suppress the genes that identify organ boundaries, BRs are essential in the formation of organ boundaries. Organ-fusion phenotypes are seen in BR-hypersensitive mutants, while enhanced organ boundaries are seen in BR-insensitive mutants. The BR-deficient dwarf1 mutant, considered as lacking BR C-6-oxidase activity in rice, displays aberrant arrangement and polar cell elongation of the leaf, stem, and other developing organs (Hong et al. 2002).

While BR is generally recognized for its growth-promoting effects, its impact on root growth is contingent upon dosage (Mussig et al. 2003). Recent investigations emphasize the necessity of a balanced influence of BR on cell division and differentiation for optimal root growth and meristem homeostasis. In the root meristem, BR fosters mitotic activity and the expression of CYCLIN D3 and CYCLIN B1 (González-García et al. 2011; Hacham et al. 2011). Conversely, elevated concentrations of BR accelerate cell differentiation and elongation, leading to a reduction in meristem size and root growth. Moreover, BR plays a role in promoting quiescent center division and columella stem cell differentiation. Notably, the expression of BRI1, using an epidermal-specific promoter, successfully restored the root growth of the bri1 mutant, underscoring the sufficiency of BR perception in the epidermis for controlling root growth and meristem size, potentially mediated by a mobile factor other than BZR1 or BZR2 (Hacham et al. 2011). Upon exogenous BR application, mutants of AtRALF (Rapid Alkalinization Factor), a peptide signal essential to cell biology that primarily controls cell expansion, exhibit markedly larger root cells and show reduced transcription levels of genes involved in BR biosynthesis (CPDs and DWFs) compared to wild-type plants (Bergonci et al. 2014). According to Bergonci et al. (2014), root cell elongation and growth may be controlled by the RALF signaling mechanism's beneficial interference with the BR signaling pathway. Additionally, BR assumes a crucial role in determining epidermal cell fate in roots, where epidermal cells differentiate into hair or non-hair cells based on their position. Disrupting BR signaling interferes with the patterning of root hair cells and the expression patterns of WEREWOLF (WER) and GLABRA2 (GL2), which are responsible for specifying epidermal cell fate (Kuppusamy et al. 2009). Therefore, the regulation of expression levels of cell-type-specific components emerges as a general mechanism through which BR exerts specific effects on diverse developmental responses. On the other hand, aberrant phloem and xylem differentiation and a rise in vascular defects are seen in loss-of-function brl1 mutants, suggesting that BR receptors are essential for both vascular differentiation and cell growth (roots and shoots). Collective information from data sources demonstrated that the BR signaling pathway was essential to plant cell differentiation and enhanced stress tolerance (Xu et al. 1994; Xia et al. 2009b; Rehman et al. 2016; Hou et al. 2017).

1.2.3 Leaves, Photosynthesis, and Stomata

Respiration, transpiration, photosynthesis, and guttation are among the vital physiological processes carried out by leaves, which function as specialized organs of plants. Leaf development is one of the many physiological and developmental processes that BRs are essential for controlling. Recent research on pepper showed that the application of BR significantly increased leaf growth, including leaf area and leaf mass per unit leaf area, and also found that the photosynthesis process increased with chlorophyll content, Fv/Fm, as well as net photosynthetic rate (Pn) with application of exogenous BR (Li et al. 2016). Previous research has consistently shown that applying BR causes important variables for plant growth under stress, like leaf area, chlorophyll content, and photosynthesis, to significantly increase (Yuan et al. 2012; Wei et al. 2015). A different study conducted recently discovered the super compact-2 mutant in cucumbers, which exhibits the typical phenotype of BR biosynthesis deficiency. The leaves of these mutants were wrinkled and dark green. Chloroplasts and stomata, the epidermal valves that permit the transfer of gases between plant leaves and the outside world, are both necessary for photosynthesis. To maximize CO_2 uptake and minimize water loss, stomata density and distribution are strictly regulated (Dong and Bergmann 2010). Several bHLH transcription factors, the mitogen-activated protein kinase (MAPK) module, members of the ERECTA family of receptor kinases, and epidermal patterning factor (EPF) family peptide signals are components of the genetically defined stomata pathway (Dong and Bergmann 2010). Due to BIN2-mediated phosphorylation of two stomatal pathway components, BR has been demonstrated to both restrict and encourage stomatal formation in leaves and hypocotyls, respectively. Depletion of BSU1 family phosphatases results in extensive stomata formation, with this phenotype being dependent on BIN2 activity (Kim et al. 2012). BIN2 phosphorylates and deactivates the MAPK kinase kinase YODA (YDA), a process that hinders stomata formation. The deactivation of BIN2 by BR induces the activation of MAPKs, thereby impeding stomatal development in leaves. In addition, BR exerts a positive influence on stomatal development in hypocotyls by phosphorylating BIN2, leading to the inactivation of the bHLH factor SPEECHLESS (SPCH) as demonstrated by Gudesblat et al. (2012). SPCH operates downstream of the MAPKs, stimulating cell division and the formation of stomata (Dong and Bergmann 2010). These findings suggest an intricate regulatory network in stomatal development involving multiple interactions between the BRI1 and ERECTA receptor kinase pathways.

Chlorophyll and photosynthesis play crucial roles in plant growth. The application of exogenous BR significantly enhances chlorophyll content and improves the photosynthetic characteristics of plants, particularly under various stresses (Anwar et al. 2018; Li et al. 2022). Chlorophyll serves as a vital parameter commonly used to gauge chloroplast development and photosynthetic activity. Its levels are highly sensitive to external stimuli (stress), leading to a reduction in total chlorophyll a, b, and carotenoid contents in leaves (Rehman et al. 2016). BR protects chlorophyll from these external stimuli and elevates its concentration under various stress conditions (Fariduddin et al. 2014b; Anwar et al. 2018; Hafeez et al. 2021; Zhang et al. 2023). Exogenous BR notably promotes chlorophyll and photosynthetic parameters, particularly under low temperature, weak light stress, and $Ca(NO_3)_2$ stress. Furthermore, BR can both regulate the combination of chlorophyll molecules (by controlling chlorophyllase activity) with membrane proteins and maintain the stability of thylakoid membranes (Xu et al. 1994). When tomato seedlings are under water stress, BRs are essential for controlling and preserving photosynthetic activity (Li et al. 2015). BR mitigates the negative effects of various stress conditions and controls the defense system by regulating the transcription levels of defense genes in cucumbers (Li et al. 2013). Additionally, under drought and heat stress in wheat, BR controls the RuBisCO activase (RCA) gene, a crucial component in photosynthesis (Degen et al. 2021). BR dramatically speeds up the process of photosynthesis and antioxidant enzyme activity (Jiang et al. 2013; Ahammed et al. 2015a; Wani et al. 2019). BR-treated seedlings exhibit increased CO_2 assimilation and quantum yield of photosystem II (PSII), along with enhanced ribulose-1,5-bisphosphate carboxylase/oxygenase (RuBisCO) activities and the expression of RuBisCO large subunit (rbcL) and RuBisCO small subunit (rbcS) genes, ultimately boosting photosynthetic capacity (Zhang et al. 2008; Xia et al. 2009a; Gururani et al. 2015). These findings highlight the positive role of BR in photosynthesis. Furthermore, transcriptome analysis of BR reveals that it enhances various chlorophylls and photosynthesis-related genes under low-temperature stress (Zhao et al. 2019). Thus, BR promotes the accumulation of chlorophylls and enhances photosynthetic capacity by regulating various enzymes, including chlorophyllase and RuBisCO, as well as transcript levels of genes involved in chlorophyll and photosynthesis under stress.

1.2.4 Nutrient Uptake

The uptake of essential nutrients such as nitrogen (N), phosphorus (P), potassium (K), calcium (Ca), magnesium (Mg), iron (Fe), manganese (Mn), copper (Cu), zinc (Zn), etc., is integral to plant growth. A higher influx of ions contributes to increased efficiency in light energy transformation, CO_2 conductance, the potential of light and dark reactions, and the overall photosynthetic rate (Talaat and Shawky 2013). A recent study highlighted that BR plays a role in increasing essential inorganic ions, decreasing toxic ions, and promoting ion homeostasis, particularly in the leaves, roots, and epicotyl of canola under salt stress (Liu et al. 2014). In the face of low temperature and weak-light stress, 24-EBL has been observed to enhance nitrogen metabolism. It boosts the activity of enzymes such as nitrite reductase (NiR), nitrate reductase (NR), glutamate synthase (GOGAT), glutamine synthetase (GS), and glutamate dehydrogenase (GDH), while also inducing photosynthetic characteristics in tomato seedlings (Shu et al. 2016). Additionally, the application of exogenous BR increases the activities of H^+-ATPase and Ca^{2+}-ATPase in both roots and leaves (Feng et al. 2019). These enzymes play a crucial role in establishing an electrochemical potential gradient to maintain ion balance in plants, thereby alleviating the adverse effects of stress.

1.2.5 Reproductive Development

A complex network of endogenous and environmental cues, such as temperature, photoperiod, BR, and GA, regulates the transition that occurs from vegetative to reproductive growth (Li et al. 2010). A floral repressor, FLOWERING LOCUS C (FLC), is expressed more frequently in *Arabidopsis* mutants with BR deficiency or insensitivity, which results in a delayed flowering phenotype (Domagalska et al. 2007). It has been discovered that REF6 (Noh et al. 2004) interacts with BZR2 (Yu et al. 2008), which regulates FLC expression negatively (Zhu et al. 2013; Bao et al. 2020). Furthermore, BR application shortens circadian rhythms, suggesting that BR may indirectly affect the flowering time through effects on the circadian clock and the photoperiod flowering pathway (Li et al. 2010). Male fertility is also decreased by defects in BR biosynthesis or signaling pathways because they cause pollen development defects and stamen shortening (Ye et al. 2010). Many of the important genes involved in the development of anthers and pollen are transcriptional targets of BZR2, and their decreased expression is correlated with these developmental defects. In conclusion, current research indicates that BR regulates sex determination in maize by inducing the stamen and suppressing the pistil development in tassels (Li and Liu 2017; Lunde et al. 2019).

1.2.6 Fruit Development

Given the growing global population, one of the main challenges facing researchers is meeting the demand for food by increasing yield and promoting yield components. However, because they severely hinder plant growth, developmental processes, and yield, biotic and abiotic stresses pose a significant obstacle to meeting the world's food demand (Oshunsanya et al. 2019; Liliane and Charles 2020). Previous research demonstrated that BR greatly increased the mustard plant's yield and yield-related components (Alam et al. 2020). Tomato growth, fruit count, and weight per plant were all markedly increased by foliar application and seed treatments of BR (Hayat et al. 2011). BR application also increased the yield of pears and yellow passion fruit (Gomes et al. 2006). Prior research has indicated that BR has a beneficial effect on pitaya quality, mango fruit growth, and fruit ripening (Khatoon et al. 2021).

1.2.7 Environmental Stresses

Abiotic and biotic stresses, particularly those caused by salinity, heat, drought, and heavy metal toxicity, are becoming more and more of an issue in agriculture and are a major risk to crop productivity (Bharti et al. 2021; Umar et al. 2021; Singh et al. 2022; Dilawari et al. 2023). Recently, stress tolerance in a range of plants has been developed through the use of plant hormones. Steroid hormones (BRs) have multifaceted impacts on plants (Ahammed et al. 2015b; Ahanger et al. 2018; Hussain et al. 2020; Basit et al. 2021b). One of the main fields of research interest has been the exogenous application of BRs to enhance plant tolerance mechanisms to different stresses (Manghwar et al. 2022). Several investigations have demonstrated how these steroidal hormones interact with other metabolic pathways and play a part in the up-regulation/downregulation of stress-related resistance genes (Table 1.1). To control a range of physiological and developmental processes in plants, BRs interact with other plant hormones, including salicylic acid, jasmonic acid, auxin, ABA, cytokinin, GA, and ET (El-Esawi 2017; Yang et al. 2019). All exogenous applications of brassinosteroid in the form of BL, 24-EBL, or 28-HBL studies have demonstrated the function of BRs in improving photosynthetic efficiency, altering nutrient metabolism, reducing transpiration rate, modulating stomatal oscillation, maintaining redox homeostasis, and influencing RuBisCO activity in response to abiotic stress or external stimuli (Fariduddin et al. 2009; Wei et al. 2015; Li et al. 2016; Rehman et al. 2016; Alyemeni and Al-Quwaiz 2016; Hu et al. 2017b; Khatoon et al. 2021).

1.3 Brassinosteroids and Hormonal Interaction

Auxin, BRs, and GA are all strongly correlated with each other, suggesting that the physiological pathways mediated by BRs are closely related to other hormonal and environmental signals (Gallego-Bartolomé et al. 2012; Li et al. 2012; Saini et al. 2015; Hu et al. 2017b; Castro-Camba et al. 2022). Therefore, controlling growth and responses to stress is greatly influenced by the complex interactions and crosstalk between BR and other plant hormones (Steber and McCourt 2001; Werner et al. 2001; Acharya and Assmann 2009;

Table 1.1 Comprehensive information of exogenous brassinosteroids application with altering metabolism to reduce the adverse effects of abiotic stress on different plant species.

Type of exogenous brassinosteroid Applied	Alone/Combine treatment	Plant	Stress	Plant development stages	Physiological changes	References
24-epibrassinolide (24-EBL)	NaCl	*Oryza sativa*	Salt	Seedling	↑Ascorbate peroxidase ↑Soluble Protein ↓Lipid peroxidation level ↓Proline	(Özdemir et al. 2004)
24-EBL	NaCl	*Brassica juncea* L.	Salt	Seedling	↑Regulating nitric oxide homeostasis ↑Osmolytes synthesis ↑Antioxidant defense systems ↑Nutrients and pigment contents	(Gupta and Seth 2023)
24-EBL		*Acacia gerrardii* Benth.	Salt	Seedling	↑Superoxide dismutase ↑Catalase ↑Enzymes of the ascorbate-glutathione pathway protect the photosynthetic electron transport chain ↑Membrane protection ↑Chlorophyll synthesis	(Abd_Allah et al. 2018)
28-homobrassinolide (HBL)		*Glycine max* L.	Drought	Seedling	↑Glyoxalase I and II, ↑Antioxidant enzyme ↑Membrane protection	(Hasan et al. 2020)

(Continued)

Table 1.1 (Continued)

Type of exogenous brassinosteroid Applied	Alone/Combine treatment	Plant	Stress	Plant development stages	Physiological changes	References
24-EBL	Rhizobium	*Vigna radiata*	Ni-stress	Seedling	↑Nodulation ↑Antioxidant enzymes (viz. catalase, peroxidase, and superoxide dismutase) activity ↑Proline synthesis	(Yusuf et al. 2012)
28-HBL	Rhizobium	*Cicer arietinum* (L.) cv. *Uday*	Cd-stress	Seedling	↑Antioxidative defense	(Hasan et al. 2008)
24-EBL	Week light condition	*Lycopersicon esculentum* Mill.	Low temperature	seedlings	↑Net photosynthetic rate (Pn), light saturation point (LSP), maximal quantum yield of PSII photochemistry (Fv/Fm), actual photochemical efficiency of PSII (ΦPSII), and photochemical quenching coefficient (q_P), ↓Intercellular CO_2 concentration (Ci), light compensation point (LCP), and apparent quantum efficiency (AQE) ↑Nitrate reductase (NR), glutamate dehydrogenase (GDH), glutamine synthetase (GS), and glutamate synthase (GOGAT) activity ↓Ammonium content ↓Nitrite reductase (NiR) activity	(Shu et al. 2016)

24-EBL and 28-HBL		*Raphanus sativus*	Water stress	Seedling	↑Superoxide dismutase, catalase, and ascorbate peroxidase activity ↑Proline content, ↓Lipid peroxidation and MDA	(Mahesh et al. 2013)
24-EBL	Spermidine	*Vigna radiata*	Salt stress and Zn-stress	Seedlings	↑Antioxidant enzyme activity ↑Proline content ↑Electrolyte leakage	(Mir et al. 2015)
28-HBL		*Brassica juncea* Czern & Coss cv. *Varuna*	Salt stress (150 mM NaCl)	Seedling	↑Nitrate reductase ↑Carbonic anhydrase activity ↑Chlorophyll content	(Siddiqui et al. 2018)
Brassinosteroid (BR)		*Eleusine coracana*	Drought	Seed, vegetative stage	↑Gas exchange parameters ↑Relative water content ↑Chlorophyll index ↑Chlorophyll fluorescence ↑Proline content ↑Catalase activity	(Mohanabharathi et al. 2019)
28-HBL		*Brassica juncea*	Cu-stress	Seedling	↑Growth ↑Photosynthetic parameters ↑Antioxidant enzymes (Superoxide Dismutase, Catalase, and Peroxidase)	(Fariduddin et al. 2009)
24-EBL		*Brassica juncea* Czern & Coss	Pb-stress	Seed and seedling	↑Antioxidant enzymes (especially Catalase)	(Soares et al. 2020)
BR		*Mentha piperita* L.	Salt stress	Seedling	↑Antioxidant enzymes ↓Membrane permeability ↓Lipid peroxidation ↑Essential oil content	(Çoban and Göktürk Baydar 2016)

(Continued)

Table 1.1 (Continued)

Type of exogenous brassinosteroid Applied	Alone/Combine treatment	Plant	Stress	Plant development stages	Physiological changes	References
24-EBL		*Triticum aestivum* L. cvs. *Omskaya 35* (O-35, drought-resistant variety) and *Salavat Yulaev* (SYu, less drought-resistant variety)	Drought stress	Seed	↑Dehydrin (low molecular weight) proteins	(Shakirova et al. 2016)
Brassinolide (BL)		*Vitis vinifera* L.	Salt stress	Seedling	↑Photosynthetic pigment concentrations ↑Ion homeostasis ↑Regulating water status ↑Antioxidant capacity ↑Leaf anatomical attributes	(El-Banna et al. 2022)
24-EBL		*Triticum aestivum* L. cvs. *HUW-468, C-306* (early and late sowing, respectively)	Heat stress	Pre-flowering stage and post-flowering stage	↑Germination percentage ↑Total leaf area ↑Total dry weight ↑Yield	(Kumari and Hemantaranjan 2018, 2019)
BR		*Brassica rapa* subsp. *pekinensis* (Lour.) Rupr.	Heat stress	Vegetative stage	↑Antioxidant enzymes	(Lee et al. 2019b)

24-EBL		*Zea mays* cv. *PR 32T83 and PR 34N24*	Salt stress	Seedling and vegetative stage	↑Growth attributes ↑Water relations ↑Antioxidant enzymes activities ↑Proline content ↓Electrolyte leakage ↓H_2O_2 ↓MDA contents ↓Na^+ and Cl^- concentrations ↑N, Ca^{2+}, K^+, and P concentrations	(Kaya et al. 2018)
24-EBL		*Hordeum vulgare* L. cv. *RD-2508* and *RD-2660* (salt tolerant-RD-2508 and salt sensitive-RD-2660)	Salt stress and K-deficiency stress	Vegetative stage	↑Antioxidant enzymes ↑Osmolytes (proline and sugar) ↑K^+ accumulation ↓Na^+ accumulation ↑K^+/Na^+ ratio	(Liaqat et al. 2020)
24-EBL	Sodium nitroprusside (SNP)	*Brassica juncea* (L.) cv. *Varuna*	Salt stress	Vegetative stage	↑Plant growth ↑Gas exchange parameters ↑Chlorophyll fluorescence ↑Nitrogen metabolism ↑K^+/Na^+ ratio ↓MDA, Proline, ABA content	(Gupta et al. 2017)
24-EBL		*Oryza sativa* cv. *Karad*	Alkaline stress	Seedlings	↑Antioxidant enzymes (Ascorbate peroxidase, catalase, and superoxide dismutase) ↓Level of MDA, H_2O_2, and Superoxide radicals	(Sharma et al. 2019)
24-EBL	Zinc	*Brassica napus* L.	Salt stress	Vegetative stage	↓Level of MDA, H_2O_2 content ↑Osmoprotectants ↑Antioxidant enzymes	(Mokari-Firuzsalari et al. 2019)

(Continued)

Table 1.1 (Continued)

Type of exogenous brassinosteroid Applied	Alone/Combine treatment	Plant	Stress	Plant development stages	Physiological changes	References
24-EBL		*Glycine max*	B-stress	Vegetative stage	↑Root tissues, specifically maximizing the vascular cylinder, metaxylem, and epidermis) thus improve nutritional status ↑Leaf anatomy (epidermis on both leaf sides, palisade parenchyma, and spongy parenchyma) to make arability of CO_2 for photosynthesis	(Pereira et al. 2022)
24-EBL	Putrescine	*Cucumis sativus* L.	Salt stress	Seedlings	↑Growth parameters ↑Antioxidant enzymes (catalase, peroxidase, and superoxide dismutase)	(Fariduddin et al. 2014a)
24-EBL		*Capsicum annuum* L.	B-stress	Seedling	↑ Water relations ↑Proline ↑Antioxidants (enzymatic and non-enzymatic) ↑Endogenous H2S and L-DES activity, ↓H_2O_2 and MDA contents ↓Membrane leakage ↑Ca^{2+}, K^{+}, and N content in leaf	(Kaya 2020)
24-EBL		*Glycine max*	Salt stress	Seedling, vegetative, flowering, and podding stage	↑Electrical conductivity of saturated soil extract (ECse) in the rhizosphere (root exudates)	(Otie et al. 2022)

BL	*Zea mays* cv. *Jingkenuo2000* (JKN2000, heat-tolerant) and *Yunuo7* (YN7, heat-sensitive)	Heat stress	Silking-stage	↑Accumulation, translocation, and remobilization of dry matter (DM) ↑Nutrients (nitrogen, phosphorus, and potassium) accumulation to grain yield	(Xu et al. 2022)
24-EBL	*Eucalyptus urophylla*	Salt stress	Seedlings	↑Photosynthetic pigments ↑Photochemical efficiency ↑Leaf anatomy with increased stomatal density, palisade parenchyma, and spongy parenchyma ↑Antioxidant system (CAT and APX)	(de Oliveira et al. 2019)
24-EBL	*Vigna unguiculata* L. cv. *BR3-Tracuateua*	Drought stress	Seedlings	↑ΦPSII, q_P and ETR ↑Antioxidant enzymes (SOD, CAT, APX and POX)	(Lima and Lobato 2017)
24-EBL	*Glycine max*	Seedlings	Seedlings	↑Root anatomy, particularly of the RET, RDT, RCT, VCD and RMD ↓ROS (superoxide and hydrogen peroxide) ↓Membrane damage (malondialdehyde and electrolyte leakage) ↑Antioxidant enzymes (SOD, CAT, APX and POX)	(Ribeiro et al. 2019)

(Continued)

Table 1.1 (Continued)

Type of exogenous brassinosteroid Applied	Alone/Combine treatment	Plant	Stress	Plant development stages	Physiological changes	References
28-HBL and 24-EBL		*Zea mays* L. var. *DKC 9106*	Salt stress	Seedling	↓Na^+ ions content ↑K+ ions content ↓Malondialdehyde content ↑Antioxidative enzymes (POD, CAT, DHAR, monodehydroascorbate reductase, MDHAR), ↑Osmoprotectants (proline, glycine betaine, mannitol, and total osmolytes content) ↑Phenolic content ↑Flavonoid content ↑DPPH content	(Rattan et al. 2020)
24-EBL		*Robinia pseudoacacia* L.	Salt stress	Seedling	↑Membrane stability ↓Na^+ levels ↑Chlorophyll content	(Yue et al. 2019)
24-EBL		*Solanum tuberosum*	Salt stress	Seedling	↑Peroxidase ↑K^+/Na^+ ratio in potato leaves ↓Leaf osmotic potential	(Kolomeichuk et al. 2020)

24-EBL	Kinetin	*Solanum lycopersicum* L. cv. *Huange 108*	Salt stress	Seed	↓Lipid peroxidation and ↓Electrolyte leakage ↓Protease and lipoxygenase ↓Redox homeostasis ↓Glutathione ↑Antioxidant system ↑Ascorbic acid ↑PSII functioning (Fv/Fm), ETR, and photochemical quenching	(Ahanger et al. 2020)
24-EBL		*Triticum aestivum* cv. *Chakwal-50* and *Faisalabad-2008*	Drought stress	Seedling	↑Osmolyte accumulation ↑Photosynthetic pigments, and ↑Gas exchange parameters ↓Electrolyte leakage ↓Malondialdehyde and hydrogen peroxide contents ↑Antioxidant enzymes (CAT, SOD, POX)	(Khan et al. 2021)
24-EBL		*Chunyou 927* (CY-927, sensitive), *Yliangyou 689* (YLY-689 tolerant)	Cr-stress	Seed	↑Antioxidant enzyme activities (SOD, POD, CAT, and APX)	(Basit et al. 2021a)

↑ improved; ↓ reduced; q_P, photochemical quenching coefficient; ΦPSII, effective quantum yield of PSII; ETR, electron transport rate; RET, root epidermis; RDT, root endodermis; RCT, root cortex; VCD, vascular cylinder; RMD, root metaxylem; ROS, reactive oxygen species; SOD, superoxide dismutase; CAT, catalase; APX, ascorbate peroxidase; POX/POD, peroxidase; DHAR, dehydroascorbate reductase; MDHAR, monodehydroascorbate reductase; DPPH, 1,1-diphenylpicrylhydrazyl.

Zhang et al. 2009; Zhu et al. 2016). Notably, earlier studies have emphasized the importance of ABA as a signaling molecule in stressed as well as unstressed plants (Zhang et al. 2009). The intricate regulation of gene expression by the interaction between BR and ABA leads to increased stress tolerance (Bulgakov and Avramenko 2020; Hussain et al. 2021; Li et al. 2021a). BR-perception (bri1-1), and ABA-insensitive (abi1, abi2) mutants are two more examples of how ABA and BR are related. This interaction controls several genes that are essential for plant development processes, such as BR biosynthesis genes CPD (constitutive photomorphogenesis and dwarfism) and DWF1 (DWARF1), as well as seed germination gene; GPA (G protein α-subunit1), along with stress-responsive genes such as SOD, POD, CAT, and GR, which support the plant defense system (Werner et al. 2001; Lapik and Kaufman 2003).

Both BR and GA growth-regulating hormones have similar effects on various developmental processes throughout the life cycle of plants (Depuydt and Hardtke 2011). To cause the DELLA family proteins to become ubiquitylated and degraded, GA attaches to its receptor, gibberellin insensitive dwarf1 (GID1), which then binds to several transcription factors, such as PIFs, to prevent their DNA binding (Hauvermale et al. 2012; Schwechheimer 2012). Therefore, GA causes DELLA to degrade, which derepresses PIFs (De Lucas et al. 2008; Hauvermale et al. 2012). To date, it was thought that the effects of BR and GA were synergistic, suggesting that their signaling pathways operated on transcriptional responses that were essentially non-overlapping. Nonetheless, recent research has shown a direct interaction as well as an interdependent relationship between the BR and GA signaling pathways (Gallego-Bartolomé et al. 2012). Furthermore, research indicates that PIFs, BZR1, and DELLA proteins directly interact to integrate BR, GA, and light signals and to regulate a core expression network that is involved in photomorphogenesis (Xue et al. 2022).

Additionally, research in the fields of genetics and physiology has demonstrated that auxin and BR work in conjunction to activate similar sets of genes in a variety of developmental contexts, including vascular bundle patterning and hypocotyl elongation (Oh et al. 2020; Mazzoni-Putman et al. 2021). BR-auxin interactions mediated potential molecular pathways are BZR1 controls a large number of genes related to auxin production, transport, and signaling (Lee et al. 2021). Furthermore, auxin raises the levels of BR by activating the genes that synthesize BR. Small auxin upregulated 15 (SAUR15) gene, which plays a role in cell elongation, can be activated by auxin hormones and BRs via these transcription factors BZR2 and ARF5 (Walcher and Nemhauser 2012; Li et al. 2012; Galstyan and Nemhauser 2019). Auxin and BR also regulate dynamic actin cytoskeleton, which plays an essential cellular part (García-González and van Gelderen 2021).

In roots and shoots, cytokinin stimulates cell division (Werner et al. 2001, 2010). The application of exogenous BR led to a decline in enzyme activity and the expression of genes encoding cytokinin oxidase, contributing significantly to an increase in cytokinin levels in wheat seedlings (Yuldashev et al. 2012). This suggests the involvement of BR in the regulation of cytokinin. Similar to a cytokinin dehydrogenase/oxidase 3 (CKX3) overexpression line in tobacco, overexpression of BRI1 in PYK10:CKX3 increased root and leaf growth. These results suggest that cytokinin and BR actively interact and play important roles in various aspects of plant growth and development (Yuldashev et al. 2012).

All plant organs contain ET, a plant hormone that is essential to plant growth and development and helps plants become more resilient to a wide range of biotic and abiotic stresses. A previous study indicated that BR plays a critical role in plants' biosynthesis of ET. Tomatoes treated with BR exhibited increased ET accumulation and biosynthesis, suggesting a major interaction between ET and BR in the context of stress tolerance responses (Zhu et al. 2016).

1.4 Conclusion

BRs are instrumental in the orchestration of critical cellular processes that underpin the very essence of a plant's existence. These phytohormones wield their influence over cell division, elongation, and differentiation throughout all stages of a plant's life cycle. Such involvement is not just beneficial but essential for the development, growth, and ultimate productivity of plants, making BRs central figures in the intricate dance of life that unfolds in the botanical world. The discovery of the chemical structure of BRs within plants and their identification signaled the beginning of their exploration. The quest to understand the roles and functions of BRs in plant growth and development began at this pivotal moment. It also marked the beginning of a new phase of research aimed at deciphering the complex network of relationships that control how BRs interact with a variety of biotic and abiotic stressors. Indeed, the remarkable versatility of BRs as regulators is underscored by their essential role in the intricate machinery of plant life. They are crucial components in the growth, development, and responses to environmental challenges, including the rigors of extreme temperatures and the desiccating grasp of drought. As these stressors continue to shape the global landscape, the functions of BRs in safeguarding plant development and orchestrating a robust response to a dynamic array of stressors have become increasingly vital. Current research on BRs has revealed a wide range of interaction domains, temporal specifics, and spatial complexities where these phytohormones function. This research information illustrated the mechanisms by which BRs interact with other hormones and play a part in the regulation of stress reactions too.

The latest advancements in our comprehension of the brassinosteroid signaling pathway have not only deepened our understanding of the physiological, cellular, and molecular mechanisms through which BRs regulate plant development – including root development, anther and pollen formation, stem elongation, vasculature differentiation, cellulose biosynthesis, and various other aspects of plant life – but also hold the potential to revolutionize our approaches to agriculture, environmental management, and the pursuit of greater resilience in a changing world.

1.5 Future Prospects

The understanding of the BR signaling pathway has advanced significantly over the past few decades with a combination of genomic, proteomic, and genetic methods. The first and currently only completely understood receptor kinase signaling pathway in plants is the

BR pathway. It is associated with numerous molecular interconnections to other signaling pathways, demonstrating a well-integrated plant regulatory network. BR's pivotal functions in regulating plant responses to light, GA, and others, in particular, strengthen the concept that it serves as a master regulator at the core of the network that regulates plant growth. Thus, a framework for creating an intricate molecular map of the plant growth regulation network has been established.

There are still a lot of important questions that need to be answered with research, such as: Why do various tissues and organs have such a wide range of BR levels? What is the mechanism by which plants use BR during typical growth and stress conditions? Are cell differentiation and morphogenesis aided by BR as a positional signal? How can different tissues and cell types respond differently to BR? What connections do other signaling pathways have with the BR pathway? Lastly, how does BR signaling interact with or influence cellular functions like cell wall integrity and expansion, vesicle trafficking, and cytoskeleton organization? To produce food and bioenergy, and to preserve the environment, we must comprehend how plants regulate their growth. The answers to these questions will help us attain this goal. Furthermore, to understand the cross-talk between BR and other phytohormones, which ultimately expands its biological activities at the transcription and post-transcriptional levels, the proteomics and metabolomics portfolio linked to BR-induced stress tolerance needs to be addressed.

References

Abd Allah, E.F., Alqarawi, A.A., Hashem, A. et al. (2018). Regulatory roles of 24-epibrassinolide in tolerance of *Acacia gerrardii* Benth to salt stress. *Bioengineered* 9: 61–71.

Acharya, B.R. and Assmann, S.M. (2009). Hormone interactions in stomatal function. *Plant Molecular Biology* 69: 451–462.

Ahammed, G.J., Li, X., Liu, A. et al. (2020). Brassinosteroids in plant tolerance to abiotic stress. *Journal of Plant Growth Regulation* 39: 1451–1464.

Ahammed, G.J., Li, X., Xia, X.-J. et al. (2015a). Enhanced photosynthetic capacity and antioxidant potential mediate brassinosteroid-induced phenanthrene stress tolerance in tomato. *Environmental Pollution* 201: 58–66.

Ahammed, G.J., Xia, X.-J., Li, X. et al. (2015b). Role of brassinosteroid in plant adaptation to abiotic stresses and its interplay with other hormones. *Current Protein and Peptide Science* 16: 462–473.

Ahanger, M.A., Ashraf, M., Bajguz, A. et al. (2018). Brassinosteroids regulate growth in plants under stressful environments and crosstalk with other potential phytohormones. *Journal of Plant Growth Regulation* 37: 1007–1024.

Ahanger, M.A., Mir, R.A., Alyemeni, M.N. et al. (2020). Combined effects of brassinosteroid and kinetin mitigates salinity stress in tomato through the modulation of antioxidant and osmolyte metabolism. *Plant Physiology and Biochemistry* 147: 31–42.

Alam, P., Kaur Kohli, S., Al Balawi, T. et al. (2020). Foliar application of 24-Epibrassinolide improves growth, ascorbate-glutathione cycle, and glyoxalase system in brown mustard (*Brassica juncea* (L.) Czern.) under cadmium toxicity. *Plants* 9: 1487.

Ali, B., Hasan, S.A., Hayat, S. et al. (2008). A role for brassinosteroids in the amelioration of aluminium stress through antioxidant system in mung bean (*Vigna radiata* L. Wilczek). *Environmental and Experimental Botany* 62: 153–159.

Alyemeni, M.N. and Al-Quwaiz, S.M. (2016). Effect of 28-homobrassinolide on the performance of sensitive and resistant varieties of *Vigna radiata*. *Saudi Journal of Biological Sciences* 23: 698–705.

Anwar, A., Liu, Y., Dong, R. et al. (2018). The physiological and molecular mechanism of brassinosteroid in response to stress: a review. *Biological Research* 51: 46.

Arteca, J.M. and Arteca, R.N. (2001). Brassinosteroid-induced exaggerated growth in hydroponically grown *Arabidopsis* plants. *Physiologia Plantarum* 112: 104–112.

Asami, T., Min, Y.K., Nagata, N. et al. (2000). Characterization of brassinazole, a triazole-type brassinosteroid biosynthesis inhibitor. *Plant Physiology* 123: 93–100.

Bajguz, A. (2011). Brassinosteroids – occurence and chemical structures in plants. In: *Brassinosteroids: A Class of Plant Hormone* (ed. S. Hayat and A. Ahmad), 1–27. Dordrecht: Springer Netherlands.

Bajguz, A., Chmur, M., and Gruszka, D. (2020). Comprehensive overview of the brassinosteroid biosynthesis pathways: substrates, products, inhibitors, and connections. *Frontiers in Plant Science* 11: 1034.

Bajguz, A. and Tretyn, A. (2003). The chemical characteristic and distribution of brassinosteroids in plants. *Phytochemistry* 62: 1027–1046.

Bao, S., Hua, C., Shen, L. et al. (2020). New insights into gibberellin signaling in regulating flowering in *Arabidopsis*. *Journal of Integrative Plant Biology* 62: 118–131.

Basit, F., Chen, M., Ahmed, T. et al. (2021a). Seed priming with brassinosteroids alleviates chromium stress in rice cultivars via improving ROS metabolism and antioxidant defense response at biochemical and molecular levels. *Antioxidants* 10: 1089.

Basit, F., Liu, J., An, J. et al. (2021b). Brassinosteroids as a multidimensional regulator of plant physiological and molecular responses under various environmental stresses. *Environmental Science and Pollution Research* 28: 44768–44779.

Bell, E.M., Lin, W., Husbands, A.Y. et al. (2012). *Arabidopsis* lateral organ boundaries negatively regulates brassinosteroid accumulation to limit growth in organ boundaries. *Proceedings of the National Academy of Sciences of the United States of America* 109: 21146–21151.

Bergonci, T., Ribeiro, B., Ceciliato, P.H.O. et al. (2014). *Arabidopsis* thaliana RALF1 opposes brassinosteroid effects on root cell elongation and lateral root formation. *Journal of Experimental Botany* 65: 2219–2230.

Bertolino, L.T., Caine, R.S., and Gray, J.E. (2019). Impact of stomatal density and morphology on water-use efficiency in a changing world. *Frontiers in Plant Science*. 10: 225.

Bharti, J., Sahil, Mehta, S. et al. (2021). Mitogen-activated protein kinase, plants, and heat stress. In: *Harsh Environment and Plant Resilience: Molecular and Functional Aspects* (ed. A. Husen), 323–354. Cham: Springer International Publishing.

Bulgakov, V.P. and Avramenko, T.V. (2020). Linking brassinosteroid and ABA signaling in the context of stress acclimation. *International Journal of Molecular Sciences* 21: 5108.

Caño-Delgado, A., Yin, Y., Yu, C. et al. (2004). BRL1 and BRL3 are novel brassinosteroid receptors that function in vascular differentiation in *Arabidopsis*. *Development* 131 (21): 5341–5351.

Castorina, G. and Consonni, G. (2020). The role of brassinosteroids in controlling plant height in *Poaceae*: a genetic perspective. *International Journal of Molecular Sciences.* 21 (4): 1191.

Castro-Camba, R., Sánchez, C., Vidal, N. et al. (2022). Interactions of gibberellins with phytohormones and their role in stress responses. *Horticulturae* 8: 241.

Catterou, M., Dubois, F., Schaller, H. et al. (2001). Brassinosteroids, microtubules and cell elongation in *Arabidopsis thaliana*. I. Molecular, cellular and physiological characterization of the *Arabidopsis* bul1 mutant, defective in the Δ^7-sterol-C5-desaturation step leading to brassinosteroid biosynthesis. *Planta* 212: 659–672.

Chakraborty, N., Sharma, P., Kanyuka, K. et al. (2015). G-protein α-subunit (GPA1) regulates stress, nitrate and phosphate response, flavonoid biosynthesis, fruit/seed development and substantially shares GCR1 regulation in *A. thaliana*. *Plant Molecular Biology* 89: 559–576.

Choudhary, S.P., Yu, J.-Q., Yamaguchi-Shinozaki, K. et al. (2012). Benefits of brassinosteroid crosstalk. *Trends in Plant Science* 17: 594–605.

Çoban, Ö. and Göktürk Baydar, N. (2016). Brassinosteroid effects on some physical and biochemical properties and secondary metabolite accumulation in peppermint (*Mentha piperita* L.) under salt stress. *Industrial Crops and Products* 86: 251–258.

Cohen, J.D. and Meudt, W.J. (1983). Investigations on the mechanism of the brassinosteroid response: I. Indole-3-acetic acid metabolism and transport. *Plant Physiology* 72: 691–694.

De Lucas, M., Daviere, J.-M., Rodríguez-Falcón, M. et al. (2008). A molecular framework for light and gibberellin control of cell elongation. *Nature* 451: 480–484.

de Oliveira, V.P., Lima, M.D.R., da Silva, B.R.S. et al. (2019). Brassinosteroids confer tolerance to salt stress in *Eucalyptus urophylla* plants enhancing homeostasis, antioxidant metabolism and leaf anatomy. *Journal of Plant Growth Regulation* 38: 557–573.

Degen, G.E., Orr, D.J., and Carmo-Silva, E. (2021). Heat-induced changes in the abundance of wheat Rubisco activase isoforms. *New Phytologist* 229: 1298–1311.

Depuydt, S. and Hardtke, C.S. (2011). Hormone signalling crosstalk in plant growth regulation. *Current Biology* 21: R365–R373.

Deslauriers, S.D. and Larsen, P.B. (2010). FERONIA is a key modulator of brassinosteroid and ethylene responsiveness in *Arabidopsis* hypocotyls. *Molecular Plant* 3: 626–640.

Dilawari, R., Mehta, S., Singh, B. et al. (2023). Chapter 11 – Switching role of hydrogen sulfide in amelioration of metal stress in plant. In: *Plant Stress Mitigators* (ed. M. Ghorbanpour and MA Shahid), 187–199. London: Academic Press.

Djemal, R., Hanin, M., and Ebel, C. (2023). Control of plant responses to salt stress: significance of auxin and brassinosteroids. In: *Making Plant Life Easier and Productive Under Salinity - Updates and Prospects* (ed. N.A. Anjum, A. Masood, P. Thangavel et al.). London: IntechOpen.

Domagalska, M.A., Schomburg, F.M., Amasino, R.M. et al. (2007). Attenuation of brassinosteroid signaling enhances FLC expression and delays flowering. *Development* 134 (15): 2841–2850.

Dong, J. and Bergmann, D.C. (2010). Stomatal patterning and development. *Current Topics in Developmental Biology* 91: 267–297.

dos Santos, T.B., Ribas, A.F., de Souza, S.G.H. et al. (2022). Physiological responses to drought, salinity, and heat stress in plants: a review. *Stresses* 2: 113–135.

El-Banna, M.F., AL-Huqail, A.A., Farouk, S. et al. (2022). Morpho-physiological and anatomical alterations of salt-affected Thompson seedless grapevine (*Vitis vinifera* L.) to brassinolide spraying. *Horticulturae* 8(7): 568.

El-Esawi, M.A. (2017). Introductory chapter: hormonal regulation in plant development and stress tolerance. In: *Phytohormones-Signaling Mech Crosstalk Plant Dev Stress Responses* (ed. M.A. El-Esawi), 1–10. Rijeka: IntechOpen.

EL Sabagh, A., Islam, M.S., Hossain, A. et al. (2022). Phytohormones as growth regulators during abiotic stress tolerance in plants. *Frontiers in Agronomy* 4: 765068.

Fàbregas, N. and Caño-Delgado, A.I. (2014). Turning on the microscope turret: a new view for the study of brassinosteroid signaling in plant development. *Physiologia Plantarum* 151: 172–183.

Fariduddin, Q., Mir, B.A., Yusuf, M. et al. (2014a). 24-epibrassinolide and/or putrescine trigger physiological and biochemical responses for the salt stress mitigation in *Cucumis sativus* L. *Photosynthetica* 52: 464–474.

Fariduddin, Q., Yusuf, M., Ahmad, I. et al. (2014b). Brassinosteroids and their role in response of plants to abiotic stresses. *Biologia Plantarum* 58: 9–17.

Fariduddin, Q., Yusuf, M., Hayat, S. et al. (2009). Effect of 28-homobrassinolide on antioxidant capacity and photosynthesis in *Brassica juncea* plants exposed to different levels of copper. *Environmental and Experimental Botany* 66: 418–424.

Feng, S., Wei, Y., and Xu, H. (2019). Study on tolerance comparison of two salvia to $CdCl_2$ stress and its physiological mechanism. *Botanical Research* 8(2): 126–136.

Gallego-Bartolomé, J., Minguet, E.G., Grau-Enguix, F. et al. (2012). Molecular mechanism for the interaction between gibberellin and brassinosteroid signaling pathways in *Arabidopsis*. *Proceedings of the National Academy of Sciences of the United States of America* 109: 13446–13451.

Galstyan, A. and Nemhauser, J.L. (2019). Auxin promotion of seedling growth via ARF5 is dependent on the brassinosteroid-regulated transcription factors BES1 and BEH4. *Plant Direct* 3: e00166.

García-González, J. and van Gelderen, K. (2021). Bundling up the role of the actin cytoskeleton in primary root growth. *Frontiers in Plant Science*. 12: 777119.

Gendron, J.M., Liu, J.-S., Fan, M. et al. (2012). Brassinosteroids regulate organ boundary formation in the shoot apical meristem of *Arabidopsis*. *Proceedings of the National Academy of Sciences of the United States of America* 109: 21152–21157.

Gomes, M.M.A., Campostrini, E., Leal, N.R. et al. (2006). Brassinosteroid analogue effects on the yield of yellow passion fruit plants (*Passiflora edulis* f. *flavicarpa*). *Scientia Horticulturae* 110: 235–240.

González-García, M.-P., Vilarrasa-Blasi, J., Zhiponova, M. et al. (2011). Brassinosteroids control meristem size by promoting cell cycle progression in *Arabidopsis* roots. *Development* 138: 849–859.

Grove, M.D., Spencer, G.F., Rohwedder, W.K. et al. (1979). Brassinolide, a plant growth-promoting steroid isolated from *Brassica napus* pollen. *Nature* 281: 216–217.

Gruszka, D., Bajguz, A., Li, Q.-F. et al. (2023). An update on brassinosteroids: homeostasis, crosstalk, and adaptation to environmental stress, Volume II. *Frontiers in Plant Science* 14: 1194566.

Gudesblat, G.E., Schneider-Pizoń, J., Betti, C. et al. (2012). SPEECHLESS integrates brassinosteroid and stomata signalling pathways. *Nature Cell Biology* 14: 548–554.

Gupta, P. and Seth, C.S. (2023). 24-Epibrassinolide regulates functional components of nitric oxide signalling and antioxidant defense pathways to alleviate salinity stress in *Brassica juncea* L. cv. Varuna. *Journal of Plant Growth Regulation* 42: 4207–4222.

Gupta, P., Srivastava, S., and Seth, C.S. (2017). 24-Epibrassinolide and sodium nitroprusside alleviate the salinity stress in *Brassica juncea* L. cv. Varuna through cross talk among proline, nitrogen metabolism and abscisic acid. *Plant and Soil* 411: 483–498.

Gururani, M.A., Mohanta, T.K., and Bae, H. (2015). Current understanding of the interplay between phytohormones and photosynthesis under environmental stress. *International Journal of Molecular Sciences* 16: 19055–19085.

Hacham, Y., Holland, N., Butterfield, C. et al. (2011). Brassinosteroid perception in the epidermis controls root meristem size. *Development* 138: 839–848.

Hafeez, M.B., Zahra, N., Zahra, K. et al. (2021). Brassinosteroids: molecular and physiological responses in plant growth and abiotic stresses. *Plant Stress* 2: 100029.

Hasan, M.M., Ali, M.A., Soliman, M.H. et al. (2020). Insights into 28-homobrassinolide (HBR)-mediated redox homeostasis, AsA–GSH cycle, and methylglyoxal detoxification in soybean under drought-induced oxidative stress. *Journal of Plant Interactions* 15: 371–385.

Hasan, S.A., Hayat, S., Ali, B. et al. (2008). 28-Homobrassinolide protects chickpea (*Cicer arietinum*) from cadmium toxicity by stimulating antioxidants. *Environmental Pollution* 151: 60–66.

Hauvermale, A.L., Ariizumi, T., and Steber, C.M. (2012). Gibberellin signaling: a theme and variations on DELLA repression. *Plant Physiology* 160: 83–92.

Hayat, S., Yadav, S., Wani, A.S. et al. (2011). Comparative effect of 28-homobrassinolide and 24-epibrassinolide on the growth, carbonic anhydrase activity and photosynthetic efficiency of *Lycopersicon esculentum*. *Photosynthetica* 49: 397–404.

Helmy, Y.I., Sawan, O.M.M., and Abd-El-Halim, S.M. (1997). Growth, yield and endogenous hormones of broad bean plants as affected by brassinosteroids. *Egyptian Journal of Horticulture* 24 (1): 109–115.

Hong, Z., Ueguchi-Tanaka, M., Shimizu-Sato, S. et al. (2002). Loss-of-function of a rice brassinosteroid biosynthetic enzyme, C-6 oxidase, prevents the organized arrangement and polar elongation of cells in the leaves and stem. *The Plant Journal* 32: 495–508.

Hou, S., Niu, H., Tao, Q. et al. (2017). A mutant in the CsDET2 gene leads to a systemic brassinosteriod deficiency and super compact phenotype in cucumber (*Cucumis sativus* L.). *Theoretical and Applied Genetics* 130: 1693–1703.

Hu, S., Sanchez, D.L., Wang, C. et al. (2017a). Brassinosteroid and gibberellin control of seedling traits in maize (*Zea mays* L.). *Plant Science* 263: 132–141.

Hu, S., Wang, C., Sanchez, D.L. et al. (2017b). Gibberellins promote brassinosteroids action and both increase heterosis for plant height in maize (*Zea mays* L.). *Frontiers in Plant Science.* 8: 1039.

Hussain, Q., Asim, M., Zhang, R. et al. (2021). Transcription factors interact with ABA through gene expression and signaling pathways to mitigate drought and salinity stress. *Biomolecules* 11: 1159.

Hussain, M.A., Fahad, S., Sharif, R. et al. (2020). Multifunctional role of brassinosteroid and its analogues in plants. *Plant Growth Regulation* 92: 141–156.

Jiang, Y., Huang, L., Cheng, F. et al. (2013). Brassinosteroids accelerate recovery of photosynthetic apparatus from cold stress by balancing the electron partitioning, carboxylation and redox homeostasis in cucumber. *Physiologia Plantarum* 148: 133–145.

Kaya, C. (2020). Role of l-cysteine desulfhydrase in epibrassinolide-induced tolerance to boron toxicity in pepper (*Capsicum annuum* L.) plants. *Journal of Plant Growth Regulation* 39: 1531–1542.

Kaya, C., Aydemir, S., Akram, N.A. et al. (2018). Epibrassinolide application regulates some key physio-biochemical attributes as well as oxidative defense system in maize plants grown under saline stress. *Journal of Plant Growth Regulation* 37: 1244–1257.

Khan, I., Awan, S.A., Ikram, R. et al. (2021). Effects of 24-epibrassinolide on plant growth, antioxidants defense system, and endogenous hormones in two wheat varieties under drought stress. *Physiologia Plantarum* 172: 696–706.

Khan, M., Rozhon, W., Bigeard, J. et al. (2013). Brassinosteroid-regulated GSK3/Shaggy-like kinases phosphorylate mitogen-activated protein (MAP) kinase kinases, which control stomata development in *Arabidopsis thaliana*. *Journal of Biological Chemistry* 288: 7519–7527.

Khatoon, F., Kundu, M., Mir, H. et al. (2021). Efficacy of foliar feeding of brassinosteroid to improve growth, yield and fruit quality of strawberry (Fragaria× ananassa Duch.) grown under subtropical plain. *Communications in Soil Science and Plant Analysis* 52: 803–814.

Kim, T.-W., Guan, S., Sun, Y. et al. (2009). Brassinosteroid signal transduction from cell-surface receptor kinases to nuclear transcription factors. *Nature Cell Biology* 11: 1254–1260.

Kim, T.-W., Michniewicz, M., Bergmann, D.C. et al. (2012). Brassinosteroid regulates stomatal development by GSK3-mediated inhibition of a MAPK pathway. *Nature* 482: 419–422.

Kim, Y., Song, J.-H., Park, S.-U. et al. (2017). Brassinosteroid-induced transcriptional repression and dephosphorylation-dependent protein degradation negatively regulate BIN2-interacting AIF2 (a BR signaling-negative regulator) bHLH transcription factor. *Plant and Cell Physiology* 58: 227–239.

Kim, S.Y., Warpeha, K.M., Huber, S.C. (2019). The brassinosteroid receptor kinase, BRI1, plays a role in seed germination and the release of dormancy by cold stratification. *Journal of Plant Physiology* 241: 153031.

Kolomeichuk, L.V., Efimova, M.V., Zlobin, I.E. et al. (2020). 24-Epibrassinolide alleviates the toxic effects of NaCl on photosynthetic processes in potato plants. *Photosynthesis Research* 146: 151–163.

Kour, J., Kohli, S.K., Khanna, K. et al. (2021). Brassinosteroid signaling, crosstalk and, physiological functions in plants under heavy metal stress. *Frontiers in Plant Science*. 12: 608061.

Krishnan, S., Azhakanandam, K., Ebenezer, G.A.I. et al. (1999). Brassinosteroids and benzylaminopurine increase yield in IR50 indica rice. *Current Science* 76: 145–147.

Kumari, A. and Hemantaranjan, A. (2018). Morpho-physiological attributes of Wheat (*Triticum aestivum* L.) genotypes as influenced by brassinosteroids under heat stress. *Journal of Pharmacognosy and Phytochemistry* 7: 2111–2115.

Kumari, A. and Hemantaranjan, A. (2019). Mitigating effects of 24-epibrassinolide on heat stress damage by shifting biochemical and antioxidant defense mechanisms in wheat (*Triticum aestivum* L.) at pre-flowering stage and post-flowering stage. *Journal of Pharmacognosy and Phytochemistry* 8: 1157–1161.

Kuppusamy, K.T., Chen, A.Y., and Nemhauser, J.L. (2009). Steroids are required for epidermal cell fate establishment in *Arabidopsis* roots. *Proceedings of the National Academy of Sciences of the United States of America* 106: 8073–8076.

Lapik, Y.R. and Kaufman, L.S. (2003). The *Arabidopsis* cupin domain protein AtPirin1 interacts with the G protein α-subunit GPA1 and regulates seed germination and early seedling development. *The Plant Cell* 15: 1578–1590.

Lee, J., Han, S., Lee, H.-Y. et al. (2019a). Brassinosteroids facilitate xylem differentiation and wood formation in tomato. *Planta* 249: 1391–1403.

Lee, J., Kim, H., Park, S. et al. (2021). Brassinosteroid-BZR1/2-WAT1 module determines the high level of auxin signalling in vascular cambium during wood formation. *New Phytologist* 230: 1503–1516.

Lee, H.J., Lee, J.H., Lee, S.G. et al. (2019b). Foliar application of biostimulants affects physiological responses and improves heat stress tolerance in Kimchi cabbage. *Horticulture, Environment, and Biotechnology* 60: 841–851.

Li, P., Chen, L., Zhou, Y. et al. (2013). Brassinosteroids-induced systemic stress tolerance was associated with increased transcripts of several defence-related genes in the phloem in *Cucumis sativus. PLoS One* 8: e66582.

Li, J. and Chory, J. (1997). A putative leucine-rich repeat receptor kinase involved in brassinosteroid signal transduction. *Cell* 90: 929–938.

Li, J., Li, Y., Chen, S. et al. (2010). Involvement of brassinosteroid signals in the floral-induction network of *Arabidopsis. Journal of Experimental Botany* 61: 4221–4230.

Li, Q. and Liu, B. (2017). Genetic regulation of maize flower development and sex determination. *Planta* 245: 1–14.

Li, Y., Ma, J., Gao, X. et al. (2022). Exogenous brassinosteroids alleviate calcium deficiency-induced tip-burn by maintaining cell wall structural stability and higher photosynthesis in mini Chinese Cabbage. *Frontiers in Plant Science* 13: 999051.

Li, Q.-F., Wang, C., Jiang, L. et al. (2012). An interaction between BZR1 and DELLAs mediates direct signaling crosstalk between brassinosteroids and gibberellins in *Arabidopsis. Science Signaling* 5: ra72.

Li, Q., Xu, F., Chen, Z. et al. (2021a). Synergistic interplay of ABA and BR signal in regulating plant growth and adaptation. Nature Plants 7: 1108–1118.

Li, J., Yang, P., Gan, Y. et al. (2015). Brassinosteroid alleviates chilling-induced oxidative stress in pepper by enhancing antioxidation systems and maintenance of photosystem II. *Physiologia Plantarum Acta Physiologiae Plantarum* 37: 222.

Li, J., Yang, P., Kang, J. et al. (2016). Transcriptome analysis of pepper (*Capsicum annuum*) revealed a role of 24-epibrassinolide in response to chilling. *Frontiers in Plant Science* 7: 1281.

Li, S., Zheng, H., Lin, L. et al. (2021b). Roles of brassinosteroids in plant growth and abiotic stress response. *Plant Growth Regulation* 93: 29–38.

Liaqat, S., Umar, S., Saffeullah, P. et al. (2020). Protective effect of 24-epibrassinolide on barley plants growing under combined stress of salinity and potassium deficiency. *Journal of Plant Growth Regulation* 39: 1543–1558.

Liliane, T.N. and Charles, M.S. (2020). Factors affecting yield of crops. In: *Agronomy: Climate Change & Food Security* (ed. S.K. Amanullah), 9–24. London: IntechOpen.

Lima, J.V. and Lobato, A.K.S. (2017). Brassinosteroids improve photosystem II efficiency, gas exchange, antioxidant enzymes and growth of cowpea plants exposed to water deficit. *Physiology and Molecular Biology of Plants* 23: 59–72.

Liu, J., Gao, H., Wang, X. et al. (2014). Effects of 24-epibrassinolide on plant growth, osmotic regulation and ion homeostasis of salt-stressed canola. *Plant Biology* 16: 440–450.

Lunde, C., Kimberlin, A., Leiboff, S. et al. (2019). Tasselseed5 overexpresses a wound-inducible enzyme, ZmCYP94B1, that affects jasmonate catabolism, sex determination, and plant architecture in maize. *Communications Biology* 2: 114.

Lv, B., Tian, H., Zhang, F. et al. (2018). Brassinosteroids regulate root growth by controlling reactive oxygen species homeostasis and dual effect on ethylene synthesis in *Arabidopsis*. *PLOS Genetics* 14: e1007144.

Mahesh, K., Balaraju, P., Ramakrishna, B. et al. (2013). Effect of brassinosteroids on germination and seedling growth of radish (*Raphanus sativus* L.) under PEG-6000 induced water stress. *American Journal of Plant Science* 4 (12): 40442.

Manghwar, H., Hussain, A., Ali, Q. et al. (2022). Brassinosteroids (BRs) role in plant development and coping with different stresses. *International Journal of Molecular Sciences* 23: 1012.

Mazzoni-Putman, S.M., Brumos, J., Zhao, C. et al. (2021). Auxin interactions with other hormones in plant development. *Cold Spring Harbor Perspectives in Biology* 13: a039990.

Mir, B.A., Khan, T.A., and Fariduddin, Q. (2015). 24-epibrassinolide and spermidine modulate photosynthesis and antioxidant systems in *Vigna radiata* under salt and zinc stress. *International Journal of Advanced Research* 3: 592–608.

Mitchell, J.W., Mandava, N., Worley, J.F. et al. (1970). Brassins—a new family of plant hormones from rape pollen. *Nature* 225: 1065–1066.

Mohanabharathi, M., Sritharan, N., Senthil, A. et al. (2019). Physiological studies for yield enhancement in finger millet under drought condition. *Journal of Pharmacognosy and Phytochemistry* 8: 3308–3312.

Mokari-Firuzsalari, S., Khomari, S., Seyed-Sharifi, R. et al. (2019). The combined influence of zinc and epibrassinolide increase tolerance to salt stress in *Brassica napus* L. *Russian Journal of Plant Physiology* 66: 240–249.

Mussig, C., Shin, G.-H., and Altmann, T. (2003). Brassinosteroids promote root growth in *Arabidopsis*. *Plant Physiology* 133: 1261–1271.

Nakaya, M., Tsukaya, H., Murakami, N. et al. (2002). Brassinosteroids control the proliferation of leaf cells of *Arabidopsis thaliana*. *Plant and Cell Physiology* 43: 239–244.

Noh, B., Lee, S.-H., Kim, H.-J. et al. (2004). Divergent roles of a pair of homologous jumonji/zinc-finger–class transcription factor proteins in the regulation of *Arabidopsis* flowering time. *The Plant Cell* 16: 2601–2613.

Nolan, T.M., Vukašinović N, Liu, D. et al. (2020). Brassinosteroids: multidimensional regulators of plant growth, development, and stress responses. *The Plant Cell* 32: 295–318.

Nomura, T., Nakayama, M., Reid, J.B. et al. (1997). Blockage of brassinosteroid biosynthesis and sensitivity causes dwarfism in garden pea. *Plant Physiology* 113: 31–37.

Oh, M.-H., Honey, S.H., and Tax, F.E. (2020). The control of cell expansion, cell division, and vascular development by brassinosteroids: a historical perspective. *International Journal of Molecular Sciences*. 21: 1743.

Oh, E., Zhu, J.-Y., Bai, M.-Y. et al. (2014). Cell elongation is regulated through a central circuit of interacting transcription factors in the *Arabidopsis* hypocotyl. *Elife* 3: e03031.

Oshunsanya, S.O., Nwosu, N.J., and Li, Y. (2019). Abiotic stress in agricultural crops under climatic conditions. In: *Sustainable Agriculture, Forest and Environmental Management* (ed. M. Jhariya, A. Banerjee, R. Meena et al.), 71–100. Singapore: Springer.

Otie, V., Ibrahim, A., Udo, I. et al. (2022). Foliarly applied 24-epibrassinolide modulates the electrical conductivity of the saturated rhizospheric soil extracts of soybean under salinity stress. *Plants* 11: 2330.

Özdemir, F., Bor, M., Demiral, T. et al. (2004). Effects of 24-epibrassinolide on seed germination, seedling growth, lipid peroxidation, proline content and antioxidative system of rice (*Oryza sativa* L.) under salinity stress. *Plant Growth Regulation* 42: 203–211.

Pandey, V. and Pati, A.K. (2016). A new candidate of catalase appears during the germination after priming in naturally aged neem (*Azadirachta indica*) seeds. *Journal of Plant Biochemistry & Physiology* 4: 1–4.

Pandey, V. and Pati, A.K. (2017). Limitation of improvement in germination by osmopriming of differentially aged non-orthodox neem (*Azadirachta indica*) seedstle. *Biochemistry and Physiology: Open Access* 6: 1–7.

Pereira, Y.C., Nascimento, A.N.D.S., Aguiar, B.T.D.S. et al. (2022). Anatomical modifications modulated by pretreatment with 24-epibrassinolide alleviate boron stress in soybean plants: valuable repercussions on nutrient contents, photosynthesis, and biomass. *Journal of Soil Science and Plant Nutrition* 22: 4533–4550.

Peres, A.L.G.L., Soares, J.S., Tavares, R.G. et al. (2019). Brassinosteroids, the sixth class of phytohormones: a molecular view from the discovery to hormonal interactions in plant development and stress adaptation. *International Journal of Molecular Sciences* 20: 331.

Perez-Borroto, L.S., Guzzo, M.C., Posada, G. et al. (2022). A brassinosteroid functional analogue increases soybean drought resilience. *Scientific Reports* 12: 11294.

Planas-Riverola, A., Gupta, A., Betegón-Putze, I. et al. (2019). Brassinosteroid signaling in plant development and adaptation to stress. *Development* 146: dev151894.

Poppenberger, B., Fujioka, S., Soeno, K. et al. (2005). The UGT73C5 of *Arabidopsis thaliana* glucosylates brassinosteroids. *Proceedings of the National Academy of Sciences* 102: 15253–15258.

Rattan, A., Kapoor, D., Kapoor, N. et al. (2020). Brassinosteroids regulate functional components of antioxidative defense system in salt stressed maize seedlings. *Journal of Plant Growth Regulation* 39: 1465–1475.

Rehman, S.U., Bilal, M., Rana, R.M. et al. (2016). Cell membrane stability and chlorophyll content variation in wheat (*Triticum aestivum*) genotypes under conditions of heat and drought. *Crop & Pasture Science* 67: 712–718.

Ribeiro, D.G.D.S., Silva, B.R.S.D., and Lobato, A.K.D.S. (2019). Brassinosteroids induce tolerance to water deficit in soybean seedlings: contributions linked to root anatomy and antioxidant enzymes. *Physiologia Plantarum Acta Physiologiae Plantarum* 41: 82.

Saini, S., Sharma, I., and Pati, P.K. (2015). Versatile roles of brassinosteroid in plants in the context of its homoeostasis, signaling and crosstalks. *Frontiers in Plant Science*. 6: 950.

Saito, M., Kondo, Y., and Fukuda, H. (2018). BES1 and BZR1 redundantly promote phloem and xylem differentiation. *Plant and Cell Physiology* 59: 590–600.

Sakakibara, M. and Mori, K. (1982). Facile synthesis of (22 R, 23 R)-homobrassinolide. *Agricultural and Biological Chemistry* 46: 2769–2779.

Sasse, J.M. (1985). The place of brassinolide in the sequential response to plant growth regulators in elongating tissue. *Physiologia Plantarum* 63: 303–308.

Schwechheimer, C. (2012). Gibberellin signaling in plants–the extended version. *Frontiers in Plant Science* 2: 107.

Shahzad, A., Ullah, S., Dar, A.A. et al. (2021). Nexus on climate change: agriculture and possible solution to cope future climate change stresses. *Environmental Science and Pollution Research* 28: 14211–14232.

Shakirova, F., Allagulova, C., Maslennikova, D. et al. (2016). Involvement of dehydrins in 24-epibrassinolide-induced protection of wheat plants against drought stress. *Plant Physiology and Biochemistry* 108: 539–548.

Sharma, P. and Bhardwaj, R. (2007). Effect of 24-epibrassinolide on seed germination, seedling growth and heavy metal uptake in *Brassica juncea* L. *General and Applied Plant Physiology* 33: 59–73.

Sharma, M., Mahajan, P., Singh, H.P. et al. (2019). 24-Epibrassinolide pre-treatment reduces alkaline-induced oxidative stress in red rice seedlings. *Environmental Science and Pollution Research* 26: 23192–23197.

Sharma, P., Pandey, V., Sharma, M.M.M. et al. (2021). A review on biosensors and nanosensors application in agroecosystems. *Nanoscale Research Letters* 16: 136.

Sharma, P., Sharma, M.M.M., Patra, A. et al. (2020). Chapter 9 – The role of key transcription factors for cold tolerance in plants. In: *Transcription Factors for Abiotic Stress Tolerance in Plants* (ed. S.H. Wani), 123–152. Waltham, MA: Academic Press.

Shu, S., Tang, Y., Yuan, Y. et al. (2016). The role of 24-epibrassinolide in the regulation of photosynthetic characteristics and nitrogen metabolism of tomato seedlings under a combined low temperature and weak light stress. *Plant Physiology and Biochemistry* 107: 344–353.

Siddiqui, H., Ahmed, K.B.M., and Hayat, S. (2018). Comparative effect of 28-homobrassinolide and 24-epibrassinolide on the performance of different components influencing the photosynthetic machinery in *Brassica juncea* L. *Plant Physiology and Biochemistry* 129: 198–212.

Singh, B., Padhy, A.K., Ambreen, H. et al. (2022). Understanding abiotic stress responses in lentil under changing climate regimes. . In: *Developing Climate Resilient Grain and Forage Legumes* (ed. U.C. Jha, H. Nayyar, S.K. Agrawal et al.), 179–204. Singapore: Springer Nature Singapore.

Singh, A.P. and Savaldi-Goldstein, S. (2015). Growth control: brassinosteroid activity gets context. *Journal of Experimental Botany* 66: 1123–1132.

Soares, T.F.S.N., dos Santos Dias, D.C.F., Oliveira, A.M.S. et al. (2020). Exogenous brassinosteroids increase lead stress tolerance in seed germination and seedling growth of *Brassica juncea* L. *Ecotoxicology and Environmental Safety* 193: 110296.

Song, Y., Wang, Y., Yu, Q. et al. (2023). Regulatory network of GSK3-like kinases and their role in plant stress response. *Frontiers in Plant Science.* 14: 1123436.

Steber, C.M. and McCourt, P. (2001). A role for brassinosteroids in germination in *Arabidopsis. Plant Physiology* 125: 763–769.

Talaat, N.B. and Shawky, B.T. (2013). 24-Epibrassinolide alleviates salt-induced inhibition of productivity by increasing nutrients and compatible solutes accumulation and enhancing antioxidant system in wheat (*Triticum aestivum* L.). *Physiologia Plantarum Acta Physiologiae Plantarum* 35: 729–740.

Thompson, M.J., Mandava, N.B., Meudt, W.J. et al. (1981). Synthesis and biological activity of brassinolide and its 22β, 23β-isomer: novel plant growth-promoting steroids. *Steroids* 38: 567–580.

Umar, O.B., Ranti, L.A., Abdulbaki, A.S. et al. (2021). Stresses in plants: biotic and abiotic. In: *Current Trends in Wheat Research* (ed. M.-R. Ansari), Ch. 7. Rijeka: IntechOpen.

Vardhini, B.V. and Rao, S.S.R. (1998). Effect of brassinosteroids on growth, metabolite content and yield of *Arachis hypogaea. Phytochemistry* 48: 927–930.

Vogler, F., Schmalzl, C., Englhart, M. et al. (2014). Brassinosteroids promote *Arabidopsis* pollen germination and growth. *Plant Reproduction* 27: 153–167.

Walcher, C.L. and Nemhauser, J.L. (2012). Bipartite promoter element required for auxin response. *Plant Physiology* 158: 273–282.

Wang, Y., Jiang, Q., Wang, X. et al. (2023). Brassinosteroid stimulates hydrogen peroxide biosynthesis and reduces the effect of cold stress. *Journal of Plant Growth Regulation* 42: 3757–3769.

Wang, R., Liu, M., Yuan, M. et al. (2016). The brassinosteroid-activated BRI1 receptor kinase is switched off by dephosphorylation mediated by cytoplasm-localized PP2A B' subunits. *Molecular Plant* 9: 148–157.

Wang, Q., Yu, F., and Xie, Q. (2020). Balancing growth and adaptation to stress: crosstalk between brassinosteroid and abscisic acid signaling. *Plant, Cell & Environment* 43: 2325–2335.

Wani, A.S., Ahmad, A., Hayat, S. et al. (2019). Epibrassinolide and proline alleviate the photosynthetic and yield inhibition under salt stress by acting on antioxidant system in mustard. *Plant Physiology and Biochemistry* 135: 385–394.

Wei, L.-J., Deng, X.-G., Zhu, T. et al. (2015). Ethylene is involved in brassinosteroids induced alternative respiratory pathway in cucumber (*Cucumis sativus* L.) seedlings response to abiotic stress. *Frontiers in Plant Science* 6: 982.

Wei, Z. and Li, J. (2016). Brassinosteroids regulate root growth, development, and symbiosis. *Molecular Plant* 9: 86–100.

Werner, T., Motyka, V., Strnad, M. et al. (2001). Regulation of plant growth by cytokinin. *Proceedings of the National Academy of Sciences* 98: 10487–10492.

Werner, T., Nehnevajova, E., Köllmer, I. et al. (2010). Root-specific reduction of cytokinin causes enhanced root growth, drought tolerance, and leaf mineral enrichment in *Arabidopsis* and tobacco. *The Plant Cell* 22: 3905–3920.

Xia, X.-J., Huang, L.-F., Zhou, Y.-H. et al. (2009a). Brassinosteroids promote photosynthesis and growth by enhancing activation of Rubisco and expression of photosynthetic genes in *Cucumis sativus*. *Planta* 230: 1185–1196.

Xia, X.-J., Wang, Y.-J., Zhou, Y.-H. et al. (2009b). Reactive oxygen species are involved in brassinosteroid-induced stress tolerance in cucumber. *Plant Physiology* 150: 801–814.

Xiong, M., Chu, L., Li, Q. et al. (2021). Brassinosteroid and gibberellin coordinate rice seed germination and embryo growth by regulating glutelin mobilization. *The Crop Journal* 9: 1039–1048.

Xu, H., Shida, A., Futatsuya, F. et al (1994). Effects of epibrassinolide and abscisic acid on sorghum plants growing under soil water deficit: ii. physiological basis for drought resistance induced by exogenous epibrassinolide and abscisic acid. *Japanese Journal of Crop Science* 63: 676–681.

Xu, Y., Zhang, X., Yang, H. et al. (2022). Effects of exogenous brassinolide application at the silking stage on nutrient accumulation, translocation and remobilization of waxy maize under post-silking heat stress. *Agriculture* 12: 572.

Xue, H., Gao, X., He, P. et al. (2022). Origin, evolution, and molecular function of DELLA proteins in plants. *The Crop Journal* 10: 287–299.

Yamagami, A., Saito, C., Nakazawa, M. et al. (2017). Evolutionarily conserved BIL4 suppresses the degradation of brassinosteroid receptor BRI1 and regulates cell elongation. *Scientific Reports* 7: 5739.

Yamamoto, R., Fujioka, S., Demura, T. et al. (2001). Brassinosteroid levels increase drastically prior to morphogenesis of tracheary elements. *Plant Physiology* 125: 556–563.

Yamauchi, D. (2007). Hormonal regulation of the expression of cysteine proteinase genes in germinated cotyledons of common bean seeds. *Plant Biotechnology* 24: 231–235.

Yang, J., Duan, G., Li, C. et al. (2019). The crosstalks between jasmonic acid and other plant hormone signaling highlight the involvement of jasmonic acid as a core component in plant response to biotic and abiotic stresses. *Frontiers in Plant Science* 10: 1349.

Ye, Q., Zhu, W., Li, L. et al. (2010). Brassinosteroids control male fertility by regulating the expression of key genes involved in *Arabidopsis* anther and pollen development. *Proceedings of the National Academy of Sciences* 107: 6100–6105.

Yin, Y., Wang, Z.-Y., Mora-Garcia, S. et al. (2002). BES1 accumulates in the nucleus in response to brassinosteroids to regulate gene expression and promote stem elongation. *Cell* 109: 181–191.

Yokota, T., Ohnishi, T., Shibata, K. et al. (2017). Occurrence of brassinosteroids in non-flowering land plants, liverwort, moss, lycophyte and fern. *Phytochemistry* 136: 46–55.

Yu, X., Li, L., Li, L. et al. (2008). Modulation of brassinosteroid-regulated gene expression by Jumonji domain-containing proteins ELF6 and REF6 in *Arabidopsis*. *Proceedings of the National Academy of Sciences* 105: 7618–7623.

Yuan, L., Shu, S., Sun, J. et al. (2012). Effects of 24-epibrassinolide on the photosynthetic characteristics, antioxidant system, and chloroplast ultrastructure in *Cucumis sativus* L. under $Ca(NO_3)_2$ stress. *Photosynthesis Research* 112: 205–214.

Yue, J., You, Y., Zhang, L. et al. (2019). Exogenous 24-epibrassinolide alleviates effects of salt stress on chloroplasts and photosynthesis in *Robinia pseudoacacia* L. Seedlings. *Journal of Plant Growth Regulation* 38: 669–682.

Yuldashev, R., Avalbaev, A., Bezrukova, M. et al. (2012). Cytokinin oxidase is involved in the regulation of cytokinin content by 24-epibrassinolide in wheat seedlings. *Plant Physiology and Biochemistry* 55: 1–6.

Yusuf, M., Fariduddin, Q., Ahmad, A. (2012). 24-Epibrassinolide modulates growth, nodulation, antioxidant system, and osmolyte in tolerant and sensitive varieties of *Vigna radiata* under different levels of nickel: a shotgun approach. *Plant Physiology and Biochemistry* 57: 143–153.

Zadvornova, Y.V., Alekseichuk, G.N., Laman, N.A. et al. (2005). Effect of brassinosteroids on activation of the cell cycle during germination of *Brassica oleracea* L. seeds. *Doklady Natsional'noi Akademii Nauk Belarusi* 49: 70–73.

Zhang, S., Cai, Z., Wang, X. (2009). The primary signaling outputs of brassinosteroids are regulated by abscisic acid signaling. *Proceedings of the National Academy of Sciences* 106: 4543–4548.

Zhang, Z., Chen, Z., Song, H., Cheng, S. (2023). From plant survival to thriving: exploring the miracle of brassinosteroids for boosting abiotic stress resilience in horticultural crops. *Frontiers in Plant Science* 14: 1218229.

Zhang, M., Zhai, Z., Tian, X. et al. (2008). Brassinolide alleviated the adverse effect of water deficits on photosynthesis and the antioxidant of soybean (*Glycine max* L.). *Plant Growth Regulation* 56: 257–264.

Zhao, M., Yuan, L., Wang, J. et al. (2019). Transcriptome analysis reveals a positive effect of brassinosteroids on the photosynthetic capacity of wucai under low temperature. *BMC Genomics* 20: 1–19.

Zhiponova, M.K., Vanhoutte, I., Boudolf, V. et al. (2013). Brassinosteroid production and signaling differentially control cell division and expansion in the leaf. *New Phytologist* 197: 490–502.

Zhu, J.-Y., Sae-Seaw, J., Wang, Z.-Y. (2013). Brassinosteroid signalling. *Development* 140: 1615–1620.

Zhu, T., Deng, X., Zhou, X. et al. (2016). Ethylene and hydrogen peroxide are involved in brassinosteroid-induced salt tolerance in tomato. *Scientific Reports* 6: 35392.

2

Brassinosteroid-mediated Regulation of Plant Biology: An Omics Approach

*Indu Sharma[1], Ashish Sharma[2], Ajay Sharma[3], Vikas Sharma[4], Pooja Sharma[5,6], Sonika Sharma[7], Rasha S. El-Serafy[8], Renu Bhardwaj[6], and Ashutosh Sharma[7]**

[1]*Department of Life Sciences, University Institute of Sciences, Sant Baba Bhag Singh University, Jalandhar, Punjab, India*
[2]*Department of Botany and Environment Studies, DAV University, Sarmastpur, Jalandhar, Punjab, India*
[3]*Department of Chemistry, Career Point University, Tikker-Kharwarian, Hamirpur, Himachal Pradesh, India*
[4]*Department of Molecular Biology and Genetic Engineering, Lovely Professional University, Phagwara, Kapurthala, Punjab, India*
[5]*Department of Microbiology, DAV University, Sarmastpur, Jalandhar, Punjab, India*
[6]*Department of Botanical and Environmental Sciences, Guru Nanak Dev University, Amritsar, Punjab, India*
[7]*Faculty of Agricultural Sciences, DAV University, Jalandhar, Punjab, India*
[8]*Horticulture Department, Faculty of Agriculture, Tanta University, Tanta, Egypt*

*Corresponding author: sharma_tosh_ashu@yahoo.co.in

2.1 Introduction

Steroidal hormones have been recorded from both animal and plant kingdoms. Mammals synthesize them in specialized glands, while in plants, each cell has the potential to produce them (Kim and Russinova 2020). Plants produce a group of plant-specific growth-promoting steroidal hormones, referred to as brassinosteroids (BRs), to regulate normal growth and development and for mitigation of various biotic and abiotic stresses. The "brassino" part of the BRs name is linked to the plant species from where they were first isolated i.e. *Brassica napus* (rapeseed plant) pollens in the year 1979 by Grove and co-workers. The first plant hormone of this category was then named (22R, 23R, 24S)-2α-3α, 22,23-tetrahydroxy-24-methyl-6,7-s-5α-cholestano-6,7-lactone or Brassinolide (BL) (Grove et al. 1979). BRs are naturally occurring, endogenous, steroidal plant growth regulators with a "5α-cholestane" ring structure, and are hydroxylated at various positions that are biologically active based on a 6-oxo functional group or lactone structure (Sharma et al. 2018a). BRs are structurally similar to androgens, estrogens, corticoids, and ecdysteroids. Due to their ubiquitous presence and important role in plant development, growth, plant stress protection, and eliciting diverse physiological responses, BRs are commonly referred to as the sixth class of plant hormones (Bajguz et al. 2020).

Brassinosteroids and Abiotic Stress: Recent Developments on Physiological and Molecular Aspects, First Edition. Edited by Anket Sharma, et al.
Published 2026 by John Wiley & Sons Ltd.

To date, more than 69 BRs have been identified from 64 plant species, including gymnosperms (6), angiosperms (41 dicots and 12 monocots), pteridophytes (1), algae (3), bryophytes (1), mosses (1), lycophytes (2), and fern species (13) (Manghwar et al. 2022; Sharma et al. 2018a). After the discovery of BRs, their exogenous applications have revealed bioactive and biostimulant action in a wide array of plants, including agricultural and horticultural crops. Various modes of BR application in plants (Figure 2.1) include seed priming, foliar spray, soil drenching, incorporation in tissue culture media, and addition in hydroponic solution (Zhang et al. 2007; Arteca and Arteca 2001; Chone et al. 2018; Hayat et al. 2012; Zheng et al. 2017). Generally, three types of BRs, C-27, C-28, and C-29, are endogenously produced through a mevalonic acid (MVA)-dependent BR-specific biosynthetic pathway. Among all BRs, BL, 24-epibrassinolide (EBL), and 28-homobrassinolide (HBL) have been widely explored for their potential in plant biology, specifically their physiological significance, and other roles for human welfare.

BRs regulate various cellular processes like cell division, expansion or elongation and differentiation, homeostasis, and a range of physiological processes such as seed germination and seedling growth, photomorphogenesis, vascular development and differentiation, xylem differentiation, photosynthesis, senescence, programmed cell death, fruit ripening and quality parameters, yield, plant stress protection and resistance, antioxidant defense system, and enzyme activation, etc. (Ali 2017; Bhardwaj et al. 2010, 2011, 2014; Gutiérrez-Villamil et al. 2024; Randhawa et al. 2010; Sharma et al. 2018a; Soares et al. 2020). Besides, BRs have also been reported to possess various medicinal and pharmacological potentials due to their antibacterial, antifungal, antiviral, antiproliferative, immunomodulatory, neuroprotective, anticancerous, antiangiogenic, and antigenotoxic properties (Bajguz et al. 2013; Bhardwaj et al. 2011; Hussain et al. 2020; Kohli et al. 2020; Ma et al. 2022). Recently, gene expression analysis revealed that EBL improved maize seedling growth through stimulation of root gravitropic response and shoot growth (Trevisan et al. 2020).

During the initial studies, BRs were mainly focused on their effects on plant growth and physiology, as well as in elucidating their chemical nature (Mitchell et al. 1970; Grove et al. 1979). However, the exact molecular mechanism and interacting partners of BR signaling were not well understood. Recent advances in omics technologies have further improved our understanding of BR-mediated responses in plants to an appreciable extent,

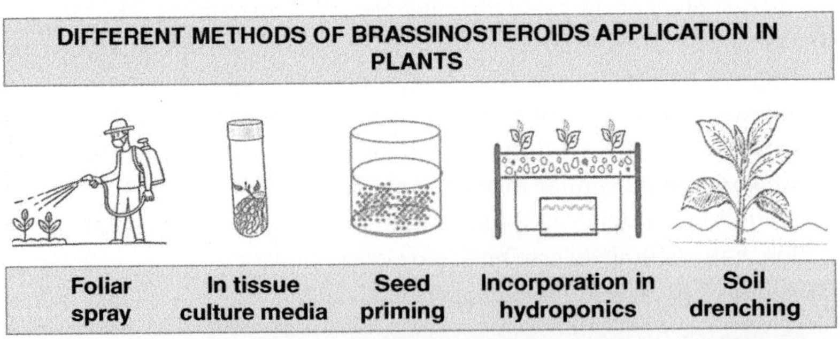

Figure 2.1 A schematic diagram representing various methods of brassinosteroid application in plants.

which further supports their much-anticipated role in improving crop productivity (Kang & Guo 2011). These "Omics" approaches (involving genomics, proteomics, phenomics, metabolomics, transcriptomics, etc.) rely on high-throughput technologies like Next Generation Sequencing (NGS), metabolite profiling, and automated/semi-automated phenotyping, together with advancements in computational capabilities that have enabled the researchers to go one step farther and analyze multi-omics data simultaneously to dissect the intricate network of BR signaling, which thereby leads to BR-mediated responses in plants. The integration of such omics technologies in comparison to the studies on a single omics type, referred to as "multi-omics," is now opening a wide landscape of opportunities to understand the intricate networking partners in generating a BR-mediated response in plants.

This chapter mainly focuses on genomics, transcriptomics, proteomics, and metabolomics approaches and their integrated use to elucidate the molecular mechanisms by which BRs influence plant growth and stress tolerance. The use of omics approaches to identify key genes, proteins, and metabolites involved in a BR-mediated response in plants and associated signaling networks has improved our current understanding with an overall aim to improve crop productivity and stress tolerance. The chapter aims to provide an up-to-date review of our current understanding of BRs' role in plant biology gained from omics and multi-omics approaches. Further, the investigation of BR-mediated responses through diverse approaches of system biology highlights the molecular signaling events of the BR biosynthetic pathway. By exploring transcript profiles, protein groups, transcription factors (TFs), phosphorylation sites of BRs, and identifying BR–receptor complexes at the plasma membrane, BRs have been found to regulate numerous growth and post-harvest characteristics in a range of plants. This chapter reviews the use of different omics approaches (i.e. use of genomics, proteomics, transcriptomics, and metabolomics, etc.) in understanding the BR-mediated regulation of various plant physiological processes.

2.2 Brassinosteroids: A Master Regulator of Plant Physiology

Phytosterols or plant sterols, belonging to the triterpene family, have a tetracyclic structure, which is similar to cholesterol (Moreau et al. 2018). These sterols stabilize lipid rafts, plasma membrane enzymes, and plant cell membranes. Phytosterols are also precursors of vital steroidal bioactive compounds like phytoecdysteroids, steroidal glycoalkaloids, steroidal saponins, and BRs. In the early BR-specific biosynthetic pathway, isopentenyl pyrophosphate (IPP) is an indirect compound involved in the cholesterol biosynthesis pathway, which is produced *via* two branches: (i) non-MVA in the lower plants and (ii) MVA in most of the higher plants (Bajguz et al. 2020). In the MVA pathway, acetyl-CoA produces 3-hydroxy-3-methylglutaryl-CoA, which is then reduced to MVA (Figure 2.2). From MVA, campesterol is derived from the sterol-specific biosynthetic pathway, and this campesterol then enters into BR-specific biosynthesis (cycloartenol- and cycloartanol-dependent) pathway in order to synthesize BL, either through the early C-22 or late C-22 oxidation branch, followed by early C-6 or late C-6 oxidation pathway. In the C28-BR biosynthetic pathway, campesterol and 22α-hydroxycampesterol are involved, whereas during C27-BR

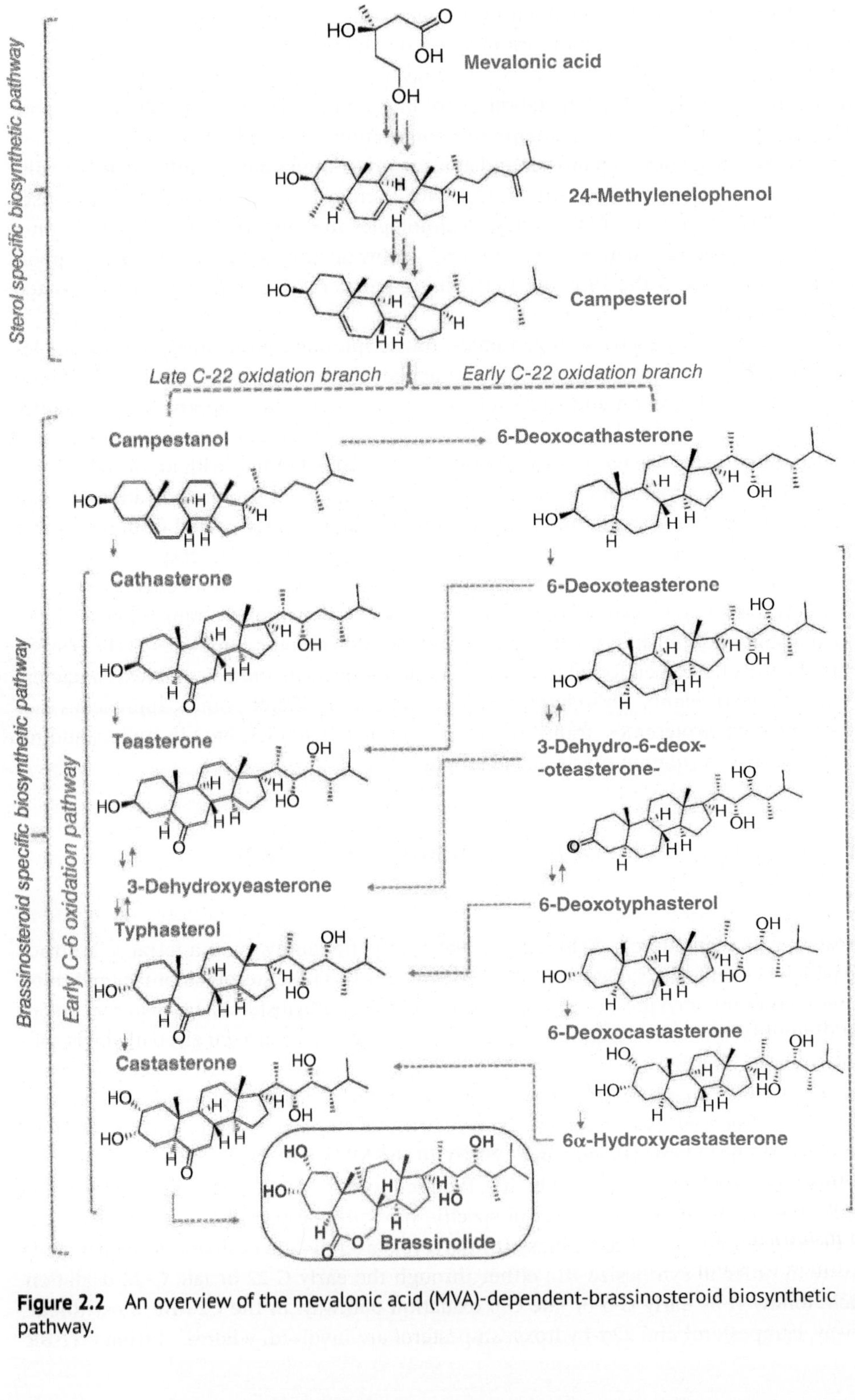

Figure 2.2 An overview of the mevalonic acid (MVA)-dependent-brassinosteroid biosynthetic pathway.

biosynthesis, cholesterol is converted to 28-norBL. Further, the C29-BR biosynthetic pathway begins with β-sitosterol, which produces 28-HBL.

A large number of researchers have reported that BRs are an important group of steroidal phytohormones that regulate the majority of physiological processes in plants. BRs play a crucial role in cell elongation and cell size (Que et al. 2018), cell expansion and cell division (Oh et al. 2020), grain length regulation (Du et al. 2023), fertility and seed filling (Wang et al. 2010), chlorophyll fluorescence and gas exchange (Pereira et al. 2019), carbohydrate metabolism (Xu et al. 2015), cellulose biosynthesis (Sánchez-Rodríguez et al. 2017; Que et al. 2018), fiber cell development and elongation (Liu et al. 2023), vascular development (Oh et al. 2020) and wood development (Lu et al. 2023), fruit quality and yield (Ghoneem et al. 2024), post-harvest processes such as regulation of ripening and senescence (Gutiérrez-Villamil et al. 2024). Exogenous BR application in fruits and vegetable crops can be a potential alternative to post-harvest technology for regulating ripening and senescence by stimulating various molecular signaling cascades (Gutiérrez-Villamil et al. 2024). Eco-friendly and non-toxic synthetic BRs are commercially available for large-scale agricultural production, which are utilized for enhancing fruit quality and quantity in horticultural crops (Ali 2017). BR treatments regulate cellulose deposition, cell elongation, and size of petioles in carrots, whereas BR-deficient mutants caused dwarfism in *Arabidopsis* and rice (Que et al. 2018).

In rice, BR signaling is regulated by plasma membrane-localized receptor kinases, OsBAKs [*Oryza sativa* BR-INSENSITIVE1 (BRI1)-ASSOCIATED RECEPTOR KINASES], where OsBZR1 (*O. sativa* BRASSINAZOLE RESISTANT 1) binds to OsBAK2 promoter, which further suppresses its expression, forming a BR feedback inhibition loop (Du et al. 2023). The *osbak2* mutant has been found to show BR-deficient phenotypes and negatively alter the accumulation of OsBZR1. Further, the study revealed that through a negative feedback loop, OsBAK2 and OsBZR1 maintain BR-homeostasis and rice grain length regulation. Apart from grain length, BR application is also reported to enhance grain size, leaf angle, and plant height in rice (Liu et al. 2021). Moreover, BRs regulate plant height, lint percentage, boll number, and yield, whereas they also delay maturity in cotton (Ghasemi Bezdi et al. 2024).

As a potential stress adaptogen, EBL is commercially sold with a brand name "Epin-Extra" for improving the growth and development of potatoes. Pavlovna et al. (2015) found that applying EBL as "Epin-Extra" increased leaf assimilation, dry matter and starch content, net photosynthesis productivity, plant stress resistance, and biochemical parameters, whereas it decreased nitrate accumulation in potato tubers. The bio-stimulant potential of this formulation in improving plant growth and development is also considered quite promising. In early and mid-season maturing sweet cherry cultivars, 28-HBL decreased the maturation period, increased fruit firmness, fruit color and stem removal forces, and advanced cherry maturity and harvest (Mandava and Wang 2015). Further, studies have also emphasized that HBL was not generally phytotoxic. In another study conducted by Lu et al. (2023), EBL was reported to be involved in cellulose and lignin synthesis, and wood formation in both interspecific hybrids and female parent of *Fraxinus mandshurica*.

Numerous studies emphasized that EBL ameliorated nickel (Ni) toxicity in radish (Sharma et al. 2011a), cadmium (Cd) and mercury (Hg) stress in radish (Sharma et al. 2012), pesticide-induced phytotoxicities (Shahzad et al. 2018), arsenic (As) stress in

Arabidopsis (Surgun-Acar and Zemheri-Navruz 2019), and lead (Pb) stress by regulating the activities of antioxidant enzymes in *Brassica juncea* seedlings (Soares et al. 2020) and *Festuca arundinacea* (Zhong et al. 2020). Of late, HBL has also been established as a safe alternative for conferring plant stress protection in diverse plants. HBL mitigated heavy metal stress (Sharma et al. 2014), including chromium (Cr) (Sharma et al. 2011b) and Cd toxicity in radish (Sharma et al. 2010) and wheat (Hayat et al. 2014), by regulating the activities and gene expression of antioxidant enzymes (Sharma et al. 2018b) as well as other abiotic stresses like temperature stress (Qazi et al. 2014; Sirhindi et al. 2017) and salt stress in *Brassica juncea* (Alyemeni et al. 2013; Kaur et al. 2018) and *Vigna radiata* (Hayat et al. 2010), drought toxicity in soyabean (Hasan et al. 2020), and chilling stress in *Cucumis sativus* (Fariduddin et al. 2011). Besides the antioxidant defense system, BRs also affect the photosynthetic machinery (Alyemeni et al. 2013; Fariduddin et al. 2011; Hayat et al. 2010; Siddiqui et al. 2018), fluorescence (Hayat et al. 2010), clastogenic aberrations (Kaur et al. 2024), physiological traits (Ahmad et al. 2017) and physio-biochemical attributes (Khan et al. 2023), redox homeostasis (Hasan et al. 2020), sugar metabolism (Kaur et al. 2024), and nutrient uptake (Khan et al. 2023) in several plants to ensure plant stress protection.

Hence, researchers have highlighted the pleiotropic impact of BRs in regulating various stress responses, plant developmental and physiological processes, including physiological, biochemical, and agronomic traits. Further genomic analysis of BR function-specific genes, their beneficial alleles, and strategies like optimizing hormone distribution would help establish BRs as important, relatively bio-safe agents, potentially useful for the next agricultural biotechnological revolution (Yang et al. 2024).

2.3 Omics Approaches to Uncover the Brassinosteroid-mediated Responses

Using a systems biology approach, BR-mediated molecular signaling events were studied in *Arabidopsis* seedlings (Clark et al. 2021). The study profiled transcripts, protein groups, and phosphorylation sites. Further, it employed a network inference pipeline to identify phosphorylation sites on BES1 (BRI1-Ethylmethylsulfone-Suppressor 1) and BRON (BRONTOSAURUS), TFs regulating cell division whose expression is modulated by BR-responsive kinases and TFs. BRON was found to be an additional gene, which is regulated during BR-mediated responses. Further, it has been found that the BR-signaling pathway is mediated by the activation of BR–receptor complexes at the plasma membrane, which are receptor-like kinase BRI1 and co-receptor BRI1-ASSOCIATED KINASE 1, i.e. BAK1 (Hwang et al. 2021). The dissociation of BRI1 inhibitor 1 and BKI1 (BRI1 KINASE INHIBITOR 1) is initiated by activating BSU1 (BRI1 SUPPRESSOR 1) and inactivating BIN2 (BRASSINOSTEROID INSENSITIVE 2) kinase. BIN2, a GSK3/SHAGGY-like kinase family member, regulates the phosphorylation status of plant-specific TFs, playing a crucial role in BR perception. GSK3 (glycogen synthase kinase 3) or SHAGGY-like kinases are highly conserved eukaryotic serine/threonine kinases that regulate various physiological processes in plants. BIN2 acts as a negative regulator of BR signaling. In the BR-signaling pathway, both BZR1 and BES1 are essential TFs, which are tightly regulated *via* phosphorylation and are controlled by multiple E3 ubiquitin ligases (Kim and Russinova 2020).

During BR-mediated gene expression, BRs inactivate BIN2, which then allows PP2A phosphatase to dephosphorylate BZR1/BES1 TFs resulting in their activation. It further activates the downstream BR-responsive gene expression in the nucleus that regulates the expression of ~4000–5000 genes (Guo et al. 2013). Various researchers explored systems biology strategies to study the molecular signaling events of BRs in the seedlings of model plant, *Arabidopsis*, to identify phosphorylation sites, TFs, and their expression through network analysis. BR-mediated gene expression and regulation of various physiological activities have been established through different omics approaches, which include phenomics, genomics, proteomics, transcriptomics, metabolomics, and integrated omics.

2.3.1 Phenomics

BRs regulate the growth, fruit ripening, and quality, post-harvest storage, crop yield, pigment concentrations, and contents of anthocyanins and phenolics in various crops (Garrido-Auñón et al. 2024). In tomato plants, application of BRs alleviated salinity stress by enhancing chlorophyll content, improving photochemical efficiency, and managing the redox imbalance (Salehi et al. 2024). In *Solidago canadensis*, 10^{-8} M foliar treatment of BR was found effective in enhancing growth parameters, plant height and stem diameter, number of leaves and leaf area, herb fresh and dry weight, flowering parameters like number of inflorescences, length of flowers and inflorescence stalks as well as antioxidant capacity, photosynthetic and total phenolic content (El-Sayed et al. 2020). Further, BR application alleviated heat stress by increasing the percentage of panicle blanking, number of filled spikelets, and F_v/F_m ratio during diverse phenological stages in heat-stressed rice plants (Pantoja-Benavides et al. 2022). The exogenous levels of BRs are also affected by natural conditions or the treatment of biostimulants. In lettuce, the application of biostimulants enhanced growth and photosynthesis performance, which was mediated through regulation of BRs (Sorrentino et al. 2022). Two BRs, namely, 3-dehydroteasterone and (22S,24R)-22-hydroxy-5α-ergostan-3-one, were observed to be down-accumulated when treated with protein hydrolysates (PH) botanical obtained from Solanaceae family (PH B) in lettuce. In grapes, 3 mg/L BR treatment enhanced metabolic activities and antioxidant enzyme activity, whereas BR at 2 mg/L concentration resulted in the highest cluster and berry weight, enhanced firmness, and reduced weight loss of berries under plastic covering induced unfavorable climatic conditions, suggesting stress protectant potential of BRs (Ghoneem et al. 2024).

In tomatoes, 24-EBL treatment alleviated the negative impact of drought by stimulating phenological characteristics such as the number of leaves plant^{-1}, fruit set percent, days to first fruit set, number of clusters plant^{-1}, diameter, and yield of fruits; and biochemical parameters including lycopene content and antioxidant defence system (Jangid and Dwivedi 2017). EBL also alleviated photosystem II (PS-II) functioning, photochemical quenching, and electron transport rate in tomatoes by improving growth, photosynthesis, and metabolism of osmolytes, antioxidants, and secondary metabolites during salinity stress (Ahanger et al. 2020). Moreover, EBL (100 nM) also enhanced the quantum yield of PS-II photochemistry, electron transport rate, water-use efficiency, and net photosynthetic rate in water-deficient soybean (Pereira et al. 2019). Also, it caused a reduction in membrane damage analyzed through malondialdehyde and electrolyte leakage studies, and accumulation of oxidant compounds such as superoxide and hydrogen peroxide to confer stress tolerance. At different ripening stages, EBL applications regulated the sugar unloading in grape berries, resulting in enhanced levels of soluble sugars in berries and decreased their contents in fruit skins (Xu et al. 2015). This study further revealed that application of EBL and brassinazole or Brz (BR-synthesis inhibitor) enhanced the activities of invertases and sucrose synthase during ripening by upregulating the mRNA levels of *VvcwINV*, which encodes sucrose metabolic invertase, and genes encoding monosaccharide *viz*., *VvHT3, VvHT4, VvHT5*, and *VvHT6*, and disaccharide *viz*., *VvSUC12* and *VvSUC27*. Both EBL and Brz enhanced BL contents by downregulating *BRASSINOSTEROID-6-OXIDASE* and *DWARF1* (*VvBR6OX1* and *VvDWF1*) genes encoding BR-biosynthetic enzymes, and upregulating *BRASSINOSTEROID INSENSITIVE 1* (*VvBRI1*) gene encoding BR receptor gene.

It has been found that in shade-intolerant plants, BR is involved in the shade-induced rapid growth, which further regulates shade-induced hypocotyl elongation in such plants (Huang et al. 2022). These hypocotyl changes are modified by BRs in association with a complex transcription regulatory network that includes PHYTOCHROME-INTERACTING FACTOR 7 (PIF7), LONG HYPOCOTYL IN FAR-RED 1 (HFR1), and PIF3-LIKE 1 (PIL1) in plant responses to shade avoidance. The overexpression or mutation of BR receptors i.e. BRL3/BRL1/BAK1 conferred drought resistance in *Arabidopsis* without penalizing growth by accumulating osmo-protectant metabolites such as proline and sugars including glucose, fructose, galactose, maltose, raffinose, and galactinol in shoots (Fàbregas et al. 2018). BRI1 is also involved in the regulation of seed thermo-inhibition responses in the endosperm of *Arabidopsis* seeds at high temperatures by inhibiting DELLA factors, which are responsible for abscisic acid (ABA) accumulation (Piskurewicz et al. 2024). Moreover, BR treatment is reported to stimulate pear bud break by positively regulating the dormancy release (Wang et al. 2024). *BZR2*, a BR-responsive TF, was further observed to inhibit gene expression of *PpyDAM3, DORMANCY-ASSOCIATED MADS-box* genes in pear, thus accelerating the pear bud break. A crucial negative regulator of the BR signaling pathway, namely, *BIN2*, regulates cellulose synthesis in *Arabidopsis* by phosphorylation of *CESA1* (*cellulose synthase A1*) in *Arabidopsis* (Sánchez-Rodríguez et al. 2017).

2.3.2 Genomics

Genomics studies an organism's entire genome, including genes, interactions, evolution, and editing, while functional genomics develops global experimental approaches for gene function analysis, ensuring rigorous data generation and public access. By exploring the BR-mediated gene expression and genomic pathways, the underlying mechanism of BRs can be understood. Ahmar and Gruszka (2022) analyzed 39 BR signaling genes, revealing their chromosomal distribution, phylogenetic relationships, and TFs. This study also predicted differences in BR-dependent expression between *Arabidopsis* and rice. Besides this, 17 gene promoters in 39 BR signaling genes have BES1 TF-binding sites. Furthermore, BRs regulate the expression of various genes to enhance drought and salt tolerance, increase productivity, and regulate growth and development under abiotic stress. The exogenous and endogenous levels of BR-mediated an array of genes, such as *DWF4, CPD, DET2, ROT3, CYP90D1*, and *CYP85A1*, which affect the expression of many essential enzymes and thus modulate diverse biochemical pathways (Miao et al. 2024). BRs regulate about

9028 genes associated with cotton fiber growth including *GhDET2, DWARF4, CPD, GhFP1, GhSMT2*, and *PAG1* (Qanmber et al. 2024). These genes are grouped according to the components, targets, and regulatory roles they play in the BR pathway, which influences the development of cotton fibers.

In *Gossypium hirsutum, GhSMO2-2*, a homologue of *Arabidopsis* Sterol C-4α Methyl Oxidase (SMO2-2), is a crucial gene involved in cotton fiber cell elongation and development (Liu et al. 2023). The gene expression of *GhSMO2-2* is enhanced by GhBES1, a TF of the BR signaling pathway. GhBES1 directly binds to the promoter region of *GhSMO2-2*, thus stimulating *GhSMO2-2* overexpression, which further enhances phytosterol synthesis, fiber cell elongation, and accumulation of some sphingolipids. Another omic-based investigation explored that BR signaling, when activated by ZnF (zinc finger proteins), ensues proteasomal destruction of TaBKI1 (*Triticum aestivum* BRI1 kinase inhibitor 1), a BR-signaling repressor TF (Song et al. 2023). Further genetic analysis revealed that the semi-dwarf trait in wheat, which results in enhanced grain yield, occurs due to the absence of *ZnF-B* (gene encoding a RING-type E3 ligase), and *ZnF-B* loss has been found to block the BR-signaling transduction pathway through stabilization of TaBKI1. Thus, *ZnF-B* is a crucial gene that modulates the BR-signaling cascade, and this may be a milestone for the future implementation of BR in the manipulation of high-yielding semi-dwarf varieties of wheat to ensure global food security.

In rice, genome editing using CRISPR/Cas9 tools created a panel of mutants within three BR-signaling gene families, namely, *GSK3/SHAGGY-LIKE KINASE1* (*GSK1*)–*GSK4*, *OsBZR1–OsBZR4*, and Kelch-like protein phosphatases, *PPKL1–PPKL3* (Liu et al. 2021). High-order mutants were created by targeting multiple genes of *GSKs* and *PPKLs* family or overlapping sequences of the *OsBZRs* family, resulting in diverse plant heights, leaf angles, and grain morphologies. Comparative phenotypic analysis revealed functional redundancy, differentiation, or dominancy among members of *GSKs, PPKLs*, and *OsBZRs* gene families. Transgenic plants overexpressing *GSK2* (*Flag-GSK2, GFP-aGSK2*, and *GFP-Myc-aGSK2*), *OsBZR1/2* (*GFP-Myc-OsBZR1, GFP-Myc-aOsBZR1*, and *Flag-aOsBZR1*), and *PPKL2* (*GFP-PPKL2* and *Flag-PPKL2*) consistently expressed economically important agronomic phenotypes including improved grain size, leaf angle, and plant height in rice. Nolan et al. (2023) employed single-cell RNA sequencing (scRNA-seq) to identify the elongating root cortex as a site for BR-mediated gene expression. BRs were found to promote a shift from proliferation to elongation, involving increased expression of cell wall-related genes. Further analysis through a tissue-specific CRISPR approach revealed loss of BR signaling in the cortex, impaired cell expansion, but had little effect on meristem cell length. *OMEOBOX FROM ARABIDOPSIS THALIANA 7* (*HAT7*) and *GT-2-LIKE 1* (*GTL1*) were identified as responsible for BR-mediated regulation of cell elongation in the cortex. Single-cell genomics identifies context-specific TFs for precise BR-mediated plant growth and stress responses, understanding how BR-signaling activates cell wall-related genes and controls cell elongation.

2.3.3 Proteomics

Proteomics studies have elucidated crucial proteins involved in BR-mediated responses and stress tolerance mechanisms by exploring the possible interactions between protein metabolites activated or inactivated during the BR-signaling cascade (Verma et al. 2021). In mung bean epicotyls, BR application re-upregulated 17 proteins and *CYP90A2* (a synthetic gene), which were downregulated by chilling stress through stimulation of methionine assimilation, ATP synthesis, cell wall construction, and stress-protective response mechanisms (Huang et al. 2006). Therefore, to study the underlying mechanisms of BR-mediated responsive role of proteins in plants, protein expression profiles during non-stressed and stressed conditions have been explored through diverse proteomic approaches (Verma et al. 2021). It has been found that a wide range of antioxidant enzymes and metabolites have been activated through complex BR-mediated signaling networks of TFs and receptors.

In *Arabidopsis*, Tang et al. (2008) employed prefractionation and two-dimensional (2D) difference gel electrophoresis (DIGE) techniques to detect early BR signaling events and response proteins. BR-induced rapid proteomic changes were detected in phosphoprotein and plasma membrane fractions. In contrast, LC-MS/MS analysis identified 19 BR-regulated plasma membrane proteins and 6 BR-regulated proteins from phosphoprotein fractions. The identified proteins showed spot shifts correlated with BR-mediated phosphorylation through BAK1 receptor kinase and the BZR1 TF. Besides this, two (Thr-507 in At1g12270 and Thr-506 in At1g62740) of BZR1 in tetratricopeptide repeat proteins, and three (Ser-62, Thr-66, and Thr-122) phosphorylation sites of BZR1 have also been identified in a phosphoenolpyruvate carboxykinase (PCK1). Overexpression of a novel BR-induced plasma protein, *viz*., DREPP (At4g20260), a developmentally regulated plasma membrane polypeptide, partially suppressed BR deficient mutant phenotypes. Identification of DREPP highlighted the BR-signal-mediated protein modifications during BR-regulated growth in plants. The 2D DIGE study by Wang et al. (2010) compared proteomic differences in *Arabidopsis* BR-insensitive and BR-deficient mutants that identified differentially expressed proteins such as OsGRP1 (a glycine-rich RNA-binding protein) and DREPP2. Their overexpression partially suppressed the dwarf phenotype of BR-insensitive mutant bri1-5, but RNA levels of *OsGRP1* were not significantly affected. The study highlighted that *OsGRP1* is regulated at the post-translational level by BRs through the downstream signal transduction pathway, which results in the promotion of cell expansion.

Recent developments in proteomic studies have transformed from traditional gel-based methods to high-throughput quantitative strategies, including the latest proteomics tools of MALDI (Matrix-Assisted Laser Desorption/Ionization) and iTRAQ (isobaric tags for relative and absolute quantification) approaches, enhancing the existing knowledge of BR-mediated regulation of diverse plant responses and biochemical mechanisms (Verma et al. 2021). In the BR-signaling pathway, new components have been identified through proteomics studies, which have further revealed the first complete pathway from receptor-like kinases (RLKs) to TFs in plants (Tang et al. 2010). The mass spectrometry (MS) analysis revealed that BR signal transduction mediated by RLK takes place primarily through phosphorylation events. In rice, Hou et al. (2017) used a label-free, MS-based phospho-proteomics and Western blot analysis to interpret the effect of phospho-sites and phosphoprotein profiles during BR-mediated signaling events. BR treatment phosphorylated 4034 phospho-sites on 1900 phosphoproteins and 1821 proteins comprising TFs and kinases *via* influencing cellulose synthesis. The dephosphorylation of another phytohormone, namely gibberellin (GA), signaling component was also found to be a crucial mechanism for BR-regulated antagonism to GA. Further, MS-based phospho-proteomics and bioinformatics analysis by Lin et al. (2015) revealed that most of the peptides identified 1104 unique phosphorylated peptides from 739 unique phosphoproteins in plants were strongly connected with known components of the BR-signaling cascade. The BR treatment significantly phosphorylated and dephosphorylated the substrates of casein kinase 2 and mitogen-activated protein kinase. The study further revealed that most phosphorylation changes were not led by BR-induced gene expression modifications, suggesting that maximum BR-signaling downstream proteins are induced by phosphorylation through kinases instead of transcriptional regulation.

In rice seeds, Li et al. (2016) reported more than 800 BR-responsive proteins through an iTRAQ-based proteomic approach comprising 88 target proteins, which shared 90% similar expression patterns. Gene ontology and string analysis further highlighted the BR-regulated highly clustered proteins involved in diverse BR-mediated biological processes including carbohydrate metabolism and protein biosynthesis, thereby revealing insights into BR-regulated seed germination molecular mechanisms. EBL treatment introduced the maximum differentially expressed proteins (DEPs), and EBL-mediated expression of O80577, Q9FJQ8, and Q9SKL2 proteins was also enhanced (Khan et al. 2019). These three BR-regulated proteins enhanced the growth and photosynthetic efficiency of tomato plants under low-temperature stress through regulation of citrate cycle pathway. In another investigation by Li et al. (2022), iTRAQ method, physiological assays, and cellular ultrastructure analysis revealed that EBL improved photosynthesis and cell structure through modification of about 346 DEPs. Out of these, 217 DEPs were upregulated, which were found to be related to photosynthetic electron transfer chain, Calvin cycle, oxidative phosphorylation, glutathione metabolism, and signaling pathway. BR treatment downregulated 129 proteins in pepper subjected to chilling stress.

In *A. thaliana*, Zhang et al. (2012) analyzed BR-mediated plant responses and metabolic pathways through integrated multidimensional protein identification technology (MudPIT)-based shotgun proteomics with a rapid sample preparation method and bioinformatics classification system. BR upregulated 60 proteins involved in diverse cellular, metabolic, and developmental processes, and cellular component organization or biogenesis, response to stimuli, and other biological regulations. The treatment of BR also upregulated 12 DEPs, comprising 5′-adenylylsulfate reductase 2, co-chaperone grpE family protein, protoporphyrinogen oxidase, and DEAD/DEAH box helicase. While some important proteins of auxin pathway, lipid metabolism, and energy storage such as auxin-binding protein 1, sulfate adenylyltransferase 4, plastid-lipid associated protein, oligosaccharyl transferase STT3 subunit family protein, and jacalin lectin family proteins were downregulated by BR application. In cotton, MS (MALDI-TOF/TOF) analysis and gene ontology annotation revealed that BR-induced resistance to *Verticillium* cotton wilt through the expression of 188 DEPs, categorized into 17 biological processes (Gao et al. 2013). The proteomic analysis further highlighted that most of these DEPs were associated with stimulus responses along with several other cellular and metabolic processes. Gene ontology reported that various identified genes were related to reactive oxygen burst, phytohormone, and secondary metabolism signaling pathways. While conferring cotton resistance to *Verticillium*, BR signaling positively regulated protein expression. The Gb14-3-3 proteins in *Arabidopsis* interacted with BIN2-phosphorylated targets in BZR1 protein, which further affected cotton resistance to *Verticillium* cotton wilt.

In apple nursery tree shoot tips, BR-mediated expression of 175 DEPs were identified *via* iTRAQ-based quantitative proteomics, and these were mostly associated with hormone signaling, cell growth, photosynthesis, and chlorophyll biosynthesis, lignin and carbohydrate metabolism, including glycolysis, citric acid cycle, and respiratory action (Zheng et al. 2018). The comparative proteomic approaches highlighted that BR regulated 52 381 peptides from 9680 proteins, including 402 DEPs, involved in the lignin biosynthesis pathway (Wang et al. 2020). Besides, BR-pretreatment and *Phytophthora sojae* infection also upregulated key enzyme expressions, leading to faster lignin accumulation in soybeans. Proteomics analysis showed exogenous BL application enhanced chlorophyll metabolism proteins and altered stress responses in tomatoes under calcium nitrate stress (Zhang et al. 2021). This BL-induced calcium nitrate stress alleviation improved photosynthetic efficiency, reduced reactive oxygen species, and promoted plant growth, emphasizing the significant application of BR in protected agriculture for managing secondary salinization. Lu et al. (2022) applied the cellular thermal shift assay (CETSA) coupled with MS to *Arabidopsis* cells and reported that GSK3s phosphorylated the auxin carrier PIN-FORMED1 and regulated its polarity, thereby highlighting a cross-talk between BR and auxin signaling.

2.3.4 Transcriptomics

Large-scale transcriptomic and proteomic profiling analyses explored the plant adaptation pathways in barley crown tissue in response to abiotic drought stress (Kuczyńska et al. 2025). These studies revealed the specific remodeling of behavior in the BR-insensitive barley *uzu1.a* mutant. The multidisciplinary approaches provided new insights into the relationships between expression of genes encoding dehydrins, protein, and phytohormone content of crown tissue in drought-stressed barley. Furthermore, some important transcriptomic approaches to uncover the BR-mediated responses in plant regulation are described as follows.

2.3.4.1 RNA Seq

RNA sequencing enables the identification of differentially expressed genes (DEGs) across the DNA. In this technique, cDNA library is prepared from isolated RNA, which is then sequenced by different sequencing platforms, such as Sanger sequencing and Illumina sequencing. By sequencing the DNA, the activity of genes involved in BR-mediated responses can be studied, and their expression can be analyzed. The RNA seq data also provide information about the biological roles of identified DEGs, which can be further analyzed to identify the pathways in which the genes are involved via pathway enrichment analysis (KEGG). Furthermore, comparative transcriptomic analysis can be used to identify novel genes involved in BR responses in control versus treated plants. For instance, comparative transcriptome analysis was carried out in the model plant *A. thaliana* to identify the BR-responsive genes, which revealed 4498 DEGs and 100 TFs that respond to BR at different stages of plant development (7- and 24-day-old seedlings) (Liu et al. 2020). Similarly, a multi-omics study profiled 26669 transcripts, 9533 proteins, and 26 617 phosphorylation sites in *Arabidopsis*. Further, they constructed Spatiotemporal Clustering and Inference of Omics Networks (SC-ION) to integrate these data, thereby unravelling the molecular signaling events of BR response in *Arabidopsis* (Clark et al. 2021).

In xylem, phloem, and leaf, DEGs were identified in *F. mandshurica* by Lu et al. (2023). Differential analyses of xylem-DEGs (X-DEGs) revealed that among 4589 DEGs, 2234 were upregulated whereas 2355 were downregulated in *F. mandshurica* xylem samples. Further analysis of xylem-specifically expressed genes (X-SEG) highlighted that regulation of DEGs and SEGs affected wood traits and gene ontology during wood development. Out of the total studied X-DEGs, 83, 63, 19, and 10 X-DEGs were involved, respectively, in cellulose biosynthesis, phenylpropanoid biosynthesis, cellulose synthase, and hemicellulose synthesis, which emphasized the active involvement of BRs in wood development.

2.3.4.2 qRT-PCR

Quantitative real-time PCR enables quantification of the expression levels of a particular gene. It involves the synthesis of cDNA from RNA followed by its PCR amplification. A fluorescent dye such as SYBR green or a fluorescent probe such as a TaqMan probe is involved, which binds with the DNA by intercalation and gives fluorescence. The intensity of fluorescence is directly proportional to the amount of amplified PCR product. A threshold value (Ct value) is used to quantify the relative expression levels. The expression levels are then compared to the internal control used in the experiment, which are the housekeeping genes (such as actin and ubiquitin). Mostly, the qRT-PCR is used to validate the results of RNA sequencing. That means the DEGs obtained (upregulated/downregulated genes) in RNA seq are confirmed by their expression levels. For example, in the functional analysis of loss-of-function mutants of a subset of genes selected from the BR transcriptome identified abiotic stress-related roles for ACID PHOSPHATASE5 (ACP5), WRKY33, JACALIN-RELATED LECTIN1-3 (JAC-LEC1-3), and a BR-RESPONSIVE-RECEPTOR-LIKE KINASE (BRRLK). The results were based on microarray expression analysis, and the accuracy of these expression analyses was confirmed by qRT-PCR (Divi et al. 2016). Similarly, in another study, transcriptomic analysis was carried out to understand the role of BR in wood development. From the identified DEGs, 16 DEGs were validated using qRT-PCR (Wang et al. 2022). In a similar study, RNA-seq was done to understand how BRs regulate the developmental processes in soybeans and explore the molecular mechanism underlying soybean early development. Of the obtained DEGs, eight randomly selected DEGs were validated for validation of RNA-seq data. The correlations between the RNA-seq and qRT-PCR data were analyzed using log2 expression levels. The results suggested that the qRT-PCR measurements were highly correlated with the RNA-seq results (R^2=0.7895). Thus, the expression patterns of the genes from the qRT-PCR results were highly consistent with the results from the RNA-seq, which indicated the fidelity and reproducibility of the RNA-seq analysis (Song et al. 2019). Likewise, a study reported that endogenous BR is a major factor promoting bamboo shoot internode elongation through physiological and transcriptomic analyses.

For the identified DEGs, qRT-PCR validation was also done (Zhang et al. 2020). Exogenous application of commercially available BR, namely, 24-EBL treatment, enhanced plant height, petioles' number, and aboveground plant weight in carrots through upregulation of *DcBRI1, DcBZR1*, and *DcBSU1* in the petioles (Que et al. 2017). The 0.5 mg/L concentration of EBL was effective in upregulating the expression profiles of *DcCYP734A1* catabolism gene, which was significantly upregulated, indicating the bioactive role of EBL in normal growth and development of carrot. Quantitative real-time PCR analysis revealed that EBL treatment induced the expression levels of *DCAR_009411* (GA signaling gene, i.e. GA5-like gene), *DCAR_019754*, and *DCAR_013973* (*cellulose synthase-like* genes) were upregulated in carrot petiole, which further suggested BR-mediated modulation of GA and cellulose synthesis that results in promotion of cell elongation in carrot petioles (Que et al. 2018). The *BES1* family, which mediates BR signaling, regulates plant stress responses in drought-stressed chickpea (Felagari et al. 2023). A genome-wide analysis of six *BES1* genes, namely, *CaBES1.1, CaBES1.2, CaBES1.3, CaBES1.4, CaBES1.5*, and *CaBES1.6*, in chickpea revealed their different expression patterns during drought stress. The transcription database revealed that *CaBES1.2* expression was downregulated under drought conditions, while qRT-PCR analysis emphasized that *CaRD26* and *CaNAC72* expression increased, which further emphasized the potential role of BRs in practical applications for sustainable agricultural strategies.

2.3.4.3 Microarray

Microarray technology allows for the simultaneous measurement of the expression levels of thousands of genes. This technique involves hybridizing cDNA (synthesized from isolated RNA) to a grid of DNA probes on a microarray chip. Fluorescent labeling of cDNA provides a signal that can be detected and quantified. Microarray analysis can identify DEGs in response to BR treatment. For example, a study on rice employed microarray analysis to identify BR-regulated genes and found numerous genes involved in stress responses, metabolism, and cell wall organization. This study revealed how BR influences diverse biological processes in rice, providing insights into BR-mediated regulatory networks (Kumar et al. 2013). Another example is in *Arabidopsis*, where microarray analysis identified a wide range of BR-responsive genes involved in various cellular processes, including cell elongation, division, and differentiation. This data helped to map out the complex regulatory networks of BR signaling in plants (Sun et al. 2010; Seyed Rahmani et al. 2021). Likewise, a study on BR-treated tomato plants used microarray analysis to identify genes involved in fruit development and ripening. The identified genes were associated with hormone signaling pathways, providing insights into BR's role in fruit physiology (Hu et al. 2020; Musser et al. 2014; Rohrmann et al. 2011). Furthermore, in a study on barley, microarray analysis was conducted to identify BR-responsive genes. The study revealed a number of genes involved in primary metabolism, stress responses, and hormone signaling pathways. These findings contributed to understanding the complex interactions between BRs and other hormonal and environmental signals in barley (Gruszka et al. 2016). Also, a microarray study on maize aimed to identify BR-responsive genes in roots and shoots. The analysis highlighted several key genes involved in cell wall biosynthesis, signaling, and transcriptional regulation, providing insights into how BRs influence growth and development in different plant tissues (Castorina et al. 2018; Jiang et al. 2018). Moreover, in *Brassica napus*, microarray analysis was used to study the effects of BRs on seedling growth. The study identified genes involved in photosynthesis, cell cycle regulation, and stress responses, showing how BRs can enhance plant growth and stress tolerance (Dhaubhadel and Krishna 2008; Sahni et al. 2016). Similarly, a study on cucumber microarray analysis was employed to investigate BR-responsive genes. The results demonstrated that BRs are likely to play a negative role in regulating Fe-deficiency-induced FRO, expression of CsFRO1 and CsIRT1, as well as Fe translocation from roots to shoots (Lin et al. 2013).

2.3.4.4 Small RNA

Small RNA sequencing (sRNA-Seq) is a powerful technique for profiling small RNAs, including microRNAs (miRNAs) and small interfering RNAs (siRNAs). These small RNAs play crucial roles in gene regulation at the post-transcriptional level, often by degrading mRNA or inhibiting its translation. sRNA-Seq can be used to uncover the roles of small RNAs in BR-mediated responses. For instance, a study in *Arabidopsis* identified several miRNAs regulated by BRs that target TFs and other regulatory proteins involved in BR-signaling pathways. This analysis provided insights into how BRs influence gene expression through small RNA-mediated mechanisms. The results suggest that BR regulates seedling development and root growth of *A. thaliana* through miR395a by suppressing GUN5 expression and its downstream signal transduction (Lin et al. 2013).

Similarly, in rice, sRNA-Seq was utilized to identify BR-responsive miRNAs, revealing their roles in regulating genes involved in growth and stress responses. These findings enhance our understanding of the regulatory layers controlled by BRs in plants (Zheng et al. 2015). Also, a study on soybeans used sRNA-Seq to identify miRNAs involved in BR-regulated early development processes (Gupta et al. 2021). The identified miRNAs were found to target genes related to cell division and expansion, highlighting their importance in BR-mediated developmental regulation. In a study on tomatoes, sRNA-Seq was used to identify BR-responsive miRNAs during fruit development. The identified miRNAs were found to target genes involved in hormone-signaling pathways and fruit ripening processes, illustrating the role of miRNAs in BR-mediated regulation of fruit physiology (Karlova et al. 2013). Similarly, a study on cotton employed sRNA-Seq to profile BR-responsive miRNAs in developing fibers. The identified miRNAs targeted genes associated with cell wall modification, stress responses, and signal transduction, providing insights into how BRs regulate fiber development at the molecular level (Kwak et al. 2009; Liu et al. 2014). In wheat, sRNA-Seq was utilized to identify miRNAs involved in BR-mediated stress responses. The study found several BR-responsive miRNAs that regulate genes involved in abiotic stress tolerance, highlighting the role of miRNAs in enhancing stress resilience in crops (Xu et al. 2022; Song et al. 2023; Wei et al. 2009). Furthermore, a study on grapevine, sRNA-Seq was used to identify BR-regulated miRNAs during berry development. The identified miRNAs were involved in regulating genes associated with fruit ripening, stress responses, and secondary metabolism, highlighting the complex regulatory networks controlled by BRs in fruit crops (Belli Kullan et al. 2015; Li et al. 2024).

2.3.5 Metabolomics

Plants produce essential building blocks, bioactive stimulants, chemicals, osmolytes, hormones, and secondary metabolites during normal processes and also for protection against stress. Metabolomics quantifies these metabolites, analyzing chemical processes in cells, tissues, and organisms. An in-depth understanding of the metabolites helps in ascertaining BR-mediated crop resilience, physiological and stress-protective responses, yield, and productivity. The function of BRs in regulating shooting in Marubakaido apple rootstock was investigated by Pereira-Netto et al. (2009). BL decreased apical dominance by altering the shoot elongation and development of primary and major lateral shoots. Higher doses of Brz200 (an inhibitor of BR production) gradually inhibited the main shoot elongation through decline in 6-deoxocathasterone, 6-deoxotyphasterol, and castasterone levels.

However, fructose, glucose, and gulonic acid levels increased in treated shoots, suggesting 28-homocastasterone-induced shoot elongation and new shoot formation. In *Arabidopsis*, hormonomic profiles of two BRs, namely, EBL and 28-norcastasterone, were identified under salt stress through ultra-high-performance liquid chromatography-tandem MS (Šimura et al. 2018). BL application upregulated heat shock-related genes, various metabolites like amino acids, poricoic acid, carbohydrates, higher accumulation of endogenous EBL and antioxidants, enhanced osmotic adjustment substances (soluble solid concentrations and soluble sugar content grapes) under high-temperature stress (Lv et al. 2022). Also, BR treatment deferred the microfibrils and middle lamella degradation.

Another investigation by McGuiness et al. (2021) examined the impact of BRs, GA, and ethylene on legume infection, nodule number, and metabolites associated with both nitrogen fixation and nutrient exchange in pea mutants. BR deficiency didn't affect the dicarboxylate levels and amino acid levels in glutamine-aspartate transamination, suggesting that although BRs influence energy and nitrogen-fixation metabolites in nodule tissue, GA is crucial for establishing a functional nodule metabolome. BR enhanced chlorophyll content, photochemical efficiency of PS-II, lowered non-photochemical quenching, stimulated antioxidant defense mechanisms by regulating activities of superoxide dismutase, catalase, ascorbate peroxidase, and reducing lipid peroxidation in tomato under salt stress (Salehi et al. 2024). Further, BR-induced salt stress tolerance is regulated by the metabolomic signatures in plants through the modulation of secondary metabolites in tomatoes. BR application significantly improved the amino acid pool, hormone and/or signal-related compounds, and secondary metabolites including phenylpropanoids and nitrogen-containing compounds to confer plant tolerance to salinity. During kiwifruit ripening, 2,4-EBL application delayed fruit firmness decline and influenced sugar metabolism pathways and flavor development (Yang et al. 2025). The application of 2,4-EBL downregulated *AdERF2* and *AdBAM3.5* expressions, slowing down starch degradation and promoting soluble sugar accumulation, potentially affecting ripe kiwifruit taste.

2.3.6 Integrated Omics

In BL-treated *Arabidopsis* seedlings, integrated multi-omic dataset analysis revealed the levels of transcripts (26 669), protein groups (9533), and phosphorylation sites (26 617), which were further integrated using the "Spatiotemporal Clustering and Inference of Omics Networks (SC-ION)," a network inference pipeline (Clark et al. 2021). The integrated omic analysis predicted S179 and S180 as putative phosphorylation sites in BES1 that are BIN2 phosphorylated. Also, BRON and ANL2 TFs were found to regulate BR-mediated cell division, as well as BR-responsive kinases also modulated gene expression of *BRON*. BL treatment induced but BRON repressed the gene expression of two cyclin genes, *viz.*, *CYCD3;1* and *CYCP4;1*, which suggested coordination between *BRON*- and BR-mediated gene expression. Another integrated transcriptomic and proteomic investigation by Wang et al. (2022) observed that the inhibition of *Arabidopsis* vascular-enriched BR-signaling receptors influences wood development in *Populus* through the regulation of mRNA and/or proteome level. Transgenic poplar lines with edited *PdBRI1* genes, orthologs of BR receptors, showed severe growth defects. The transcriptome and proteome study found the differential expression of genes, *pdbri1-1;2;3;6*, was related to wood development, emphasizing the significant role of BRs in procambial cell division and xylem differentiation in *Populus*. Recently, Montes et al. (2022) discovered a system-wide link between BRs and Target of Rapamycin Complex (TORC) in plant growth and stress responses through quantification of transcripts, proteins, and phosphorylation sites in wild-type *A. thaliana* and mutants with altered levels of BIN2. The integrated multi-omics data established a predictive network defining molecular interactions between BR- or TORC-regulated growth and autophagy.

In maize, integrated analysis of transcriptomic, proteomic, and phosphoproteomic highlighted the BR-mediated regulation of photosynthesis antenna proteins, carotenoid biosynthesis, mitogen-activated protein kinase (MAPK, a family of serine/threonine protein kinases) signaling, and phytohormone signal transduction (Li et al. 2023). BR-mediated upregulation of gene expression was observed to be dose-dependent, which further regulated 42 and 186 TFs responses, thus providing insights into the photosynthetic response. Further proteomic and metabolomic analyses in maize revealed that BR-biosynthesis and signal transduction pathways were regulated during embryonic callus induction from immature embryo (Ge et al. 2017). BR-induced chilling stress protection was analyzed in tomato fruits through transcriptomic, metabolomic, and proteomic methods (Bai et al. 2021). BR induced chilling stress protection by inhibiting cell wall and membrane lipid damage, regulating the expression of *lipoxygenase* (*LOX*) gene involved in α-linolenic acid metabolism pathway and several DEPs, and improving the JA-coenzyme A (JA-CoA) synthesis in tomato fruits. Integrated omics revealed that artificial aging treatment improved rice seed longevity through regulation of BR-signaling and suppression of JA-signaling, which stimulated a signaling cascade ensuring the accumulation of enzymatic and non-enzymatic antioxidants for efficient detoxification of aging-induced reactive oxygen species (Gupta et al. 2024). The genes and metabolites involved in the BR-biosynthetic pathway were found to be enhanced in the somatic embryogenesis of Darjeeling tea, which was established through integrated metabolome and transcriptome analysis (Awon et al. 2024). In BR-biosynthesis, six metabolites were accumulated in heart-shaped embryos, globular embryos, and embryogenic callus of tea. The globular embryo had the highest concentration of 6-deoxocastasterone, and the highest expression of the *brassinosteroid-6-oxidase-1* (*BRD1*) gene. Additionally, *BRD1* and castasterone were significantly upregulated in globular embryos, and these genes were positively correlated with *LEAFY*, *WOX3*, *FUS3*, and *LEA5-D* genes.

2.4 Questions Unanswered

Although an array of physiological, biochemical, molecular, and economically important agronomic attributes of crops has been affected by BR application, large-scale BR applications remain limited due to their complex effects on plants (Tong and Chu 2018). Further, genomics and molecular strategies identifying specific BR components for the breeding desirable traits would help in improving crop performance and yield. Moreover, targeting crops to enhance their agronomic trait diversity *via* editing BR-related family genes for screening desired plants may further facilitate commercial BR application. The BR-mediated TFs, TF-binding sites, and gene promoters, revealing up- or downregulation, are necessary for plant growth and diversifying organs, with differences predicted between *Arabidopsis* and rice have been investigated (Ahmar and Gruszka 2022). However, further research is required to fully understand the BR-responsive regulatory processes and interdependencies influencing rice development and stress response. Recent research has emphasized on the significant role of BRs in controlling various environmental stressors in plants by interacting with other phytohormones, yet little is known about how they work to withstand stress and interact with other hormones (Miao et al. 2024). The regulatory network of BRs and possible reasons for increased agricultural output should be investigated in future studies. Moreover, uncovering the redundancy in cotton fiber development and its impact on yield and quality is a challenging question (Qanmber et al. 2024). Most research is conducted in controlled laboratory conditions, leaving a gap between findings and field conditions. Thus, understanding BR responses without penalizing yield is crucial for improving cotton fiber quality. BR-mediated responses have been generally analyzed through BR mutant-based studies, which may suggest varied responses depending on BR dose, experimental design, errors, or experimental conditions. Therefore, it is still unclear if BRs have a broad impact on plants under normal conditions or different stresses or if they act as particular regulators of some particular stress or regulate only a few responses.

2.5 Conclusion and Future Prospects

BRs regulate plant biology by regulating processes like cell signaling, metabolism, redox homeostasis, growth and developmental processes, and stress mechanisms. Due to their role in enhancing quality and crop productivity as well as immense medicinal and pharmacological potentials, BRs are being researched for their potential in agricultural practices. Large-scale BR applications are limited due to their complex effects on plants. Recent advances in omics technologies have improved our understanding of the role of BRs in improving crop productivity. Further, genomics and molecular strategies can identify BR-mediated components for breeding desirable traits and improving crop performance. By exploring the precise molecular and biochemical mechanisms of BR action in regulating plant biology, bio-safe BRs can be employed in various agro-industries to enhance global food security and promote sustainable farming methods, reducing reliance on harmful pesticides and agrochemicals. Further research is needed to understand BR-responsive processes and interactions.

Conflict of Interest

None declared.

Acknowledgments

The first author (IS) and corresponding author (AS) acknowledge Sant Baba Bhag Singh University and DAV University respectively for necessary infrastructural facilities.

References

Ahanger, M.A., Mir, R.A., Alyemeni, M.N. et al. (2020). Combined effects of brassinosteroid and kinetin mitigates salinity stress in tomato through the modulation of antioxidant and osmolyte metabolism. *Plant Physiology and Biochemistry* 147: 31–42.

Ahmad, H., Hayat, S., Ali, M. et al. (2017). Regulation of growth and physiological traits of cucumber (*Cucumis sativus* L.) through various levels of 28-homobrassinolide under salt stress conditions. *Canadian Journal of Plant Science* 98 (1): 132–140.

Ahmar, S. and Gruszka, D. (2022). In-silico study of brassinosteroid signaling genes in rice provides insight into mechanisms which regulate their expression. *Frontiers in Genetics* 13: 953458.

Ali, B. (2017). Practical applications of brassinosteroids in horticulture-some field perspectives. *Scientia Horticulturae* 225: 15–21.

Alyemeni, M.N., Hayat, S., Wijaya, L. et al. (2013). Foliar application of 28-homobrassinolide mitigates salinity stress by increasing the efficiency of photosynthesis in *Brassica juncea*. *Acta Botanica Brasilica* 27: 502–505.

Arteca, J.M. and Arteca, R.N. (2001). Brassinosteroid-induced exaggerated growth in hydroponically grown *Arabidopsis* plants. *Physiologia Plantarum* 112 (1): 104–112.

Awon, V.K., Dutta, D., Banerjee, S. et al. (2024). Integrated metabolomics and transcriptomics analysis highlight key pathways involved in the somatic embryogenesis of Darjeeling tea. *BMC Genomics* 25 (1): 207.

Bai, C., Zheng, Y., Watkins, C.B. et al. (2021). Revealing the specific regulations of brassinolide on tomato fruit chilling injury by integrated multi-omics. *Frontiers in Nutrition* 8: 769715.

Bajguz, A., Bajguz, A.J., and Tryniszewska, E.A. (2013). Recent advances in medicinal applications of brassinosteroids, a group of plant hormones. *Studies in Natural Products Chemistry* 40: 33–49.

Bajguz, A., Chmur, M., and Gruszka, D. (2020). Comprehensive overview of the brassinosteroid biosynthesis pathways: substrates, products, inhibitors, and connections. *Frontiers in Plant Science* 11: 1034.

Belli Kullan, J., Lopes Paim Pinto, D., Bertolini, E. et al. (2015). miRVine: a microRNA expression atlas of grapevine based on small RNA sequencing. *BMC Genomics* 16: 1–23.

Bhardwaj, R., Arora, N., Uppal, P. et al. (2011). Prospects of brassinosteroids in medicinal applications. In: *Brassinosteroids: A Class of Plant Hormone*, 439–458.

Bhardwaj, R., Sharma, I., Arora, N. et al. (2010). *Regulation of Oxidative Stress by BRs in Plants. Oxidative Stress: Role of Antioxidants in Plants*. New Delhi: Stadium Press Pvt. Ltd., 215–231.

Bhardwaj, R., Sharma, I., Kapoor, D. et al. (2014). Brassinosteroids: improving crop productivity and abiotic stress tolerance. *Physiological Mechanisms and Adaptation Strategies in Plants Under Changing Environment* 2: 161–187.

Castorina, G., Persico, M., Zilio, M. et al. (2018). The maize lilliputian1 (lil1) gene, encoding a brassinosteroid cytochrome P450 C-6 oxidase, is involved in plant growth and drought response. *Annals of Botany* 22 (2): 227–238.

Chone, R.M.S., Rocha, D.I., Monte-Bello, C.C. et al. (2018). Brassinosteroid increases the cytokinin efficiency to induce direct somatic embryogenesis in leaf explants of *Coffea arabica* L. (Rubiaceae). *Plant Cell, Tissue and Organ Culture (PCTOC)* 135: 63–71.

Clark, N.M., Nolan, T.M., Wang, P. et al. (2021). Integrated omics networks reveal the temporal signaling events of brassinosteroid response in *Arabidopsis*. *Nature Communications* 12 (1): 5858.

Dhaubhadel, S. and Krishna, P. (2008). Identification of differentially expressed genes in brassinosteroid-treated *Brassica napus* seedlings. *Journal of Plant Growth Regulation* 27: 297–308.

Divi, U.K., Rahman, T., and Krishna, P. (2016). Gene expression and functional analyses in brassinosteroid-mediated stress tolerance. *Plant Biotechnology Journal* 14 (1): 419–432.

Du, H., Yong, R., Zhang, J. et al. (2023). OsBAK2/OsSERK2 expression is repressed by *OsBZR1* to modulate brassinosteroid response and grain length in rice. *Journal of Experimental Botany* 74 (17): 4978–4993.

El-Sayed, I.M., Salim, R.G., El-Haggar, E.F. et al. (2020). Molecular characterization and positive impact of brassinosteroids and chitosan on *Solidago canadensis* cv. Tara characteristics. *Horticulturae* 6 (4): 100.

Fàbregas, N., Lozano-Elena, F., Blasco-Escámez, D. et al. (2018). Overexpression of the vascular brassinosteroid receptor BRL3 confers drought resistance without penalizing plant growth. *Nature Communications* 9 (1): 4680.

Fariduddin, Q., Yusuf, M., Chalkoo, S. et al. (2011). 28-homobrassinolide improves growth and photosynthesis in *Cucumis sativus* L. through an enhanced antioxidant system in the presence of chilling stress. *Photosynthetica* 49: 55–64.

Felagari, K., Bahramnejad, B., Siosemardeh, A. et al. (2023). A comparison of the physiological traits and gene expression of brassinosteroids signaling under drought conditions in two chickpea cultivars. *Agronomy* 13 (12): 2963.

Gao, W., Long, L., Zhu, L.F. et al. (2013). Proteomic and virus-induced gene silencing (VIGS) analyses reveal that gossypol, brassinosteroids, and jasmonic acid contribute to the resistance of cotton to *Verticillium dahliae*. *Molecular & Cellular Proteomics* 12 (12): 3690–3703.

Garrido-Auñón, F., Puente-Moreno, J., García-Pastor, M.E. et al. (2024). Brassinosteroids: an innovative compound family that could affect the growth, ripening, quality, and postharvest storage of fleshy fruits. *Plants* 13 (21): 3082.

Ge, F., Hu, H., Huang, X. et al. (2017). Metabolomic and proteomic analysis of maize embryonic callus induced from immature embryo. *Scientific Reports* 7 (1): 1004.

Ghasemi Bezdi, K., Ghorbani Nasrabad, G., Faghani, E. et al. (2024). Evaluating the effect of different growth regulators on yield and morphological traits of three upland (*Gossypium hirsutum* L.) cotton cultivars. *Iranian Journal of Cotton Researches* 12 (1): 83–100.

Ghoneem, G.M., Khalil, R., Yusuf, M. et al. (2024). Brassinosteroids enhance the tolerance, production, and fruit quality of covered table grape vines. *New Zealand Journal of Crop and Horticultural Science*, 1–20. https://doi.org/10.1080/01140671.2024.2398053

Grove, M.D., Spencer, G.F., Rohwedder, W.K. et al. (1979). Brassinolide, a plant growth-promoting steroid isolated from *Brassica napus* pollen. *Nature* 281 (5728): 216–217.

Gruszka, D., Janeczko, A., Dziurka, M. et al. (2016). Barley brassinosteroid mutants provide an insight into phytohormonal homeostasis in plant reaction to drought stress. *Frontiers in Plant Science* 7: 1824.

Guo, H., Li, L., Aluru, M. et al. (2013). Mechanisms and networks for brassinosteroid regulated gene expression. *Current Opinion in Plant Biology* 16 (5): 545–553.

Gupta, K., Mishra, S.K., Gupta, S. et al. (2021). Functional role of miRNAs: key players in soybean improvement. *Phyton* 90 (5): 1339.

Gupta, R., Min, C.W., Cho, J.H. et al. (2024). Integrated "-omics" analysis highlights the role of brassinosteroid signaling and antioxidant machinery underlying improved rice seed longevity during artificial aging treatment. *Plant Physiology and Biochemistry* 206: 108308.

Gutiérrez-Villamil, D.A., Magnitskiy, S., and Balaguera-López, H.E. (2024). Physiological and molecular functions of brassinosteroids during fruit development, ripening, and postharvest damage of horticultural products: A review. *Postharvest Biology and Technology 214*: 112984.

Hasan, M.M., Ali, M.A., Soliman, M.H. et al. (2020). Insights into 28-homobrassinolide (HBR)-mediated redox homeostasis, AsA–GSH cycle, and methylglyoxal detoxification in soybean under drought-induced oxidative stress. *Journal of Plant Interactions* 15 (1): 371–385.

Hayat, S., Alyemeni, M.N., and Hasan, S.A. (2012). Foliar spray of brassinosteroid enhances yield and quality of *Solanum lycopersicum* under cadmium stress. *Saudi Journal of Biological Sciences* 19 (3): 325–335.

Hayat, S., Hasan, S.A., Yusuf, M. et al. (2010). Effect of 28-homobrassinolide on photosynthesis, fluorescence and antioxidant system in the presence or absence of salinity and temperature in *Vigna radiata*. *Environmental and Experimental Botany* 69 (2): 105–112.

Hayat, S., Khalique, G., Wani, A.S. et al. (2014). Protection of growth in response to 28-homobrassinolide under the stress of cadmium and salinity in wheat. *International Journal of Biological Macromolecules* 64: 130–136.

Hou, Y., Qiu, J., Wang, Y. et al. (2017). A quantitative proteomic analysis of brassinosteroid-induced protein phosphorylation in rice (*Oryza sativa* L.). *Frontiers in Plant Science* 8: 514.

Hu, S., Liu, L., Li, S. et al. (2020). Regulation of fruit ripening by the brassinosteroid biosynthetic gene *SlCYP90B3* via an ethylene-dependent pathway in tomato. *Horticulture Research* 7: 163.

Huang, B., Chu, C.H., Chen, S.L. et al. (2006). A proteomics study of the mung bean epicotyl regulated by brassinosteroids under conditions of chilling stress. *Cellular & Molecular Biology Letters* 11: 264–278.

Huang, S., Yang, C., and Li, L. (2022). Unraveling the dynamic integration of auxin, brassinosteroid and gibberellin in early shade-induced hypocotyl elongation. *Phenomics* 2 (2): 119–129.

Hussain, M.A., Fahad, S., Sharif, R. et al. (2020). Multifunctional role of brassinosteroid and its analogues in plants. *Plant Growth Regulation* 92: 141–156.

Hwang, H., Ryu, H., and Cho, H. (2021). Brassinosteroid signaling pathways interplaying with diverse signaling cues for crop enhancement. *Agronomy* 11 (3): 556.

Jangid, K.K. and Dwivedi, P. (2017). Physiological and biochemical changes by nitric oxide and brassinosteroid in tomato (*Lycopersicon esculentum* Mill.) under drought stress. *Acta Physiologiae Plantarum* 39: 1–10.

Jiang, L., Ball, G., Hodgman, C. et al. (2018). Analysis of gene regulatory networks of maize in response to nitrogen. *Genes* 9 (3): 151.

Kang, Y.Y. and Guo, S.R. (2011). Role of brassinosteroids on horticultural crops. In: *Brassinosteroids: A Class of Plant Hormone* 269–288.

Karlova, R., van Haarst, J.C., Maliepaard, C. et al. (2013). Identification of microRNA targets in tomato fruit development using high-throughput sequencing and degradome analysis. *Journal of Experimental Botany* 64 (7): 1863–1878.

Kaur, H., Kaur, G., Sirhindi, G. et al. (2024). Exploring the role of 28-homobrassinolide in regulation of temperature induced clastogenic aberrations and sugar metabolism of *Brassica juncea* L. *Plant Physiology and Biochemistry* 214: 108893.

Kaur, H., Sirhindi, G., Bhardwaj, R. et al. (2018). 28-homobrassinolide regulates antioxidant enzyme activities and gene expression in response to salt-and temperature-induced oxidative stress in *Brassica juncea*. *Scientific Reports* 8 (1): 8735.

Khan, M., Ahmed, S., Yasin, N.A. et al. (2023). 28-homobrassinolide primed seed improved lead stress tolerance in *Brassica rapa* L. through modulation of physio-biochemical attributes and nutrient uptake. *Plants* 12 (20): 3528.

Khan, T.A., Yusuf, M., Ahmad, A. et al. (2019). Proteomic and physiological assessment of stress sensitive and tolerant variety of tomato treated with brassinosteroids and hydrogen peroxide under low-temperature stress. *Food Chemistry* 289: 500–511.

Kim, E.J. and Russinova, E. (2020). Brassinosteroid signalling. *Current Biology* 30 (7): R294–R298.

Kohli, S.K., Bhardwaj, A., Bhardwaj, V. et al. (2020). Therapeutic potential of brassinosteroids in biomedical and clinical research. *Biomolecules* 10 (4): 572.

Kuczyńska, A., Michałek, M., Ogrodowicz, P. et al. (2025). Disorders in brassinosteroids signal transduction triggers the profound molecular alterations in the crown tissue of barley under drought. *PLoS One* 20 (2): e0318281.

Kumar, K., Kumar, M., Kim, S.R. et al. (2013). Insights into genomics of salt stress response in rice. *Rice* 6: 1–15.

Kwak, P.B., Wang, Q.Q., Chen, X.S. et al. (2009). Enrichment of a set of microRNAs during the cotton fiber development. *BMC Genomics* 10: 1–11.

Li, H., He, X., Gao, Y. et al. (2023). Integrative analysis of transcriptome, proteome, and phosphoproteome reveals potential roles of photosynthesis antenna proteins in response to brassinosteroids signaling in maize. *Plants* 12 (6): 1290.

Li, J., Sohail, H., Nawaz, M.A. et al. (2022). Physiological and proteomic analyses reveals that brassinosteroids application improves the chilling stress tolerance of pepper seedlings. *Plant Growth Regulation* 96(2): 1–15.

Li, Q.F., Xiong, M., Xu, P. et al. (2016). Dissection of brassinosteroid-regulated proteins in rice embryos during germination by quantitative proteomics. *Scientific Reports* 6 (1): 34583.

Li, W.F., Zhou, Q., Ma, Z.H. et al. (2024). Regulatory mechanism of GA3 application on grape (*Vitis vinifera* L.) berry size. *Plant Physiology and Biochemistry* 210: 108543.

Lin, L.L., Hsu, C.L., Hu, C.W. et al. (2015). Integrating phosphoproteomics and bioinformatics to study brassinosteroid-regulated phosphorylation dynamics in *Arabidopsis*. *BMC Genomics* 16: 1–17.

Lin, L.L., Wu, C.C., Huang, H.C. et al. (2013). Identification of microRNA 395a in 24-epibrassinolide-regulated root growth of *Arabidopsis thaliana* using microRNA arrays. *International Journal of Molecular Sciences* 14 (7): 14270–14286.

Liu, D., Yu, Z., Zhang, G. et al. (2021). Diversification of plant agronomic traits by genome editing of brassinosteroid signaling family genes in rice. *Plant Physiology* 187 (4): 2563–2576.

Liu, F., Wei, T., Wang, Q. et al. (2023). *GhSMO2-2* is regulated by brassinosteroid signal and involved in cotton fiber elongation *via* influencing phytosterol and sphingolipid biosynthesis. *Industrial Crops and Products* 205: 117527.

Liu, N., Tu, L., Tang, W. et al. (2014). Small RNA and degradome profiling reveals a role for mi RNA s and their targets in the developing fibers of *Gossypium barbadense*. *The Plant Journal* 80 (2): 331–344.

Liu, X., Yang, H., Wang, Y., Zhu, Z., Zhang, W., & Li, Jet al. (2020). Comparative transcriptomic analysis to identify brassinosteroid response genes. *Plant Physiology* 184 (2): 1072–1082.

Lu, H., Chen, M., Fu, M. et al. (2023). Brassinosteroids affect wood development and properties of *Fraxinus mandshurica*. *Frontiers in Plant Science* 14: 1167548.

Lu, Q., Zhang, Y., Hellner, J. et al. (2022). Proteome-wide cellular thermal shift assay reveals unexpected cross-talk between brassinosteroid and auxin signaling. *Proceedings of the National Academy of Sciences of the United States of America* 119 (11): e2118220119.

Lv, J., Dong, T., Zhang, Y. et al. (2022). Metabolomic profiling of brassinolide and abscisic acid in response to high-temperature stress. *Plant Cell Reports* 41 (4): 935–946.

Ma, F., An, Z., Yue, Q. et al. (2022). Effects of brassinosteroids on cancer cells: a review. *Journal of Biochemical and Molecular Toxicology* 36 (6): e23026.

Mandava, B. and Wang, Y. (2015). Effect of brassinosteroids on cherry maturation, firmness and fruit quality. In: *III Balkan Symposium on Fruit Growing 1139*, 451–458.

Manghwar, H., Hussain, A., Ali, Q. et al. (2022). Brassinosteroids (BRs) role in plant development and coping with different stresses. *International Journal of Molecular Sciences* 23 (3): 1012.

McGuiness, P.N., Reid, J.B., and Foo, E. (2021). The influence of ethylene, gibberellins and brassinosteroids on energy and nitrogen-fixation metabolites in nodule tissue. *Plant Science* 305: 110846.

Miao, R., Li, C., Liu, Z. et al. (2024). The role of endogenous brassinosteroids in the mechanisms regulating plant reactions to various abiotic stresses. *Agronomy* 14 (2): 356.

Mitchell, J.W., Mandava, N., Worley, J.F. et al. (1970). Brassins—A new family of plant hormones from rape pollen. *Nature* 225 (5237): 1065–1066.

Montes, C., Wang, P., Liao, C.Y. et al. (2022). Integration of multi-omics data reveals interplay between brassinosteroid and target of Rapamycin complex signaling in *Arabidopsis*. *New Phytologist* 236 (3): 893–910.

Moreau, R.A., Nyström, L., Whitaker, B.D. et al. (2018). Phytosterols and their derivatives: Structural diversity, distribution, metabolism, analysis, and health-promoting uses. *Progress in Lipid Research* 70: 35–61.

Musser, R.O., Hum-Musser, S.M., Gallucci, M. et al. (2014). Microarray analysis of tomato plants exposed to the nonviruliferous or viruliferous whitefly vector harboring pepper golden mosaic virus. *Journal of Insect Science* 14 (1): 230.

Nolan, T.M., Vukašinović, N., Hsu, C.W. et al. (2023). Brassinosteroid gene regulatory networks at cellular resolution in the *Arabidopsis* root. *Science* 379 (6639): eadf4721.

Oh, M.H., Honey, S.H., and Tax, F.E. (2020). The control of cell expansion, cell division, and vascular development by brassinosteroids: a historical perspective. *International Journal of Molecular Sciences* 21 (5): 1743.

Pantoja-Benavides, A.D., Garces-Varon, G., and Restrepo-Díaz, H. (2022). Foliar cytokinins or brassinosteroids applications influence the rice plant acclimatization to combined heat stress. *Frontiers in Plant Science* 13: 983276.

Pavlovna, U.I., Nikolaevna, K.N., Vladimirovna, D.E. et al. (2015). Brassinosteroids as a factor of photosyntetic activity increase of improved potatoes. *Biosciences Biotechnology Research Asia* 12 (2): 1481–1485.

Pereira, Y.C., Rodrigues, W.S., Lima, E.J.A. et al. (2019). Brassinosteroids increase electron transport and photosynthesis in soybean plants under water deficit. *Photosynthetica* 57 (1): 181–191.

Pereira-Netto, A.B., Roessner, U., Fujioka, S. et al. (2009). Shooting control by brassinosteroids: metabolomic analysis and effect of brassinazole on *Malus prunifolia*, the Marubakaido apple rootstock. *Tree Physiology* 29 (4): 607–620.

Piskurewicz, U., Glauser, G., and Lopez-Molina, L. (2024). Endospermic brassinosteroids moderate seed thermoinhibition responses in *Arabidopsis thaliana*. *New Phytologist* 241 (6): 2320–2325. https://doi.org/10.1111/nph.19491

Qanmber, G., Liu, Z., Li, F. et al. (2024). Brassinosteroids in cotton: orchestrating fiber development. *New Phytologist* 244 (5): 1732–1741.

Qazi, F., Mohammad, Y., Mahmooda, B. et al. (2014). 28-homobrassinolide protects photosynthetic machinery in Indian mustard under high temperature stress. *Journal of Stress Physiology & Biochemistry* 10 (1): 181–194.

Que, F., Khadr, A., Wang, G.L. et al. (2018). Exogenous brassinosteroids altered cell length, gibberellin content, and cellulose deposition in promoting carrot petiole elongation. *Plant Science* 277: 110–120.

Que, F., Wang, G.L., Xu, Z.S. et al. (2017). Transcriptional regulation of brassinosteroid accumulation during carrot development and the potential role of brassinosteroids in petiole elongation. *Frontiers in Plant Science* 8: 1356.

Randhawa, N., Sharma, I., and Bhardwaj, R. (2010). Effect of 24-epibrassinolide and 28-homobrassinolide on some biochemical parameters in *Raphanus sativus* L. plants under chromium stress. *Seed Science and Biotechnology* 4 (1): 28–32.

Rohrmann, J., Tohge, T., Alba, R. et al. (2011). Combined transcription factor profiling, microarray analysis and metabolite profiling reveals the transcriptional control of metabolic shifts occurring during tomato fruit development. *The Plant Journal* 68 (6): 999–1013.

Sahni, S., Prasad, B.D., Liu, Q. et al. (2016). Overexpression of the brassinosteroid biosynthetic gene *DWF4* in *Brassica napus* simultaneously increases seed yield and stress tolerance. *Scientific Reports* 6 (1): 28298.

Salehi, H., Miras-Moreno, B., Tolisano, C. et al. (2024). Brassinosteroids alleviate salinity stress in tomato through the modulation of photosynthetic efficiency, antioxidant machinery, and metabolomic profile. *Antioxidant Machinery, and Metabolomic Profile*. http://dx.doi.org/10.2139/ssrn.4752002

Sánchez-Rodríguez, C., Ketelaar, K., Schneider, R. et al. (2017). BRASSINOSTEROID INSENSITIVE2 negatively regulates cellulose synthesis in *Arabidopsis* by phosphorylating cellulose synthase 1. *Proceedings of the National Academy of Sciences of the United States of America* 114 (13): 3533–3538.

Seyed Rahmani, R., Shi, T., Zhang, D. et al. (2021). Genome-wide expression and network analyses of mutants in key brassinosteroid signaling genes. *BMC Genomics* 22 (1): 465.

Shahzad, B., Tanveer, M., Che, Z. et al. (2018). Role of 24-epibrassinolide (EBL) in mediating heavy metal and pesticide induced oxidative stress in plants: a review. *Ecotoxicology and Environmental Safety* 147: 935–944.

Sharma, I., Bhardwaj, R., Gautam, V. et al. (2018a). Brassinosteroids: occurrence, structure and stress protective activities. In: *Frontiers in Natural Product Chemistry*, 4: 204–239. Bentham Science Publishers.

Sharma, I., Pati, P.K., and Bhardwaj, R. (2010). Regulation of growth and antioxidant enzyme activities by 28-homobrassinolide in seedlings of *Raphanus sativus* L. under cadmium stress. *Indian Journal of Biochemistry & Biophysics* 47: 172–177.

Sharma, I., Pati, P.K., and Bhardwaj, R. (2011a). Effect of 24-epibrassinolide on oxidative stress markers induced by nickel-ion in *Raphanus sativus* L. *Acta Physiologiae Plantarum* 33: 1723–1735.

Sharma, I., Pati, P.K., and Bhardwaj, R. (2011b). Effect of 28-homobrassinolide on antioxidant defence system in *Raphanus sativus* L. under chromium toxicity. *Ecotoxicology* 20: 862–874.

Sharma, I., Sharma, A., Pati, P. et al. (2018b). Brassinosteroids reciprocates heavy metals induced oxidative stress in radish by regulating the expression of key antioxidant enzyme genes. *Brazilian Archives of Biology and Technology* 61: e18160679.

Sharma, N., Hundal, G. S., Sharma, I. et al. (2012). Effect of 24-epibrassinolide on protein content and activities of glutathione-S-transferase and polyphenol oxidase in *Raphanus sativus* L. plants under cadmium and mercury metal stress. *Terrestrial and Aquatic Environmental Toxicology* 6 (1): 1–7.

Sharma, N., Hundal, G.S., Sharma, I. et al. (2014). 28-Homobrassinolide alters protein content and activities of glutathione-s-transferase and polyphenol oxidase in *Raphanus sativus* L. plants under heavy metal stress. *Toxicology International* 21 (1): 44.

Siddiqui, H., Ahmed, K.B.M., and Hayat, S. (2018). Comparative effect of 28-homobrassinolide and 24-epibrassinolide on the performance of different components influencing the photosynthetic machinery in *Brassica juncea* L. *Plant Physiology and Biochemistry* 129: 198–212.

Šimura, J., Antoniadi, I., Široká, J. et al. (2018). Plant hormonomics: multiple phytohormone profiling by targeted metabolomics. *Plant Physiology* 177 (2): 476–489.

Sirhindi, G., Kaur, H., Bhardwaj, R. et al. (2017). 28-Homobrassinolide potential for oxidative interface in *Brassica juncea* under temperature stress. *Acta Physiologiae Plantarum* 39: 1–10.

Soares, T.F.S.N., dos Santos Dias, D.C.F., Oliveira, A.M.S. et al. (2020). Exogenous brassinosteroids increase lead stress tolerance in seed germination and seedling growth of *Brassica juncea* L. *Ecotoxicology and Environmental Safety* 193: 110296.

Song, L., Chen, W., Yao, Q.et al. (2019). Genome-wide transcriptional profiling for elucidating the effects of brassinosteroids on *Glycine max* during early vegetative development. *Scientific Reports* 9 (1): 16085.

Song, L., Liu, J., Cao, B. et al. (2023). Reducing brassinosteroid signalling enhances grain yield in semi-dwarf wheat. *Nature* 617 (7959): 118–124.

Sorrentino, M., Panzarová, K., Spyroglou, I. et al. (2022). Integration of phenomics and metabolomics datasets reveals different mode of action of biostimulants based on protein hydrolysates in *Lactuca sativa* L. and *Solanum lycopersicum* L. under salinity. *Frontiers in Plant Science* 12: 808711.

Sun, Y., Fan, X.Y., Cao, D.M.et al. (2010). Integration of brassinosteroid signal transduction with the transcription network for plant growth regulation in *Arabidopsis*. *Developmental Cell* 19 (5): 765–777.

Surgun-Acar, Y. and Zemheri-Navruz, F. (2019). 24-Epibrassinolide promotes arsenic tolerance in *Arabidopsis thaliana* L. by altering stress responses at biochemical and molecular level. *Journal of Plant Physiology* 238: 12–19.

Tang, W., Deng, Z., and Wang, Z.Y. (2010). Proteomics shed light on the brassinosteroid signaling mechanisms. *Current Opinion in Plant Biology* 13 (1): 27–33.

Tang, W., Deng, Z., Oses-Prieto, J.A. et al. (2008). Proteomics studies of brassinosteroid signal transduction using prefractionation and two-dimensional DIGE. *Molecular & Cellular Proteomics* 7 (4): 728–738.

Tong, H. and Chu, C. (2018). Functional specificities of brassinosteroid and potential utilization for crop improvement. *Trends in Plant Science* 23 (11): 1016–1028.

Trevisan, S., Forestan, C., Brojanigo, S. et al. (2020). Brassinosteroid application affects the growth and gravitropic response of maize by regulating gene expression in the roots, shoots and leaves. *Plant Growth Regulation* 92 (1): 117–130.

Verma, A., Ansari, M.W., Singh, H. et al. (2021). Proteomics for brassinosteroid signalling: understanding brassinosteroids mediated stress responses through advanced proteomics. *Plant Gene* 26: 100282.

Wang, C., Liu, N., Geng, Z. et al. (2022). Integrated transcriptome and proteome analysis reveals brassinosteroid-mediated regulation of cambium initiation and patterning in woody stem. *Horticulture Research* 9: uhab048.

Wang, F., Bai, M.Y., Deng, Z. et al (2010). Proteomic study identifies proteins involved in brassinosteroid regulation of rice growth. *Journal of Integrative Plant Biology* 52 (12): 1075–1085.

Wang, X., Wei, J., Wu, J. et al. (2024). Transcription factors BZR2/MYC2 modulate brassinosteroid and jasmonic acid crosstalk during pear dormancy. *Plant Physiology* 194 (3): 1794–1814.

Wang, X., Zhao, Z., Guo, N. et al. (2020). Comparative proteomics analysis reveals that lignin biosynthesis contributes to brassinosteroid-mediated response to *Phytophthora sojae* in soybeans. *Journal of Agricultural and Food Chemistry* 68 (19): 5496–5506.

Wei, B., Cai, T., Zhang, R. et al. (2009). Novel microRNAs uncovered by deep sequencing of small RNA transcriptomes in bread wheat (*Triticum aestivum* L.) and *Brachypodium distachyon* (L.) Beauv. *Functional & Integrative Genomics* 9: 499–511.

Xu, F., Xi, Z.M., Zhang, H. et al. (2015). Brassinosteroids are involved in controlling sugar unloading in *Vitis vinifera* 'Cabernet Sauvignon' berries during véraison. *Plant Physiology and Biochemistry* 94: 197–208.

Xu, H., Sun, H., Dong, J., et al. (2022). The brassinosteroid biosynthesis gene TaD11–2A controls grain size and its elite haplotype improves wheat grain yields. *Theoretical and Applied Genetics* 135 (8): 2907–2923.

Yang, Y., Chen, M., Li, Z. et al. (2025). Transcriptome and metabolome analyses reveal the effects of brassinolide treatment on the ripening and flavor development of postharvest kiwifruit. *Postharvest Biology and Technology* 222: 113380.

Yang, Y., Chu, C., Qian, Q. et al. (2024). Leveraging brassinosteroids towards the next Green Revolution. *Trends in Plant Science* 29 (1): 86–98.

Zhang, S., Hu, J., Zhang, Y. et al. (2007). Seed priming with brassinolide improves lucerne (*Medicago sativa* L.) seed germination and seedling growth in relation to physiological changes under salinity stress. *Australian Journal of Agricultural Research* 58 (8): 811–815.

Zhang, Y., Chen, H., Li, S. et al. (2021). Comparative physiological and proteomic analyses reveal the mechanisms of brassinolide-mediated tolerance to calcium nitrate stress in tomato. *Frontiers in Plant Science* 12: 724288.

Zhang, Y., Liu, S., Dai, S.Y. et al. (2012). Integration of shot-gun proteomics and bioinformatics analysis to explore plant hormone responses. In: *BMC Bioinformatics*, 13: 1–11. BioMed Central.

Zhang, Z., Yang, X., Cheng, L. et al. (2020). Physiological and transcriptomic analyses of brassinosteroid function in moso bamboo (*Phyllostachys edulis*) seedlings. *Planta* 252: 1–13.

Zheng, C., Zhao, L., Wang, Y. et al. (2015). Integrated RNA-Seq and sRNA-Seq analysis identifies chilling and freezing responsive key molecular players and pathways in tea plant (*Camellia sinensis*). *PLoS One* 10 (4): e0125031.

Zheng, L., Ma, J., Zhang, L. et al. (2018). Revealing critical mechanisms of BR-mediated apple nursery tree growth using iTRAQ-based proteomic analysis. *Journal of Proteomics* 173: 139–154.

Zheng, Y., Xu, B., Ren, K. et al. (2017). Impact of soil drench and foliar spray of 24-epibrassinolide on the growth, yield, and quality of field-grown *Moringa oleifera* in Southwest China. *Journal of Plant Growth Regulation* 36: 931–941.

Zhong, W., Xie, C., Hu, D. et al. (2020). Effect of 24-epibrassinolide on reactive oxygen species and antioxidative defense systems in tall fescue plants under lead stress. *Ecotoxicology and Environmental Safety* 187: 109831.

3

Regulation of Plant Primary Metabolism by Brassinosteroids Under Abiotic Stress

*Slimane Mokrani[1,2] and El-hafid Nabti[2]**

[1]University Mustapha Stambouli, Department of Agronomy, Laboratory of Research on Biological Systems and Geomantic, Mascara, Algeria
[2]Laboratoire de Maitrise des Energies Renouvelables, Faculté des Sciences de la Nature et de la Vie, Université de Bejaia, Bejaia, Algeria

*Corresponding author: nabtielhafid1977@yahoo.com

3.1 Introduction

Plant secondary metabolites are typically described as molecules that have no basic involvement in the maintenance of life processes in plants. However, they are vital for the plant to interact with its environment for adaptation and protection (Akula and Ravishankar 2011). Plants use a variety of signaling molecules, including hormones, to mediate their responses to stresses (Guo et al. 2019). Stresses, whether abiotic or biotic, have a significant impact on agricultural profitability, and global farming is facing the challenge of increasing efficiency in order to meet the growing dietary demand of the rising population (Parihar et al. 2015). Furthermore, it has been estimated that abiotic stresses may significantly affect 70% of the output of staple support crops (Kaur et al. 2008; Mantri et al. 2012). Research has demonstrated that plant responses to a certain stress are mediated by a combination of hormones and signaling molecules, rather than a single hormone (Choudhary et al. 2012; Planas-Riverola et al. 2019). At the 13th Annual Meeting of International Plant Growth Substances in 1998, brassinosteroids (BRs) were confirmed as the sixth class of plant hormones. Previously known plant hormones include auxin (IAA), gibberellin (GA), cytokinin (CTK), abscisic acid (ABA), and ethylene (ETH) (Clouse and Sasse 1998). This family of phytohormones is represented by more than 70 substances, which have been isolated or discovered from more than 100 plant species, ranging from algae to angiosperms, indicating their widespread distribution across the plant kingdom (Zullo and Adam 2002).

BRs are plant steroid hormones that were initially discovered roughly 40 years ago. Approximately 60 similar chemicals have been discovered (Xia et al. 2009a; Choudhary et al. 2012). In 1979, Brassica napus pollen was used to isolate the plant growth-promoting steroid brassinolide (BL) (Grove et al. 1979). The most physiologically active component among all the BRs that have been discovered so far is known as BL, and it is present in a

Brassinosteroids and Abiotic Stress: Recent Developments on Physiological and Molecular Aspects, First Edition. Edited by Anket Sharma, et al.
Published 2026 by John Wiley & Sons Ltd.

wide variety of plant species (Kim et al. 2008). Considerable study on BRs has been focused on BL, a C28 complex molecule with an S-methyl group at C24 of its 5α-ergostane structure side chain (Ashraf et al. 2010). Several functions are played by the plant steroid hormones known as BRs in the growth and development of plants. In plants, phenotypic defects and impaired physiological processes are caused by BR deficits. Numerous studies demonstrate how beneficial BRs are for plants' responses to abiotic stresses such as pesticides, heavy metals, heat, cold, drought, and salinity (Ahammed et al. 2020). Throughout their life cycle, BRs help plants adjust to changes in their environment, maintain homeostasis, and control growth and development (Li et al. 2020).

Recent developments in the accessibility of biological resources and methodologies have produced the most useful methods for BRs de novo synthesis, translocation, and signaling pathways (Ahmad et al. 2018). These crucial genes are controlled by BRs because they encode essential proteins and enzymes that are required to prevail in the battle against stress. When plants are growing normally, BRs are usually not very active, but under stressful environments, they demonstrate their usefulness (Chaudhuri et al. 2022). It is now one of the most well-studied hormonal pathways in plants due to recent advancements in the use of contemporary genetic, biochemical, and proteomic tools along with a wide range of biological resources that are accessible. These tools have allowed for a thorough investigation of the key regulatory elements in BR signaling networks (Sharma et al. 2017). BR-induced stress tolerance to various environmental stimuli has been the subject of several investigations. Numerous environmental stressors cause the plant system to generate more free radicals and other oxidative species (Yaqoob et al. 2022). BRs increase the synthesis of secondary metabolites such as forskolin, phenolics, and shikonin (Swamy and Rao 2011). The majority of these secondary metabolites are reactive oxygen species (ROS) scavengers, which help the plant mitigate the impacts of diverse abiotic challenges (Yaqoob et al. 2022).

Numerous research teams have previously proven that BRs regulate a number of important physiological and biochemical functions. *Arabidopsis thaliana*, the model plant, has been the basis of several recent discoveries and offers a possible platform for BR experiments. Plants possess an intricate system of remarkable signaling networks that enable them to perceive and react to their constantly shifting surroundings (Chaudhuri et al. 2022). Experienced researchers have shown that when the membrane-bound steroid receptor BRI1 binds to BR, a series of cytoplasmic signaling events occurs, ultimately resulting in the expression of genes linked to BR. In this context, it was well-observed that the main roots of *A. thaliana* exhibit precise signals that respond to stress as an adaptive mechanism (Planas-Riverola et al. 2019). Researchers are focusing more and more on BR-associated stress-responsive signaling pathways as a means of understanding how plants might adapt to both biotic and abiotic stress (Anwar et al. 2018; Nolan et al. 2017). The chapter covers the advances in different aspects related to the regulation of plant primary metabolism by BRs under abiotic stress.

3.2 Primary Metabolism Regulation

Plant growth and development are significantly influenced by the plant hormone BR, which also regulates the responses of plants to stress. From a mechanical perspective, this dual capability of BR is fascinating, because it presents a workable way to stabilize agricultural production in the face of changing climate circumstances. BR integrates several stress-related traits even under non-stress settings, which are subsequently complemented by a dynamic stress response under unfavorable conditions (Divi et al. 2016). Previous research has demonstrated that BRs interact with a variety of plant morphophysiological processes, including those regulated by salicylic acid, abscisic acid, IAA, GA, ETH, CTK, and jasmonic acid (Ahmad et al. 2018). Several biological processes in plants depend on endogenous BR control (Tanaka et al. 2005). The endogenous level of BR in plants is actively maintained by the production, translocation, and metabolism of BR, which are essential components of metabolic balance. Furthermore, BRs are master regulators of plant disease resistance and defensive responses to pathogen assault, but further study is necessary to fully understand the mechanisms behind these responses, which are currently unclear and even controversial (Albrecht et al. 2012; Nahar et al. 2013).

Subsequent physiological and genetic investigations have shown that BRs have additional functions in regulating a wide range of processes, including source/sink relationships, seed germination, photosynthesis, senescence, flowering, and responses to biotic and abiotic stresses, despite the fact that BRs were first distinguished by their ability to promote growth (Deng et al. 2007). Primary metabolite stress tolerance is primarily attained by the low molecular weight antioxidants ascorbate and glutathione, as well as the traditional antioxidant system, which includes ascorbate peroxidase, superoxide dismutase, catalase, glutathione reductase (GR), and monodehydroascorbate reductase. On the other hand, a variety of secondary metabolites, such as tannins, flavonoids, hydroxycinnamate esters, phenolics, and lignins that are synthesized in plants, help to achieve the second level of defense and are highly effective in scavenging ROS (Yaqoob et al. 2022).

3.2.1 Respiration

From a biochemical perspective, nighttime respiration is an amphibolic, multistep, enzymatically controlled process that generates ATP through the oxidation of glucose produced during photosynthesis. During glycolysis, glucose is first broken down into pyruvate. This oxidized pyruvate then forms acetyl-CoA, which releases a CO_2 molecule (Sharma et al. 2022). Environmental factors have an influence on growth respiration (Rg) and maintenance respiration (Rm), which are the two functional components of dark respiration (Rd) (Amthor 1986; Lambers 1985; Slot and Kitajima 2015). The quantity of photoassimilates respired to supply energy for the synthesis of more biomass is known as growth respiration(Amthor et al. 2019). O_2 availability is necessary for plant respiration and mitochondrial energy production. Consequently, reduced O_2 concentrations have a negative impact on mitochondrial activity. Hypoxia, or low oxygen concentrations, causes cellular stress because it reduces the synthesis of ATP, depletes energy stores, and accumulates metabolic intermediates (Jethva et al. 2022).

In *Nicotiana benthamiana*, exogenously given BL, the most active BR, stimulated, but brassinazole (BRZ), a BR biosynthesis inhibitor, decreased alternative respiration and AOX1 expression. The BR-induced alternative respiratory route was hampered by chemical scavenging of H_2O_2 and virus-induced gene silencing (VIGS) of *NbRBOHB*. *NbAOX1* promoter analysis further supported this conclusion. Furthermore, reduced plant resilience to environmental stresses and reduced BR-induced stress tolerance resulted from AOX activity reduction by chemical treatment or a VIGS-based approach (Deng et al. 2015). Under salt stress, the rate of respiration in lettuce roots and shoots decreased (Zapata et al. 2007) to counteract the impact of stress. This is the plant's protective reaction. Compared to other important ions, NaCl ions have more mobility as they are fully dissociated into Na^+ and Cl^- ions. They obstruct the path for them, resulting in an ion deficit that forces the plants to breathe more quickly than usual. Application of BR returns respiration to normal (Siddiqi and Husen 2021).

3.2.2 Photosynthesis

Numerous studies indicate that BRs benefit plants' photosynthetic apparatus in both stressed and unstressed environments. By considerably increasing PSII's photochemical efficiency, quantum efficiency, and photochemical quenching coefficient, exogenous administration of BRs is known to reduce photoinhibition (Xia et al. 2009b; Ahammed et al. 2015). Under stress, BRs also significantly increase the intercellular CO_2 concentration, transpiration rate, stomatal conductance, and net photosynthetic rate (Farooq et al. 2010; Hayat et al. 2012). BRs increase the amount of chlorophyll and decrease the activity of the enzyme chlorophyllase, which is in charge of breaking down chlorophyll pigment while under abiotic stress (Bajguz 2011; Hayat et al. 2012; Sharma et al. 2015). Abiotic stresses of many kinds significantly impair the mechanism of photosynthesis (Sharma et al. 2019b; Dey et al. 2022) via adjusting the chemical processes and chlorophyll biosynthesis mediated by photosystems I (PSI) and II (PSII) (Sharma et al. 2020).

As a reflection of their dynamic responsibilities, BR treatments under regular conditions also increase net photosynthetic rate. In model plants *Brassica* and *Arabidopsis*, BR-insensitive mutants had been identified (Clouse et al. 1996; Caño-Delgado et al. 2004), and numerous developmental abnormalities were present in the plants, including dwarfism, late blooming, deep green foliage, and male sterility (Chakraborty et al. 2015; Hou et al. 2017). In particular, BRs boost the early activity of RuBiSCO to enhance CO_2 uptake in plants and the gene expression of carbon fixation (Xia et al. 2006; Sharma et al. 2019a). Through an increase in chlorophyll content, BR improves the plant's ability to withstand a variety of stresses. This, in turn, results in an increase in photosynthetic capacity, an improvement in the antioxidant system, an increase in enzymatic activity, and an upregulation of stress response genes, including ascorbate peroxide (APX), GR, superoxide (SOD), and peroxide (POD) (Shu et al. 2016; Yuan et al. 2012b). However, the way that BRs affect pigments and photosynthesis varies greatly. The method of administration of BRs and the connection between exogenously administered BRs and the basal chlorophyll content in different crop species determine the degree of an impact they have on the content of chlorophyll (Gomes 2011).

3.2.3 Flowering

The ability of a plant species to reproduce through seeds assures its survival; therefore, flowering is closely regulated by a confluence of environmental and developmental signals that are combined and transmitted by ad hoc regulatory pathways called flowering pathways (Amasino and Michaels 2010). Plant damage sustained during flowering might result in permanent reproductive failure (Sunkar et al. 2007). Thus, changes in flowering time are frequently seen in stressful conditions as a coping strategy (Wada and Takeno 2010). Abiotic stress has a variety of consequences on plants, from qualitative and quantitative modifications in the types of proteins that are synthesized in crops to interference with the development of flower buds and the pollination process (Oshunsanya et al. 2019). The plant often grows vegetatively for a longer time in N-hazardous circumstances, delaying flowering and fruiting (Hodges 2014; Ashoka et al. 2017). Since they mature more slowly than wild types, BR-insensitive and -deficient mutants are sometimes called late-flowering (Clouse 2008; Zhao and Li 2012). In contrast, plants that are overaccumulating BRs have longer petioles and hypocotyls as well as greater height (Rozhon et al. 2019).

Ren et al. (2022) examined the thermotolerance of a number of transgenic plants and mutants of *Arabidopsis* with hyperactive or lacking BR biosynthesis/signaling in a systematic way. They demonstrated the critical function BIN2 plays in regulating the control of BR signaling-induced plant heat shock (HS) responses. BIN2 was produced by HS, which caused an accumulation of the plant and sped up the life cycle of *Arabidopsis* plants by encouraging early flowering and diminishing their tolerance to heat. These findings provide insight into a unique method by which plants stimulate early flowering in order to adapt to heat stress in their natural environments. They also clarify why *Arabidopsis* plants become more thermotolerant when treated with BR. Since their development is slowed, BR-insensitive and -deficient mutants are sometimes referred to as late flowering (Clouse 2008; Zhao and Li 2012). In contrast, plants that are overaccumulating BRs have longer petioles and hypocotyls as well as higher heights (Rozhon et al. 2019).

3.2.4 Seed Germination

Growth of seedlings and seed germination are crucial factors in a good agricultural production system. During germination, seeds are extremely sensitive and readily harmed by minimal outside stimuli (stress) (Lapik and Kaufman 2003). External environmental conditions have a substantial impact on the germination stage of seeds, which is highly susceptible to abiotic stresses such as salinity, drought, and extreme temperatures (Awan et al. 2023). The assessment of abiotic stress tolerance during seed germination may take into consideration changes in the germination markers, such as the germination index (GI), germination potential (GP), and germination rate (GR) (Zhang et al. 2020). Various traditional techniques have been employed to mitigate the harmful impacts of abiotic stresses on seed germination and radicle emergence (Seleiman et al. 2021).

When seeds germinate, BRs have the ability to initiate the cell cycle (Zadvornova et al. 2005). In *A. thaliana*, BR transcription modulates the genes and pathways that control the development of the seed and ovule, regulating seed development and impacting seed size, weight, and quantity (Huang et al. 2013; Jiang et al. 2013). Furthermore, under moderate nitrogen deficiency, BRs control the root nitrogen foraging response in *A. thaliana*. The main root elongates in response to moderate nitrogen lack due to the action of the Brassinosteroid signaling kinase 3 (*BSK3*) gene (Jia et al. 2019). In addition, Chakma et al. (2021) reported that of cotton seeds primed with BR (24-epibrassinolide, or 24-EBL) alone or in conjunction with other hormones (abscisic acid, auxin, and gibberellic acid) revealed that BR promoted germination under no stress as well as under salinity and heat stress conditions, while other hormones were ineffective under stress conditions. BR also promoted cotyledon opening and the development of lateral roots in germinated seedlings. The ability of BR to positively impact seedling growth across different stress conditions suggests that priming cotton seeds with BR may help in early and successful establishment of seedlings, which may benefit the plant through its lifecycle. Whereas other hormones were ineffective under stress, BR facilitated germination in the presence of heat, salt, and both. In seedlings that germinated, BR further promoted the growth of lateral roots and cotyledon opening. More than 800 BR-responsive proteins were found in a recent proteomic study of rice seeds that were germinating, with 88 of those proteins exhibiting high confidence (Li et al. 2016). Many of these proteins were linked to the metabolism of carbohydrates and proteins, which is compatible with the reactivation of metabolism and the production of proteins during germination. It was shown earlier that BR affects the stability and synthesis of the translational machinery components, which has a significant impact on protein synthesis under stress (Dhaubhadel et al. 2002; Divi et al. 2015).

3.2.5 Senescence

Senescence is a term that mainly relates to the aging process and is connected to decay, death, or diminished fertility with age. It comes from the Latin word "senēscere," which means to grow weak, become exhausted, and be in a decline (Caswell and Salguero-Gómez 2013), but it is actually a very widespread concept for plants (Miryeganeh 2021). Abiotic stresses such as inappropriate temperature lead to early senescence in plants, which offsets normal processes and reduces plant productivity by 50% on average (Sağlam Çağ 2019). Abiotic stresses accelerate plant senescence, resulting in decreased yields and photosynthetic losses. Source–sink interactions are put to the test by abiotic stress-induced senescence, which also significantly lowers crop yields. While the mechanisms governing stress-induced senescence are poorly understood, the processes connected to plant senescence are extensively investigated (Sade et al. 2018).

Numerous abiotic stimuli have been explored in relation to the reaction of osmoprotectants in plant species to heat stress (Kothari and Lachowiec 2021). The effects of high temperatures are also mitigated by the induction of osmoprotectants, which are facilitated by BR. Heat stress causes cellular membranes to become unstable, which speeds up membrane damage and electrolyte leakage, ultimately resulting in cell death and senescence (Akter and Islam 2017). Brassinosteroids induced senescence in detached cucumber seedling cotyledons (Zhao et al. 1990) and leaves of mung bean seedlings (He et al. 1996).

3.2.6 Plant Growth

A crucial function for BRs in plant development has been demonstrated by molecular genetics investigations of BR-deficient and BR-insensitive mutants. These studies have also led to the identification and characterization of other BR signaling components (Vert et al. 2005; Belkhadir et al. 2006). Microarray investigations have identified a vast number of BR-regulated genes, most of which are linked to aspects of plant growth and development such as hormone production, cytoskeleton construction, and cell wall modification (Vert et al. 2005). The rice BR-deficient dwarf1 mutant displays defective leaf, stem, and other developmental organ structures as well as polar cell elongation due to the absence of BR C-6-oxidase activity (Hong et al. 2002). Maintaining a balance between plant growth and development and stress tolerance has several advantages. Plants must reproduce and grow in order to resist harsh environmental conditions (Raza 2020; Hafeez et al. 2021a; Saddiq et al. 2021). Although BRs have the ability to promote the growth of thin primary branched roots, crown roots, and primary and secondary branched roots (Sasse 2003). BRs effectively improved growth characteristics, proline and metabolite levels, and reduced oxidative damage in bitter melon under salt stress (Souza et al. 2017). Recently, it has been shown that BIN2-mediated BR signaling interacts with the autophagy system to control plant survival and growth under conditions of starvation and drought. When BES1 homodimers became less available, BR-biosynthesis de-depressed. At higher temperatures, the BES1-PIF4 (BES1-PHYTOCHROME INTERACTING FACTOR 4) complex promotes thermomorphogenesis (Hafeez et al. 2021b).

3.2.7 Nutrient Uptake

Another crucial factor in plant development is the uptake of vital nutrients (N, P, K, Ca, Mg, Fe, Mn, Cu, Zn, etc.). Enhanced light energy transformation efficiency, CO_2 conductivity, light-dark response potential, and photosynthetic rate are all made possible by higher ion intake (Talaat 2013; Song et al. 2016). Plant growth regulators, or PGRs, have the potential to improve crop quality characteristics, including nutrient efficiency and resistance to abiotic stress (Colla and Rouphael 2015). PGRs have been demonstrated to increase nutrition activity and boost the effectiveness of nutrient exporters in cell membranes, both of which have been linked to increased plant nutrient efficiency (De Pascale et al. 2017). BRs tend to communicate the distribution of nutrients at the rhizosphere level for improved development and high yield in agricultural plants. They also control numerous physiological processes, such as root growth under nutrient deficiencies in nitrogen, phosphorus, and boron. But some of the intake of mineral nutrients can be partially offset by phytohormones, such as BRs (Praveena et al. 2020).

The exogenous administration of BRs, either spraying, priming, or feeding in conjunction with nutrients, supports the systems involved in stress tolerance (Fariduddin et al. 2009; Nolan et al. 2017). According to Liu et al. (2014), when canola leaves, roots, and epicotyls were subjected to salt stress, BR enhanced the absorption of necessary inorganic ions, reduced the buildup of harmful ions, and fostered ion homeostasis. The activities of nitrite reductase, nitrate reductase, glutamate dehydrogenase, glutamate synthase, and glutamine synthase in tomato seedlings were all elevated by the exogenous administration of 24-EBL (Shu et al. 2016). Additionally, BR treatment increases the activities of H+-ATPase and Ca2+-ATPase in the roots and leaves, which are in charge of controlling the electrochemical gradient and preserving ionic equilibrium to withstand the challenges of stressful conditions (Shu et al. 2016). Nonetheless, exogenous BR treatment makes plants more sensitive to iron shortage, indicating that BR has two roles for assisting plants to tolerate nutritional deficiencies (Wang et al. 2012). Furthermore, Janeczko et al. (2010) reported that under salt stress, *Triticum aestivum*'s nutrient absorption was significantly increased by the foliar application of EBL. The K/Na ratio can be stabilized and absorption maximized by nutrients like calcium and potassium. Additionally, they observed that the exogenous

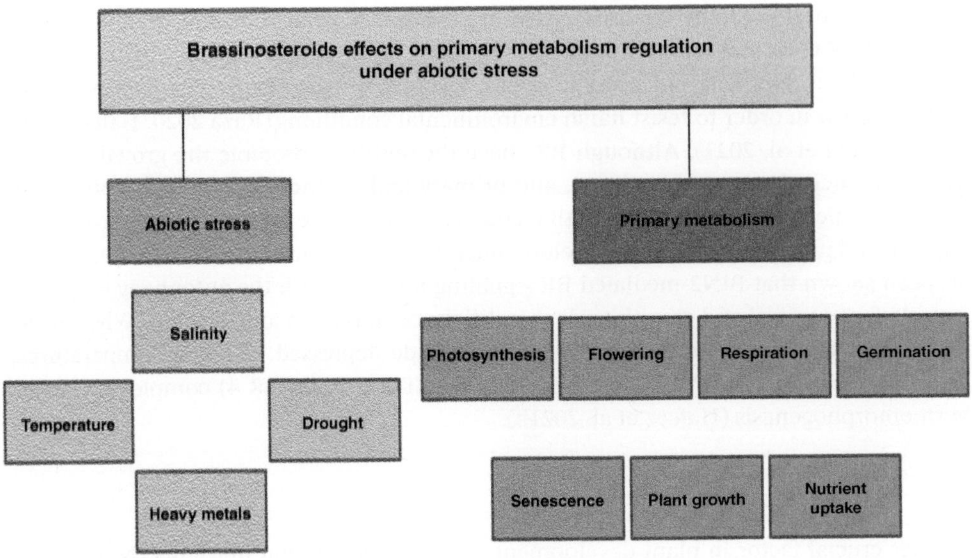

Figure 3.1 Control of physiological mechanisms implicating brassinosteroids under abiotic constraints.

administration of EBL (0.01 M) by foliar or soaking enhanced the absorption of potassium, magnesium, and calcium while lowering iron and salt levels in wheat.

Figure 3.1 summarizes the control of physiological mechanisms implicating brassinosteroids under abiotic constraints. The brassinosteroid effects on metabolism regulation include photosynthesis, flowering, respiration, germination, senescence, plant growth, and nutrient uptake under different abiotic stresses like salinity, temperature, drought, and heavy metals.

3.3 Brassinosteroids Mediation and Abiotic Stress Tolerance Regulation

The fundamental mechanisms of BR-induced stress tolerance are mostly unexplained. The key transcription factor Brassinazole-Resistant1 (BZR1), which regulates the transcription of BR-responsive genes in the nucleus, is activated by phosphorylation processes that follow BR receptors' detection of BR on the cell surface (Ahammed et al. 2020). Membrane-localized receptors detect BR, and cytosolic regulators downstream translate BR-mediated signals to the nucleus, where they trigger the transcription of BR-responsive genes that promote cell division (Belkhadir and Jaillais 2015). They operate in combination with other hormones to control how the plant responds to abiotic stresses such as salinity, temperature changes, and drought. Exogenous BR administration to plants can successfully boost agricultural production, stimulate plant development, and raise the tolerance of plants to abiotic stresses (Divi and Krishna 2009).

Certain phenotypes are exhibited by plants with defects in BR biosynthesis or signal transduction pathways, such as low fertility, short stature, round and curled leaves, short petioles, and short pedicels. These mutants' cotyledons opened without an apical hook, and their hypocotyls were short when they were developed in the dark (Choe 2004). The biochemical analysis of these mutations has made it easier to confirm previously established pathways and identify previously unidentified stages (Choe 2006). The general consensus is that BRs use transcription factors to initiate a phosphorylation cascade reaction, which is how they transmit signals to the nucleus, where they activate the transcription of BR-responsive genes and promote cell development (Li et al. 2020). Recently, it has been demonstrated that BRs influence autophagy, a cell's self-destructive mechanism that plants utilize to regulate responses to stresses, which is implicated in plant responses to nitrogen (N) deficiency (Wang et al. 2019).

Several approaches were used to mitigate the negative effects of stress, and the use of plant growth regulators has been discussed as a more reasonable alternative. In this context, it has been extensively reported that BRs and related chemicals stabilize the effects of a number of abiotic stresses, such as heavy metals and salt (Ahmad et al. 2018). The last 10 years have seen a clear establishment of BR's function in plant responses to abiotic stress, but the mechanisms in which BR interacts with other stress-related hormones and their communication pathways to impart stress tolerance are seldom documented (Divi et al. 2010).

It should be highlighted that a variety of parameters, such as dosage, plant species, development stage, growth conditions (with or without stress), types of stress, duration of stress, and their interaction with other hormones, growth regulators, and signaling molecules, all significantly influence the actions of BR (Yin et al. 2019). Moreover, BR helps plants become more resistant to a variety of abiotic stresses, such as drought, heat, salt, low and high temperatures, $Ca(NO_3)_2$, and heavy metal toxicity (Anuradha and Rao 2001; Yuan et al. 2012a). ROS are produced when plants are subjected to environmental stresses. Proteins, nucleic acids, and lipids are among the cellular macromolecules that these activated oxygen species frequently oxidize, a process that compromises the fundamental viability of the cell. Plants have developed a complex mechanism of defense against free radicals ROS in order to prevent the build-up of these molecules and to ensure their own survival. Various enzymatic and nonenzymatic chemicals are created as part of the antioxidant defense system to mitigate the harmful effects of environmental stresses (Bartwal et al. 2013). Higher expression of stress signal genes, such as *heat shock protein (hsp)* genes like *RD29A* and *ERD10*, is often connected with an increase in BR (Dhaubhadel et al. 1999; Kagale et al. 2007), indicating that plants treated with BR have a greater stress tolerance, partially attributed to enhanced expression of stress-responsive genes (Divi et al. 2010).

Figure 3.2 illustrates some brassinosteroid mechanisms involved in the plant response to abiotic stress.

3.3.1 Salinity

Currently, salt affects around 800 million hectares of land worldwide, making up about 6% of the planet's total land area (Rengasamy 2002). Salinity is becoming an increasingly significant issue that impacts agriculture and soils globally. Many fundamental metabolic processes are disturbed when salt is present in plant cells, which has a detrimental

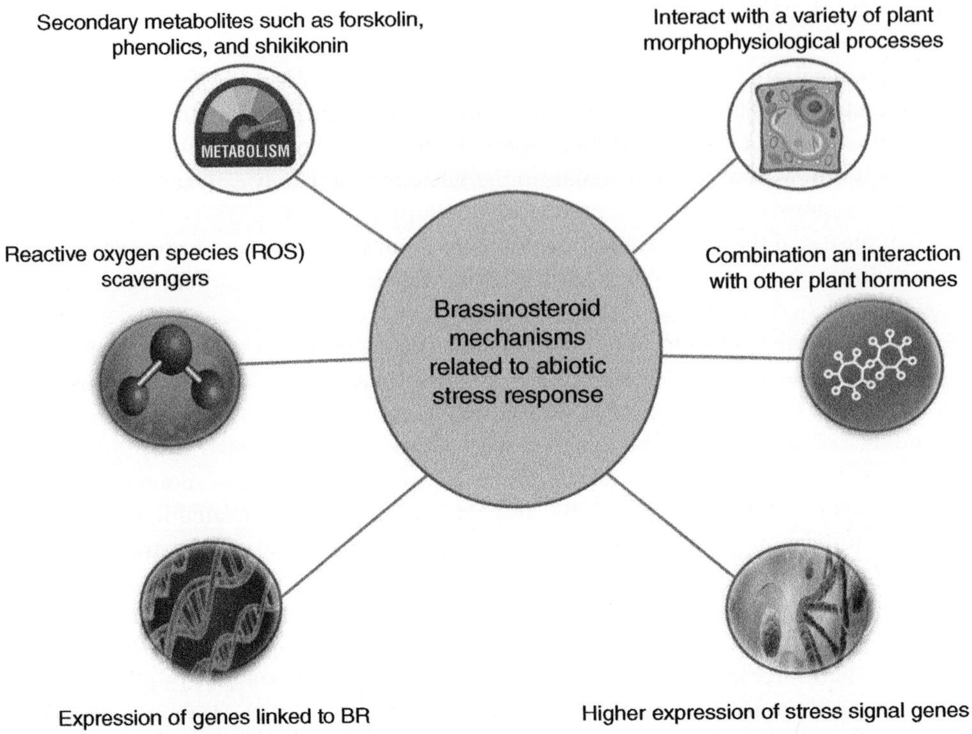

Figure 3.2 Brassinosteroid mechanisms involved in the plant response to abiotic stress.

impact on plant growth and development (Hameed et al. 2021). It has an impact on every biological function, including lipid metabolism, protein synthesis, energy metabolism, disturbance of cellular homeostasis, photosynthesis, mRNA processing, transcription, and amino acid biosynthesis (Tseng et al. 2007; Koussevitzky et al. 2007; Munns and Tester 2008; Yan et al. 2013; Abdelhamid et al. 2020). In addition, the adverse consequences of salinity encompass ion accumulation-induced toxicity, osmotic stress, nutritional imbalance and deficiency, increased cell-turgor loss, decreased leaf water potential, disrupted physiological and biochemical processes, and excessive production of ROS (Munns and Tester 2008).

BRs are important for the salt response of plants (Tanveer et al. 2018). Proline is a necessary compatible solute that helps preserve the equilibrium of REDOX, detoxify ROS, and maintain protein structure (Han et al. 2014). Using chickpeas and mung beans as model organisms, it was found that BRs might enhance the activity of antioxidant enzymes by promoting the accumulation of proline (Hayat et al. 2010). Combining two wheat cultivars with varying salt tolerances and treating their leaves with EBL, Shahbaz et al. (2008) found that the two kinds of antioxidant systems were impacted. Salt stress resulted in an increase in SOD, CAT, and POD activities. It has been demonstrated that BRs may decrease the negative impacts of salinity in a variety of plant species, such as rapeseed (*Brassica juncea*), eggplant (*Solanum melongena*), pepper (*Capsicum annuum*), cucumber (*Cucumis sativus*), common bean (*Phaseolus vulgaris*), maize (*Zea mays*), black locust (*Robinia *pseudoacacia* L.), and *Arabidopsis* (*A. thaliana*) (Hayat et al. 2010; Yuan et al. 2012b; Yue et al. 2018). Also, Özdemir et al. (2004) reported that compared to seedlings treated with NaCl alone, EBL treatment of salt-sensitive rice cv. IR-28 seeds increased seedling development, reduced lipid damage, and lowered proline accumulation. Moreover, Kagale et al. (2007) showed that EBL can reduce the inhibition of NaCl-induced germination of *Brassica napus* and *A. thaliana* seeds. In another investigation, Kutay and Venkatesan (2022) reported that the outcomes of reducing salinity in the Vigna radiata L. Co (Gg) 7 variety were assessed by foliar application of brassinosteroid at several doses (0.5, 0.75, and 1.0 μM) and basal treatment with 75 mM of NaCl. However, in salt and non-salinity stress settings, 0.75 μM brassinosteroid concentration was beneficial in enhancing growth, chlorophyll, protein, and antioxidant enzyme activities. 1.0 μM brassinosteroid concentration enhanced proline content.

3.3.2 Drought

At least 40% of agricultural losses globally are caused by drought, and the percentage is rising sharply as a result of climate change (Boyer 1982). As one of the most severe abiotic stressors known to date, it reduces the production of photosynthetic pigments, stomatal conductance, and photosynthesis machinery, which in turn causes a drop in plant development (Hasanuzzaman et al. 2012). Research has demonstrated that following BR treatment, water content and potential increased, enhancing the survival of plants in low-water environments (Yuan et al. 2010).

Yuan et al. (2010) found that EBL treatment may help tomato seedlings under drought stress by lowering water loss, which is in line with other research (Kagale et al. 2007; Li et al. 2008). The effects of exogenously applied BRs on genotypes that are drought-tolerant or sensitive should be compared in order to identify any changes that the BRs have caused. This is especially important for the sensitive genotypes, since the tolerant genotypes should be less affected by drought. This should be comparable to what happens when BRs are administered exogenously to plants that are subjected to mild, moderate, or less severe drought: plants that are more severely stressed by drought always respond more favorably to BRs (Naveen et al. 2021).

Fàbregas et al. (2018) indicated that BRL3 receptors located in the vascular system provide drought tolerance without negatively impacting the overall development of plants, in contrast to mutations in the ubiquitously distributed BRI1 receptor pathway that exhibit increased resilience to drought at the price of growth. Systematic investigations show that in response to drought stress, the vascular BRL3 receptor pathway preferentially favors overall plant development by triggering the production and mobilization of osmoprotectant compounds, primarily proline and carbohydrates, in the roots. Plants are more adapted to drought stress when BRs are applied exogenously and when the BR biosynthesis enzyme DWF4 is overexpressed (Kagale et al. 2007; Sahni et al. 2016), whereas repression of the BRI1 receptor leads to phenotypes resistant to drought (Ye et al. 2017). Intriguingly, following BR perception and upstream of the Bin2 (Brassinosteroid-Insensitive 2) kinase, ABA signaling suppresses the BR signaling pathway (Zhang et al. 2009; Gui et al. 2016), and the overlapping transcriptional regulation of many BR- and ABA-regulated genes mediates additional downstream interaction (Chung et al. 2014).

3.3.3 Temperature

One of the main environmental elements influencing plant development is temperature. Plants that are exposed to nonfreezing temperatures below 12 °C for an extended amount of time – beyond the crucial period–will experience "low-temperature damage," a physiological malfunction (Seydpour and Sayyari 2015). Stress from high temperatures can also seriously harm a plant's growth and development as well as its physiological processes (Sun et al. 2019; Zhuang et al. 2020). Employing heat-sensitive rice (IR36) as the source material, Chen et al. (2019) found that when applied to juvenile ears at 40 °C, sucrose transporter genes *OsSUT1*, *OsSUT2*, and *OsSUT4* might be expressed frequently. Additionally, the use of EBL could lower the amount of superoxide anion present, hence minimizing damage to the cell membrane. In another study on rice (*Oryza sativa* L.), the fresh weight of the shoots and roots and the leaf area were lower at a temperature of 40/30 °C (seven days) than they were in the control group, which had been cultivated at the optimal temperature. 24-EBL (10–8 M) was shown to decrease this impact (Thussagunpanit et al. 2015a). In field studies on rice carried out in the hot season (February to May 2012, Thailand), 24-EBL enhanced seed mass by 35% and the quantity of filled seeds per panicle by 16% (Thussagunpanit et al. 2015b). It seems reasonable that plants lacking in BR or experiencing abnormalities with BR signaling would be more vulnerable to high temperatures if exogenous BR treatment boosts resistance to such stress. Interestingly, the study carried out on BR-defective barley plants by Sadura et al. (2019) showed that all barley plants with BR deficiencies or abnormalities in their perception of BR were considerably more resistant to high temperatures (38 °C and 45 °C) than their wild-type Bowman and Delisa cultivars following acclimation at 27 °C (seven days).

On the one hand, plants exposed to below-freezing conditions had higher survival rates when BR was applied (Kim et al. 2010). Frost tolerance was reduced in *A. thaliana* mutants lacking in different stages of BR signaling, such as the weak *bri1-301* allele and the strong *bri1-1* allele, as well as in lines of *BIN2* and its homolog *ASKtheta* overexpression. Plants with constitutively active BR signaling, such as the higher-order *bin2-3* bil1 bil2 mutant and the *BRI1* overexpression line *35S:BRI1- GFP*, exhibited greater resistance to frost damage (Eremina et al. 2016; Li et al. 2017). Chilling damage affects a plant's capacity to withstand low temperatures by causing a loss of membrane fluidity, which can lead to membrane leakage, even though it is obviously not related to the formation of ice crystals (Verslues et al. 2006) and it seems that these occurrences can shield BRs from their harmful consequences. Plants that are susceptible to chilling, such as *Cucumis sativus* (cucumber), *Solanum lycopersicum* (tomato), *Oryza sativa* (rice), *Zea mays* (maize), and the cold-hardy plant *A. thaliana*, all performed better in the cold when BR treatment was applied (Xia et al. 2009b; Aghdam et al. 2012).

3.3.4 Heavy Metals

In the Periodic Table, heavy metals mostly belong to the transition elements. The elements classified as heavy metals have a specific weight greater than 5 g cm^3 (Leonard et al. 2004). Additionally classified as heavy metals are those that have a density of at least 5 times that of water (Stohs and Bagchi 1995). There are two types of heavy metals: non-essential (Cd, Ni, As, Hg, Pb) and essential (Mo, Mn, Cu, Ni, Fe, Zn) (Bharti and Sharma 2022). Plants in metal-contaminated environments show changes in their metabolism, decreased rate of growth, and accumulation of metals (Asati et al. 2016). Both essential and non-essential heavy metals often have similar harmful effects on plants, which include reduced biomass accumulation, chlorosis, growth and photosynthetic inhibition, altered water balance and nutrient absorption, and senescence, which eventually results in plant death (Singh et al. 2016).

The accumulation of radioactive elements and heavy metals in plants can be minimized by using BRs, which have the capacity to control the absorption of ions into plant cells. Additionally, the harmful effects and symptoms caused by an excess of heavy metals are alleviated by BRs (Bajguz and Hayat 2009). BRs also play a crucial role in maintaining cellular ion homeostasis and restricting metal accumulation in plants. The application of BRs offers a promising strategy to mitigate the toxic effects of V and Cr stress on plant growth and development, contributing to improved crop productivity and environmental sustainability (Ahammed et al. 2013). Hypothesized mechanism via which BRs control plants' ability to withstand high metal stress is the inhibition of lipid degradation, resulted from the overproduction of reactive oxygen species under stress conditions, and increases the activity of antioxidants. They also have the ability to promote phytochelatin synthesis. BRs administered exogenously improve resistance to heavy metals and oxidative damage. Accompanying it is the formation of H_2O_2 and the enhancement of antioxidant enzymes, which scavenge excess ROS (Rajewska et al. 2016).

Kour et al. (2021) highlighted in depth the molecular process of BR production, signaling, transcriptional networks' function in plants' response to heavy metal stress, and the manner in which BRs interact with other phytohormones to avert heavy metal damage. BRs were used to decrease the amount of hazardous and radioactive metals that had been collected in plant components. The donor groups of BRs may prevent metal ions from depositing in cells by generating soluble complexes with them. Cadmium stress significantly inhibited the formation of biomass in tobacco seedlings (Ahammed et al. 2013). For instance, Sharma et al. (2007) reported that pre-sowing treatments of 28-homobrassinolide lowered the uptake of zinc and enhanced the activities of enzymes and protein concentration of *Brassica juncea* seedlings. Bajguz and Hayat (2009) noted that BRs can reduce the harmful effects caused by an excess of heavy metals. Furthermore, Hayat et al. (2007) reported that the use of HBL indicates that *Brassica juncea* is experiencing less Cd stress. They observed a decrease in the growth, chlorophyll content, nitrate reductase activity, carbonic anhydrase activity, nitrate, and sugar content of the plants under Cd stress. However, the use of HBL has eliminated the harmful impact. It decreased the activities of peroxidase and ascorbic acid oxidase while increasing proline accumulation and oxidative enzyme activity.

Table 3.1 shows some protective effects of BRs in plant development under abiotic stress responses.

3.4 Challenges and Opportunities

Numerous research teams have previously proven that BRs regulate a number of important physiological and biochemical functions. Several recent findings have come from the model plant *A. thaliana*, which is also an appropriate platform for BR experiments. A complex web of amazing signaling networks allows plants to perceive and react to their constantly changing surroundings. Since stress nowadays has a more complicated nature, the

Table 3.1 Protective effects of brassinosteroids in plant development under abiotic stress responses.

Plant species	Abiotic stress	BRs dose	Protective effects	References
Wheat (*Triticum aestivum* L.)	Salinity	0.0125, 0.025, and 0.0375 mg L^{-1}	Increased plant Biomass	Shahbaz and Ashraf (2007)
Lulo (*Solanum quitoense* L. cv. *septentrionale*	Drought	4 and 8 mL L^{-1}	Sprays enhanced the photochemical efficiency of PSII, and plant growth	Castañeda-Murillo et al. (2022)
Oilseed Rape (*Brassica napus* L.)	Salinity	10^{-10} M	Stem growth was fully restored, plant assimilation area increased by as much as 67–76%, fresh and dry weight largely recovered (up to 85–92% of the control values), and diminished the inhibitory effect of NaCl on photosynthetic pigments	Efimova et al. (2014)
Wheat (*Triticum aestivum* L.)	Drought	1, 1.5, and 2 μM	Improved the plant height, spike length, and the dry weights of roots, shoots, and grains	Khan et al. (2021a)
Eggplant (*Solanum melongena* L.)	High-temperature	0.05–0.2 μM	Promoted the plant growth and alleviated high-temperature-induced inhibition of photosynthesis	Wu et al. (2014)
Tomato (*Lycopersicon esculentum*)	Low-temperature	0 or 10–8 M	Accelerated the antioxidative enzymes and proline content	Khan et al. (2015)
Mung bean (*Vigna radiata* L.)	Heavy metals (zinc)	0.1, 0.2, and 0.4 PPM	Enhanced germination percentage, speed of germination, and seedling growth	Kumar et al. (2023)

active involvement of several sensors for signal perception and transmission is avoided (Chaudhuri et al. 2022). BRs can interact with ABA, ETH, and other plant hormones to collectively control a range of abiotic stressors, as demonstrated by several researches. Nevertheless, the biosynthesis method has not been well investigated; currently, there is very little technology available to artificially produce BRs. The development of affordable and effective BRs will therefore probably become an increasingly important field of study as biotechnology advances (Li et al. 2021).

However, recent climate-related problems have been shown to be constraints that have put the sustainability of the crop types introduced during the Green Revolution in jeopardy. Various sophisticated instruments and methods can streamline numerous intricacies in this investigation. Significantly improved instruments, accurate translatomics (Vragović et al. 2015), and fluorescent-activated cell sorting (FACS) (Vilarrasa-Blasi et al. 2014), can effectively map BR responses, and once that is done, it can track how BRs interact with other phytohormones in a particular tissue under different stress conditions. Under harmful stress circumstances, BR distribution and activity should be monitored using highly sophisticated instruments and procedures (Chaudhuri et al. 2022).

In order to improve abiotic stress tolerance, there are still a lot of transcription factors that have to be discovered about their interactions with the various parts of the BRs signal transduction relay. Examining the functions of BRs at the cellular and molecular levels will be difficult. BRs' diverse roles indicate and clarify how they interact with other plant growth hormones that affect target gene response. Further backing for this field might come from combining genome editing technologies with forward genetics. To support the BR cascade mechanism, its elements, targeted genes, related transcriptional factors, and proteins, more investigation will be required in the future (Bhandari and Nailwal 2020). To further increase our comprehension of the intricate web of regulatory processes involving these phytohormones, more data on the relationships between BRs, photosynthesis, and radiation are also required. Additionally, the understanding of these steroids' physical and chemical characteristics encourages us to view them as a powerful stress reliever and a very promising, environmentally friendly way to increase agricultural output (Fariduddin et al. 2014). The high cost of BRs is one of the main obstacles to their wider use in the fields. Nevertheless, recent developments in the chemical synthesis of BRs and their analogs have given rise to commercially viable strategies that have made large-scale uses for raising yields quite accessible to farmers (Fariduddin et al. 2014).

3.5 Recommendations and Expected Future Aspects

The potential of BR in stress resistance has been demonstrated in a number of studies; however, its interactions with ion homeostasis and transporters (N, P, K, Ca, Mg, etc.) have received little attention, which might be useful in the discovery and functional characterization of different nutrient transporters. Furthermore, the investigation of BR's function in yield has received very little attention. Future research is thus required to fully comprehend the process and investigate how BR functions in growth and development in response to environmental stimuli (Anwar et al. 2018). To support the BR cascade mechanism, its elements, targeted genes, related transcriptional factors, and proteins more thorough study will be required in the future. The process behind the roles that BRs play in plant systems, including their communication during cell division, cell wall regeneration, morphogenesis, cytoskeleton structure, etc., remains largely unknown (Bhandari and Nailwal 2020).

Although BR controls hundreds of genes, it remains unknown which other signaling proteins and transcription factors interact with BZR1 and BES1 to improve stress tolerance. To learn more about how plants react to environmental challenges, it will be fascinating to investigate the function of the BR signaling pathway, hormone interactions, and crosstalk at the organ, tissue, and cell levels (Anwar et al. 2018; Basit et al. 2021). The field of genetic engineering has grown significantly in prominence over time. Targeted and modified genes linked to BRs have already yielded the desired feature of stress resistance. Nevertheless, a thorough investigation is required to identify more similar genes that might be targeted in the future (Bhandari and Nailwal 2020). Hopefully, much more information will be added to the current body of literature as the research continues. As previously mentioned, these steroids cause a variety of physiomorphological changes in plants by interacting with the genome. Moreover, they do not cause pests to coevolve, which adds to our arsenal of 21st-century plant protection strategies (Khan et al. 2021b).

3.6 Conclusion

Several studies have been carried out to determine the mechanisms by which BRs contribute to the regulation of primary plant metabolism under abiotic stress. These studies were carried out at different levels, molecular and genetic, treating the contributions of BRs under various abiotic stresses. In addition, these studies attempt to identify the different signaling pathways involving these hormones, as well as their interaction with other plant hormones. Nevertheless, research on BRs must continue in order to deepen our understanding of these mechanisms and to establish all related aspects so that we can effectively use large-scale agrarian hormones as effective hormones for the mitigation of different abiotic stresses.

References

Abdelhamid, M.T., Sekara, A., Pessarakli, M. et al. (2020). New approaches for improving salt stress tolerance in rice. In: *Rice Research for Quality Improvement: Genomics and Genetic Engineering* (ed. A. Roychoudhury), 247–268. Springer: Singapore.

Aghdam, M.S., Asghari, M., Farmani, B. et al. (2012). Impact of postharvest brassinosteroids treatment on PAL activity in tomato fruit in response to chilling stress. *Scientia Horticulturae* 144: 116–120.

Ahmad, F., Singh, A., and Kamal, A. (2018). Crosstalk of brassinosteroids with other phytohormones under various abiotic stresses. *Journal of Applied Biology and Biotechnology* 6: 56–62.

Ahammed, G.J., Choudhary, S.P., Chen, S. et al. (2013). Role of brassinosteroids in alleviation of phenanthrene cadmium co-contamination-induced photosynthetic inhibition and oxidative stress in tomato. *Journal of Experimental Botany* 64: 199–213.

Ahammed, G.J., Li, X., Xia, X.J. et al. (2015). Enhanced photosynthetic capacity and antioxidant potential mediate brassinosteriod-induced phenanthrene stress tolerance in tomato. *Environmental Pollution* 201: 58–66.

Ahammed, G.J., Li, X., Liu, A. et al. (2020). Brassinosteroids in plant tolerance to abiotic stress. *Journal of Plant Growth Regulation* 39: 1451–1464.

Akter, N. and Islam, M.R. (2017). Heat stress effects and management in wheat: a review. *Agronomy for Sustainable Development* 37: 1–17.

Akula, R. and Ravishankar, G.A. (2011). Influence of abiotic stress signals on secondary metabolites in plants. *Plant Signaling & Behavior* 6: 1720–1731.

Albrecht, C., Boutrot, F., Segonzac, C. et al. (2012). Brassinosteroids inhibit pathogen-associated molecular pattern-triggered immune signaling independent of the receptor kinase BAK1. *Proceedings of the National Academy of Sciences of the United States of America* 109: 303–308.

Amasino, R.M. and Michaels, S.D. (2010). The timing of flowering. *Plant Physiology* 154: 516–520.

Amthor, J.S. (1986). Evolution and applicability of a whole plant respiration model. *Journal of Theoretical Biology* 122: 473–490.

Amthor, J.S., Bar-Even, A., Hanson, A.D. et al. (2019). Engineering strategies to boost crop productivity by cutting respiratory carbon loss. *Plant Cell* 31: 297–314.

Anuradha, S. and Rao, S.S.R. (2001). Effect of brassinosteroids on salinity stress induced inhibition of seed germination and seedling growth of rice (*Oryza sativa* L.). *Plant Growth Regulation* 33: 151–153.

Anwar, A., Liu, Y., Dong, R. et al. (2018). The physiological and molecular mechanism of brassinosteroid in response to stress: a review. *Biological Research* 51: 1–15.

Asati, A., Pichhode, M., and Nikhil, K. (2016). Effect of heavy metals on plants: an overview. *International Journal of Applied Innovation in Engineering and Management* 5: 56–66.

Ashoka, P., Meena, R.S., Kumar, S. et al. (2017). Green nanotechnology is a key for ecofriendly agriculture. *Journal of Cleaner Production* 142: 4440–4441.

Ashraf, M., Akram, N.A., Arteca, R.N. et al. (2010). The physiological, biochemical and molecular roles of brassinosteroids and salicylic acid in plant processes and salt tolerance. *Critical Reviews in Plant Sciences* 29: 162–190.

Awan, S.A., Khan, I., and Yang, F. (2023). Pre-treatment of melatonin enhances the seed germination responses and physiological mechanisms of soybean (*Glycine max* L.) under abiotic stresses. *Frontiers in Plant Science* 14: 1149873.

Bajguz, A. (2011). Suppression of *Chlorella vulgaris* growth by cadmium, lead, and copper stress and its restoration by endogenous brassinolide. *Archives of Environmental Contamination and Toxicology* 60: 406–416.

Bajguz, A. and Hayat, S. (2009). Effects of brassinosteroids on the plant responses to environmental stresses. *Plant Physiology and Biochemistry* 47: 1–8.

Bartwal, A., Mall, R., Lohani, P. et al. (2013). Role of secondary metabolites and brassinosteroids in plant defense against environmental stresses. *Journal of Plant Growth Regulation* 32: 216–232.

Basit, F., Liu, J., An, J. et al. (2021). Brassinosteroids as a multidimensional regulator of plant physiological and molecular responses under various environmental stresses. *Environmental Science and Pollution Research* 28: 44768–44779.

Belkhadir, Y., Wang, X., and Chory, J. (2006). Brassinosteroid signaling pathway. *Science's STKE* 364:cm 4.

Belkhadir, Y. and Jaillais, Y. (2015). The molecular circuitry of brassinosteroid signaling. *New Phytologist* 206: 522–540.

Bhandari, S. and Nailwal, T.K. (2020). Role of brassinosteroids in mitigating abiotic stresses in plants.*Biologia* 75: 2203–2230.

Bharti, R. and Sharma, R. (2022). Effect of heavy metals: an overview. *Materials Today: Proceedings* 51: 880–885.

Boyer, J.S. (1982). Plant productivity and environment. *Science* 218: 443–448.

Caño-Delgado, A., Yin, Y., Yu, C. et al. (2004). BRL1 and BRL3 are novel brassinosteroid receptors that function in vascular differentiation in *Arabidopsis*. *Development* 131: 5341–5351.

Castañeda-Murillo, C.C., Rojas-Ortiz, J.G., Sánchez-Reinoso, A.D. et al. (2022). Foliar brassinosteroid analogue (DI-31) sprays increase drought tolerance by improving plant growth and photosynthetic efficiency in lulo plants. *Heliyon* 8: e08977.

Caswell, H. and Salguero-Gómez, R. (2013). Age, stage and senescence in plants. *Journal of Ecology* 101: 585–595.

Chakma, S.P., Chileshe, S.M., Thomas, R. et al. (2021). Cotton seed priming with brassinosteroid promotes germination and seedling growth. *Agronomy* 11: 566.

Chakraborty, N., Sharma, P., Kanyuka, K. et al. (2015). G-protein α-subunit (GPA1) regulates stress, nitrate and phosphate response, flavonoid biosynthesis, fruit/seed development and substantially shares GCR1 regulation in *Arabidopsis thaliana*. *Plant Molecular Biology* 89: 559–576.

Chaudhuri, A., Halder, K., Abdin, M.Z. et al. (2022). Abiotic stress tolerance in plants: brassinosteroids navigate competently. *International Journal of Molecular Sciences* 23: 14577.

Chen, Y.H., Wang, Y.L., Zhu, D.F. et al. (2019). Study on the mechanism of exogenous brassinolide in alleviating high temperature damage during panicle differentiation of rice. *Chinese Journal of Rice Science* 33: 457–466.

Choe, S. (2004). Brassinosteroid biosynthesis and metabolism. In: *Plant Hormones: Biosynthesis, Signal Transduction* (ed. P.J. Davies), 156–178. Dordrecht, The Netherlands: Kluwer Academic Publishers.

Choe, S. (2006). Brassinosteroid biosynthesis and inactivation. *Physiologia Plantarum* 126: 539–548.

Choudhary, S.P., Yu, J.-Q., Yamaguchi-Shinozaki, K. et al. (2012). Benefits of brassinosteroid crosstalk. *Trends in Plant Science* 17: 594–605.

Chung, Y., Kwon, S.I., and Choe, S. (2014). Antagonistic regulation of *Arabidopsis* growth by brassinosteroids and abiotic stresses. *Molecules and Cells* 37: 795.

Clouse, S.D. (2008). The molecular intersection of brassinosteroid-regulated growth and flowering in *Arabidopsis*. *Proceedings of the National Academy of Sciences of the United States of America* 105: 7345–7346.

Clouse, S.D., Langford, M., and McMorris, T.C. (1996). A brassinosteroid-insensitive mutant in *Arabidopsis* exhibits multiple defects in growth and development. *Plant Physiology* 11: 671–678.

Clouse, S.D. and Sasse, J.M. (1998). Brassinosteroids: essential regulators of plant growth and development. *Annual Review of Plant Biology* 49: 427–451.

Colla, G. and Rouphael, Y. (2015). Biostimulants in horticulture. *Scientia Horticulturae* 196: 1–134.

Deng, Z., Zhang, X., Tang, W. et al. (2007). A proteomics study of brassinosteroid response in *Arabidopsis*. *Molecular & Cellular Proteomics* 6: 2058–2071.

Deng, X.G., Zhu, T., Zhang, D. W. et al. (2015). The alternative respiratory pathway is involved in brassinosteroid-induced environmental stress tolerance in *Nicotiana benthamiana*. *Journal of Experimental Botany* 66: 6219–6232.

De Pascale, S., Rouphael, Y., and Colla, G. (2017). Plant biostimulants: innovative tool for enhancing plant nutrition in organic farming. *European Journal of Horticultural Science* 82: 277–285.

Dey, P., Datta, D., Pattnaik, D. et al. (2022). Physiological, biochemical, and molecular adaptation mechanisms of photosynthesis and respiration under challenging environments. In: *Plant Perspectives to Global Climate Changes* (ed. T. Aftab and A. Roychoudhury), 79–100. Academic Press.

Dhaubhadel, S., Browning, K.S., Gallie, D.R. et al. (2002). Brassinosteroid functions to protect the translational machinery and heat-shock protein synthesis following thermal stress. *The Plant Journal* 29: 681–691.

Dhaubhadel, S., Chaudhary, S., Dobinson, K.F. et al. (1999). Treatment with 24-epibrassinolide, a brassinosteroid, increases the basic thermotolerance of *Brassica napus* and tomato seedlings. *Plant Molecular Biology* 40: 333–342.

Divi, U.K. and Krishna, P. (2009). Brassinosteroid: a biotechnological target for enhancing crop yield and stress tolerance. *New Biotechnology* 26: 131–136.

Divi, U.K., Rahman, T., and Krishna, P. (2010). Brassinosteroid-mediated stress tolerance in *Arabidopsis* shows interactions with abscisic acid, ethylene and salicylic acid pathways. *BMC Plant Biology* 10: 1–14.

Divi, U.K., Rahman, T., and Krishna, P. (2015). Gene expression and functional analyses in brassinosteroid-mediated stress tolerance. *Plant Biotechnology Journal* 14: 419–432.

Divi, U.K., Rahman, T., and Krishna, P. (2016). Gene expression and functional analyses in brassinosteroid-mediated stress tolerance. *Plant Biotechnology Journal* 14: 419–432.

Efimova, M.V., Savchuk, A.L., Hasan, J.A.K. et al. (2014). Physiological mechanisms of enhancing salt tolerance of oilseed rape plants with brassinosteroids. *Russian Journal of Plant Physiology* 61: 733–743.

Eremina, M., Unterholzner, S.J., Rathnayake, A.I. et al. (2016) Brassinosteroids participate in the control of basal and acquired freezing tolerance of plants. *Proceedings of the National Academy of Sciences of the United States of America* 113: E5982–E5991.

Fàbregas, N., Lozano-Elena, F., Blasco-Escámez, D. et al. (2018). Overexpression of the vascular brassinosteroid receptor BRL3 confers drought resistance without penalizing plant growth. *Nature Communications* 9: 4680.

Fariduddin, Q., Khanam, S., Hasan, S.A. et al. (2009). Effect of 28-homobrassinolide on the drought stress-induced changes in photosynthesis and antioxidant system of *Brassica juncea* L. *Acta Physiologiae Plantarum* 31: 889–897.

Fariduddin, Q., Yusuf, M., Ahmad, I. et al. (2014).Brassinosteroids and their role in response of plants to abiotic stresses. *Biologia Plantarum* 58: 9–17.

Farooq, M., Wahid, A., Lee, D.J. et al. (2010). Comparative time course action of the foliar applied glycinebetaine, salicylic acid, nitrous oxide, brassinosteroids and spermine in improving drought resistance of rice. *Journal of Agronomy and Crop Science* 196: 336–345.

Gomes, M.M.A. (2011). Physiological effects related to brassinosteroid application in plants. In: *Brassinosteroids: A Class of Plant Hormones* (ed. S. Hayat and A. Ahmad), 193–242. Dordrecht: Springer.

Grove, M.D., Spencer, G.F., Rohwedder, W.K. et al. (1979). Brassinolide, a plant growth-promoting steroid isolated from *Brassica napus* pollen. *Nature* 281: 216–217.

Gui, J., Zheng, S., Liu, C. et al. (2016). OsREM4. 1 interacts with OsSERK1 to coordinate the interlinking between abscisic acid and brassinosteroid signaling in rice. *Developmental Cell* 38: 201–213.

Guo, D.L., Wang, Z.G., Li, Q. et al. (2019). Hydrogen peroxide treatment promotes early ripening of Kyoho grape. *Australian Journal of Grape and Wine Research* 25: 357–362.

Hafeez, M.B., Raza, A., Zahra, N. et al. (2021a). Gene regulation in halophytes in conferring salt tolerance. In: *Handbook of Bioremediation* (ed. M. Hasanuzzaman and M.N.V. Prasad), 341–370. New York: Academic Press.

Hafeez, M.B., Zahra, N., Zahra, K. et al. (2021b). Brassinosteroids: molecular and physiological responses in plant growth and abiotic stresses. *Plant Stress* 2: 100029.

Hameed, A., Ahmed, M.Z., Hussain, T. et al. (2021). Effects of salinity stress on chloroplast structure and function. *Cells* 10: 2023.

Han, G.L., Wang, M.J., Yuan, F. et al. (2014). The CCCH zinc finger protein gene AtZFP1 improves salt resistance in *Arabidopsis thaliana*. *Plant Molecular Biology* 86: 237–253.

Hasanuzzaman, M., Hossain, M.A., da Silva, J.A. et al. (2012). Plant response and tolerance to abiotic oxidative stress: antioxidant defense is a key factor. In: *Crop Stress and Its Management: Perspectives and Strategies* (ed. B. Venkateswarlu, A.K. Shanker, and C. Shanker et al.), 261–315. Dordrecht: Springer.

Hayat, S., Ali, B., Hasan, S.A. et al. (2007). Brassinosteroid enhanced the level of antioxidants under cadmium stress in *Brassica juncea*. *Environmental and Experimental Botany* 60: 33–41.

Hayat, S., Hasan, S.A., Yusuf, M. et al. (2010). Effect of 28-homobrassinolide on photosynthesis, fluorescence and antioxidant system in the presence or absence of salinity and temperature in *Vigna radiata*. *Environmental and Experimental Botany* 69: 105–112.

Hayat, S., Alyemeni, M., and Hasan, S. (2012). Foliar spray of brassinosteroid enhances yield and quality of *Solanum lycopersicum* under cadmium stress. *Saudi Journal of Biological Sciences* 19: 325–335.

He, Y.J., Xu, R.J., and Zhao, Y.J. (1996). Enhancement of senescence by epibrassinolide in leaves of mung bean seedlings. *Acta Phytophysiologica Sinica* 22: 58–62.

Hodges, S.C. (2014).*Soil Fertility Basics*, pp. 1–75. Raleigh: North Carolina State University Press.

Hong, Z., Ueguchi-Tanaka, M., Shimizu-Sato, S. et al. (2002). Loss-of-function of a rice brassinosteroid biosynthetic enzyme, C-6 oxidase, prevents the organized arrangement and polar elongation of cells in the leaves and stem. *The Plant Journal* 32: 495–508.

Hou, S., Niu, H., Tao, Q. et al. (2017). A mutant in the CsDET2 gene leads to a systemic brassinosteriod deficiency and super compact phenotype in cucumber (*Cucumis sativus* L.). *Theoretical and Applied Genetics* 130: 1693–1703.

Huang, H.Y., Jiang, W.B., Hu, Y.W. et al. (2013). BR signal influences *Arabidopsis* ovule and seed number through regulating related genes expression by BZR1. *Molecular Plant* 6: 456–469.

Janeczko, A., Biesaga-Koscielniak, J., Oklestkova, J. et al. (2010). Role of 24-epibrassinolide in wheat production: physiological effects and uptake. *Journal of Agronomy and Crop Science* 196: 311–321.

Jethva, J., Schmidt, R.R., Sauter, M. et al. (2022). Try or die: dynamics of plant respiration and how to survive low oxygen conditions. *Plants* 11: 205.

Jia, Z., Giehl, R.F., Meyer, R.C. et al. (2019). Natural variation of BSK3 tunes brassinosteroid signaling to regulate root foraging under low nitrogen. *Nature Communications* 10: 2378.

Jiang, W.B., Huang, H.Y., Hu, Y.W. et al. (2013). Brassinosteroid regulates seed size and shape in *Arabidopsis*. *Plant Physiology* 162: 1965–1977.

Kagale, S., Divi, U.K., Krochko, J.E. et al. (2007). Brassinosteroid confers tolerance in *Arabidopsis thaliana* and *Brassica napus* to a range of abiotic stresses. *Planta* 225: 353–364.

Kaur, G., Kumar, S., Nayyar, H. et al. (2008). Cold stress injury during the pod-filling phase in chickpea (*Cicer arietinum* L.): effects on quantitative and qualitative components of seeds. *Journal of Agronomy and Crop Science* 194: 457–464.

Khan, I., Awan, S.A., Ikram, R. et al. (2021a). Effects of 24-epibrassinolide on plant growth, antioxidants defense system, and endogenous hormones in two wheat varieties under drought stress. *Physiologia Plantarum* 172: 696–706.

Khan, T.A., Fariduddin, Q., and Yusuf, M. (2015). Lycopersicon esculentum under low temperature stress: an approach toward enhanced antioxidants and yield. *Environmental Science and Pollution Research* 22: 14178–14188.

Khan, E.A., Upadhyay, T.K., Prajapat, R.K. et al. (2021b). Revisiting brassinosteroids signaling in plants: current advances and challenges. In: *Brassinosteroids in Plant Developmental Biology and Stress Tolerance* (ed. G.J. Ahammed and A. Sharma), 15–41. New York: Academic Press.

Kim, B.K., Fujioka, S., Takatsuto, S. et al. (2008). Castasterone is a likely end product of brassinosteroid biosynthetic pathway in rice. *Biochemical and Biophysical Research Communications* 374: 614–619.

Kim, S.Y., Kim, B.H., Lim, C.J. et al. (2010). Constitutive activation of stress-inducible genes in a brassinosteroid-insensitive 1 (bri1) mutant results in higher tolerance to cold. *Physiologia Plantarum* 138: 191–204.

Kothari, A. and Lachowiec, J. (2021). Roles of brassinosteroids in mitigating heat stress damage in cereal crops. *International Journal of Molecular Sciences* 22: 2706.

Kour, J., Kohli, S.K., Khanna, K. et al. (2021). Brassinosteroid signaling, crosstalk and, physiological functions in plants under heavy metal stress. *Frontiers in Plant Science* 12: 608061.

Koussevitzky, S., Nott, A., Mockler, T.C. et al. (2007). Signals from chloroplasts converge to regulate nuclear gene expression. *Science*. 316: 715–719.

Kumar, N., Sharma, V., Kaur, G. et al. (2023). Brassinosteroids as promoters of seedling growth and antioxidant activity under heavy metal zinc stress in mung bean (*Vigna radiata* L.). *Frontiers in Microbiology* 14: 1259103.

Kutay, Z.A. and Venkatesan, A. (2022). Mitigation of salinity induced effects by foliar application of brassinosteroids in *Vigna radiata* L. Co (Gg) 7 variety. *Indian Journal of Natural Sciences* 12: 38748–38757.

Lambers, H. (1985). Respiration in intact plants and tissues: its regulation and dependence on environmental factors, metabolism and invaded organisms. In: *Higher Plant Cell Respiration*

(Encyclopedia of Plant Physiology). New Ser. Vol. 18 (ed. R. Douce and D.A. Day), 418–473. Berlin: Springer.

Lapik, Y.R. and Kaufman, L.S. (2003). The *Arabidopsis* cupin domain protein AtPirin1 interacts with the G protein α-subunit GPA1 and regulates seed germination and early seedling development. *Plant Cell* 15: 1578–1590.

Leonard, S.S., Harris, G.K., and Shi, X. (2004). Metal-induced oxidative stress and signal transduction. *Free Radical Biology and Medicine* 37: 1921–1942.

Li, K.R., Wang, H.H., Han, G. et al. (2008). Effects of brassinolide on the survival, growth and drought resistance of *Robinia pseudoacacia* seedlings under water-stress. *New Forests* 35: 255–266.

Li, Q.F., Xiong, M., Xu, P. et al. (2016). Dissection of brassinosteroid-regulated proteins in rice embryos during germination by quantitative proteomics. *Scientific Reports* 6: 34583.

Li, C., Zhang, S., and Wang, X. (2017). Novel signaling interface constituted with membrane receptor-like kinases emerged from the study of interaction and transphosphorylation of BRI1 and BAK1. *Current Topics in Medicinal Chemistry* 17: 2393–2400.

Li, S., Zheng, H., Lin, L. et al. (2020). Roles of brassinosteroids in plant growth and abiotic stress response. *Plant Growth Regulation* 93: 29–38.

Liu, J., Gao, H., Wang, X. et al. (2014). Effects of 24-epibrassinolide on plant growth, osmotic regulation and ion homeostasis of salt stressed canola. *Plant Biology* 16: 440–450.

Mantri, N., Patade, V., Penna, S. et al. (2012). Abiotic stress responses in plants: present and future. In: *Abiotic Stress Responses in Plants: Metabolism, Productivity and Sustainability* (ed. P. Ahmad and M.N.V. Prasad), 1–19. New York: Springer.

Miryeganeh, M. (2021). Senescence: the compromised time of death that plants may call on themselves. *Genes* 12: 143.

Munns, R. and Tester, M. (2008). Mechanisms of salinity tolerance. *Annual Review of Plant Biologyogy* 59: 651–681.

Nahar, K., Kyndt, T., Hause, B. et al. (2013). Brassinosteroids suppress rice defense against root-knot nematodes through antagonism with the jasmonate pathway. *Molecular Plant-Microbe Interactions* 26: 106–115.

Naveen, N., Kumari, N., Avtar, R. et al. (2021). Evaluation of effect of brassinolide in *Brassica juncea* leaves under drought stress in field conditions. *Horticulturae* 7: 514.

Nolan, T., Chen, J., and Yin, Y. (2017). Cross-talk of brassinosteroid signalling in controlling growth and stress responses. *Biochemical Journal* 474: 2641–2661.

Oshunsanya, S.O., Nwosu, N.J., and Li, Y. (2019). Abiotic stress in agricultural crops under climatic conditions. In: *Sustainable Agriculture, Forest and Environmental Management* (ed. M.K. Jhariya, A. Banerjee, R.S. Meena et al.), 71–100. Singapore: Springer Singapore.

Özdemir, F., Bor, M., Demiral, T. et al. (2004). Effects of 24-epibrassinolide on seed germination, seedling growth, lipid peroxidation, proline content and antioxidative system of rice (*Oryza sativa* L.) under salinity stress. *Plant Growth Regulation* 42: 203–211.

Parihar, P., Singh, S., Singh, R. et al. (2015). Effect of salinity stress on plants and its tolerance strategies: a review. *Environmental Science and Pollution Research International* 22: 4056–4075.

Planas-Riverola, A., Gupta, A., Betegoń-Putze, I. et al. (2019). Brassinosteroid signaling in plant development and adaptation to stress. *Development* 146: dev151894.

Praveena, J., Dash, S., Behera, L. et al. (2020). Brassinosteroids: a multifunctional phytohormone of plant development and stress responses. *Current Journal of Applied Science and Technology* 39: 174–196.

Rajewska, I., Talarek, M., and Bajguz, A. (2016). Brassinosteroids and response of plants to heavy metals action. *Frontiers in Plant Science* 7: 197845.

Raza, A. (2020). Eco-physiological and biochemical responses of rapeseed (*Brassica napus* L.) to abiotic stresses: consequences and mitigation strategies. *Journal of Plant Growth Regulation* 40: 1368–1388.

Ren, H., Wu, X., Zhao, W. et al. (2022). Heat shock-induced accumulation of the glycogen synthase kinase 3-like kinase Brassinosteroid Insensitive 2 promotes early flowering but reduces thermotolerance in *Arabidopsis*. *Frontiers in Plant Science* 13: 838062.

Rengasamy, P. (2002). Transient salinity and subsoil constraints to dryland farming in Australian sodic soils: an overview. *Australian Journal of Experimental Agriculture* 42: 351–61.

Rozhon, W., Akter, S., Fernandez, A. (2019). Inhibitors of brassinosteroid biosynthesis and signal transduction. *Molecules* 24: 4372.

Saddiq, M.S., Iqbal, S., Hafeez, M.B. et al. (2021). Effect of salinity stress on physiological changes in winter and spring wheat. *Agronomy* 11: 1193.

Sade, N., del Mar Rubio-Wilhelmi, M., Umnajkitikorn, K. et al. (2018). Stress-induced senescence and plant tolerance to abiotic stress. *Journal of Experimental Botany* 69: 845–853.

Sadura, I., Pociecha, E., Dziurka, M. et al. (2019). Mutations in the HvDWARF, HvCPD and HvBRI1 genes-involved in brassinosteroid biosynthesis/signalling: altered photosynthetic efficiency, hormonal homeostasis and tolerance to high/low temperatures in barley. *Journal of Plant Growth Regulation* 38: 1062–1081.

Sağlam Çağ, S. (2019). Brassinosteroids and senescence. In: *Brassinosteroids: Plant Growth and Development* (ed. S. Hayat, M. Yusuf, R. Bhardwaj et al.), 169–184. Singapore: Springer.

Sahni, S., Prasad, B.D., Liu, Q. et al. (2016). Overexpression of the brassinosteroid biosynthetic gene DWF4 in *Brassica napus* simultaneously increases seed yield and stress tolerance. *Scientific Reports* 6: 28298.

Sasse, J.M. (2003). Physiological actions of brassinosteroids: an update. *Journal of Plant Growth Regulation* 22: 276–288.

Seleiman, M.F., Al-Suhaibani, N., Ali, N. et al. (2021). Drought stress impacts on plants and different approaches to alleviate its adverse effects. *Plants* 10: 259.

Seydpour, F. and Sayyari, M. (2015). Chilling injury in cucumber seedlings amelioration by methyl salicylate. *International Journal of Vegetable Science* 22: 432–441.

Shahbaz, M. and Ashraf, M. (2007). Influence of exogenous application of brassinosteroid on growth and mineral nutrients of wheat (*Triticum aestivum* L.) under saline conditions. *Pakistan Journal of Botany* 39: 513.

Shahbaz, M., Ashraf, M., and Athar, H.U.R. (2008). Does exogenous application of 24-epibrassinolide ameliorate salt induced growth inhibition in wheat (*Triticum aestivum* L.)? *Plant Growth Regulation* 55: 51–64.

Sharma, P., Bhardwaj, R., Arora, N. et al. (2007). Effect of 28-homobrassinolide on growth, zinc metal uptake and antioxidative enzyme activities in *Brassica juncea* L. seedlings. *Brazilian Journal of Plant Physiology* 19: 203–210.

Sharma, I., Bhardwaj, R., and Pati, P.K. (2015). Exogenous application of 28-Homobrassinolide modulates the dynamics of salt and pesticides induced stress responses in an elite rice variety Pusa Basmati-1. *Journal of Plant Growth Regulation* 34: 509–518.

Sharma, I., Kaur, N., and Pati, P.K. (2017). Brassinosteroids: a promising option in deciphering remedial strategies for abiotic stress tolerance in rice. *Frontiers in Plant Science* 8: 301539.

Sharma, A., Kumar, V., Shahzad, B. et al. (2020). Photosynthetic response of plants under different abiotic stresses: a review. *Journal of Plant Growth Regulation* 39: 509–531.

Sharma, A., Shahzad, B., Kumar, V. et al. (2019b). Phytohormones regulate accumulation of osmolytes under abiotic stress. *Biomolecules* 9: 285.

Sharma, A., Yuan, H., Kumar, V. et al. (2019a). Castasterone attenuates insecticide induced phytotoxicity in mustard. *Ecotoxicology and Environmental Safety* 179: 50–61.

Sharma, N., Thakur, M., Suryakumar, P. et al. (2022). 'Breathing Out' under heat stress—respiratory control of crop yield under high temperature. *Agronomy* 12: 806.

Shu, S., Tang, Y., Yuan, Y. et al. (2016). The role of 24-epibrassinolide in the regulation of photosynthetic characteristics and nitrogen metabolism of tomato seedlings under a combined low temperature and weak light stress. *Plant Physiology and Biochemistry* 107: 344–353.

Siddiqi, K.S. and Husen, A. (2021). Significance of brassinosteroids and their derivatives in the development and protection of plants under abiotic stress. *Biologia* 76: 2837–2857.

Singh, S., Parihar, P., Singh, R. et al. (2016). Heavy metal tolerance in plants: role of transcriptomics, proteomics, metabolomics, and ionomics. *Frontiers in Plant Science* 6: 165395.

Slot, M. and Kitajima, K. (2015). General patterns of acclimation of leaf respiration to elevated temperatures across biomes and plant types. *Oecologia* 177: 885–900.

Song, Y.L., Dong, Y.J., Tian, X.Y. et al. (2016). Role of foliar application of 24-epibrassinolide in response of peanut seedlings to iron deficiency. *Biologia Plantarum* 60: 1–14.

Souza, L.A., Monteiro, C.C., Carvalho, R.F. et al. (2017). Dealing with abiotic stresses: an integrative view of how phytohormones control abiotic stress-induced oxidative stress. *Theoretical and Experimental Plant Physiology* 29: 109–127.

Stohs, S.J. and Bagchi, D. (1995). Oxidative mechanisms in the toxicity of metal ions. *Free Radical Biology and Medicine* 18: 321–336.

Sun, X., Lin, L., and Sui, N. (2019). Regulation mechanism of microRNA in plant response to abiotic stress and breeding. *Molecular Biology Reports* 46: 1447–1457.

Sunkar, R., Chinnusamy, V., Zhu, J. et al. (2007). Small RNAs as big players in plant abiotic stress responses and nutrient deprivation. *Trends in Plant Science* 12: 301–309.

Swamy, K.N. and Rao, S.S.R. (2011). Effect of brassinosteroids on the performance of coleus (*Coleus forskohlii*). *Journal of Herbs, Spices & Medicinal Plants* 17: 12–20.

Talaat, N.B. (2013). 24-Epibrassinolide alleviates salt-induced inhibition of productivity by increasing nutrients and compatible solutes accumulation and enhancing antioxidant system in wheat (*Triticum aestivum* L.). *Acta Physiologiae Plantarum* 35: 729–740.

Tanaka, K., Asami, T., Yoshida, S. et al. (2005). Brassinosteroid homeostasis in *Arabidopsis* is ensured by feedback expressions of multiple genes involved in its metabolism. *Plant Physiology* 138: 1117–1125.

Tanveer, M., Shahzad, B., Sharma, A. et al. (2018). 24-Epibrassinolide; an active brassinolide and its role in salt stress tolerance in plants: a review. *Plant Physiology and Biochemistry* 130: 69–79.

Thussagunpanit, J., Jutamanee, K., Kaveeta, L. et al. (2015a). Comparative effects of brassinosteroid and brassinosteroid mimic on improving photosynthesis, lipid peroxidation and rice seed set under heat stress. *Journal of Plant Growth Regulation* 34: 320–331.

Thussagunpanit, J., Jutamanee, K., Sonjaroon, W. et al. (2015b). Effects of brassinosteroid and brassinosteroid mimic on photosynthetic efficiency and rice yield under heat stress. *Photosynthetica* 53: 312–320.

Tseng, M.J., Liu, C.W., and Yiu, J.C. (2007). Enhanced tolerance to sulfur dioxide and salt stress of transgenic Chinese cabbage plants expressing both superoxide dismutase and catalase in chloroplasts. *Plant Physiology and Biochemistry* 45: 822–833.

Verslues, P.E., Agarwal, M., Katiyar-Agarwal, S. et al. (2006). Methods and concepts in quantifying resistance to drought, salt and freezing, abiotic stresses that affect plant water status. *The Plant Journal* 45: 523–539.

Vert, G., Nemhauser, J.L., Geldner, N. et al. (2005). Molecular mechanisms of steroid hormone signaling in plants. *Annual Review of Cell and Developmental Biology* 21: 177–201.

Vilarrasa-Blasi, J., González-García, M.P., Frigola, D. et al. (2014). Regulation of plant stem cell quiescence by a brassinosteroid signaling module. *Developmental Cell* 30: 36–47.

Vragović K, Selaa, A., Friedlander-Shani, L. et al. (2015). Translatomeanalyses capture of opposing tissue-specific brassinosteroid signals orchestrating root meristem differentiation. *Proceedings of the National Academy of Sciences of the United States of America* 112: 923–928.

Wada, K.C. and Takeno, K. (2010). Stress-induced flowering. *Plant Signaling & Behavior* 5: 944–947.

Wang, B., Li, Y., and Zhang, W.H. (2012a). Brassinosteroids are involved in response of cucumber (*Cucumis sativus*) to iron deficiency. *Annals of Botany* 110: 681–688.

Wang, Y., Cao, J.J., Wang, K.X. et al. (2019). BZR1 mediates brassinosteroid-induced autophagy and nitrogen starvation tolerance in tomato. *Plant Physiology* 179: 671–685.

Wu, X., Yao, X., Chen, J. et al. (2014). Brassinosteroids protect photosynthesis and antioxidant system of eggplant seedlings from high-temperature stress. *Acta Physiologiae Plantarum* 36: 251–261.

Xia, X.J., Huang, Y.Y., Wang, L. et al. (2006). Pesticides-induced depression of photosynthesis was alleviated by 24-epibrassinolide pretreatment in *Cucumis sativus* L. *Pesticide Biochemistry and Physiology* 86: 42–48.

Xia, X.J., Huang, L.F., Zhou, Y.H. et al. (2009a). Brassinosteroids promote photosynthesis and growth by enhancing activation of Rubisco and expression of photosynthetic genes in *Cucumis sativus* L. *Planta* 230: 1185.

Xia, X.J., Wang, Y.J., Zhou, Y.H. et al. (2009b). Reactive oxygen species are involved in brassinosteroid-induced stress tolerance in cucumber. *Plant Physiology* 150: 801–814.

Yan, K., Shao, H., Shao, C. et al. (2013). Physiological adaptive mechanisms of plants grown in saline soil and implications for sustainable saline agriculture in coastal zone. *Acta Physiologiae Plantarum* 35: 2867–2878.

Yaqoob, U., Jan, N., Raman, P.V. et al. (2022). Crosstalk between brassinosteroid signaling, ROS signaling and phenylpropanoid pathway during abiotic stress in plants: does it exist? *Plant Stress* 4: 100075.

Ye, H., Liu, S., Tang, B. et al. (2017). RD26 mediates crosstalk between drought and brassinosteroid signalling pathways. *Nature Communications* 8: 14573.

Yin, W., Dong, N., Niu, M. et al. (2019). Brassinosteroid-regulated plant growth and development and gene expression in soybean. *The Crop Journal* 7: 411–418.

Yuan, G.F., Jia, C.G., Li, Z. et al. (2010). Effect of brassinosteroids on drought resistance and abscisic acid concentration in tomato under water stress. *Scientia Horticulturae* 126: 103–108.

Yuan, L., Shu, S., Sun, J. et al. (2012a). Effects of 24-epibrassinolide on the photosynthetic characteristics, antioxidant system, and chloroplast ultrastructure in *Cucumis sativus* L. under $Ca(NO_3)_2$ stress. *Photosynthesis Research* 112: 205–214.

Yuan, L., Yuan, Y., Du, J. et al. (2012b). Effects of 24-epibrassinolide on nitrogen metabolism in cucumber seedlings under $Ca(NO_3)_2$ stress. *Plant Physiology and Biochemistry* 61: 29–35.

Yue, J., You, Y., Zhang, L. et al. (2018). Exogenous 24-epibrassinolide alleviates effects of salt stress on chloroplasts and photosynthesis in *Robinia pseudoacacia* L. seedlings. *Journal of Plant Growth Regulation* 38: 669–682.

Zadvornova, Y., Alekseichuk, G., Laman, N. et al. (2005). Effect of brassinosteroids on activation of the cell cycle during germination of *Brassica oleracea* L. seeds. *Doklady Natsional'noi Akademii Nauk Belarusi* 49: 70–73.

Zapata, P.J., Botella, M.A., Petrel, M.T. et al. (2007). Responses of ethylene biosynthesis to saline stress in seedlings of eight plant species. *Plant Growth Regulation* 53: 97–106.

Zhang, S., Cai, Z., and Wang, X. (2009). The primary signaling outputs of brassinosteroids are regulated by abscisic acid signaling. *Proceedings of the National Academy of Sciences of the United States of America* 106: 4543–4548.

Zhang, M., He, S., Qin, B. et al. (2020). Exogenous melatonin reduces the inhibitory effect of osmotic stress on antioxidant properties and cell ultrastructure at germination stage of soybean. *PLoS One* 15: e0243537.

Zhao, Y.J., Xu, R.J., and Cuo, W.H. (1990). Inhibitory effects of abscisic acid on epibrassinolide-induced senescence of detached cotyledons in cucumber seedlings. *Chinese Science Bulletin* 35: 928–931.

Zhao, B. and Li, J. (2012). Regulation of brassinosteroid biosynthesis and inactivation[F]. *Journal of Integrative Plant Biology* 54: 746–759.

Zhuang, K.Y., Gao, Y.Y., Liu, Z.B. et al. (2020). WHIRLY1 regulates HSP21.5A expression to promote thermotolerance in tomato. *Plant Cell Physiology* 61: 169–177.

Zullo, M.A.T. and Adam, G. (2002). Brassinosteroid phytohormones: structure, bioactivity and applications. *Brazilian Journal of Plant Physiology* 14: 143–181.

4

Regulation of Plant Secondary Metabolism by Brassinosteroids Under Abiotic Stress

Dhriti Sharma[1,2], Bharat Kapoor[3], and Dhriti Kapoor[4*]*

[1]*Department of Bioengineering and Biosciences, Lovely Professional University, Jalandhar, Punjab, India*
[2]*Sidharth Govt. College, Nadaun, Hamirpur, Himachal Pradesh, India*
[3]*Department of Tourism and Hospitality, Guru Nanak Dev University, Amritsar, Punjab, India*
[4]*School of Biological and Environmental Sciences, Shoolini University, Solan, Himachal Pradesh, India*

*Corresponding authors: dhriti405@gmail.com, bharatkapoor22@gmail.com

4.1 Introduction

The life of plants is greatly influenced by the prevailing environmental conditions. If a particular environmental factor exceeds in its intensity and persists for a longer time span, then it turns stressful and incompatible with the normal process of plant development. This can negatively affect plant growth, productivity, and even the quality of produce. Extremes of pH, temperature, water, salts, and radiations, together with contamination of heavy metals and pesticides, can be named as a few major examples of abiotic stress-inducing factors (Begum et al. 2019; Ali et al. 2021; Riaz et al. 2022). Plants respond to these stresses by bringing about changes in their metabolic affairs at the expense of internal energy, thereby upsetting the different growth-related stages (Iqbal and Ansari 2020). Endurance or sensitivity of the plants to these stresses is quite an intricate affair and causes a drastic drop in the crop yield worldwide (Etesami and Jeong 2018). This scenario will become more critical with the enhancement in desertification of land, salinization of water and soil, water scarcity, and rise in ecological defiling.

One of the major metabolic byproducts of abiotic stress is surplus production of reactive oxygen species (ROS), leading to oxidative damage and degradation of macromolecules like lipids, proteins, and nucleic acids (El-Beltagi et al. 2020; Medina et al. 2021).

As a part of stress response and survival strategy, plants synthesize several compounds called metabolites, namely phenolics, amino acids, compatible solutes, hormones, and antioxidants, which play a pivotal role in plant defense and confer stress tolerance. These metabolites are either primary or secondary in nature, generally classified on the basis of their synthesis, occurrence, and function. Secondary metabolites are synthesized to meet specific requirements, and that too from modification of either substrates or synthetic pathways of primary metabolites (Pott et al. 2019). The classical system, comprised of enzymatic

Brassinosteroids and Abiotic Stress: Recent Developments on Physiological and Molecular Aspects,
First Edition. Edited by Anket Sharma, et al.
Published 2026 by John Wiley & Sons Ltd.

and non-enzymatic antioxidants, falls under the category of primary metabolites to achieve stress tolerance, whereas secondary metabolites mainly include phenolics, flavonoids, tannins, stilbenes, lignin, and hydroxycinnamate esters, etc., which scavenge ROS in stressed plants. In order to get adapted to the changes in the environment, secondary metabolites are genetically coded and prominently synthesized by plants (Rastegari et al. 2019).

Furthermore, plant growth-regulating hormones play the role of chemical messengers that govern different developmental and physiological aspects in plant metabolism. With the help of various extrinsic and intrinsic stimuli, these hormones modulate signal transduction mechanisms to make plants capable of combating various stresses (Sarkar and Sadhukhan 2022). One such phytohormone is brassinosteroid (BR), which constitutes a significant part of the defense network of plants against challenging and stressful environmental circumstances. First discovered inside *Brassica* pollens, it has now been found in almost every other plant part, ranging from roots and stems to leaves, flowers, pollens, and seeds, besides being reported in the entire spectrum of plant diversity encompassing lower to higher plants, especially the angiospermic ones (Yokota et al. 2017; Zullo and Bajguz 2019).

Applying BRs exogenously to combat stress conditions has become a promising strategy to realize the goal of enhancing crop yield sustainably together with protecting the plants ecologically (Shahzad et al. 2018; Baghel et al. 2019). Functional efficacy of BRs in influencing the different aspects of growth has been recorded against a diverse array of stressful conditions as created by salinity (Larré et al. 2014, temperature extremes (Gornik and Lahuta 2017), water deficit/excess (Shahana et al. 2015; Mahesh et al. 2013), and heavy metals – Al (Madhan et al. 2014; Abdullahi et al. 2002), As (Raghu et al. 2014), Cd (Piacentini et al. 2022; Anusha et al. 2015; Hasan et al. 2008; Janeczko et al. 2005), Cr (Sharma et al. 2011b), Cu (Yadav et al. 2016), Ni (Lukatkin et al. 2013), and Pb (Rao and Raghu 2016; Anuradha and Rao 2009).

For being able to battle out these abiotic stresses, BRs not only make the growth and photosynthetic attributes better in stressed plants but also upregulate the content of metabolites. Considering the stress-protective role of both the BRs and secondary metabolites, it becomes all the more important to elucidate their correlation under adverse environmental conditions. In this chapter, a comprehensive insight into the regulation of secondary metabolism by the application of BRs in plants subjected to abiotic stresses has been dealt with. For this, different types of secondary metabolites and regulation of their functioning under the influence of BRs in abiotically stressed plants have been reviewed.

4.2 Secondary Metabolites Involved in Abiotic Stress Tolerance

Biological importance of secondary metabolites has been neglected for a long time, but their role in attracting pollinators, screening against UV radiation, preventing herbivory, and toxicity has compelled the scientific world to designate them as essential for the existence of plants (Ahmed et al. 2017). Researchers in the field of organic chemistry have extensively worked out the structural and functional details of secondary metabolites since the 1950s (Mauseth 2014). And the findings helped in revealing that secondary metabolites play a key role in underlying defense mechanisms in plants against abiotic and biotic stresses (Croteau et al. 2000; Srivastava et al. 2021).

Some major secondary metabolites involved in conferring abiotic stress tolerance in plants can be enumerated as below:

1. **Phenolics:** These are of paramount significance among all the metabolites on account of their ubiquitous occurrence in different plant parts and their key role in providing resistance against UV radiation and pathogen-induced stresses (Michalak 2006). Phenols and other polyphenolic groups (phenolic acid, flavonoids, anthocyanins, lignans, and stilbenoids) constitute the class of secondary metabolites with varied physiological roles. The presence of aromatic ring with hydroxyl (OH) or methoxy ($-OCH_3$) side groups makes them effective electron or proton donors, thereby conferring them with antioxidant potential, and these are found to be far superior in mitigating oxidative damage than the non-enzymatic antioxidants like vitamin E (tocopherol) and vitamin C (ascorbic acid) (Sakihama et al. 2002; Bolda et al. 2011). Furthermore, phenols (especially flavonoids) are mainly concentrated in vacuolar and apoplastic regions of foliar and floral parts, where they reduce the fluidity to bring about the stabilization of cellular membranes, leading to a decrease in diffusion of free radicals and lipid peroxidation thereof (Apel and Hirt 2004). Flavonoids like quercetin and luteolin glycosides are noted to quench singlet oxygen formed in response to blue light-induced stress (Agati et al. 2009).

   Their production is enhanced under stressful conditions as a response of plants' defense, and by either donating electrons or protons, they form phenoxy radicals and mitigate abiotic stresses via acting as H_2O_2 quenchers or metal chelators (Gill and Tuteja 2010). Flavonoids have been reported to perform quenching of ROS either inside the plant part or in proximity to the site where they are generated, for instance, in the case of the chloroplast, ROS scavenging by flavonoids occurs on the outer surface of the chloroplast (Mullineaux and Karpinski 2002; Agati et al. 2009). Anthocyanins, on the other hand, get accumulated in the vacuolar regions of mesophyll cells to perform sequestering of ROS, like superoxide anions and hydrogen peroxide (Kytridis and Manetas 2006).
2. **Osmolytes:** Also termed as compatible solutes, these water-soluble organic metabolites possess a low molecular weight and their concentration rises in intracellular regions to protect plants from various environmental stresses (Bajguz and Hayat 2009; Ma et al. 2010). Amino acids (proline), betaines, polyamines, polyols, sugars (trehalose), and sugar alcohols (mannitol and sorbitol) are the most common examples of these compatible solutes. Stocking up these solutes is advantageous for plants in response to water scarcity or salinity excess because stress tolerance is provided to the plants by them without upsetting the cellular mechanisms. The extent of their intracellular accumulation under abiotic stress makes the plants differentially tolerant or sensitive to stress (Obata and Fernie 2012). These neutral molecules act as a shield to safeguard proteins and various other biological membranes when subjected to challenging conditions affecting the cellular metabolic status. These solutes bridge the osmotic gap between the cytosol and the surrounding regions of the cell (Wani et al. 2013).

3. **Phytohormones:** These are the low-concentration chemicals of a regulatory nature that have a major say over a varied array of developmental and functional processes inside plants. They are intricately involved in crosstalk with each other and regulate signal transduction cascades in response to stressful circumstances (Raza et al. 2019; Sarkar and Sadhukhan 2022). Auxins, gibberellins, cytokinins, ethylene, and abscisic acid constitute five principal phytohormones, and BRs, jasmonates, salicylates, strigolactones, etc., are some of the new additions to this list (Pons 2020). These phytohormones, especially ABA, jasmonates, and salicylates, create a defensive network to battle out various environmental stress factors (Kumudini et al. 2018). Gaining deeper insights into the interactive mechanisms existing between these phytohormones and signaling cascades of defense could result in discovering novel arenas of resistance mechanisms in plants (Nunes da Silva et al. 2022).

Chemical nature of different secondary metabolites and their functional efficacy against abiotic stresses have been summarized in Table 4.1.

4.3 Brassinosteroids: An Up-and-coming Phytohormone to Battle Out Abiotic Stresses

Versatility in BR applicability contributes to making this hormone a favorable option for improving plant growth and development (Vardhini 2016; Yusuf et al. 2017). They possess steroidal nature, plant specificity, and are sensed by receptor kinases of cell surfaces belonging to leucine-rich repeat family of receptors i.e. BRI1 (BRASSINOSTEROID INSENSITIVE 1), which collaborates with BAK1 (BRI1 ASSOCIATED RECEPTOR KINASE 1), a co-receptor, consequently getting involved in a cascade of phosphorylation and dephosphorylation events for sending transduction signals to the nuclear region so that the expression levels of a variety of genes regulating functional aspects of plant development could be modulated (Sharma et al. 2013a; Nakamura et al. 2017). BRs are operational on different fronts in the life cycle of plants, ranging from germination of seeds, division and elongation of cells, light-induced morphogenesis, together with xylem differentiation (Vriet et al. 2012; Zhang et al. 2014).

However, recent years have witnessed the elevation in their prominence in the agricultural domain whereby crop quality and yield are improved by them under both stressful and stress-free environmental conditions (Zhang et al. 2014; Marakli and Gozukirmizi 2016; Liu et al. 2017). Taking into account the scope and significance of their agricultural applications, more concerted research efforts have been made lately just to discern the active mechanism behind their metabolic and signaling strategies (Belkhadir and Jaillais 2015). Although multiple works strongly suggest their efficacy in amelioration of abiotic stresses (Zhang et al. 2014; Ahammed et al. 2015; Sharma et al. 2015, 2023; Hegazi et al. 2017; Wang et al. 2017), yet comprehensive account of inherent mode of their working in doing so is still needed to be deciphered. This might be because of the integrated involvement of BR signals with various other networks of signaling when it comes to alleviation of stresses (Divi and Krishna 2010; Ahammed et al. 2015; Divi et al. 2016). One such possible mechanism could be enacted by BRs by modulating the components of secondary metabolome to mitigate abiotic stresses, which is discussed in detail hereinafter.

Table 4.1 Secondary metabolites and their role in conferring abiotic stress tolerance.

Name of the secondary metabolite	Chemical nature of the secondary metabolite	Role against abiotic stress(es)	References
A) Phenolics (Phenols and polyphenols)	Phenolic acid	Enhances its own endogenous production and abiotic stress tolerance thereof, due to upregulation of major synthetic genes like PAL and CHS in phenylpropanoid pathway; high pH stress is mitigated by escalated secretions of phenolic acid in epidermal and stelar regions of roots, which in turn promotes absorption of ions and decelerates Fe-deficiency-induced reactions in rice plants.	Li et al. (2021) and Dehghanian et al. (2022)
	Flavonoids	Strengthens antioxidant defense system; measures of different flavonoids, namely cyanidin, kaempferol, and quercetin glycosides are elevated as observed a) in *Oryza sativa* plants in response to heat and salinity stress, and b) *Brassica rapa* and *Brassica oleracea* var. *acephala* in response to water, temperature, and light extremes. Flavonoid glucosinolates too strengthen the defense system of plants by making them tolerant to abiotic stresses; Another flavonoid i.e. anthocyanins can scavenge ROS and perform photoprotection against stress.	Kovinich et al. (2014); Nakabayashi et al. (2014); Rechner et al. (2017); Jan et al. (2021); Ljubej et al. (2021); Rao et al. (2021); and Laoué et al. (2022)
	Lignans	Effective as antioxidants; enhancement in endogenous production as observed in *Nicotiana tabacum* plants to protect against chilling stress and in *Sesamum indicum* under water scarcity; modulation in extent of lignification in *Glycine max* roots by upregulating genes like DIR2 and peroxidases to make them tolerant to Mn toxicity.	Zhou et al. (2018); Kermani et al. (2019); and Šamec et al. (2021)
	Stilbenoids	Biosynthetic process of stilbenoids is enhanced to mitigate oxidative damage, brings about upregulation of antioxidant defense mechanism as observed in different varieties of *Vitis vinifera* under UV and ozone stress.	Versari et al. (2001); Chong et al. (2009); and Valletta et al. (2021)
B) Osmolytes	Glycine betaine	Osmotic protection, maintenance of structural integrity (proteins, enzymes, and cell membranes), photosynthetic fortification, ROS scavenging through upregulation of AsA-GSH cycle and modulation of expression of stress-related genes.	Yang et al. (2007b); Hoque et al. (2007, 2008)

(*Continued*)

Table 4.1 (Continued)

Name of the secondary metabolite	Chemical nature of the secondary metabolite	Role against abiotic stress(es)	References
	Proline	Direct detoxification of ROS, maintenance of water balance and redox homeostasis, modulation in the enzymatic action of antioxidants, enhancement in its own concentration via upregulating the expression of Δ1-pyrroline-5-carboxylic synthase (PCS5) genes while lowering down the expression levels of proline dehydrogenase (PDH) to combat salt and water stress; promotion of cellular signaling.	Kishor et al. (2005); Verbruggen and Hermans (2008); Yang et al. (2010); de Campos et al. (2011); Liang et al. (2013); and Qamar et al. (2015)
	Polyamines	Stabilization of macromolecular components and regulation of environmental stimuli i.e. biotic and abiotic stressors; enhanced production of antioxidants and modulation of auxin (IAA) and abscisic acid (ABA) pathways by binding to brassinosteroids to reinforce the tolerance against toxicity; Alteration in genetic and enzymatic control of ethylene biosynthesis together with enhancement in the production of different polyamines make the plants to combat abiotic stress better as seen in tobacco.	Kusano et al. (2007); Bitrián et al. (2012); Choudhary et al. (2012); and Naing et al. (2021)
C) Phytohormones	Abscisic acid	Commonly designated as the stress hormone, makes plants capable of perennating harsh environmental conditions; causes stomatal closure, involved in architectural changes and root development during water scarcity; synthesis of osmoprotectants, protective proteins, and dehydrins in response to abiotic stresses.	Guimaraes and Jueneman (2008); Atkinson and Urwin (2012); Fernando and Schroeder (2016); Negin and Moshelion (2016); Soma et al. (2021); Hossain et al. (2022); and Kuromori et al. (2022)
	Auxin	Regulation of plant growth under stress; control of interactions between stressors of both biotic and abiotic nature; transcription of main genes of auxin response as reported in *Arabidopsis*, soybean, and rice; essential for making plants more tolerant to salinity and heavy metal toxicity.	Ahmad et al. (2010); Casanova-Sáez et al. (2021); Quamruzzaman et al. (2021); Kou et al. (2022); Mathur et al. (2022); and Mishra et al. (2022)
	Ethylene	Modulation of endogenous measures in response to abiotic stress-inducing factors; crosstalk with other hormones, improvement of stress tolerance in higher concentrations.	Jha et al. (2021) and Li et al. (2022)

Cytokinins	Make the plants capable to combat salt and water shortage-induced stress; exhibit antagonistic behavior to ABA and break dormancy in seeds; promote apical dominance with reduction in concentration and impart drought resistance.	Wani et al. (2016) and He et al. (2018)
Gibberellins	Protection against osmotic stress; involved in intermutual talks with other hormones to strengthen the defense network of plants against abiotic stresses.	Xie et al. (2019); Fenn and Giovannoni (2021); and Mujtaba et al. (2021)
Salicylic acid	Enhancement of antioxidant potential of plants in low quantities; regulation of heat shock proteins, chaperones, and genes encoding for the synthesis of secondary metabolites, together with ABA, controls drought responses in plants.	Verbruggen and Hermans (2008); Gupta et al. (2017); Ahmad et al. (2019); Lefevere et al. (2020); and Guo et al. (2021)
Brassinosteroids	Mitigates a variety of abiotic stresses via regulation of expression of stress marker genes, for instance, cold-responsive genes and heat shock proteins, thus making plants tolerant to temperature extremes; upkeeps the photosynthetic machinery, promotes chlorophyll biosynthesis, and enzymatic actions essential for abiotic stress tolerance.	Bajguz and Hayat (2009); Wani et al. (2016); Bhuyan et al. (2020); and Hossain et al. (2022)
Jasmonates	Promotion of tolerance against abiotic stresses (drought, irradiation, salinity and freezing); enhancement in endogenous measures inside the roots in response to salt stress; external supplementation in the form of methyl jasmonate mitigates salt stress along with heavy metal toxicity via activating antioxidant machinery and accumulation of phytochelatins as reported in case of cadmium and copper.	Ryu and Cho (2015); Pál et al. (2018); Jan et al. (2019); Romero-Puertas et al. (2019); Taheri et al. (2020); and Wang et al. (2020)
Strigolactones	Mars the production of ROS and MDA; elevates antioxidant metabolism; makes the foliar chlorophyll content and thereby photosynthetic efficiency better; maintains transpiration balance during abiotic stress amelioration	Aliche et al. (2020); Rehman et al. (2021); Bhoi et al. (2021); Brewer (2022); and Kleman and Matsusova (2023)

4.4 Regulatory Role of Brassinosteroids over Secondary Metabolites Under Abiotic Stress

Stimulation of stress-responsive genes by BRs enables them to regulate the biosynthesis of various secondary metabolites and then get engaged in crosstalk with them to confer stress tolerance, which has been elaborated as follows:

4.4.1 BRs and Phenolics

Phenolics are major secondary metabolites, chemically termed as phenylpropanoids and contain a carbon skeleton of phenylalanine with a number of structural variants involved in the protection, support, and subsistence of plants (Dixon and Paiva 1995). Several reports demonstrate the employment of phytohormones to increase the level of phenolic compounds in plants, such as the enhanced content of anthocyanin in carrot (Narayan et al. 2005). Ahammed et al. (2013) suggested that an enhanced level of phenolics by the application of phytohormone BR is possibly due to a change in the levels of polyphenols, which results in enhanced antioxidative defense ability of plants (Yadav et al. 2016).

4.4.1.1 Phenylpropanoid Mechanism

Plants synthesize phenolic compounds via shikimate pathway either from PEP i.e. phosphoenolpyruvic acid (an intermediate of Embden–Meyerhof–Parnas (EMP) pathway/glycolysis) or erythrose-4-phosphate (an intermediate of the hexose monophosphate shunt or pentose phosphate pathway) whereby they get reduced to form DAHP i.e. 3-deoxy-D-heptulosonate 7-phosphate along with an inorganic phosphate by DAHP synthase enzyme, which undergoes subsequent metabolic modifications, ultimately resulting in the biosynthesis of shikimate (Fraser and Chapple 2011). Shikimate is then subjected to phosphorylation, which gives rise to EPSP i.e. 5-enol pyruvyl-shikimate-3-phosphate by the catalytic action of EPSP synthase. This compound is then transformed into chorismate with the enzymatic action of chorismate synthase involving its 1-4 trans-elimination. This sequence of reactions for the generation of various secondary metabolites is briefly depicted in Figure 4.1. Vogt in 2010 had noted that a complex network of biocatalysts is essential in synthesizing chorismate from a varied array of precursor molecules.

Chorismate is then utilized differently to form three principal amino acids of aromatic nature, namely phenylalanine (Phe), tryptophan (Trp), and tyrosine (Tyr). Out of these three, the formation of Phe and Tyr takes place from chorismate through its conversion into prephenate by chorismate mutase, while Trp gets synthesized from chorismate after transforming the latter into anthranilate in the presence of anthranilate synthase. The catalytic presence of prephenate aminotransferase transaminates prephenate into arogenate from where the synthesis of Phe and Trp vary and arogenate then finally gives rise to phenylalanine by taking help of arogenate dehydratase enzyme. Next occurs the biosynthesis of phenylpropanoid via initially deaminating phenylalanine into trans-cinnamic acid + NH_3 by PAL i.e. phenylalanine ammonia-lyase whereafter NH_3 gets re-added to phenylalanine to continue metabolism of phenylpropanoids (Fraser and Chapple 2011). Isoforms of phenylalanine ammonia lyase (PAL) enzyme namely, PAL1–PAL4 with variable extent of expression in *Arabidopsis* have been recognized (Raes et al. 2003). Stems are

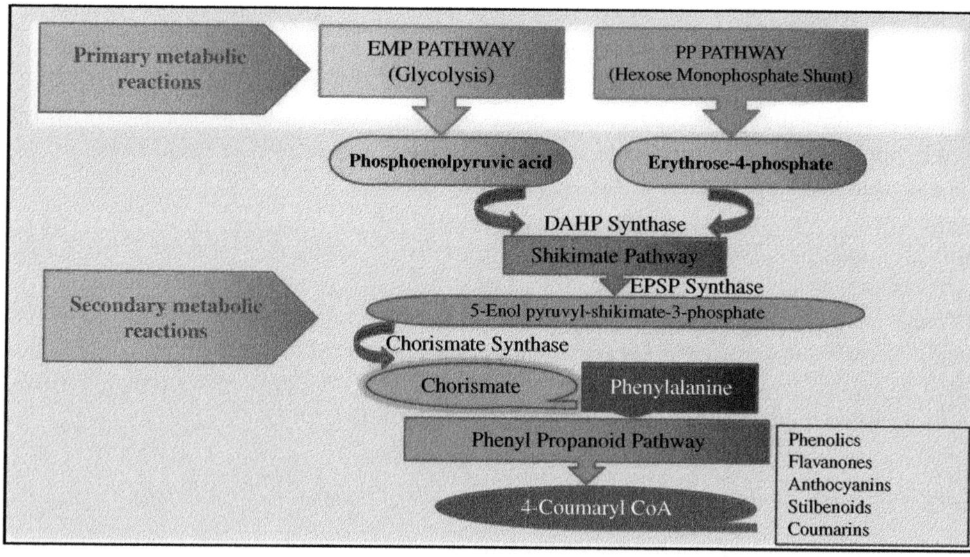

Figure 4.1 Generalized scheme of phenyl propanoid pathway for the formation of secondary metabolites.

reported to have a basal degree of PAL3 gene expression, while later developmental stages in plants exhibited elevation in expression extent of PAL1 and PAL2 along with PAL4 in *Arabidopsis* (Raes et al. 2003; Rohde et al. 2004). Numerous findings further suggested toward the essential role of PAL1–PAL2 genes in encoding principal PAL enzymes participating in the phenylpropanoid mechanism (Raes et al. 2003).

4.4.1.2 Crosstalk Between Brassinosteroids and Phenylpropanoid Mechanism

Intricate crosstalks exist between the phenylpropanoid mechanism and signaling of BRs as observed in grapes (*Vitis vinifera*), where exogenous treatment promoted the levels of secondary metabolites, namely anthocyanins and phenolics (Zhu et al. 2013). The role of 24-epibrassinolide (EBL) (an active form of BRs) in the regulation of phenylpropanoid mechanism is substantiated by elevated measures of secondary metabolites (total phenolics, anthocyanins, tannins, and flavonoids) when grapevine (var. Yan 73 and Cabernet Sauvignon) was subjected to 24-EBL spray (0.1 and 0.4 mg L^{-1}). The reason being that all the rate-limiting biocatalysts of the phenylpropanoid mechanism, PAL (phenylalanine ammonia lyase) and UFGT (UDP-glucose: flavonoid 3-O-glucosyltransferase), are upregulated by 24-EBL application (Zhang et al. 2015). On similar lines, external supplementation of 24-EBL promotes the secondary metabolite production in the form of catechins and theanines, the products of the phenylpropanoid mechanism, in tea (*Camellia sinensis*) without changing the levels of caffeine (Ahammed et al. 2013; Luan et al. 2013).

Phytotoxic effects of phenanthrene (PHE) have been ameliorated by 24-EBL treatment via promoting the enzymatic action of antioxidants, PAL, and UFGT to enhance the levels of secondary metabolites (Ahammed et al. 2012). Tomato plants exogenously supplemented with 24-EBL (0.1 μM), exhibited overexpression in enzymes of secondary metabolism, namely PAL, CAD (cinnamyl alcohol dehydrogenase), and SKDH (shikimate dehydrogenase) by 27, 39, and 127%, respectively, when correlated to untreated plants of control (Ahammed et al. 2013). This enhancement in the enzymatic action of PAL by BR application consequently accelerates the successive steps of the phenylpropanoid pathway, thereby speeding up the synthesis of secondary metabolites (Dixon and Paiva 1995).

Biosynthesis of another significant secondary metabolite, lignin, is also regulated by BRs as supported by a drastic reduction in lignin content in DIM1 (Delta (24)-sterol reductase) mutants of *Arabidopsis*, which fails to perform BR synthesis (Hossain et al. 2012). Furthermore, involvement of BRs in lignin biosynthesis and development of secondary cell wall has been worked out in *Arabidopsis*, where modulation in expression levels of synthetic regulators of lignin (MYB transcriptional factors) is brought about by VND6 and 7 (Vascular-Related NAC-Domain 6 & 7), which are in turn upregulated by BES 1-transcriptional factor of BR signaling (Ohtani and Demura 2019).

4.4.2 BRs and Osmolytes

Another common line of defense against abiotic stresses has been reported to be the amassing of compatible solutes or osmolytes (betaines, proline, polyamines, sugar alcohols, fructans, and myo-inositol, etc.) as documented in rice plants (Kumar et al. 2013). These solutes perform osmoprotection and make the plants stress-tolerant by performing ROS scavenging, maintaining cell turgidity by osmotic balance, together with upholding the integrity of cellular constituents.

Several studies have substantiated the role of BRs in inducing the accumulation of different osmolytes in response to abiotic stresses, which is held accountable for improving stress survival strategy in plants, and it is duly summarized in Figure 4.2 (Sharma et al. 2012, 2013b;

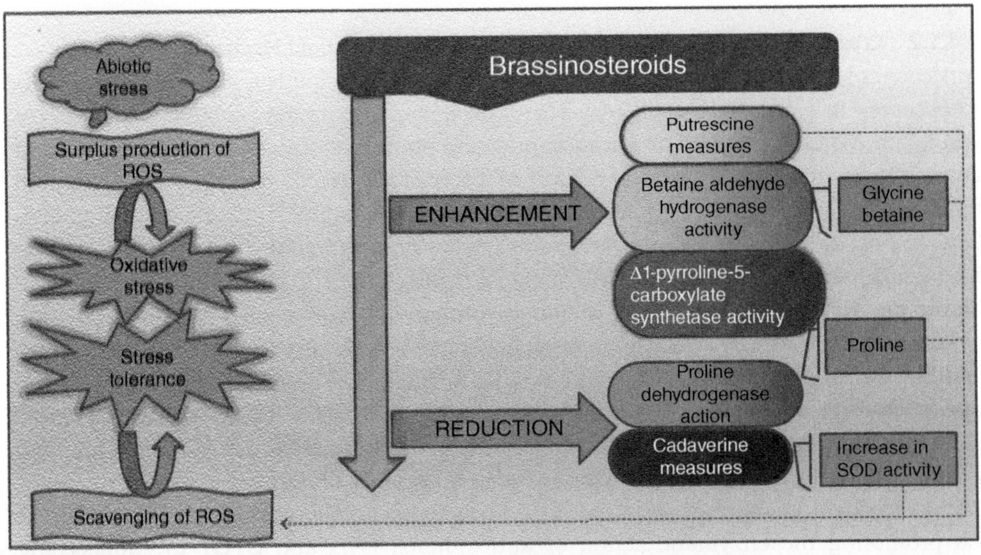

Figure 4.2 Overview of BR-mediated regulation of osmolytes in response to various abiotic stresses.

Kumar et al. 2013; Kohli et al. 2018). Signaling of BRs regulates a variety of physiological procedures in stress-exposed plants, and BRL3 (a constituent member of the family of BR receptors) is held together responsible for modulating the biosynthetic process of all the major osmoprotectants (Fàbregas et al. 2018). Elevation in expression levels of BRL3 in drought-stressed *Arabidopsis* plants has been documented, which in turn boosts the stock of key osmolytes, thereby making these plants tolerant against this stress (Fàbregas et al. 2018).

4.4.2.1 Brassinosteroids and Proline

Proline plays the role of a molecular chaperone by scavenging free radicals along with the stabilization of cytosolic redox status (Kido et al. 2013; Kumar et al. 2013). Multiple reports exhibited the enhancement of proline synthesis under the influence of BRs. Treatment of 24-EBL in Cr- and Cu-stressed radish (*Raphanus sativus*) and Cu-stress-exposed cucumber (*Cucumis sativus*) brought about a notable elevation in the endogenous measures of proline (Choudhary et al. 2010; Sharma et al. 2011a; Fariduddin et al. 2013). Foliar supplementation of 24-EBL also exhibited promotion of proline content in drought-stressed *Salvia miltiorrhiza* and *Capsicum annuum* var. *frutescens* plants (Zhu et al. 2014; Gao et al. 2016; Khamsuk et al. 2018). Subjection of *Vigna radiata* to salinity stress boosted proline levels, and co-application of brassinolide and salt led to amassing of proline in even higher quantities (Lalotra et al. 2017). Accumulation of this osmoprotectant could be due to upregulated expression of genes encoding for its biosynthesis (Lalotra et al. 2017). Peach (*Prunus persica*) fruits, when stored at a low temperature, have been reported to be protected from chilling stress by enhanced proline metabolism upon receiving 24-EBL treatment. Promotion of the enzymatic action of P5CS (Δ1-pyrroline-5-carboxylate synthetase) and downregulation of proline dehydrogenase enzyme have been responsible for enhanced measures of proline in these fruits (Gao et al. 2016).

Research studies further revealed that this enzyme encoding proline biosynthesis i.e. P5CS1 is also upregulated under salinity stress, which leads to accumulation of proline and elevation in the extent of acclimatization in response to this stress (Yamada et al. 2005), whereas without P5CS1, stress-mediated proline synthesis gets impaired, making plants hypersensitive to salt stress (Szekely et al. 2008). Additionally, the impaired genes of BRs biosynthesis and signaling, namely det-2 and bin-2, in mutants induce hypersensitivity to salinity stress, which has been connected to downregulated expression of P5CS1 and low measures of proline, thus confirming the significant role played by BRs in proline synthesis and accumulation to deal with various stresses (Zeng et al. 2010).

In addition to the positive impact of exogenous BR supplementation on the intrinsic levels of proline, coupled external application of BRs with proline can also mitigate a variety of abiotic stresses as observed in *Brassica juncea* (cv. RH30 and Varuna) under salt stress whereby not only the salinity-induced toxic effects are ameliorated but also the photosynthetic potential and yield were enhanced (Wani et al. 2019).

4.4.2.2 Brassinosteroids and Glycine Betaine

Another major osmolyte – Glycine betaine (GB) – is able to keep its form at neutral pH, making it highly water-soluble and capable of water retention and maintenance of protein structure (Kurepin et al. 2013, 2015, 2017). Additionally, its stress-ameliorative role has been corroborated in many studies with escalated levels of its accumulation in abiotic-stress-exposed plants. In most of the instances, its accumulation takes place in the photosynthetic apparatus of young leaves from which it gets translocated via phloem to the sites where its need arises (Park et al. 2007; Hattori et al. 2009). Thus, the production of GB inside the chloroplasts is responsible for providing tolerance against different stresses. Hence, the stress tolerance ability of GB is usually attributed to chloroplast-produced GB (Kurepin et al. 2015). Besides the maintenance of photosynthetic machinery (chloroplast and especially Photosystem II), GB maintains macromolecules (proteins, lipids, enzymes, etc.) required for maintaining electron transport on either side of thylakoid membranes and is specifically produced to deal with stressful conditions and sustain plant growth and productivity (Hattori et al. 2009; Adams et al. 2013; Kurepin et al. 2013).

Exogenous treatment of BRs has a synergistic impact on the accumulation of GB, which in turn boosts the stress tolerance efficiency of plants. Notable elevation in levels of GB was observed in 24-EBL supplemented salt-stressed tolerant and sensitive varieties of pea and in Cr-stress-exposed radish plants at their seven-day-old stage, leading to respective amelioration of salt and Cr toxicity (Choudhary et al. 2011; Shahid et al. 2014). Another research work centered on tomato plants subjected to salt-stress, confirmed the stress ameliorative role of 24-EBL where pre-soaking of tomato seeds in 24-EBL prior to their sowing in salt-rich soil, brought about betterment in accumulation of this osmoprotectant, when contrasted against untreated seeds grown in saline conditions (Ahmad et al. 2018). In a very similar fashion, 24-EBL-mediated enhancement in content and accumulation extent of GB in Cd-stress-exposed pea seedlings escalated the stress tolerance efficiency (Jan et al. 2018). The possible reason behind this BR-prompted GB biosynthesis is the improvement in the enzymatic action of BADH (betaine aldehyde dehydrogenase), which plays a prominent role in GB formation from choline (Qayyum et al. 2007; Rattan et al. 2014).

4.4.2.3 Brassinosteroids and Polyamines

Nitrogen-rich aliphatic compounds – polyamines are polycationic by nature and their molecular weight is low (Sengupta et al. 2016). Their notable role in the mitigation of abiotic stress is quite evident from different research works, whereby their endogenous levels are modulated (Liu et al. 2015). The three key polyamines, namely Put (putrescine), Spd (spermidine), together with Spe (spermine) exhibited a significant boost in their measures in the plants that are exposed to various abiotic stresses (Yang et al. 2007a). Impact of BRs on the levels of polyamines in abiotically stressed plants has been noticed, which suggests an underlying interactive role between these two, which brings about stress tolerance. However, even after attaining significant advancement over time, more explorative research is still required to comprehend this crosstalk between BRs and polyamines. A recent study revealed the promotion in endogenous measures of free PA in abiotically and biotically stressed plants upon supplementation with BRs, which alleviated Cu prompted oxidative damage (Guo et al. 2017; Fahad et al. 2019).

In addition to this, BR supplementation fixes the required levels of spermidine for optimal plant growth and that of putrescine to confer stress tolerance while decrementing the measures of cadaverine responsible for generation of an oxidative burst to mitigate metal toxicity (Nagajyoti et al. 2010). Moreover, the co-application of Cu with BRs reduces the content of cadaverine, resulting in modulation the enzymatic action of SOD essential for stress resistance (Dugardeyn and Van Der Straeten 2008). As per another research finding, Cu-stress-exposed seven-day-old radish seedlings exhibited changes in polyamine levels, and the stress was ameliorated by supplementing BRs, which further altered the content of polyamines (Choudhary et al. 2010. Copper stress was further found to be mitigated in radish plants by the co-supplementation of BRs with Spd, which was possibly on account of modulation of expression levels of polyamine enzymes providing stress tolerance (Choudhary et al. 2012). It affirms the main role of communication between BRs and PA in providing the plants' stress tolerance, which can be efficiently put to use in the upkeep of modern-day agribusiness while unraveling the attributes and translational factors connected with the signaling pathways of both BRs and PA.

4.4.2.4 Brassinosteroids and Sugars/Sugar Alcohols

Among all the different types of osmoprotectants, sugars are the largest contributors in the sustenance of osmotic potential, especially in response to osmotic stress (Parvaiz and Satyawati 2008). Enhancement in measures of various sugars has been reported in response to water or salinity-induced stress despite the decrease in photosynthetic carbon reduction and assimilation (Murakeozy et al. 2003). A key role is attributed to these soluble sugars in maintaining osmotic balance, together with ROS scavenging. The soluble sugars have a major role in the maintenance of osmotic homeostasis as well as scavenging of ROS and are further noted to be dependent on BRs for growth-related responses, specifically under low light stress (Parvaiz and Satyawati 2008; Zhang and He 2015). The study executed on *Arabidopsis* revealed the repression of sugar-incited hypocotyl elongation-regulating genes via brassinazole (an inhibitor of BRs biosynthesis) (Zhang et al. 2015). The findings further ascertained the elevation in transcript levels of BR-prompted response factors, namely BES1 and BZR1, in response to subjection to exogenous sugar treatments. This established that exogenously applied sugar in plants growing under dark conditions might assist in keeping the levels of BZR1 protein high (Zhang et al. 2015).

4.4.3 Brassinosteroids and Phytohormones

The interaction of BRs with other phytohormones in stress-exposed plants has been reported to boost their tolerance potential (Zhang et al. 2009). Several research findings revealed the role of exogenously supplemented BRs in promoting endogenous measures of JA (jasmonic acid), SA (salicylic acid), and ethylene, whereby the intricate involvement of BRs with these phytohormones can be confirmed to bring about BR-mediated stress tolerance through signaling pathways (Wu et al. 2017). Exogenous application of BRs showed a synergistic effect in incrementing the endogenous measures of different hormones (Yuan et al. 2010). Anwar et al. (2018) documented the regulatory role of exogenous BRs in accumulating IAA, ABA, BR, and a form of cytokinin i.e. ZR (Zeatin riboside), while decreasing JA, GA4, and another naturally occurring cytokinin i.e. iPA (isopentenyl adenosine), when exposed to chilling stress in cucumber. Different pathways involving BRs in signal transduction can be concluded to play a principal role in various metabolic, transcriptional, and signal transducing activities, which bring about notable resistance against a varied array of stresses and keep plants protected from injury. Individual involvement of BRs with these phytohormones has been elaborated as follows:

4.4.3.1 BRs and ABA

The well-known stress hormone, ABA, has been linked with inhibition of seed germination and promotion of seed dormancy whereas BRs are responsible for counteracting the action of ABA in stimulating the germination of seeds (Steber and McCourt 2001). Both of them have antagonistic roles in germination and post-germination development of seeds (Hu and Yu, 2014). While ABA hinders these processes, BRs boost them. Lately, evaluation of mutants hinted at a synergistic association of these two hormones in inciting certain reactions involving production of H_2O_2, quality articulation of RBOH1 (respiratory burst oxidase homolog 1), action of NADPH oxidase along with arbitration to impart tolerance against thermal and salt stress (Zhou et al. 2014).

As per the earlier findings, ABA has a role of signal molecule in both normal and stressed plants but its intermutual talk with BRs controls the genetic expression of different genes to confer resistance against abiotic stresses which suggests its role in not only the developmental process of plants including seed germination [GPA (G protein α-subunit1); biosynthesis of BRs (DWF1 (DWARF1) and CPD (CONSTITUTIVE PHOTOMORPHOGENESIS AND DWARFISM) but also in the upregulation of antioxidant defense system (Werner et al. 2010; Nishiyama et al. 2011; Zhu et al. 2016). Additionally, mutants of BR perception i.e. bril-1, and ABA insensitivity i.e. abil and abi2 substantiate the underlying interaction between these two hormones. More detailed studies have also revealed the key role of ABA biosynthesis in reinforcing stress tolerance signaled by BR-inducted pathways.

4.4.3.2 BRs and Auxins

Different stages of plant development and growth have been reported to be controlled by the intermutual talk of BRs with auxin (Hao et al. 2013; Saini et al. 2013; Chaiwanon and Wang 2015). Impact of this crosstalk is evident in various physiological processes like elongation of hypocotyl and development of roots, but the role of their interaction in managing stress-induced responses still remains undefined (Kissoudis et al. 2014). BRX (BRAVIS RADIX) plays a mediatory role in maintaining the required level of BRs in *Arabidopsis* and is stimulated by auxin, confirming the role of BRX at the junction of an internal orbit in *Arabidopsis* (Mouchel et al. 2006). Interestingly, BR biosynthetic genes like CPD and DWF4 are also regulated by BRX, hinting at an underlying linkage between biosynthesis of BRs and signaling of auxin (Tanaka et al. 2005). Additionally, the interaction between BRs and auxin for regulating developmental phases of plant growth is also evident from the collaboration mechanism existing between BIN2 and a member of the transcription controlling group- ARF2 (auxin response factors) (Vert et al. 2008). Although the relationship of BRs with auxins is quite efficient in regulating formative stages of plant development, nonetheless, the need for in-depth evaluations is still there to fathom the details of their crosstalk, which is linked to the provision of stress tolerance.

4.4.3.3 BRs and Jasmonic Acid

To boost tolerance against abiotic stresses, the synergistic relationship between BRs and JA plays a major role in regulating plant growth. Levels of JA have been found to be promoted by BRs in rice plants in response to stress, which undeniably indicates the crosstalk existing between them (Kitanaga et al. 2006). Surprisingly, curbing of JA-mediated stocking up of anthocyanins by BRs (brassinazole) in *Arabidopsis* has been further handled by motioning of BRs on the JA route (Peng et al. 2011). The expression levels of biosynthetic and signaling genes of JA exhibited downregulation if they received low BR focus, but the same were upregulated upon receiving high focus of BR.

These outcomes were further substantiated via external foliar supplementation of JA, which mediated the downregulation of genes of both BR synthesis and signaling in rice, such as OsBRI1 and OsDWF4, confirming their counteractive nature (Nahar et al. 2013). Moreover, it has also been observed that biosynthesis of BRs is regulated by improvement in JA precursor i.e. 12-oxo-phytodienoic destructive, which consequently links initiation of BR and JA pathways (Nahar et al. 2013).

4.4.3.4 BRs and Ethylene

Ethylene, being present in all the plant parts, regulates different phases of plant growth and development including the provision of stress tolerance. BRs have been reported to play a principal role in the biosynthetic mechanism of ethylene and its accumulation, whereby S-adenosylmethionine (SAM) (formed from methionine/ethylene precursor in the presence of enzyme methionine adenosyltransferases) is transformed first into 1-aminocyclopropane-1-carboxylic acid (ACC) by ACC synthase and then into ethylene by getting enrichment from BR signaling pathway as reported in tomato (Choudhary et al. 2012; Zhu et al. 2016). As per another research finding, BRs remarkably upregulated the extent of transcription of genes involved in biosynthetic signaling of ethylene, namely ACS1, ACS2, ACS3 (1-aminocyclopropane-1-carboxylate synthesis1, 2 and 3), ACC (1-aminocyclopropane-1-carboxylic acid), ACO1, ACO2 (1-aminocyclopropane-1-carboxylate oxidase 1 and 2), along with AOX (alternative oxidase) in *Cucumis sativus* (cucumber) under stressful conditions (Wei et al. 2015). These studies hint at underlying crosstalk between ethylene and BRs in plants to make them stress tolerant.

4.4.3.5 BRs and Cytokinin

Cytokinin plays a key role in promoting cell division of different plant parts (Werner et al. 2001, 2010). Exogenous treatment of BRs in wheat seedlings exhibited a boost in measures of cytokinin via downregulation of genes encoding for the enzymatic action of cytokinin oxidase, confirming the regulatory role of BRs in relation to cytokinin (Yuldashev et al. 2012). In case of mutant PYK10:CKX3, overexpression of gene BRI1 promoted growth of roots and leaves in comparison to wild type, which is similar to overexpression of CKX3 (CYTOKININ DEHYDROGENASE/OXIDASE 3) (CKX3) in *Nicotiana tabacum* (tobacco), suggesting active involvement of both cytokinin and BRs intermutual tasks to regulate different aspects of developmental processes in plants (Nishiyama et al. 2011).

4.4.3.6 BRs and Salicylic Acid

SA has an important role in alleviating salt stress along with other types of abiotic stresses (Ahmad et al. 2017). Interactive talks between BRs and SA have been documented to ameliorate the stresses of both biotic and abiotic nature. For instance, their crosstalk rendered plants tolerant to salinity stress when facilitated by other phytohormones (Divi et al. 2010). The gene NPR1, which is upregulated by SA, is essentially linked with 24-EBL-incited enhancement in tolerance against heat and salt-induced stress in *Arabidopsis* plants (Divi et al. 2010).

4.4.3.7 BRs and Nitric Oxide

Nitric oxide (NO) executes the indispensable role in various physiological strategies involving plant development, improvement, and their tolerance against abiotic stresses (Zheng et al. 2009; Zafra et al. 2010). While dealing with water stress and higher levels of ABA, this hormone performs signal transduction and makes plants more resilient to water scarcity (Neill et al. 2003; Sun et al. 2007; Farooq et al. 2009). In this context, the effect of NO on BR-prompted ABA accumulation to give rise to BR-incited stress tolerance against oxidative damage has been noticed. Both the exogenous and endogenous measures of BRs prolonged the stay of NO in mesophyll cells of *Zea mays* leaves, which lowered the BR-promoted ABA aggregation, hinting at the reliance of ABA amassing on BR-mediated NO production. These results clearly indicate the role played by the instigation of BRs in generating NO to intercede with the biosynthetic process of ABA, which in turn upgrades the resilience of plants to water stress-induced oxidative damage (Ahmad et al. 2018).

4.5 Conclusion

A dramatic improvement in comprehending the significance of metabolic modulations to combat adverse ecological conditions has been observed in recent years. However, stress management in plants is quite a complex mechanism that is centered around multiple levels of networking among different metabolites. Major secondary metabolites involved in conferring stress tolerance include different phenolics, osmolytes, and plant growth regulators. A complex interactive network has been reported to exist between BRs and secondary metabolites whereby the former regulates various developmental activities, including germination of seeds, development of roots, elongation of stems, closure of stomata, together with stress management.

Multiple BR signaling pathways modulate the measures of secondary metabolites and engage them in intricate crosstalks to improve plants' responses to different abiotic stresses. This stress regulatory role of BRs enables the secondary metabolites to alleviate oxidative stress via scavenging ROS, improve the photosynthetic and antioxidant efficiency, stabilize the plasma membrane, and upregulate gene expression. In spite of scientific advancements in unraveling the mystery behind BRs and secondary metabolome, a number of queries still need to be addressed with regard to the developmental phases of plants and the metabolic status of cells during these phases, which possess a direct connection with stress-induced responses during challenging environmental conditions. Solving them through discerning metabolic profiling, omics methods, and hormonal crosstalking would help in furthering the knowledge regarding the interaction of BRs with metabolites of defense signaling, which could possibly lead to the engineering of stress-resistant plants.

References

Abdullahi, B.A., Gu, X.G., Gan, Q.L. et al. (2002). Brassinolide amelioration of aluminum toxicity in mungbean seedling growth. *Journal of Plant Nutrition* 26 (9): 1725–1734.

Adams, W.W., Muller, O., Cohu, C.M. et al. (2013). May photoinhibition be a consequence, rather than a cause, of limited plant productivity? *Photosynthesis Research* 117: 31–44.

Agati, G., Stefano, G., Biricolti, S. et al. (2009). Mesophyll distribution of 'antioxidant' flavonoid glycosides in *Ligustrum vulgare* leaves under contrasting sunlight irradiance. *Annals of Botany* 104 (5): 853–861.

Ahammed, G.J., Li, X., Xia, X.J. et al. (2015). Enhanced photosynthetic capacity and antioxidant potential mediate brassinosteriod-induced phenanthrene stress tolerance in tomato. *Environmental Pollution* 201: 58–66.

Ahammed, G.J., Zhang, S., Shi, K. et al. (2012). Brassinosteroid improves seed germination and early development of tomato seedling under phenanthrene stress. *Plant Growth Regulation* 68 (1): 87–96.

Ahammed, G.J., Zhou, Y.H., Xia, X.J. et al. (2013). Brassinosteroid regulates secondary metabolism in tomato towards enhanced tolerance to phenanthrene. *Biologia Plantarum* 57 (1): 154–158.

Ahmad, P., Abd Allah, E.F., Alyemeni, M.N. et al. (2018). Exogenous application of calcium to 24-epibrassinosteroid pre-treated tomato seedlings mitigates NaCl toxicity by modifying ascorbate–glutathione cycle and secondary metabolites. *Scientific Reports* 8: 13515.

Ahmad, P., Jaleel, C.A., Salem, M.A. et al. (2010). Roles of enzymatic and nonenzymatic antioxidants in plants during abiotic stress. *Critical Reviews in Biotechnology* 30 (3): 161–175.

Ahmad, R., Manzoor, M., Muhammad, H.M.D. et al. (2023). Exogenous melatonin spray enhances salinity tolerance in *Zizyphus germplasm*: a brief theory. *Life* 13 (2): 493.

Ahmad, F., Singh, A., and Kamal, A. (2017). Ameliorative effect of salicylic acid in salinity stressed *Pisum sativum* by improving growth parameters, activating photosynthesis and enhancing antioxidant defense system. *Bioscience Biotechnology Research Communications* 10: 481–489.

Ahmad, F., Singh, A., and Kamal, A. (2018). Crosstalk of brassinosteroids with other phytohormones under various abiotic stresses. *Journal of Applied Biology and Biotechnology* 6 (1): 56–62.

Ahmad, F., Singh, A., and Kamal, A. (2019). Salicylic acid–mediated defense mechanisms to abiotic stress tolerance. In: *Plant Signaling Molecules* (ed. M. Iqbal R. Khan, P.S. Reddy, A. Ferrante et al.), 355–369. Woodhead Publishing.

Ahmed, E., Arshad, M., Khan, M.Z. et al. (2017). Secondary metabolites and their multidimensional prospective in plant life. *Journal of Pharmacognosy and Phytochemistry* 6 (2): 205–214.

Ali, M., Afzal, S., Parveen, A. et al. (2021). Silicon mediated improvement in the growth and ion homeostasis by decreasing Na^+ uptake in maize (*Zea mays* L.) cultivars exposed to salinity stress. *Plant Physiology and Biochemistry* 158: 208–218.

Aliche, E.B., Screpanti, C., De Mesmaeker, A. et al. (2020). Science and application of strigolactones. *New Phytologist* 227 (4): 1001–1011.

Anuradha, S., Rao, R., and Eeta, S. (2011). Amelioration of lead toxicity in radish (*Raphanus sativus* L.) plants by brassinolide. *Journal of Applied Biological Sciences* 5 (3).

Anusha, P., Taheniyat, S., and Sujatha, E. (2015). Germination, growth and oxidative responses of maize seedlings to salt and/or cadmium stress: mechanism of toxicity amelioration by 28 homobrassinolide. *International Journal of Current Microbiology and Applied Sciences* 4 (9): 442–453.

Anwar, A., Bai, L., Miao, L. et al. (2018). 24-Epibrassinolide ameliorates endogenous hormone levels to enhance low-temperature stress tolerance in cucumber seedlings. *International Journal of Molecular Sciences* 19 (9): 2497.

Apel, K. and Hirt, H. (2004). Reactive oxygen species: metabolism, oxidative stress, and signal transduction. *Annual Review of Plant Biology* 55 (1): 373–399.

Atkinson, N.J. and Urwin, P.E. (2012). The interaction of plant biotic and abiotic stresses: from genes to the field. *Journal of Experimental Botany* 63 (10): 3523–3543.

Baghel, M., Nagaraja, A., Srivastav, M. et al. (2019). Pleiotropic influences of brassinosteroids on fruit crops: a review. *Plant Growth Regulation* 87: 375–388.

Bajguz, A. and Hayat, S. (2009). Effects of brassinosteroids on the plant responses to environmental stresses. *Plant Physiology and Biochemistry* 47 (1): 1–8.

Begum, N., Qin, C., Ashraf, M. et al. (2019). Role of arbuscular mycorrhizal fungi in plant growth regulation: implications in abiotic stress tolerance. *Frontiers in Plant Science* 10: 466052.

Belkhadir, Y. and Jaillais, Y. (2015). The molecular circuitry of brassinosteroid signaling. *New Phytologist* 206 (2): 522–540.

Bhoi, A., Yadu, B., Chandra, J. et al. (2021). Contribution of strigolactone in plant physiology, hormonal interaction and abiotic stresses. *Planta* 254 (2): 28.

Bhuyan, M.B., Hasanuzzaman, M., Parvin, K. et al. (2020). Nitric oxide and hydrogen sulfide: two intimate collaborators regulating plant defense against abiotic stress. *Plant Growth Regulation* 90: 409–424.

Bitrián, M., Zarza, X., Altabella, T. et al. (2012). Polyamines under abiotic stress: metabolic crossroads and hormonal crosstalks in plants. *Metabolites* 2 (3): 516–528.

Bolda, V.V., Botau, D., Szôllôsi, R. et al. (2011). Studies on elemental composition and antioxidant capacity in callus cultures and native plants of *Vaccinium myrtillus* L. local populations. *Acta Biologica Szegediensis* 55 (2): 255–259.

Brewer, P.B. (2022). Emerging roles of Strigolactones in plant responses toward biotic stress. In: *Emerging Plant Growth Regulators in Agriculture* (ed. T. Aftab and M. Naeem), 205–214. Academic Press.

Casanova-Sáez, R., Mateo-Bonmatí, E., and Ljung, K. (2021). Auxin metabolism in plants. *Cold Spring Harbor Perspectives in Biology* 13 (3): a039867.

Chaiwanon, J. and Wang, Z.Y. (2015). Spatiotemporal brassinosteroid signaling and antagonism with auxin pattern stem cell dynamics in *Arabidopsis* roots. *Current Biology* 25 (8): 1031–1042.

Chong, J., Poutaraud, A., and Hugueney, P. (2009). Metabolism and roles of stilbenes in plants. *Plant Science* 177 (3): 143–155.

Choudhary, S.P., Bhardwaj, R., Gupta, B.D. et al. (2010). Epibrassinolide induces changes in indole-3-acetic acid, abscisic acid and polyamine concentrations and enhances antioxidant potential of radish seedlings under copper stress. *Physiologia Plantarum* 140 (3): 280–296.

Choudhary, S.P., Kanwar, M., Bhardwaj, R. et al. (2011). Epibrassinolide ameliorates Cr (VI) stress via influencing the levels of indole-3-acetic acid, abscisic acid, polyamines and antioxidant system of radish seedlings. *Chemosphere* 84: 592–600.

Choudhary, S.P., Kanwar, M., Bhardwaj, R. et al. (2012). Chromium stress mitigation by polyamine-brassinosteroid application involves phytohormonal and physiological strategies in *Raphanus sativus* L. *PLoS One* 7: e33210.

Croteau, R., Kutchan, T.M., and Lewis, N.G. (2000). Natural products (secondary metabolites). *Biochemistry and Molecular Biology of Plants* 24: 1250–1319.

de Campos, M.K.F., de Carvalho, K., de Souza, F.S. et al. (2011). Drought tolerance and antioxidant enzymatic activity in transgenic 'Swingle' citrumelo plants over-accumulating proline. *Environmental and Experimental Botany* 72 (2): 242–250.

Dehghanian, Z., Habibi, K., Dehghanian, M. et al. (2022). Reinforcing the bulwark: unravelling the efficient applications of plant phenolics and tannins against environmental stresses. *Heliyon* 8 (3): e09094.

Divi, U.K. and Krishna, P. (2010). Brassinosteroids confer stress tolerance. In: *Plant Stress Biology: From Genomics to Systems Biology* (ed. H. Hirt), 119–135. Weinheim: Wiley-VCH.

Divi, U.K., Rahman, T., and Krishna, P. (2010). Brassinosteroid-mediated stress tolerance in *Arabidopsis* shows interactions with abscisic acid, ethylene and salicylic acid pathways. *BMC Plant Biology* 10: 1–14.

Divi, U.K., Rahman, T., and Krishna, P. (2016). Gene expression and functional analyses in brassinosteroid-mediated stress tolerance. *Plant Biotechnology Journal* 14 (1): 419–432.

Dugardeyn, J. and Van Der Straeten, D. (2008). Ethylene: fine-tuning plant growth and development by stimulation and inhibition of elongation. *Plant Science* 175: 59–70.

El-Beltagi, H.S., Sofy, M.R., Aldaej, M.I. et al. (2020). Silicon alleviates copper toxicity in flax plants by up-regulating antioxidant defense and secondary metabolites and decreasing oxidative damage. *Sustainability* 12 (11): 4732.

Etesami, H. and Jeong, B.R. (2018). Silicon (Si): review and future prospects on the action mechanisms in alleviating biotic and abiotic stresses in plants. *Ecotoxicology and Environmental Safety* 147: 881–896.

Fàbregas, N., Lozano-Elena, F., Blasco-Escámez, D. et al. (2018). Overexpression of the vascular brassinosteroid receptor BRL3 confers drought resistance without penalizing plant growth. *Nature Communications* 9 (1): 4680.

Fahad, S., Rehman, A., Shahzad, B. et al. (2019). Rice responses and tolerance to metal/metalloid toxicity. In: *Advances in Rice Research for Abiotic Stress Tolerance* (ed. M. Hasanuzzaman, M. Fujita, K. Nahar et al), 299–312. Cambridge, UK: Woodhead Publishing.

Fariduddin, Q., Khalil, R.R., Mir, B.A. et al. (2013). 24-Epibrassinolide regulates photosynthesis, antioxidant enzyme activities and proline content of *Cucumis sativus* under salt and/or copper stress. *Environmental Monitoring and Assessment* 185: 7845–7856.

Farooq, M., Wahid, A., and Basra, S.M.A. (2009). Improving water relations and gas exchange with brassinosteroids in rice under drought stress. *Journal of Agronomy and Crop Science* 195 (4): 262–269.

Fenn, M.A., and Giovannoni, J.J. (2021). Phytohormones in fruit development and maturation. *The Plant Journal* 105 (2): 446–458.

Fernando, V.D., and Schroeder, D.F. (2016). Role of ABA in *Arabidopsis* salt, drought, and desiccation tolerance. In: *Abiotic and Biotic Stress in Plants-Recent Advances and Future Perspectives* (ed. A.K. Shanker and C. Shanker). IntechOpen.

Fraser, C.M. and Chapple, C. (2011). The phenylpropanoid pathway in Arabidopsis. *The Arabidopsis Book/American Society of Plant Biologists* 9: e0152.

Gao, H., Zhang, Z., Lv, X. et al. (2016). Effect of 24-epibrassinolide on chilling injury of peach fruit in relation to phenolic and proline metabolisms. *Postharvest Biology and Technology* 111: 390–397.

Gill, S.S. and Tuteja, N. (2010). Reactive oxygen species and antioxidant machinery in abiotic stress tolerance in crop plants. *Plant Physiology and Biochemistry* 48 (12): 909–930.

Górnik, K. and Lahuta, L.B. (2017). Application of phytohormones during seed hydropriming and heat shock treatment on sunflower (*Helianthus annuus* L.) chilling resistance and changes in soluble carbohydrates. *Acta Physiologiae Plantarum* 39: 1–12.

Guimaraes, E. and Jueneman, E. (2008). *The Global Partnership Initiative for Plant Breeding Capacity Building (GIPB)* (No. IAEA-CN--167).

Guo, H., Chen, H., Hong, C. et al. (2017). Exogenous malic acid alleviates cadmium toxicity in *Miscanthus sacchariflorus* through enhancing photosynthetic capacity and restraining ROS accumulation. *Ecotoxicology and Environmental Safety* 141: 119–128.

Guo, Q., Li, X., Niu, L. et al. (2021). Transcription-associated metabolomic adjustments in maize occur during combined drought and cold stress. *Plant Physiology* 186: 677–695.

Gupta, A., Hisano, H., Hojo, Y. et al. (2017). Global profiling of phytohormone dynamics during combined drought and pathogen stress in *Arabidopsis thaliana* reveals ABA and JA as major regulators. *Scientific Reports* 7 (1): 4017.

Hao, J., Yin, Y., and Fei, S.Z. (2013). Brassinosteroid signaling network: implications on yield and stress tolerance. *Plant Cell Reports* 32: 1017–1030.

Hasan, S.A., Hayat, S., Ali, B. and Ahmad, A. (2008). 28-homobrassinolide protects chickpea (*Cicer arietinum*) from cadmium toxicity by stimulating antioxidants. *Environmental Pollution* 151 (1): 60–66.

Hattori, T., Mitsuya, S., Fujiwara, T. et al. (2009). Tissue specificity of glycinebetaine synthesis in barley. *Plant Science* 176: 112–118.

He, M., He, C.-Q., and Ding, N.-Z. (2018). Abiotic stresses: general defenses of land plants and chances for engineering multistress tolerance. *Frontiers in Plant Science* 9: 1771.

Hegazi, A.M., El-Shraiy, A.M., and Ghoname, A.A. (2017). Mitigation of salt stress negative effects on sweet pepper using arbuscular mycorrhizal fungi (AMF), *Bacillus megaterium* and brassinosteroids (BRs). *Gesunde Pflanzen* 69 (2): 91–102.

Hoque, M.A., Banu, M.N.A., Nakamura, Y. et al. (2008). Proline and glycinebetaine enhance antioxidant defense and methylglyoxal detoxification systems and reduce NaCl-induced damage in cultured tobacco cells. *Journal of Plant Physiology* 165: 813–824.

Hoque, M.A., Banu, M.N.A., Okuma, E. et al. (2007). Exogenous proline and glycinebetaine increase NaCl-induced ascorbate–glutathione cycle enzyme activities, and proline improves salt tolerance more than glycinebetaine in tobacco Bright Yellow-2 suspension-cultured cells. *Journal of Plant Physiology* 164 (11): 1457–1468.

Hossain, Z., McGarvey, B., Amyot, L. et al. (2012). DIMINUTO 1 affects the lignin profile and secondary cell wall formation in Arabidopsis. *Planta* 235 (3): 485–498.

Hossain, A., Pamanick, B., Venugopalan, V.K. et al. (2022). Emerging roles of plant growth regulators for plants adaptation to abiotic stress–induced oxidative stress. In: *Emerging Plant Growth Regulators in Agriculture* (ed. T. Aftab and M. Naeem), 1–72. Academic Press.

Hu, Y. and Yu, D. (2014). BRASSINOSTEROID INSENSITIVE2 interacts with ABSCISIC ACID INSENSITIVE5 to mediate the antagonism of brassinosteroids to abscisic acid during seed germination in *Arabidopsis*. *The Plant Cell* 26 (11): 4394–4408.

Iqbal, M.S. and Ansari, M.I. (2020). Microbial bioinoculants for salt stress tolerance in plants. In: *Microbial Mitigation of Stress Response of Food Legumes* (ed. N. Amaresan, S. Murugesan, K. Kumar et al.). 155–163. Boca Raton, FL: CRC Press.

Jan, S., Alyemeni, M.N., Wijaya, L. et al. (2018). Interactive effect of 24-epibrassinolide and silicon alleviates cadmium stress via the modulation of antioxidant defense and glyoxalase systems and macronutrient content in *Pisum sativum* L. seedlings. *BMC Plant Biology* 18: 146.

Jan, R., Khan, M.A., Asaf, S. et al. (2019). Metal resistant endophytic bacteria reduces cadmium, nickel toxicity, and enhances expression of metal stress related genes with improved growth of *Oryza sativa*, via regulating its antioxidant machinery and endogenous hormones. *Plants* 8: 363.

Jan, R., Kim, N., Khan, M.A. et al. (2021). Enhanced flavonoid accumulation reduces combined salt and heat stress through regulation of transcriptional and hormonal mechanisms. *Frontiers in Plant Science* 12: 796956.

Janeczko, A., Koscielniak, J., Pilipowicz, M. et al. (2005). Protection of winter rape photosystem 2 by 24- epibrassinolide under cadmium stress. *Photosynthetica* 43: 293–298.

Jha, C.K., Sharma, P., Shukla, A. et al. (2021). Microbial enzyme, 1-aminocyclopropane-1-carboxylic acid (ACC) deaminase: an elixir for plant under stress. *Physiological and Molecular Plant Pathology* 115: 101664.

Kermani, S.G., Saeidi, G., Sabzalian, M.R. et al. (2019). Drought stress influenced sesamin and sesamolin content and polyphenolic components in sesame (*Sesamum indicum* L.) populations with contrasting seed coat colors. *Food Chemistry* 289: 360–368.

Khamsuk, O., Sonjaroon, W., Suwanwong, S. et al. (2018). Effects of 24-epibrassinolide and the synthetic brassinosteroid mimic on chili pepper under drought. *Acta Physiologiae Plantarum* 40: 1–12.

Kido, E.A., Ferreira Neto, J.R., Silva, R.L. et al. (2013). Expression dynamics and genome distribution of osmoprotectants in soybean: identifying important components to face abiotic stress. *BMC Bioinformatics* 14: 1–11.

Kishor, P.K., Sangam, S., Amrutha, R.N. et al. (2005). Regulation of proline biosynthesis, degradation, uptake and transport in higher plants: its implications in plant growth and abiotic stress tolerance. *Current Science* 88: 424–438.

Kissoudis, C., van de Wiel, C., Visser, R.G. et al. (2014). Enhancing crop resilience to combined abiotic and biotic stress through the dissection of physiological and molecular crosstalk. *Frontiers in Plant Science* 5: 90871.

Kitanaga, Y., Jian, C., Hasegawa, M. et al. (2006). Sequential regulation of gibberellin, brassinosteroid, and jasmonic acid biosynthesis occurs in rice coleoptiles to control the transcript levels of anti-microbial thionin genes. *Bioscience, Biotechnology, and Biochemistry* 70 (10): 2410–2419.

Kleman, J. and Matusova, R. (2023). Strigolactones: current research progress in the response of plants to abiotic stress. *Biologia* 78 (2): 307–318.

Kohli, S.K., Handa, N., Bali, S. et al. (2018). Modulation of antioxidative defense expression and osmolyte content by co-application of 24-epibrassinolide and salicylic acid in Pb exposed Indian mustard plants. *Ecotoxicology and Environmental Safety* 147: 382–393.

Kou, X., Zhao, X., Wu, B. et al. (2022). Auxin response factors are ubiquitous in plant growth and development, and involved in crosstalk between plant hormones: a review. *Applied Sciences* 12 (3): 1360.

Kovinich, N., Kayanja, G., Chanoca, A. et al. (2014). Not all anthocyanins are born equal: distinct patterns induced by stress in *Arabidopsis*. *Planta* 240: 931–940.

Kumar, K., Kumar, M., Kim, S.R. et al. (2013). Insights into genomics of salt stress response in rice. *Rice* 6: 1–15.

Kumudini, B.S., Jayamohan, N.S., Patil, S.V. et al. (2018). Primary plant metabolism during plant–pathogen interactions and its role in defense. In: *Plant Metabolites and Regulation Under Environmental Stress* (ed. P. Ahmad, M.A. Ahanger, V.P. Singh et al.), 215–229. New York, NY: Academic Press.

Kurepin, L.V., Ivanov, A.G., Zaman, M. et al. (2015). Stress-related hormones and glycinebetaine interplay in protection of photosynthesis under abiotic stress conditions. *Photosynthesis Research* 126: 221–235.

Kurepin, L.V., Ivanov, A.G., Zaman, M. et al. (2017). Interaction of glycine betaine and plant hormones: protection of the photosynthetic apparatus during abiotic stress. In: *Photosynthesis: Structures, Mechanisms, and Applications* (ed. H.J.M. Hou, M.M. Najafpour, G.F. Moore et al.), 185–202. Berlin, Germany: Springer.

Kurepin, L.V., Ozga, J.A., Zaman, M. et al. (2013). The physiology of plant hormones in cereal, oilseed and pulse crops. *Prairie Soils Crops* 6: 7–23.

Kuromori, T., Fujita, M., Takahashi, F. et al. (2022). Inter-tissue and inter-organ signaling in drought stress response and phenotyping of drought tolerance. *The Plant Journal* 109 (2): 342–358.

Kusano, T., Yamaguchi, K., Berberich, T. et al. (2007). Advances in polyamine research in 2007. *Journal of Plant Research* 120: 345–350.

Kytridis, V.P. and Manetas, Y. (2006). Mesophyll versus epidermal anthocyanins as potential in vivo antioxidants: evidence linking the putative antioxidant role to the proximity of oxy-radical source. *Journal of Experimental Botany* 57 (10): 2203–2210.

Lalotra, S., Hemantaranjan, A., Kumar, S. et al. (2017). Effect of brassinosteroid (Brassinolide) on seedling traits, morphology and metabolism in mungbean under salinity stress. *Annual Research & Review in Biology* 12 (4): 1–8.

Laoué, J., Fernandez, C., and Ormeño, E. (2022). Plant flavonoids in mediterranean species: a focus on flavonols as protective metabolites under climate stress. *Plants* 11 (2): 172.

Larré, C.F., Marini, P., Moraes, C.L. et al. (2014). Influência do 24-epibrassinolídeo na tolerância ao estresse salino em plântulas de arroz. *Semina: Ciências Agrárias* 35 (1): 67–76.

Lefevere, H., Bauters, L., and Gheysen, G. (2020). Salicylic acid biosynthesis in plants. *Frontiers in Plant Science* 11: 521987.

Li, D., Mou, W., Van de Poel, B. et al. (2022). Something old, something new: conservation of the ethylene precursor 1-amino-cyclopropane-1-carboxylic acid as a signaling molecule. *Current Opinion in Plant Biology* 65: 102116.

Li, S.M., Zheng, H.X., Zhang, X.S. et al. (2021). Cytokinins as central regulators during plant growth and stress response. *Plant Cell Reports* 40: 271–282.

Liang, X., Zhang, L., Natarajan, S.K. et al. (2013). Proline mechanisms of stress survival. *Antioxidants & Redox Signaling* 19 (9): 998–1011.

Liu, J.-H., Wang, W., Wu, H. et al. (2015). Polyamines function in stress tolerance: from synthesis to regulation. *Frontiers in Plant Science* 6: 827.

Liu, J., Zhang, D., Sun, X. et al. (2017). Structure-activity relationship of brassinosteroids and their agricultural practical usages. *Steroids* 124: 1–17.

Ljubej, V., Radojčić Redovniković, I., Salopek-Sondi, B. et al. (2021). Chilling and freezing temperature stress differently influence glucosinolates content in *Brassica oleracea* var. acephala. *Plants* 10 (7): 1305.

Luan, L.Y., Zhang, Z.W., Xi, Z.M. et al. (2013). Brassinosteroids regulate anthocyanin biosynthesis in the ripening of grape berries. *South African Journal of Enology and Viticulture* 34 (2): 196–203.

Lukatkin, A.S., Kashtanova, N.N., and Duchovskis, P. (2013). Changes in maize seedlings growth and membrane permeability under the effect of epibrassinolide and heavy metals. *Russian Agricultural Sciences* 39 (4): 307–310.

Ma, L., Pegram, L., Record Jr, M.T. et al. (2010). Preferential interactions between small solutes and the protein backbone: a computational analysis. *Biochemistry* 49 (9): 1954–1962.

Madhan, M., Mahesh, K., and Rao, S.S. (2014). Effect of 24-epibrassinolide on aluminium stress induced inhibition of seed germination and seedling growth of *Cajanus cajan* (L.) Millsp. *International Journal of Multidisciplinary and Current Research* 2: 286–290.

Mahesh, K., Balaraju, P., Ramakrishna, B. et al. (2013). Effect of brassinosteroids on germination and seedling growth of radish (*Raphanus sativus* L.) under PEG-6000 induced water stress. *American Journal of Plant Sciences* 04 (12): 2305–2313.

Marakli, S. and Gozukirmizi, N. (2016). Abiotic stress alleviation with brassinosteroids in plant roots. In: *Abiotic and Biotic Stress in Plants-Recent Advances and Future Perspectives* (ed. A.K. Shanker and C. Shanker), 373–394. Rijeka: IntechOpen.

Mathur, P., Tripathi, D.K., Baluska, F. et al. (2022). Auxin-mediated molecular mechanisms of heavy metal and metalloid stress regulation in plants. *Environmental and Experimental Botany*196: 104796. doi: 10.1016/j.envexpbot.2022.104796.

Mauseth, J.D. (2014). *Botany: An Introduction to Plant Biology*. Burlington, MA: Jones & Bartlett Publishers.

Medina, E., Kim, S.H., Yun, M. et al. (2021). Recapitulation of the function and role of ROS generated in response to heat stress in plants. *Plants* 10 (2): 371.

Michalak, A. (2006). Phenolic compounds and their antioxidant activity in plants growing under heavy metal stress. *Polish Journal of Environmental Studies*, 15 (4): 523–530.

Mishra, B.S., Sharma, M., and Laxmi, A. (2022). Role of sugar and auxin crosstalk in plant growth and development. *Physiologia Plantarum*, 174 (1): e13546.

Mouchel, C.F., Osmont, K.S., and Hardtke, C.S. (2006). BRX mediates feedback between brassinosteroid levels and auxin signalling in root growth. *Nature* 443 (7110): 458–461.

Mujtaba, M., Sharif, R., Ali, Q. et al. (2021). Biopolymer based nanofertilizers applications in abiotic stress (drought and salinity) control. In: *Advances in Nano-Fertilizers and Nano-Pesticides in Agriculture* (ed. S. Jogaiah, H.B. Singh, L.F. Fraceto et al.), 85–110. Woodhead Publishing.

Mullineaux, P. and Karpinski, S. (2002). Signal transduction in response to excess light: getting out of the chloroplast. *Current Opinion in Plant Biology* 5 (1): 43–48.

Murakeozy, E.P., Nagy, Z., Duhaze, C. et al. (2003). Seasonal changes in the levels of compatible osmolytes in three halophytic species of inland saline vegetation in Hungary. *Journal of Plant Physiology* 160: 395–401.

Nagajyoti, P.C., Lee, K.D., and Sreekanth, T.V.M. (2010). Heavy metals, occurrence and toxicity for plants: a review. *Environmental Chemistry Letters* 8: 199–216.

Nahar, K., Kyndt, T., Hause, B. et al. (2013). Brassinosteroids suppress rice defense against root-knot nematodes through antagonism with the jasmonate pathway. *Molecular Plant-Microbe Interactions* 26 (1): 106–115.

Naing, A.H. and Jeong, H.Y. (2021). Overexpression of 1-Aminocyclopropane-1-carboxylic acid deaminase (acdS) gene in *Petunia hybrida* improves tolerance to abiotic stresses. *Frontiers in Plant Science* 12: 737490.

Nakabayashi, R., Yonekura-Sakakibara, K., Urano, K. et al. (2014). Enhancement of oxidative and drought tolerance in *Arabidopsis* by overaccumulation of antioxidant flavonoids. *The Plant Journal* 77 (3): 367–379.

Nakamura, A., Tochio, N., Fujioka, S. et al. (2017). Molecular actions of two synthetic brassinosteroids, iso-carbaBL and 6-deoxoBL, which cause altered physiological activities between *Arabidopsis* and rice. *PLoS One* 12 (4): e0174015.

Narayan, M.S., Thimmaraju, R., and Bhagyalakshmi, N. (2005). Interplay of growth regulators during solid-state and liquid-state batch cultivation of anthocyanin producing cell line of *Daucus carota*. *Process Biochemistry* 40 (1): 351–358.

Negin, B. and Moshelion, M. (2016). The evolution of the role of ABA in the regulation of water-use efficiency: from biochemical mechanisms to stomatal conductance. *Plant Science* 251: 82–89.

Neill, S.J., Desikan, R., and Hancock, J.T. (2003). Nitric oxide signalling in plants. *New Phytologist* 159 (1): 11–35.

Nishiyama, R., Watanabe, Y., Fujita, Y. et al. (2011). Analysis of cytokinin mutants and regulation of cytokinin metabolic genes reveals important regulatory roles of cytokinins in drought, salt and abscisic acid responses, and abscisic acid biosynthesis. *The Plant Cell* 23 (6): 2169–2183.

Nunes da Silva, M., Carvalho, S.M., Rodrigues, A.M. et al. (2022). Defence-related pathways, phytohormones and primary metabolism are key players in kiwifruit plant tolerance to *Pseudomonas syringae* pv. actinidiae. *Plant, Cell & Environment* 45 (2): 528–541.

Obata, T. and Fernie, A.R. (2012). The use of metabolomics to dissect plant responses to abiotic stresses. *Cellular and Molecular Life Sciences* 69: 3225–3243.

Ohtani, M. and Demura, T. (2019). The quest for transcriptional hubs of lignin biosynthesis: beyond the NAC-MYB-gene regulatory network model. *Current opinion in biotechnology* 56: 82–87.

Pál, M., Janda, T., and Szalai, G. (2018). Interactions between plant hormones and thiol-related heavy metal chelators. *Plant Growth Regulation* 85: 173–185.

Park, E.J., Jeknić, Z., Pino, M.T. et al. (2007). Glycinebetaine accumulation is more effective in chloroplasts than in the cytosol for protecting transgenic tomato plants against abiotic stress. *Plant Cell Environ.* 30: 994–1005.

Parvaiz, A. and Satyawati, S. (2008). Salt stress and phyto-biochemical responses of plants-a review. *Plant, Soil and Environment* 54: 89.

Peng, Z., Han, C., Yuan, L. et al. (2011). Brassinosteroid enhances jasmonate-induced anthocyanin accumulation in *Arabidopsis* seedlings. *Journal of Integrative Plant Biology* 53 (8): 632–640.

Piacentini, D., Della Rovere, F., D'Angeli, S. et al. (2022). Convergence between development and stress: Ectopic xylem formation in Arabidopsis hypocotyl in response to 24-epibrassinolide and cadmium. *Plants* 11 (23): 3278.

Pons, S. (2020). *Les hormones dans la symbiose mycorhizienne: étude de la production et des effets d'hormones végétales par les champignons endomycorhiziens* (Doctoral dissertation, Université Paul Sabatier-Toulouse III).

Pott, D.M., Osorio, S., and Vallarino, J.G. (2019). From central to specialized metabolism: an overview of some secondary compounds derived from the primary metabolism for their role in conferring nutritional and organoleptic characteristics to fruit. *Frontiers in Plant Science* 10: 454686.

Qamar, A., Mysore, K., and Senthil-Kumar, M. (2015). Role of proline and pyrroline-5-carboxylate metabolism in plant defense against invading pathogens. *Frontiers in Plant Science* 6: 144732.

Qayyum, B., Shahbaz, M., and Akram, N.A. (2007). Interactive effect of foliar application of 24-epibrassinolide and root zone salinity on morpho-physiological attributes of wheat (*Triticum aestivum* L.). *International Journal of Agriculture and Biology* 9: 584–589.

Quamruzzaman, M., Manik, S.N., Shabala, S. et al. (2021). Improving performance of salt-grown crops by exogenous application of plant growth regulators. *Biomolecules* 11 (6): 788.

Raes, J., Rohde, A., Christensen, J.H. et al. (2003). Genome-wide characterization of the lignification toolbox in Arabidopsis. *Plant Physiology* 133 (3): 1051–1071.

Raghu, K., Mahesh, K., Divya Sri, N. et al. (2014). Effect of brassinosteroids on the seed germination and seedling growth of radish (*Raphanus sativus* L.) under arsenic toxicity stress. *International Journal of Development Research* 4: 1929–1933.

Rao, S.Q., Chen, X.Q., Wang, K.H. et al. (2021). Effect of short-term high temperature on the accumulation of glucosinolates in *Brassica rapa*. *Plant Physiology and Biochemistry* 161: 222–233.

Rao, S.S.R. and Raghu, K. (2016). Amelioration of lead toxicity by 24-epibrassinolide in rose scented geranium [Pelargonium graveolens (L.) Herit], a potential agent for phytoremediation. *Journal of Medicinal Plants* 4 (6): 06–08.

Rastegari, A.A., Yadav, A.N., and Yadav, N. (2019). Genetic manipulation of secondary metabolites producers. In: *New and Future Developments in Microbial Biotechnology and Bioengineering* (ed. V.K. Gupta and A. Pandey), 13–29. Elsevier.

Rattan, A., Kapoor, D., Kapoor, N. et al. (2014). Application of brassionsteroids reverses the inhibitory effect of salt stress on growth and photosynthetic activity of Zea mays plants. *International Journal of Theoretical and Applied Sciences* 6 (2): 13.

Raza, A., Mehmood, S.S., Tabassum, J. et al. (2019). Targeting plant hormones to develop abiotic stress resistance in wheat. *Wheat Production in Changing Environments: Responses, Adaptation and Tolerance* (ed. M. Hasanuzzaman, K. Nahar, and M. Hossain), 557–577. Singapore: Springer.

Rechner, O., Neugart, S., Schreiner, M. et al. (2017). Can narrow-bandwidth light from UV-A to green alter secondary plant metabolism and increase Brassica plant defenses against aphids? *PLoS One* 12 (11): e0188522.

Rehman, N.U., Li, X., Zeng, P. et al. (2021). Harmony but not uniformity: role of strigolactone in plants. *Biomolecules* 11 (11): 1616.

Riaz, M., Kamran, M., Rizwan, M. et al. (2022). Foliar application of silica sol alleviates boron toxicity in rice (Oryza sativa) seedlings. *Journal of Hazardous Materials* 423: 127175.

Rohde, A., Morreel, K., Ralph, J. et al. (2004). Molecular phenotyping of the pal1 and pal2 mutants of Arabidopsis thaliana reveals far-reaching consequences on phenylpropanoid, amino acid, and carbohydrate metabolism. *The Plant Cell* 16 (10): 2749–2771.

Romero-Puertas, M.C., Terrón-Camero, L.C., Pelaez-Vico, M.A. et al. (2019). Reactive oxygen and nitrogen species as key indicators of plant responses to Cd stress. *Environmental and Experimental Botany* 161: 107–119.

Ryu, H. and Cho, Y.G. (2015). Plant hormones in salt stress tolerance. *Journal of Plant Biology* 58: 147–155.

Saini, S., Sharma, I., Kaur, N. et al. (2013). Auxin: a master regulator in plant root development. *Plant Cell Reports* 32: 741–757.

Sakihama, Y., Cohen, M.F., Grace, S.C. et al. (2002). Plant phenolic antioxidant and prooxidant activities: phenolics-induced oxidative damage mediated by metals in plants. *Toxicology* 177 (1): 67–80.

Šamec, D., Karalija, E., Šola, I. et al. (2021). The role of polyphenols in abiotic stress response: the influence of molecular structure. *Plants* 10 (1): 118.

Sarkar, A.K. and Sadhukhan, S. (2022). Imperative role of trehalose metabolism and trehalose-6-phosphate signaling on salt stress responses in plants. *Physiologia Plantarum* 174 (1): e13647.

Sengupta, A., Chakraborty, M., Saha, J. et al. (2016). Polyamines: osmoprotectants in plant abiotic stress adaptation. In: *Osmolytes and Plants Acclimation to Changing Environment: Emerging Omics Technologies* (ed. N. Iqbal, R. Nazar, and N.A. Khan), 97–127. Berlin, Germany: Springer.

Shahana, T., Rao, P.A., Ram, S.S. et al. (2015). Mitigation of drought stress by 24-epibarassinolide and 28-homobrassinolide in pigeon pea seedlings. *International Journal of Multidisciplinary and Current Research* 3: 904–911.

Shahid, M.A., Balal, R.M., Pervez, M.A. et al. (2014). Treatment with 24-epibrassinolide mitigates NaCl-induced toxicity by enhancing carbohydrate metabolism, osmolyte accumulation, and antioxidant activity in *Pisum sativum*. *Turkish Journal of Botany* 38: 511–525.

Shahzad, B., Tanveer, M., Che, Z. et al. (2018). Role of 24-epibrassinolide (EBL) in mediating heavy metal and pesticide induced oxidative stress in plants: a review. *Ecotoxicology and Environmental Safety* 147: 935–944.

Sharma, I., Bhardwaj, R., and Pati, P.K. (2012). Mitigation of adverse effects of chlorpyrifos by 24-epibrassinolide and analysis of stress markers in a rice variety Pusa Basmati-1. *Ecotoxicology and Environmental Safety* 85: 72–81.

Sharma, I., Bhardwaj, R., and Pati, P.K. (2013a). Stress modulation response of 24-epibrassinolide against imidacloprid in an elite indica rice variety Pusa Basmati-1. *Pesticide Biochemistry and Physiology* 105 (2): 144–153.

Sharma, I., Bhardwaj, R., and Pati, P.K. (2015). Exogenous application of 28-homobrassinolide modulates the dynamics of salt and pesticides induced stress responses in an elite rice variety Pusa Basmati-1. *Journal of Plant Growth Regulation* 34: 509–518.

Sharma, D., Bhardwaj, S., Raza, A. et al. (2023). Alleviatory effects of silicon and 24-epibrassinolide in modulation of growth, osmolytes, metabolites, antioxidant defense system, and gene expression in lead-exposed fenugreek (*Trigonella foenum-graecum* L.) plants. *Agronomy* 13 (7): 1884.

Sharma, I., Kaur, N., Saini, S. et al. (2013b). Emerging dynamics of brassinosteroids research. In: *Biotechnology: Prospects and Applications* (ed. R. Salar, S. Gahlawat, P. Siwach et al.), 3–17. New Delhi: Springer.

Sharma, I., Pati, P.K., and Bhardwaj, R. (2011a). Effect of 24-epibrassinolide on oxidative stress markers induced by nickel-ion in *Raphanus sativus* L. *Acta Physiologiae Plantarum* 33: 1723–1735.

Sharma, I., Pati, P.K., and Bhardwaj, R. (2011b). Effect of 28-homobrassinolide on antioxidant defence system in *Raphanus sativus* L. under chromium toxicity. *Ecotoxicology* 20: 862–874.

Soma, F., Takahashi, F., Yamaguchi-Shinozaki, K. et al. (2021). Cellular phosphorylation signaling and gene expression in drought stress responses: ABA-dependent and ABA-independent regulatory systems. *Plants* 10 (4): 756.

Srivastava, A.K., Mishra, P., and Mishra, A.K. (2021). Effect of climate change on plant secondary metabolism: an ecological perspective. In: *Evolutionary Diversity as a Source for Anticancer Molecules* (ed. A.K. Srivastava, V.K. Kannaujiya, R.K. Singh et al.), 47–76. Academic Press.

Steber, C.M. and McCourt, P. (2001). A role for brassinosteroids in germination in *Arabidopsis*. *Plant physiology* 125 (2): 763–769.

Sun, B., Jing, Y., Chen, K. et al. (2007). Protective effect of nitric oxide on iron deficiency-induced oxidative stress in maize (*Zea mays*). *Journal of Plant Physiology* 164 (5): 536–543.

Székely, G., Ábrahám, E., Cséplö, A. et al. (2008). Duplicated P5CS genes of Arabidopsis play distinct roles in stress regulation and developmental control of proline biosynthesis. *Plant Journal* 53: 11–28.

Taheri, Z., Vatankhah, E., and Jafarian, V. (2020). Methyl jasmonate improves physiological and biochemical responses of *Anchusa italica* under salinity stress. *South African Journal of Botany* 130: 375–382.

Tanaka, K., Asami, T., Yoshida, S. et al. (2005). Brassinosteroid homeostasis in *Arabidopsis* is ensured by feedback expressions of multiple genes involved in its metabolism. *Plant Physiology* 138 (2): 1117–1125.

Valletta, A., Iozia, L.M., and Leonelli, F. (2021). Impact of environmental factors on stilbene biosynthesis. *Plants* 10 (1): 90.

Vardhini, B.V. (2016). Enhancement of vegetables and fruits growth and yield by application of brassinosteroids under abiotic stresses: a review. *Plant-Environment Interaction: Responses and Approaches to Mitigate Stress* (ed. M.M. Azooz and P. Ahmad), 124–140. Chichester: John Wiley & Sons, Ltd.

Verbruggen, N. and Hermans, C. (2008). Proline accumulation in plants: a review. *Amino Acids* 35: 753–759.

Versari, A., Parpinello, G.P., Tornielli, G.B. et al. (2001). Stilbene compounds and stilbene synthase expression during ripening, wilting, and UV treatment in grape cv. Corvina. *Journal of Agricultural and Food Chemistry* 49 (11): 5531–5536.

Vert, G., Walcher, C.L., Chory, J. et al. (2008). Integration of auxin and brassinosteroid pathways by Auxin Response Factor 2. *Proceedings of the National Academy of Sciences* 105 (28): 9829–9834.

Vogt, T. (2010). Phenylpropanoid biosynthesis. *Molecular Plant* 3 (1): 2–20.

Vriet, C., Russinova, E., and Reuzeau, C. (2012). Boosting crop yields with plant steroids. *The Plant Cell* 24 (3): 842–857.

Wang, J., Song, L., Gong, X. et al. (2020). Functions of jasmonic acid in plant regulation and response to abiotic stress. *International Journal of Molecular Sciences* 21 (4): 1446.

Wang, X., Teng, Y., Zhang, N. et al. (2017). Rhizobial symbiosis alleviates polychlorinated biphenyls-induced systematic oxidative stress via brassinosteroids signaling in alfalfa. *Science of the Total Environment* 592: 68–77.

Wani, A.S., Ahmad, A., Hayat, S. et al. (2019). Epibrassinolide and proline alleviate the photosynthetic and yield inhibition under salt stress by acting on antioxidant system in mustard. *Plant Physiology and Biochemistry* 135: 385–394.

Wani, S.H., Kumar, V., Shriram, V. et al. (2016). Phytohormones and their metabolic engineering for abiotic stress tolerance in crop plants. *The Crop Journal* 4 (3): 162–176.

Wani, S.H., Singh, N.B., Haribhushan, A. et al. (2013) Compatible solute engineering in plants for abiotic stress tolerance-role of glycine betaine. *Current Genomics* 14: 157–165. doi: 10.2174/1389202911314030001.

Werner, T., Motyka, V., Strnad, M. et al. (2001). Regulation of plant growth by cytokinin. *Proceedings of the National Academy of Sciences* 98 (18): 10487–10492.

Werner, T., Nehnevajova, E., Köllmer, I. et al. (2010). Root-specific reduction of cytokinin causes enhanced root growth, drought tolerance, and leaf mineral enrichment in Arabidopsis and tobacco. *The Plant Cell* 22 (12): 3905–3920.

Wu, W., Zhang, Q., Ervin, E.H. et al. (2017). Physiological mechanism of enhancing salt stress tolerance of perennial ryegrass by 24-epibrassinolide. *Frontiers in Plant Science* 8: 1017.

Xie, Z., Nolan, T.M., Jiang, H. et al. (2019). AP2/ERF transcription factor regulatory networks in hormone and abiotic stress responses in *Arabidopsis*. *Frontiers in Plant Science* 10: 437723.

Yadav, P., Kaur, R., Kohli, S.K. et al. (2016). Castasterone assisted accumulation of polyphenols and antioxidant to increase tolerance of *B. juncea* plants towards copper toxicity. *Cogent Food & Agriculture* 2 (1): 1276821.

Yamada, M., Morishita, H., Urano, K. et al. (2005). Effects of free proline accumulation in petunias under drought stress. *Journal of Experimental Botany* 56 (417): 1975–1981.

Yang, S., Vanderbeld, B., Wan, J. et al. (2010). Narrowing down the targets: towards successful genetic engineering of drought-tolerant crops. *Molecular Plant* 3 (3): 469–490.

Yang, X., Wen, X., Gong, H. et al. (2007b). Genetic engineering of the biosynthesis of glycinebetaine enhances thermotolerance of photosystem II in tobacco plants. *Planta* 225: 719–733.

Yang, J., Zhang, J., Liu, K. et al. (2007a). Involvement of polyamines in the drought resistance of rice. *Journal of Experimental* 58: 1545–1555.

Yokota, T., Ohnishi, T., Shibata, K. et al. (2017). Occurrence of brassinosteroids in non-flowering land plants, liverwort, moss, lycophyte and fern. *Phytochemistry* 136: 46–55.

Yuan, G.F., Jia, C.G., Li, Z. et al. (2010). Effect of brassinosteroids on drought resistance and abscisic acid concentration in tomato under water stress. *Scientia Horticulturae* 126 (2): 103–108.

Yuldashev, R., Avalbaev, A., Bezrukova, M. et al. (2012). Cytokinin oxidase is involved in the regulation of cytokinin content by 24-epibrassinolide in wheat seedlings. *Plant Physiology and Biochemistry* 55: 1–6.

Yusuf, M., Khan, T.A., and Fariduddin, Q. (2017). Brassinosteroids: physiological roles and its signalling in plants. In: *Stress Signaling in Plants: Genomics and Proteomics Perspective, Volume 2* (ed. M. Sarwat, A. Ahmad, M. Abdin et al.), 241–260. Cham: Springer.

Zafra, A., Rodríguez-García, M.I., and Alché, J.D.D. (2010). Cellular localization of ROS and NO in olive reproductive tissues during flower development. *BMC Plant Biology* 10: 1–14.

Zeng, H., Tang, Q., and Hua, X. (2010). *Arabidopsis* brassinosteroid mutants det2-1 and bin2-1 display altered salt tolerance. *Journal of Plant Growth Regulation* 29: 44–52.

Zhang, C., Bai, M.Y., and Chong, K. (2014). Brassinosteroid-mediated regulation of agronomic traits in rice. *Plant Cell Reports* 33: 683–696.

Zhang, Y. and He, J. (2015). Sugar-induced plant growth is dependent on brassinosteroids. *Plant Signaling and Behavior* 10: e1082700.

Zhang, Y., Liu, Z., Wang, J. et al. (2015). Brassinosteroid is required for sugar promotion of hypocotyl elongation in *Arabidopsis* in darkness. *Planta* 242: 881–893.

Zhang, S., Wei, Y., Lu, Y. et al. (2009). Mechanisms of brassinosteroids interacting with multiple hormones. *Plant Signaling & Behavior* 4 (12): 1117–1120.

Zheng, C., Jiang, D., Liu, F. et al. (2009). Exogenous nitric oxide improves seed germination in wheat against mitochondrial oxidative damage induced by high salinity. *Environmental and Experimental Botany* 67 (1): 222–227.

Zhou, P., Li, Q., Liu, G. et al. (2018). Integrated analysis of transcriptomic and metabolomic data reveals critical metabolic pathways involved in polyphenol biosynthesis in Nicotiana tabacum under chilling stress. *Functional Plant Biology* 46 (1): 30–43.

Zhou, J., Wang, J., Li, X. et al. (2014). H_2O_2 mediates the crosstalk of brassinosteroid and abscisic acid in tomato responses to heat and oxidative stresses. *Journal of Experimental Botany* 65 (15): 4371–4383.

Zhu, T., Deng, X., Zhou, X. et al. (2016). Ethylene and hydrogen peroxide are involved in brassinosteroid-induced salt tolerance in tomato. *Scientific Reports* 6 (1): 35392.

Zhu, J., Lu, P., Jiang, Y. et al. (2014). Effects of brassinosteroid on antioxidant system in *Salvia miltiorrhiza* under drought stress. *Journal of Research in Agriculture and Animal Science* 2: 1–6.

Zhu, Y., Zuo, M., Liang, Y. et al. (2013). MAP65-1a positively regulates H_2O_2 amplification and enhances brassinosteroid-induced antioxidant defence in maize. *Journal of Experimental Botany* 64 (12): 3787–3802.

Zullo, M.A.T. and Bajguz, A. (2019). The brassinosteroids family–structural diversity of natural compounds and their precursors. In: *Brassinosteroids: Plant Growth and Development* (ed. S. Hayat, M. Yusuf, R. Bhardwaj et al.), 1–44. Singapore: Springer Singapore.

5

Brassinosteroids Improve Resilience Against Nutritional Stress: Dissecting the Responses in Crops

Allan Klynger da Silva Lobato, Roberto Rivelino do Nascimento Barbosa, Leidy Alves dos Santos, Lorene Bianca Araújo Tadaiesky, and Ynglety Cascaes Pereira*

Núcleo de Pesquisa Vegetal Básica e Aplicada, Universidade Federal Rural da Amazônia. Paragominas, Pará, Brazil

*Corresponding author: allanllobato@yahoo.com.br

5.1 Introduction

Knowledge of plant nutrition is one of the main factors for increasing agricultural production, along with water availability. Understanding the dynamics of nutrients and their role in the soil/plant system makes it possible to enhance plant growth and development. Nutrients are essential chemical elements to meet vital needs throughout the plant cycle. In addition to oxygen (O_2), hydrogen (H), and carbon (C), which are acquired from the atmosphere and water, there are nutrients required by the plant through the soil, which can be enriched through fertilization. They are separated into macronutrients, which are absorbed in higher concentrations – such as nitrogen (N), phosphorus (P), potassium (K), calcium (Ca), magnesium (Mg), and sulfur (S) – and micronutrients, which are absorbed in smaller quantities, such as iron (Fe), boron (B), molybdenum (Mo), chlorine (Cl), nickel (Ni), copper (Cu), zinc (Zn), and manganese (Mn). Among these nutrients, Mg, Fe, and B are metals of great importance for plant development, and in this chapter, we will discuss their importance.

5.2 Iron and Its Functions in Plants

Iron (Fe), an essential nutrient for plants, is necessary for physiological processes in plants (Kim and Guerinot 2007). The primary function of Fe in plants is to be an enzymatic component, where most participate in oxidation processes (Kobayashi and Nishizawa 2012). The enzymes that act in the transfer of electrons use Fe as a cofactor of choice, being these enzymes involved in a variety of reversible redox reactions (Fourcroy et al. 2014), participate in the route of chlorophyll synthesis (Abadía et al. 2011), in the photosynthesis

Brassinosteroids and Abiotic Stress: Recent Developments on Physiological and Molecular Aspects, First Edition. Edited by Anket Sharma, et al.
Published 2026 by John Wiley & Sons Ltd.

(Liu et al. 2017), DNA synthesis, respiration (Briat et al. 2015), biosynthesis of phytoriums (gibberellic acid, jasmonic acid and ethylene) in the production and the elimination of reactive oxygen species (EROs) (Dixon and Stockwell 2014).

Once absorbed by the roots, the iron is loaded into the xylem and translocated to the upper part through the transpiratory flow (Becker and Asch 2005). Iron is oxidized when released into the xylem vessels and transported as complexes with organic acids, especially citrate, the primary metal chelator in the xylem (Kobayashi and Nishizawa 2012). Iron absorption by mesophyll cells also requires a reduction step after the citrate molecule releases the ferric ion, suggesting the presence of a specific Fe^{2+} transporter in the foliar cell plasmalemma. The electron donor for iron reduction is NADPH, and this process appears to be strongly induced by the increase in the NADPH/$NADP^+$ ratio resulting from the photochemical step of photosynthesis. Once in the mesophyll, iron can be stored in the vacuoles or immobilized by the ferritin protein, which occurs mainly in plastids (Zancani et al. 2004).

5.3 Fe Deficiency, Symptoms, and Impacts on Metabolism

One of the most important factors limiting agricultural production worldwide is Fe deficiency. Plants exposed to soils with low Fe content generally present interveinal chlorosis in the leaves, causing yellowing and the leaf may suffer from whitish chlorosis due to the lack of pigmentation from the chlorophyll molecule, since Fe has a catalytic role in the synthesis of chlorophyll, resulting in a decrease in the photosynthetic rates of plants, leading to reduced growth and development of the plant, affecting crop yield (Connorton et al. 2017; Zhang et al. 2019).

Fe is a poorly mobile element in the plant; deficiency symptoms appear in the youngest leaves. Its distribution in the phloem is limited, even when the nutrient is discharged via xylem it is reduced due to the formation of insoluble compounds contributing to the accumulation of Fe in the older leaves and roots and in the younger leaves the appearance of chlorosis (Jeong et al. 2017).

Fe deficiency can induce the secretion of phenolic compounds, especially coumarins. Coumarin can accelerate the mobilization and reduction of Fe, a strategy to avoid the impairment of plant metabolism in the absence of Fe (Connorton et al. 2017; Stringlis et al. 2019). In addition to chlorotic symptoms and coumarin synthesis, Fe deficiency in a more severe degree can lead to total yellowing with the presence of necrotic zones at the ends, which will result in premature leaf abscission, enabling complete defoliation (Mahender et al. 2019).

5.4 Occurrence of Iron Toxicity and Implications for Plant Performance

Iron is an abundant element in nature and corresponds to 5% of the Earth's crust. However, the majority of the iron present in the soil is unavailable to the plants, forming insoluble complexes in the presence of oxygen under neutral or alkaline pH (Chen et al. 2015). The amount of iron in the soil solution required for plant growth ranges from 10^{-9} to 10^{-4} M;

however, anoxic soils with acidic pH increase the amount of iron available, contributing to plants' excessive absorption of this element (Guerinot and Yi 1994).

Fe toxicity is caused by large Fe^{2+} concentrations in the soil solution and, consecutively, by high iron levels in the plant tissues. The iron concentration, which precedes soil submersion, rarely exceeds 0.1 mg L^{-1}, but can reach, in acid soils, approximately 600 mg L^{-1}. However, in extreme cases, values up to 5000 mg L^{-1} were already detected (Harmsen and van Breemen 1975). The Fe^{2+} concentrations in the soil solution capable of affecting production generally range from 10 to >2000 mg L^{-1} (Becker and Asch 2005).

Production losses associated with iron toxicity are estimated at 15–20% (Schmidt et al. 2013). Some authors extend this production loss to 30% (Becker and Asch 2005; Liu et al. 2016). However, the literature has already described total production losses caused by an excess of this element (Audebert and Sahrawat 2000). This amplitude of production losses in crop plants occurs due to various factors that interfere with the cultivation, such as the cultivar used, the plant's development stage, iron levels present in the solution, and crop management (Audebert and Fofana 2009).

Cultivars or/and varieties with different levels of tolerance to iron toxicity and agronomic practices such as alternative planting, adequate water management, and application of fertilizers have been developed and used in areas of occurrence of iron excess toxicity (Schmidt et al. 2013; Suriyagoda et al. 2017). In this context, the most efficient practice is the choice of resistant genotypes. However, none of these options is applicable or efficient because the environment's diversity and excess iron toxicity can occur (Becker and Asch 2005). Thus, knowledge about the impact of excess iron on the physiology of crop plants is necessary.

Excess iron in the shoot can cause phytotoxicity and necrotic spots on the leaves, browning of the roots, and inhibition of plant growth, and cationic imbalance, causing nutritional disorder (Audebert and Fofana 2009). Among the essential nutrients, a decrease in the absorption of P, K, Ca, and Mg has been reported in sensitive plants to Fe (Silveira et al. 2007). In addition, iron toxicity also potentiates oxidative stress, with increased production of reactive oxygen species (ROS), such as hydrogen peroxide, superoxide, and hydroxyl radical, the latter considered more toxic to the cell (Gill and Tuteja 2010).

To support growth and prevent cellular iron toxicity, plants rely on the ability to store and remove iron (Briat et al. 2007). In rice, two ferritin class proteins are reported, FER1 and FER2. Ferritin acts as a mechanism intensely involved in the tolerance of this metal since it incorporates Fe in its structure, minimizing its reactivity (Majerus et al. 2007).

5.5 Magnesium and Its Essentiality for Higher Plants

Magnesium (Mg) is an essential element in plants; it is a constituent of the chlorophyll molecule and acts in phosphorylation, aggregation of ribosome subunits (Zhang et al. 2022), translocation of photoassimilates, and the activation of multiple enzymes, such as glutathione synthetase, phosphoenolpyruvate (PEP) carboxylase and Ribulose 1,5 bisphosphate oxidase/carboxylase – Rubisco (Verbruggen and Hermans 2013). Therefore, crop growth and productivity are highly affected by Mg deficiency in areas of intensive agricultural production (He et al. 2023).

The absorbed Mg is transported in the plant as Mg^{2+}, accumulating in its young tissues due to its high mobility (Waters 2011). The main targets of Mg deficiency are photosynthesis and sugar transport from the source organs to the drains (Farhat et al. 2016). Mg is required for chlorophyll's synthesis and structural stability (Albus et al. 2012). In the photochemical phase of photosynthesis, the synthesis of ATP in chloroplasts is increased by the extreme supply of Mg, as this process has an absolute requirement for Mg as a binding component between ADP and the ATPase enzyme. In the chemical phase of photosynthesis, the modulation of the Rubisco enzyme requires Mg to increase the enzyme's affinity for its substrate (CO_2) and the rate of new enzyme synthesis (Cakmak and Yazici 2010).

5.6 Visual Aspects and Interference in Physiological Processes Caused by Mg Deficiency

Mg-deficient plants have lower root growth and root/shoot ratio because the loading of carbohydrates in the phloem is an active process dependent on the consumption of ATP, which also requires Mg. The accumulation of non-structural sugars in the leaves and the change in the carbohydrate partition observed under Mg deficiency are attributed to the inhibition of phloem loading (Hawkesford et al. 2012).

Plants with Mg deficiency affect the productivity and quality of agricultural production and are a common situation in several agricultural plantations in the biosphere (Verbruggen and Hermans 2013). The classic symptoms of deficiency of this nutrient are interveinal chlorosis in older leaves, surrounding the veins of the leaf blade. This chlorosis can progress to younger leaves, due to the high mobility of Mg in the plant (Marschner 2012; Farhat et al. 2014). The leaves remain green, but the veins may present yellow, tan, or reddish coloration, because the element is part of the constitution of the tetrapyrrolic center of chlorophyll (Verbruggen and Hermans 2013).

In general, leaves begin to become chlorotic due to decreasing Mg concentrations, plant growth begins to be compromised due to the impediment in the accumulation of sugars, leading to the accumulation of starch and the activation of antioxidant metabolism before any noticeable effect during photosynthesis (Hermans et al. 2004; Chaudhry et al. 2021).

With Mg deficiency in plants, there is a reduction in the synthesis of chlorophyll, which is the pigment of great importance for capturing light that results in the triggering of redox reactions in the photochemical stage. If there is no chlorophyll, there is no production of energy in the form of ATP or of reducing agent in the form of NADPH, which are essential for carboxylation reactions for the production of triose phosphate and consequent starch and sucrose (Hariadi and Shabala 2004). Furthermore, Mg deficiency can compromise the enzymatic activity of Rubisco in reducing CO_2 (Ridolfi and Garrec 2000). If there is a reduction in the production of photoassimilates, the drains become compromised. For example, roots that do not receive sugars will have reduced growth, which may compromise the absorption of water and nutrients. If the amount of water in the plant is reduced, this may imply a decrease in stomatal conductance, preventing the flow of gas exchange and consequently causing losses in productivity (Hermans and Verbruggen 2005).

The accumulation of sugars and amino acids in the source organs generally inhibits their growth, causing a chlorotic appearance and reduced photosynthetic rates. It may be due to structural impairment and destabilization in the phloem tissues, decreased drainage force compromising the load of photoassimilates in the phloem (Hermans et al. 2006). The accumulation of starch confers attenuation of root growth, and therefore the accumulation of carbon in the source leaves is the main change induced by Mg deficiency in plants (Chaudhry et al. 2021).

5.7 Overproduction of Reactive Oxygen Species Induced by Mg Excess

Stress due to high concentrations of Mg is not every day. Still, serpentine soils and semi-arid regions can have a high Mg: Ca ratio and thus affect the growth and development of crops, as high concentrations of Mg^{2+} restrict water absorption by the roots, causing high osmolarity and ionic toxicity, and as a con-sequence, oxidative stress, decreased nutrient availability, especially Ca^{2+}, K^{+}, and Mn^{2+} and reductions in the growth rate (Parihar et al. 2015).

Excessive presence of Mg in the medium has toxic effects on roots of Kalidium capsicum, and roots and shoots of common bean (*Phaseolus vulgaris* L.). Excess Mg in the soil decreases the biomass of shoots and roots in these plants, indicating a toxic effect of high Mg concentrations on plants (Kant et al. 2008). High Mg concentrations in plants can induce the production of ROS (Mittler 2017). Elevated ROS production leads to cell damage due to oxidative stress (Bhattacharjee 2019). Some of the effects of increased ROS synthesis include increased production of malondialdehyde (MDA), reduced soluble sugar, and reduced leaf chlorophyll content (Huang et al. 2019).

Plants develop a series of cellular responses to combat the negative effects of ROS. One of these responses includes the production of free proline, which plays an important role in osmotic adjustment and resistance to environmental stress (Per et al. 2017; Lamichhane et al. 2023).

In rice plants, excess Mg can cause restrictions on biomass and plant growth by decreasing chlorophyll concentrations and influencing CO_2 carboxylation. Therefore, in agricultural management, the Mg level in the soil/plant must always be in the appropriate conditions for crops (Lamichhane et al. 2023).

5.8 Boron Is an Essential Micronutrient for Vascular Plants

Boron (B) is an essential micronutrient for vascular plants, and its availability in the soil represents a limiting factor for world agricultural production (Kobayashi et al. 2018). B deficiency can be triggered by inadequate soil conditions, such as high pH and low organic matter content, and environmental factors, such as drought, high temperature, and light (Atique-ur-Rehman et al. 2018). Another problem, B is susceptible to leaching in soil because it is highly soluble and mobile, which can easily lead to plant deficiency after excessive rainfall (Goli et al. 2019). On the other hand, arid and semi-arid regions can occur problems related to the toxicity of this micronutrient (Nawaz et al. 2020), which B can be taken to the surface layers of the soil due to the combined actions linked to the evaporation process of the soil water and capillary method (Yau and Ryan 2008).

B has structural and metabolic functions in plants, which are intrinsically related to calcium in cell wall development, acting in protein synthesis, sugar transport, respiration, and carbohydrate metabolism (Reid 2014; Shireen et al. 2018). Additionally, the mobility and permeability of this nutrient can vary by species (Wimmer and Eichert 2013). B deficiency reflects mainly on meristematic tissues (Hänsch and Mendel 2009), affecting root growth and the development of seeds, flowers, and fruits (Herrera-Rodríguez et al. 2010), also limiting absorption and transport of water and nutrients, due to damage to the root structures, including the xylem vessels (Li et al. 2017b). On the other hand, B excess is associated with oxidative stress, in which B toxicity causes the overproduction of ROS, such as superoxide (O_2^-) and hydrogen peroxide (H_2O_2), causing an imbalance in metabolism, affecting the division and expansion of the plant cells (Sakamoto et al. 2011; Farghaly et al. 2021).

5.9 B Deficiency Impairs the Gas Exchange and Production

One of the characteristic signs (Liu et al. 2017) of B deficiency in plants is reduced plant growth, since B is not mobile in the phloem (Lehto et al. 2010). Plants deficient in this nutrient show reduced root growth, translocation of photoassimilates, impaired carbohydrate metabolism, nucleic acid synthesis, and pollen tube growth. B deficiency instigates changes in root morphology, resulting in an increase in the leaf/root ratio and/or thickening of the root cell wall due to the accumulation of cellulose, amino acids, and lignin. The sum of all these metabolic changes implies a complete inhibition of nutrient absorption and transport (Wimmer and Eichert 2013).

In B-deficient plants, the pool of antioxidant enzymes (superoxide dismutase, peroxidase, catalase, and ascorbate peroxidase) is reduced, disrupting ROS controls and causing the plant to undergo photooxidation. There is a reduction in photosynthetic rates, transpiration rates, and stomatal conductance. The accumulation of ROS through B deficiency in plants can cause lipid peroxidation, plasmolysis, problems in flower organogenesis, and pollen tube sterility (Hua et al. 2017; Zhou et al. 2017).

B deficiency has negative impacts on the production and quality of crops of some plant species. In fruit trees of the Rosaceae family, such as apples and pears, B deficiency causes the flowers to become wilted with black or brown coloration, and clear areas appear in the pulp of the pseudofruits that are full of water, turning brown and leading to dehydration. The surface and external pulp become rough, brownish or reddish, with deep cracks, and there is deformation of the pseudofruit (Geilfus 2019). B is of great importance for vegetative growth, reproductive development, production, and quality of crops. Recent scientific research has brought to light that B in adequate concentrations in the plant can be an effective technique to reduce the sensitivity of agricultural and forestry crops to abiotic stresses, such as water deficit, salinity, or heavy metal toxicity (García-Sánchez et al. 2020; Roquette et al. 2023).

5.10 B Toxicity Increases the Stress Indicators

Boron toxicity as well as deficiency results in the accumulation of ROS (Kaya et al. 2020), slows down growth, takes away part of the water present in the root, as a visible symptom, chlorosis on older leaves (Kayıhan et al. 2017; Farghaly et al. 2021). Studying the influence of boron on Capsicum annuum plants, described a reduction in root growth and restrictions in potassium (K) absorption in plants subjected to high boron concentrations.

El-Shazoly et al. (2019) studying the tolerance of *Triticum aestivum* to boron toxicity, verified that the excess promoted a significant reduction in the growth of the plants selected by the dry mass and fresh mass, and even that the adverse effects on the development were direct as it should be of boron in the plant.

5.11 Brassinosteroids, Molecules that Positively Regulate Plant Metabolism

Plant hormones are naturally occurring regulatory molecules that act as chemical signals to regulate plant morphogenesis. Brassinosteroids (BRs) are part of the group of plant hormones that promote cell elongation and division, stimulate enzymes involved in proton pumps, protein, and nucleic acid synthesis, and induce changes in the composition of some amino acids in proteins. This phytoregulator also acts in antioxidant activation and gas exchange, allowing higher photosynthetic rates (Vriet et al. 2012; Yuan et al. 2012; Oliveira et al. 2019).

BRs constitute a class of approximately 70 polyhydroxysteroid derivatives that appear to be distributed throughout the plant kingdom (Siddiqui et al. 2018). The identification of endogenous plant steroid compounds has resulted in almost 30 years of research to identify new substances capable of promoting growth in pollen extracts from different plant species (Fariduddin et al. 2013).

24-Epibrassinolide (EBR) is a type of BR that regulates several biochemical and physiological reactions in plants, including leaves, roots, flowers, and seeds (Merwad et al. 2018; Jan et al. 2018). This phytohormone occurs in all parts of the plant and is being used as a methodology to reduce the effects of abiotic stresses such as water deficit, high and low temperatures, soil salinity, and toxicity of chemical elements (Larré et al. 2014). These hormones are being used in agriculture as an alternative to alleviate physiological and biochemical disturbances caused by abiotic stresses obtained through ROS.

Studies indicate that BRs give tolerance to plants to abiotic stresses, stimulating the activity of antioxidant enzymes and promoting the maintenance of the photosynthetic rate (Ramakrishna and Rao 2014; Ahanger et al. 2018). Research by Santos et al. (2020) revealed that 24-EBR mitigated oxidative stress induced by low and high Zn concentrations in *Glycine max*. Likewise, Rodrigues et al. (2020) found that the exogenous application of EBR (100 nM) promoted the maintenance of photosynthetic pigments (*Chl a, Chl b, total Chl*, and *Car*) and less accumulation of EROs (O^{2-} and H_2O_2) in soybean plants (*Glycine max*) with deficiency and toxicity of Mn.

Hypothesis of this chapter considers that the inadequate supply of Fe^{2+}, Mg^{2+}, and B (deficiency or toxicity) causes deleterious effects on antioxidant metabolism, gas exchange, and photosynthetic pigments, possibly affecting plant anatomy and growth. On the other hand, EBR represents a biodegradable substance that efficiently modulates tolerance in plants exposed to deficiency or toxicity.

In this sense, EBR can be a possible solution to mitigate the damage caused by deficiencies and excess of these nutrients in plants because this steroid has a set of actions linked to increases in nutrient content (Yuan et al. 2015), EROs (Cunha et al. 2020), and biomass stimulation (Pereira et al. 2019). This chapter reports data on the role of EBR as

Table 5.1 Fe contents in rice plants treated with EBR and exposed to Fe toxicity.

Fe supply	EBR	Fe in the root (μg g DM^{-1})	Fe in the shoot (μg g DM^{-1})
Control	−	547.1 ± 18.7c	62.2 ± 2.1c
Control	+	559.7 ± 18.3c	62.9 ± 4.1c
Toxicity	−	2257.3 ± 48.7a	275.6 ± 15.4a
Toxicity	+	1864.0 ± 41.9b	166.1 ± 15.6b

Fe = Iron. Columns with different letters indicate significant differences from the Scott–Knott test ($P < 0.05$). Values correspond to means from five repetitions and standard deviations (Tadaiesky et al. 2020).

an attenuator of oxidative stress associated with an inadequate supply of nutrients and the possible repercussions on plant metabolism.

5.12 Plants Exposed to Toxicity Had Less Fe After EBR

The root and shoot Fe contents of Fe-toxic rice plants were dramatically increased (Table 5.1). Plants treated with $FeCl_2$ (6250 μM Fe) and EBR (10 nM) showed significant decreases of 17% and 40% in root and shoot, respectively, compared to steroid-free treatment.

Increases in Fe contents of rice roots and shoots confirm the toxicity symptoms shown in Figure 5.2. Treatment with 10 nM EBR attenuated the negative impact, reducing Fe contents. Higher phytochelatin (PC) synthesis can be attributed to the detoxification of heavy metals by peptides (Anwar et al. 2018). Peptides chelate metal ions in the cytoplasm of cells to detoxify heavy metals (Bajguz and Hayat 2009; Bajguz 2010).

5.13 EBR Slows Chlorophyll Breakdown and Boosts Photosynthesis

Low and high Mg supply reduced Chl *a*, Chl *b*, Total Chl, and Car values (Table 5.2) and increased Chl *a*/Chl *b* and Total Chl/Car ratios. Chl *b*, Total Chl, and Car increased 32%, 7%, and 33% under low Mg with EBR. Compared to the same treatment without EBR, Chl *a*, Chl *b*, Total Chl, and Car increased 1%, 31%, 9%, and 53% with high Mg + EBR. In this sequence, low and high Mg with EBR decreased Chl *a*/Chl *b* EBR by 24% and 19%. The Total Chl/Car ratio showed 20% and 29% decreases in low and high Mg coupled with EBR.

Low and high Mg affected photosystem II (Table 5.2). Plants treated with 100 nM EBR and soft Mg showed 9% and 9% increases in Φ_{PSII} and ETR, respectively, and 3% and 6% decreases in EXC and ETR/P_N values compared to the same treatment without EBR. High Mg with EBR resulted in 43%, 27%, and 50% increases in Φ_{PSII}, q_P, and ETR, and reductions in NPQ (19%) and EXC (6%), compared to the same treatment without EBR.

Gas exchange was negatively affected by low and high Mg (Table 5.2). However, applying EBR to plants with low Mg increased P_N, WUE, and P_N/C_i by 16%, 23%, and 120%,

Table 5.2 Photosynthetic pigments, chlorophyll fluorescence, and gas exchange in EBR-sprayed soybean subjected to varied Mg supply.

EBR	Mg supply	Chl *a* (mg g^{-1} FM)	Chl *b* (mg g^{-1} FM)	Total Chl (mg g^{-1} FM)	Car (mg g^{-1} FM)	Ratio Chl *a*/Chl *b*	Ratio total Chl/Car
–	Low	6.37 ± 0.37 Ba	1.93 ± 0.14 Cb	8.30 ± 0.25 Ba	0.27 ± 0.01 Cb	3.30 ± 0.20 Aa	31.18 ± 1.65 Aa
–	Control	8.46 ± 0.36 Ab	3.26 ± 0.13 Ab	11.72 ± 0.45 Ab	0.49 ± 0.02 Ab	2.60 ± 0.20 Ba	24.24 ± 1.70 Ba
–	High	8.35 ± 0.26 Aa	2.79 ± 0.06 Bb	11.14 ± 0.34 Ab	0.38 ± 0.03 Bb	3.00 ± 0.14 Aa	29.57 ± 2.49 Aa
+	Low	6.38 ± 0.87 Ca	2.54 ± 0.17 Ca	8.92 ± 0.39 Ca	0.36 ± 0.04 Ca	2.51 ± 0.30 Ab	24.81 ± 1.58 Ab
+	Control	9.88 ± 0.18 Aa	5.00 ± 0.13 Aa	14.88 ± 0.31 Aa	0.74 ± 0.06 Aa	1.98 ± 0.02 Bb	20.30 ± 1.23 Bb
+	High	8.45 ± 0.21 Ba	3.65 ± 0.59 Ba	12.10 ± 0.33 Ba	0.58 ± 0.02 Ba	2.42 ± 0.21 Ab	20.96 ± 1.36 Bb
EBR	**Mg supply**	Φ_{PSII}	q_P	**NPQ**	**ETR (μmol m^{-2} s^{-1})**	**EXC (μmol m^{-2} s^{-1})**	**ETR/P_N**
–	Low	0.23 ± 0.01 Aa	0.33 ± 0.01 Aa	0.62 ± 0.03 Ba	34.9 ± 0.9 Ab	0.70 ± 0.01 Ba	2.17 ± 0.10 Ba
–	Control	0.24 ± 0.01 Ab	0.33 ± 0.01 Ab	0.59 ± 0.05 Ba	36.1 ± 1.8 Ab	0.70 ± 0.01 Ba	1.99 ± 0.13 Ba
–	High	0.07 ± 0.01 Bb	0.15 ± 0.01 Bb	0.83 ± 0.05 Aa	10.3 ± 0.5 Bb	0.82 ± 0.02 Aa	6.63 ± 0.40 Ab
+	Low	0.25 ± 0.01 Ba	0.33 ± 0.02 Ba	0.62 ± 0.04 Aa	38.1 ± 2.1 Ba	0.68 ± 0.02Ba	2.05 ± 0.11 Ba
+	Control	0.31 ± 0.01 Aa	0.45 ± 0.03 Aa	0.48 ± 0.03 Bb	46.4 ± 2.8 Aa	0.62 ± 0.02 Cb	2.36 ± 0.28 Ba
+	High	0.10 ± 0.01 Ca	0.19 ± 0.01 Ca	0.67 ± 0.03 Ab	15.5 ± 1.1 Ca	0.77 ± 0.02 Ab	7.63 ± 0.56 Aa
EBR	**Mg supply**	**P_N (μmol m^{-2} s^{-1})**	**E (mmol m^{-2} s^{-1})**	**g_s (mol m^{-2} s^{-1})**	**C_i (μmol mol^{-1})**	**WUE (μmol mmol^{-1})**	**P_N/C_i (μmol m^{-2} s^{-1} Pa^{-1})**
–	Low	16.0 ± 0.6 Bb	2.28 ± 0.07 Ba	0.21 ± 0.01 Ba	241 ± 6 Ba	7.06 ± 0.21 Ab	0.066 ± 0.003 Bb
–	Control	18.1 ± 0.6 Ab	2.47 ± 0.09 Aa	0.24 ± 0.01 Aa	224 ± 6 Ca	7.38 ± 0.56 Ab	0.081 ± 0.004 Ab
–	High	1.5 ± 0.1 Cb	1.70 ± 0.06 Ca	0.11 ± 0.01 Ca	378 ± 8 Aa	0.92 ± 0.08 Ba	0.004 ± 0.001 Ca
+	Low	18.5 ± 0.4 Ba	2.14 ± 0.08 Aa	0.12 ± 0.01 Bb	128 ± 6 Bb	8.70 ± 0.41 Aa	0.145 ± 0.011 Ba
+	Control	19.7 ± 0.6 Aa	2.19 ± 0.10 Ab	0.15 ± 0.01 Ab	100 ± 6 Cb	9.01 ± 0.33 Aa	0.197 ± 0.013 Aa
+	High	2.0 ± 0.1 Ca	1.82 ± 0.06 Ba	0.12 ± 0.01 Ba	353 ± 11 Ab	1.12 ± 0.19 Ba	0.005 ± 0.001 Ca

Chl *a* = Chlorophyll *a*; Chl *b* = Chlorophyll *b*; Total chl = Total chlorophyll; Car = Carotenoids; Φ_{PSII} = Effective quantum yield of PSII photochemistry; q_P = Photochemical quenching coefficient; NPQ = Nonphotochemical quenching; ETR = Electron transport rate; EXC = Relative energy excess at the PSII level; ETR/P_N = Ratio between the electron transport rate and net photosynthetic rate; P_N = Net photosynthetic rate; E = Transpiration rate; g_s = Stomatal conductance; C_i = Intercellular CO_2 concentration; WUE = Water-use efficiency; P_N/C_i = Carboxylation instantaneous efficiency. Columns with distinct capital letters (low, control, and high Mg supply) and lowercase letters (with and without EBR) show significant differences from the Scott–Knott test ($P < 0.05$). Means ± SD, $n = 5$ (Santos et al. 2023).

respectively, and decreased C_i by 47%. In the high Mg + EBR treatment, P_N, E, g_s, WUE, and P_N/C_i increased by 33%, 7%, 9%, 22%, and 25%, respectively, while C_i decreased by 7%.

This study found that EBR reduced oxidative damages from O_2^- and H_2O_2 in plants subjected to Mg^{2+} stress, delaying the negative effects on photosynthetic pigments (Chl *a*, Chl *b*, Chl Total, and Car). The PSII showed increased efficiency, as evidenced by increased PSII and ETR, indicating that the EBR enhanced the thylakoid membrane's structural organization, maximizing pigment performance and energy distribution (Dobrikova et al. 2014). Mg deficit and oversupply reduced photosynthetic pigments. Mg^{2+} stress (high and low) can affect several physiological processes related to chlorophyll biosynthesis, plant growth, and development because Mg occupies the central position of the porphyrin ring of chlorophyll molecules and activates or regulates several kinases, such as ATPases and RuBisCO. Mg^{2+} stacks the chloroplast granum, which reduces plant photosynthetic efficacy in low-Mg^{2+} environments (Ceppi et al. 2012).

Benefits of EBR include higher Φ_{PSII}, q_P, and ETR values, indicating enhanced photosystem II efficiency and electron transfer during photochemical processes. This steroid increases fluorescence dissipation, allowing electrons to flow into chloroplasts (Jiang et al. 2012), which may boost ATP and NADPH production for CO_2 fixation in the Calvin cycle (Kumari et al. 2017). EBR spray reduced NPQ and EXC in plants supplemented with low and high Mg, indicating that EBR reduced heat dissipation and increased quantum fluorescence output (Nath et al. 2013). This is fascinating since the EBR reduced energy losses (heat) from high light energy, which may cause photooxidation and damage to the photosystem. EBR and low Mg treatment decreased ETR/P_N, indicating that this steroid improves photochemical electron use and reduces the use of alternative electron drains like photorespiration and Mehler reactions in Mg deficiency.

EBR reduced gas exchange side effects from low and high Mg supplements. In this context, plants exposed to soft Mg and sprayed with EBR (100 nM) had increases in P_N and P_N/C_i, and WUE, confirming EBR's effects on photosynthetic machinery and CO_2 uptake. The increase in P_N and reduction of C_i suggest higher RuBisCO enzyme activity, the main enzyme involved in CO_2 fixation in the Calvin cycle. Due to reduced CO_2 fixation, Mg-deficient plants had lower P_N, E, g_s, and WUE, resulting in wasted electron buildup. When chloroplasts absorb more energy, they produce ROS, which photo-oxidize and destroy chlorophyll and chloroplast membranes. Li et al. (2017a) found that *Citrus sinensis* seedlings deficient in Mg (0 mM $MgSO_4$ for 16 weeks) had significantly lower g_s, RuBisCO, and CO_2 assimilation in lower leaves. In three *Solanum lycopersicum* genotypes (dwarf mutant deficient in EBR biosynthesis, dwarf line efficient in EBR biosynthesis, and wild plants). Li et al. (2016) found that high endogenous EBR levels increased P_N, E, g_s, and P_N/C_i, and probably RuBisCO enzyme carboxylation and regeneration in vivo compared to mutant deficient in EBR and wild genotypes.

In plants overexposed to Mg + EBR (100 nM), gas exchange was maximized, with increases in P_N, E, g_s, WUE, P_N/C_i. These results can be explained by the positive effects on stomatal performance, as shown by the increases in g_s, SD, and SI. An increase in g_s aids CO_2 absorption and may promote RuBisCO activity (P_N/C_i and C_i reduction). WUE increased due to P_N and E increases in high Mg and EBR-sprayed plants. WUE is caused by the interaction between P_N and E and is linked to the stomatal system that regulates gas exchange and water usage, boosting photosynthetic efficiency. Hayat et al. (2012) found that foliar spraying two EBR analogues (28-homobrassinolide and 24-EBR at 10^{-8} M) in *Lycopersicon esculentum* plants subjected to Cd toxicity increased photosynthesis, namely P_N, g_S, E, and WUE.

5.14 Plants with Low Boron Levels, EBR Reduces Oxidative Damage

Low and high B supply increased stress markers (Figure 5.1). Plants sprayed with EBR had 1%, 2%, and 26% lower O_2^- levels in low, control, and high B than those without EBR. To H_2O_2, EBR sprayed in plants experienced declines of 2%, 14%, and 10% in low, control, and

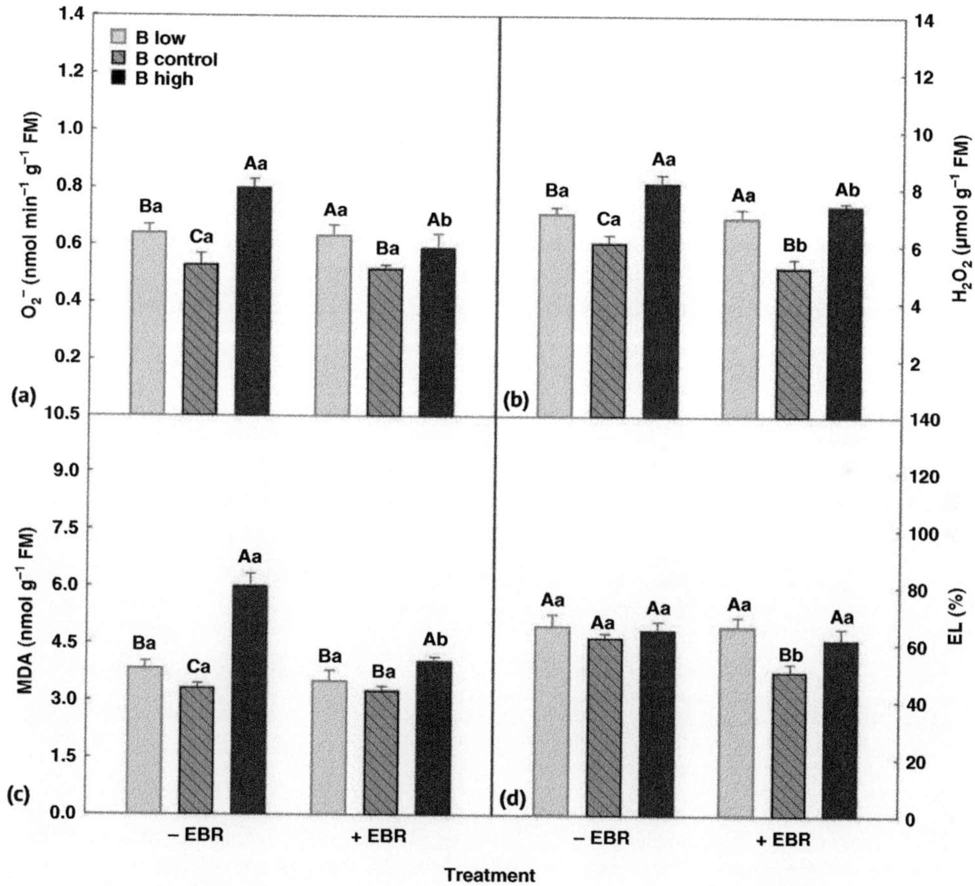

Figure 5.1 Superoxide (O_2^-), hydrogen peroxide (H_2O_2), malondialdehyde (MDA), and electrolyte leakage (EL) in soybean plants sprayed with EBR and subjected to varied B supplies. Columns with distinct capital letters (low, control, and high B supply under equal EBR) and lowercase letters (with and without EBR) show significant differences from the Scott–Knott test ($P < 0.05$). Means ± SD, $n = 5$. *Source*: Pereira et al. (2022)/Springer Nature.

high B treatments, respectively. EBR-pretreated plants reduced MDA by 9%, 3%, and 33% in low, control, and high B treatments. In EL (plants with EBR) in low, control, and high B supply decreased by 1%, 19%, and 5%, respectively.

High B stress increased O_2^-, H_2O_2, MDA, and EL, although EBR mitigated these effects. Previous antioxidant enzyme activity enhanced with EBR spray. The SOD at the start of the defense process converts O_2^- to H_2O_2, which is then degraded by the CAT, APX, and POX (Zhang et al. 2008; Dalyan et al. 2018). Ahammed et al. (2013) discovered that *Solanum lycopersicum* plants subjected to EBR and four polychlorinated biphenyl doses (0.0, 0.4, 2.0, and 10.0 μg L^{-1}) showed decreased O_2^-, H_2O_2, and MDA levels, comparable to our study. Ogweno et al. (2008) discovered reduced H_2O_2 and MDA levels in *Lycopersicon esculentum* plants treated with EBR under standard (25/18 °C) and high (40/30 °C) temperature conditions and three EBR concentrations (0.01, 0.1, 1.0 mg L^{-1}).

5.15 Fe Toxicity on Biomass Was Mitigated by EBR

The growth was greatly reduced by Fe toxicity (Figure 5.2). Interestingly, EBR-treated control plants increased SDM, RDM, and TDM by 40%, 24%, and 33%, respectively (Figure 5.3). Fe toxicity with 10 nM EBR optimized RDM (36%) and TDM (35%).

Steroids reduced Fe toxicity's impact on plant biomass (RDM and TDM). EBR improves growth by increasing tissue anatomical characteristics, which reduces Fe toxicity (RAA and LAA), increases nutrient absorption and accumulation (RMD and VCD), and maintains homeostasis with other metals of equal valence like Mg, Mn, Cu, and Zn. Quinet et al. (2012)

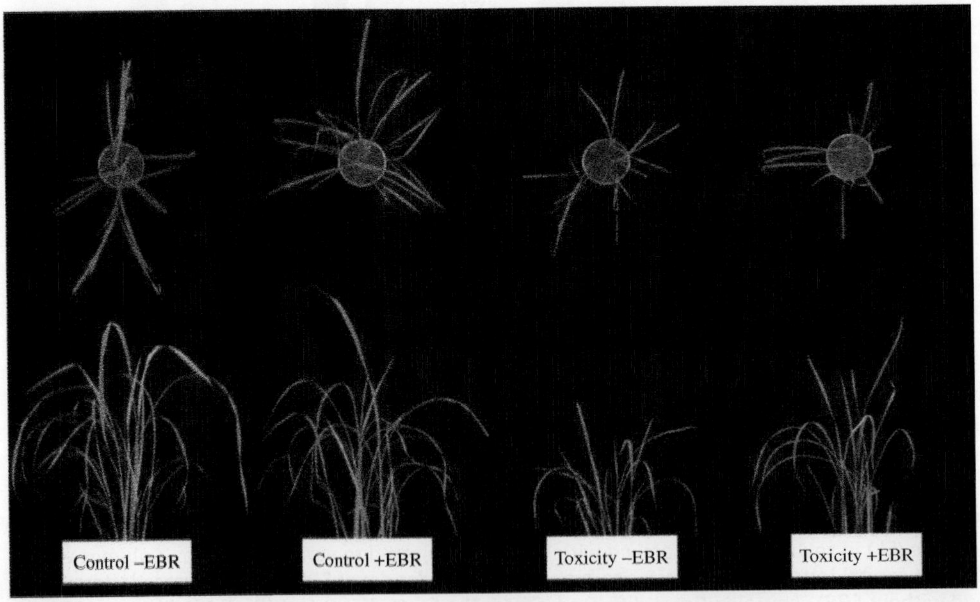

Figure 5.2 Rice plants treated with EBR and exposed to Fe toxicity. *Source:* Tadaiesky et al. (2020)/John Wiley & Sons.

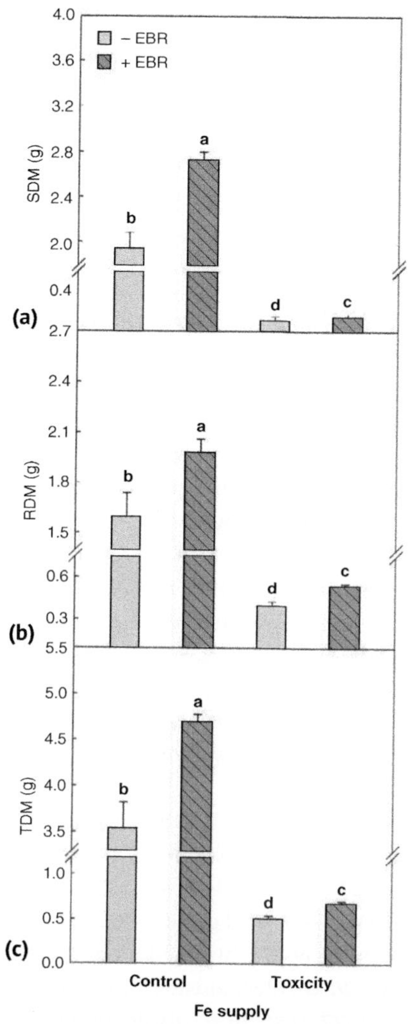

Figure 5.3 Shoot dry matter, root dry matter, and total dry matter in rice plants treated with EBR and subjected to Fe toxicity. Different lettered bars show significant differences from the Scott–Knott test ($P < 0.05$). Bars show averages and standard deviations from five replications. *Source*: Tadaiesky et al. (2020)/John Wiley & Sons.

found 31% and 26% reductions in shoot and root gene expression and morphology of Oryza sativa subjected to high Fe (125 mg L^{-1} $FeSO_4$ for three weeks) compared to the control.

5.16 Conclusion

Our chapter revealed that 24-EBR alleviated the adverse effects of Fe, Mg, and B (deficiency or toxicity) in plants, thereby reducing element contents during excess or toxicity. Additionally, electron transfer and gas exchange were maximized in plants overexposed to

Mg and pretreated with EBR (100 nM), enhancing photosystem II efficiency. EBR reduced oxidative damages from superoxide and hydrogen peroxide in plants subjected to boron stress, reducing negative impacts on metabolism. Therefore, these results indicate that EBR can improve the tolerance of plants exposed to Fe, Mg, and B (deficiency or toxicity).

References

Abadía, J., Vázquez, S., Rellán-Álvarez, R. et al. (2011). Towards a knowledge-based correction of iron chlorosis. *Plant Physiology and Biochemistry* 49: 471–482. https://linkinghub.elsevier.com/retrieve/pii/S0981942811000428

Ahammed, G.J., Ruan, Y.P., Zhou, J. et al. (2013). Brassinosteroid alleviates polychlorinated biphenyls-induced oxidative stress by enhancing antioxidant enzymes activity in tomato. *Chemosphere* 90: 2645–2653.

Ahanger, M.A., Ashraf, M., Bajguz, A. et al. (2018). Brassinosteroids regulate growth in plants under stressful environments and crosstalk with other potential phytohormones. *Journal of Plant Growth Regulation* 37: 1007–1024. http://dx.doi.org/10.1007/s00344-018-9855-2.

Albus, C.A., Salinas, A., Czarnecki, O. et al. (2012). LCAA, a novel factor required for magnesium protoporphyrin monomethylester cyclase accumulation and feedback control of aminolevulinic acid biosynthesis in tobacco. *Plant Physiology* 160: 1923–1939. https://academic.oup.com/plphys/article/160/4/1923/6109576

Anwar, A., Liu, Y., Dong, R. et al. (2018). The physiological and molecular mechanism of brassinosteroid in response to stress: a review. *Biological Research* 51: 1–15. https://doi.org/10.1186/s40659-018-0195-2.

Atique-ur-Rehman, F.M., Rashid, A., Nadeem, F. et al. (2018). Boron nutrition of rice in different production systems. A review. *Agronomy for Sustainable Development* 38: 25. http://link.springer.com/10.1007/s13593-018-0504-8

Audebert, A. and Fofana, M. (2009). Rice yield gap due to iron toxicity in West Africa. *Journal of Agronomy and Crop Science* 195: 66–76. https://doi.org/10.1111/j.1439-037X.2008.00339.x.

Audebert, A. and Sahrawat, K.L. (2000). Mechanisms for iron toxicity tolerance in lowland rice. *Journal of Plant Nutrition* 23: 1877–1885. http://dx.doi.org/10.1080/01904160009382150.

Bajguz, A. (2010). An enhancing effect of exogenous brassinolide on the growth and antioxidant activity in *Chlorella vulgaris* cultures under heavy metals stress. *Environmental and Experimental Botany* 68: 175–179. http://linkinghub.elsevier.com/retrieve/pii/S0098847209002433

Bajguz, A. and Hayat, S. (2009). Effects of brassinosteroids on the plant responses to environmental stresses. *Plant Physiology and Biochemistry* 47: 1–8. http://dx.doi.org/10.1016/j.plaphy.2008.10.002.

Becker, M. and Asch, F. (2005). Iron toxicity in rice - conditions and management concepts. *Journal of Plant Nutrition and Soil Science* 168: 558–573. https://doi.org/10.1002/jpln.200520504.

Bhattacharjee, S. (2019). *Reactive Oxygen Species in Plant Biology*. New Delhi: Springer.

Briat, J.F., Curie, C., and Gaymard, F. (2007). Iron utilization and metabolism in plants. *Current Opinion in Plant Biology* 10: 276–282. https://doi.org/10.1016/j.pbi.2007.04.003.

Briat, J.F., Dubos, C., and Gaymard, F. (2015). Iron nutrition, biomass production, and plant product quality. *Trends in Plant Science* 20: 33–40. https://dx.doi.org/10.1016/j.tplants.2014.07.005.

Cakmak, I. and Yazici, A.M. (2010). Magnesium: a forgotten element in crop production. *Better Crops* 94: 23–25.

Ceppi, M.G., Oukarroum, A., Çiçek, N. et al. (2012). The IP amplitude of the fluorescence rise OJIP is sensitive to changes in the photosystem I content of leaves: a study on plants exposed to magnesium and sulfate deficiencies, drought stress and salt stress. *Physiologia Plantarum* 144: 277–288. http://doi.wiley.com/10.1111/j.1399-3054.2011.01549.x.

Chaudhry, A.H., Nayab, S., Hussain, S.B. et al. (2021). Current understandings on magnesium deficiency and future outlooks for sustainable agriculture. *International Journal of Molecular Sciences* 22: 1–18.

Chen, L., Ding, C., Zhao, X. et al. (2015). Differential regulation of proteins in rice (*Oryza sativa* L.) under iron deficiency. *Plant Cell Reports* 34: 83–96. https://doi.org/10.1007/s00299-014-1689-1.

Connorton, J.M., Balk, J., and Rodríguez-Celma, J. (2017). Iron homeostasis in plants-a brief overview. *Metallomics* 9: 813–823.

Cunha, L.F.S., Oliveira, V.P., Nascimento, A.W.S. et al. (2020). Leaf application of 24-epibrassinolide mitigates cadmium toxicity in young *Eucalyptus urophylla* plants by modulating leaf anatomy and gas exchange. *Physiologia Plantarum* 173 (1): 67–87. https://onlinelibrary.wiley.com/doi/abs/10.1111/ppl.13182

Dalyan, E., Yüzbaşıoğlu, E., and Akpınar, I. (2018). Effect of 24-epibrassinolide on antioxidative defence system against lead-induced oxidative stress in the roots of *Brassica juncea* L. seedlings. *Russian Journal of Plant Physiology* 65: 570–578. http://link.springer.com/10.1134/S1021443718040118

Dixon, S.J. and Stockwell, B.R. (2014). The role of iron and reactive oxygen species in cell death. *Nature Chemical Biology* 10: 9–17.

Dobrikova, A.G., Vladkova, R.S., Rashkov, G.D. et al. (2014). Effects of exogenous 24-epibrassinolide on the photosynthetic membranes under non-stress conditions. *Plant Physiology and Biochemistry* 80: 75–82. http://dx.doi.org/10.1016/j.plaphy.2014.03.022.

El-Shazoly, R.M., Metwally, A.A., and Hamada, A.M. (2019). Salicylic acid or thiamin increases tolerance to boron toxicity stress in wheat. *Journal of Plant Nutrition* 42: 702–722. https://doi.org/10.1080/01904167.2018.1549670.

Farghaly, F.A., Salam, H.K., Hamada, A.M. et al. (2021). The role of benzoic acid, gallic acid and salicylic acid in protecting tomato callus cells from excessive boron stress. *Scientia Horticulturae* 278: 109867. https://doi.org/10.1016/j.scienta.2020.109867.

Farhat, N., Elkhouni, A., Zorrig, W. et al. (2016). Effects of magnesium deficiency on photosynthesis and carbohydrate partitioning. *Acta Physiologiae Plantarum* 38: 145. http://link.springer.com/10.1007/s11738-016-2165-z

Farhat, N., Rabhi, M., Krol, M. et al. (2014). Starch and sugar accumulation in Sulla carnosa leaves upon Mg^{2+} starvation. *Acta Physiologiae Plantarum* 36: 2157–2165.

Fariduddin, Q., Khalil, R.R.A.E., Mir, B.A. et al. (2013). 24-Epibrassinolide regulates photosynthesis, antioxidant enzyme activities and proline content of *Cucumis sativus* under salt and/or copper stress. *Environmental Monitoring and Assessment* 185: 7845–7856.

Fourcroy, P., Sisó-Terraza, P., Sudre, D. et al. (2014). Involvement of the ABCG37 transporter in secretion of scopoletin and derivatives by *Arabidopsis* roots in response to iron deficiency. *New Phytologist* 201: 155–167.

García-Sánchez, F., Simón-Grao, S., Martínez-Nicolás, J.J. et al. (2020). Multiple stresses occurring with boron toxicity and deficiency in plants. *Journal of Hazardous Materials* 397: 122713. https://doi.org/10.1016/j.jhazmat.2020.122713.

Geilfus, C.-M. (2019). Plant secondary compounds. In: *Controlled Environment Horticulture*, 19–33. Cham: Springer.

Gill, S.S. and Tuteja, N. (2010). Reactive oxygen species and antioxidant machinery in abiotic stress tolerance in crop plants. *Plant Physiology and Biochemistry* 48: 909–930. https://linkinghub.elsevier.com/retrieve/pii/S0981942810001798

Goli, E., Hiemstra, T., and Rahnemaie, R. (2019). Interaction of boron with humic acid and natural organic matter: experiments and modeling. *Chemical Geology* 515: 1–8. https://doi.org/10.1016/j.chemgeo.2019.03.021.

Guerinot, M.L. and Yi, Y. (1994). Iron: nutritious, noxious, and not readily available. *Plant Physiology* 104: 815–820.

Hänsch, R. and Mendel, R.R. (2009). Physiological functions of mineral micronutrients (Cu, Zn, Mn, Fe, Ni, Mo, B, Cl). *Current Opinion in Plant Biology* 12: 259–266.

Hariadi, Y. and Shabala, S. (2004). Screening broad beans (*Vicia faba*) for magnesium deficiency. II. Photosynthetic performance and leaf bioelectrical responses. *Functional Plant Biology* 31: 539–549.

Harmsen, K. and van Breemen, N. (1975). Translocation of iron in acid sulfate soils: II. Production and diffusion of dissolved ferrous iron. *Soil Science Society of America Journal* 39: 1148–1153.

Hawkesford, M., Horst, W., Kichey, T. et al. (2012). Functions of macronutrients. In: *Marschner's Mineral Nutrition of Higher Plants*, 135–189. Elsevier. http://dx.doi.org/10.1016/B978-0-12-384905-2.00006-6.

Hayat, S., Alyemeni, M.N., and Hasan, S.A. (2012). Foliar spray of brassinosteroid enhances yield and quality of *Solanum lycopersicum* under cadmium stress. *Saudi Journal of Biological Sciences* 19: 325–335. http://dx.doi.org/10.1016/j.sjbs.2012.03.005.

He, D., Chen, X., Zhang, Y. et al. (2023). Magnesium is a nutritional tool for the yield and quality of oolong tea (*Camellia sinensis* L.) and reduces reactive nitrogen loss. *Scientia Horticulturae* 308: 111590. https://linkinghub.elsevier.com/retrieve/pii/S0304423822007087

Hermans, C., Hammond, J.P., White, P.J. et al. (2006). How do plants respond to nutrient shortage by biomass allocation? *Trends in Plant Science* 11: 610–617.

Hermans, C., Johnson, G.N., Strasser, R.J. et al. (2004). Physiological characterisation of magnesium deficiency in sugar beet: acclimation to low magnesium differentially affects photosystems I and II. *Planta* 220: 344–355.

Hermans, C. and Verbruggen, N. (2005). Physiological characterization of Mg deficiency in *Arabidopsis thaliana*. *Journal of Experimental Botany* 56: 2153–2161.

Herrera-Rodríguez, M., González-Fontes, A., Rexach, J. et al. (2010). Role of boron in vascular plants and response mechanisms to boron stresses. *Plant Stress* 4: 115–122.

Hua, Y., Feng, Y., Zhou, T. et al. (2017). Genome-scale mRNA transcriptomic insights into the responses of oilseed rape (*Brassica napus* L.) to varying boron availabilities. *Plant and Soil* 416: 205–225.

Huang, H., Ullah, F., Zhou, D.X. et al. (2019). Mechanisms of ROS regulation of plant development and stress responses. *Frontiers in Plant Science* 10: 1–10.

Jan, S., Alyemeni, M.N., Wijaya, L. et al. (2018). Interactive effect of 24-epibrassinolide and silicon alleviates cadmium stress via the modulation of antioxidant defense and glyoxalase systems and macronutrient content in *Pisum sativum* L. seedlings. *BMC Plant Biology* 18: 146. https://bmcplantbiol.biomedcentral.com/articles/10.1186/s12870-018-1359-5

Jeong, J., Merkovich, A., Clyne, M. et al. (2017). Directing iron transport in dicots: regulation of iron acquisition and translocation. *Current Opinion in Plant Biology* 39: 106–113.

Jiang, Y., Cheng, F., Zhou, Y. et al. (2012). Hydrogen peroxide functions as a secondary messenger for brassinosteroids-induced CO_2 assimilation and carbohydrate metabolism in *Cucumis sativus*. *Journal of Zhejiang University Science B - Biomedicine & Biotechnology* 13: 811–823. http://www.pubmedcentral.nih.gov/articlerender.fcgi?artid=3468824&tool=pmcentrez&rendertype=abstract

Kant, C., Aydin, A., and Turan, M. (2008). Ameliorative effect of hydro gel substrate on growth, inorganic ions, proline, and nitrate contents of bean under salinity stress. *Journal of Plant Nutrition* 31: 1420–1439.

Kaya, C., Sarıoğlu, A., Ashraf, M. et al. (2020). Gibberellic acid-induced generation of hydrogen sulfide alleviates boron toxicity in tomato (*Solanum lycopersicum* L.) plants. *Plant Physiology and Biochemistry* 153: 53–63. https://doi.org/10.1016/j.plaphy.2020.04.038.

Kayıhan, C., Öz, M.T., Eyidoğan, F. et al. (2017). Physiological, biochemical, and transcriptomic responses to boron toxicity in leaf and root tissues of contrasting wheat cultivars. *Plant Molecular Biology Reporter* 35: 97–109. http://link.springer.com/10.1007/s11105-016-1008-9

Kim, S.A. and Guerinot, M.L. (2007). Mining iron: iron uptake and transport in plants. *FEBS Letters* 581: 2273–2280. http://doi.wiley.com/10.1016/j.febslet.2007.04.043

Kobayashi, M., Miyamoto, M., Matoh, T. et al. (2018). Mechanism underlying rapid responses to boron deprivation in *Arabidopsis* roots. *Soil Science and Plant Nutrition* 64: 106–115. https://doi.org/10.1080/00380768.2017.1416670.

Kobayashi, T. and Nishizawa, N.K. (2012). Iron uptake, translocation, and regulation in higher plants. *Annual Review of Plant Biology* 63: 131–152.

Kumari, J., Udawat, P., Dubey, A.K. et al. (2017). Overexpression of SbSI-1, a nuclear protein from *Salicornia brachiata* confers drought and salt stress tolerance and maintains photosynthetic efficiency in transgenic tobacco. *Frontiers in Plant Science* 8: 1215. http://www.ncbi.nlm.nih.gov/pubmed/28751902%0Ahttp://www.pubmedcentral.nih.gov/articlerender.fcgi?artid=PMC5508026%0Ahttp://journal.frontiersin.org/article/10.3389/fpls.2017.01215/full

Lamichhane, S., Tarpley, L., and Dou, F. (2023). Impact of excess magnesium salt supply on rice yield, physiological response, and grain mineral content. *Sustainability (Switzerland)* 15: 15741.

Larré, C.F., Marini, P., Moraes, C.L. et al. (2014). Influência do 24-epibrassinolídeo na tolerância ao estresse salino em plântulas de arroz. *Semina: Ciencias Agrarias* 35: 67–76.

Lehto, T., Ruuhola, T., and Dell, B. (2010). Boron in forest trees and forest ecosystems. *Forest Ecology and Management* 260: 2053–2069.

Li, X., Guo, X., Zhou, Y. et al. (2016). Overexpression of a brassinosteroid biosynthetic gene dwarf enhances photosynthetic capacity through activation of Calvin cycle enzymes in tomato. *BMC Plant Biology* 16: 33. http://www.biomedcentral.com/1471-2229/16/33

Li, C.-P., Qi, Y.-P., Zhang, J. et al. (2017a). Magnesium-deficiency-induced alterations of gas exchange, major metabolites and key enzymes differ among roots, and lower and upper leaves of *Citrus sinensis* seedlings. *Tree Physiology* 37: 1564–1581. https://academic.oup.com/treephys/article/37/11/1564/3867961

Li, M., Zhao, Z., Zhang, Z. et al. (2017b). Effect of boron deficiency on anatomical structure and chemical composition of petioles and photosynthesis of leaves in cotton (*Gossypium hirsutum* L.). *Scientific Reports* 7: 1–10. http://dx.doi.org/10.1038/s41598-017-04655-z.

Liu, H., Soomro, A., Zhu, Y. et al. (2016). QTL underlying iron and zinc toxicity tolerances at seedling stage revealed by two sets of reciprocal introgression populations of rice (*Oryza sativa* L.). *Crop Journal* 4: 280–289. http://dx.doi.org/10.1016/j.cj.2016.05.007.

Liu, H., Zhang, C., Wang, J. et al. (2017). Influence and interaction of iron and cadmium on photosynthesis and antioxidative enzymes in two rice cultivars. *Chemosphere* 171: 240–247. http://dx.doi.org/10.1016/j.chemosphere.2016.12.081.

Mahender, A., Swamy, B., Anandan, A. et al. (2019). Tolerance of iron-deficient and -toxic soil conditions in rice. *Plants* 8: 1–34. http://www.mdpi.com/2223-7747/8/2/31

Majerus, V., Bertin, P., Swenden, V. et al. (2007). Organ-dependent responses of the african rice to short-term iron toxicity: ferritin regulation and antioxidative responses. *Biologia Plantarum* 51: 303–312.

Marschner, P. (2012). *Marschner's Mineral Nutrition of Higher Plants*. Elsevier Science.

Merwad, A.R.M.A., Desoky, E.S.M., and Rady, M.M. (2018). Response of water deficit-stressed Vigna unguiculata performances to silicon, proline or methionine foliar application. *Scientia Horticulturae* 228: 132–144. http://dx.doi.org/10.1016/j.scienta.2017.10.008.

Mittler, R. (2017). ROS are good. *Trends in Plant Science* 22: 11–19.

Nath, K., Jajoo, A., Poudyal, R.S. et al. (2013). Towards a critical understanding of the photosystem II repair mechanism and its regulation during stress conditions. *FEBS Letters* 587: 3372–3381. http://dx.doi.org/10.1016/j.febslet.2013.09.015.

Nawaz, M., Ishaq, S., Ishaq, H. et al. (2020). Salicylic acid improves boron toxicity tolerance by modulating the physio-biochemical characteristics of maize (*Zea mays* L.) at an early growth stage. *Agronomy* 10: 2013. https://www.mdpi.com/2073-4395/10/12/2013

Ogweno, J.O., Song, X.S., Shi, K. et al. (2008). Brassinosteroids alleviate heat-induced inhibition of photosynthesis by increasing carboxylation efficiency and enhancing antioxidant systems in *Lycopersicon esculentum*. *Journal of Plant Growth Regulation* 27: 49–57. http://link.springer.com/10.1007/s00344-007-9030-7

Oliveira, V.P., Lima, M.D.R., Silva, B.R.S. et al. (2019). Brassinosteroids confer tolerance to salt stress in *Eucalyptus urophylla* plants enhancing homeostasis, antioxidant metabolism and leaf anatomy. *Journal of Plant Growth Regulation* 38: 557–573. http://link.springer.com/10.1007/s00344-018-9870-3

Parihar, P., Singh, S., Singh, R. et al. (2015). Effect of salinity stress on plants and its tolerance strategies: a review. *Environmental Science and Pollution Research* 22: 4056–4075.

Per, T.S., Khan, N.A., Reddy, P.S. et al. (2017). Approaches in modulating proline metabolism in plants for salt and drought stress tolerance: phytohormones, mineral nutrients and transgenics. *Plant Physiology and Biochemistry* 115: 126–140. http://dx.doi.org/10.1016/j.plaphy.2017.03.018.

Pereira, Y.C., Nascimento, A.N.S., Aguiar, B.T.S. et al. (2022). Anatomical modifications modulated by pretreatment with 24-epibrassinolide alleviate boron stress in soybean plants: valuable repercussions on nutrient contents, photosynthesis, and biomass. *Journal of Soil Science and Plant Nutrition* 22: 4533–4550. https://doi.org/10.1007/s42729-022-01053-x.

Pereira, Y.C., Rodrigues, W.S., Lima, E.J.A. et al. (2019). Brassinosteroids increase electron transport and photosynthesis in soybean plants under water deficit. *Photosynthetica* 57: 181–191. https://doi.org/10.32615/ps.2019.029.

Quinet, M., Vromman, D., Clippe, A. et al. (2012). Combined transcriptomic and physiological approaches reveal strong differences between short- and long-term response of rice (*Oryza sativa*) to iron toxicity. *Plant Cell and Environment* 35: 1837–1859.

Ramakrishna, B. and Rao, S.S.R. (2014). Foliar application of brassinosteroids alleviates adverse effects of zinc toxicity in radish (*Raphanus sativus* L.) plants. *Protoplasma* 252: 665–677. http://link.springer.com/10.1007/s00709-014-0714-0

Reid, R. (2014). Understanding the boron transport network in plants. *Plant and Soil* 385: 1–13.

Ridolfi, M. and Garrec, J.P. (2000). Consequences of an excess Al and a deficiency in Ca and Mg for stomatal functioning and net carbon assimilation of beech leaves. *Annals of Forest Science* 57: 209–218.

Rodrigues, W.S., Pereira, Y.C., de Souza, A.L.M. et al. (2020). Alleviation of oxidative stress induced by 24-epibrassinolide in soybean plants exposed to different manganese supplies: upregulation of antioxidant enzymes and maintenance of photosynthetic pigments. *Journal of Plant Growth Regulation* 39: 1425–1440. http://link.springer.com/10.1007/s00344-020-10091-7

Roquette, J.G., Ortega-Rodriguez, D.R., Portal-Cahuana, L.A. et al. (2023). Environmental forensics evaluation of residual soybean sludge using trees of Brazilian savannah. *Environmental Nanotechnology, Monitoring and Management* 20: 100814.

Sakamoto, T., Inui, Y.T., Uraguchi, S. et al. (2011). Condensin II alleviates DNA damage and is essential for tolerance of boron overload stress in *Arabidopsis*. *The Plant Cell* 23: 3533–3546. https://academic.oup.com/plcell/article/23/9/3533-3546/6097767

Santos, L.A., Batista, B.L., and Lobato, A.K.S. (2023). 24-Epibrasinolide delays chlorophyll degradation and stimulates the photosynthetic machinery in magnesium-stressed soybean plants. *Journal of Plant Growth Regulation* 42: 183–198. https://doi.org/10.1007/s00344-021-10539-4.

Santos, L.R., da Silva, B.R.S., Pedron, T. et al. (2020). 24-Epibrassinolide improves root anatomy and antioxidant enzymes in soybean plants subjected to zinc stress. *Journal of Soil Science and Plant Nutrition* 20: 105–124.

Schmidt, F., Fortes, M.Á., Wesz, J. et al. (2013). Impacto do manejo da água na toxidez por ferro no arroz irrigado por alagamento. *Revista Brasileira de Ciencia do Solo* 37: 1226–1235.

Shireen, F., Nawaz, M., Chen, C. et al. (2018). Boron: functions and approaches to enhance its availability in plants for sustainable agriculture. *International Journal of Molecular Sciences* 19: 1856. http://www.mdpi.com/1422-0067/19/7/1856

Siddiqui, H., Ahmed, K.B.M., and Hayat, S. (2018). Comparative effect of 28-homobrassinolide and 24-epibrassinolide on the performance of different components influencing the photosynthetic machinery in *Brassica juncea* L. *Plant Physiology and Biochemistry* 129: 198–212. https://linkinghub.elsevier.com/retrieve/pii/S0981942818302444

Silveira, V.C., Oliveira, A.P., Sperotto, R.A. et al. (2007). Influence of iron on mineral status of two rice (*Oryza sativa* L.) cultivars. *Brazilian Journal of Plant Physiology* 19: 127–139. http://www.scielo.br/scielo.php?script=sci_arttext&pid=S1677-04202007000200005&lng=en&tlng=en

Stringlis, I.A., De Jonge, R., and Pieterse, C.M.J. (2019). The age of coumarins in plant-microbe interactions. *Plant and Cell Physiology* 60: 1405–1419.

Suriyagoda, L.D.B., Sirisena, D.N., Somaweera, K.A.T.N. et al. (2017). Incorporation of dolomite reduces iron toxicity, enhances growth and yield, and improves phosphorus and potassium nutrition in lowland rice (*Oryza sativa* L). *Plant and Soil* 410: 299–312.

Tadaiesky, L.B.A., Silva, B.R.S., Batista, B.L. et al. (2020). Brassinosteroids trigger tolerance to iron toxicity in rice. *Physiologia Plantarum* 171: 371–387. https://doi.org/10.1111/ppl.13230.

Verbruggen, N. and Hermans, C. (2013). Physiological and molecular responses to magnesium nutritional imbalance in plants. *Plant and Soil* 368: 87–99. http://link.springer.com/10.1007/s11104-013-1589-0

Vriet, C., Russinova, E., and Reuzeau, C. (2012). Boosting crop yields with plant steroids. *The Plant Cell* 24: 842–857. http://www.plantcell.org/cgi/doi/10.1105/tpc.111.094912

Waters, B.M. (2011). Moving magnesium in plant cells. *The New Phytologist* 190: 510–3. http://doi.wiley.com/10.1111/j.1469-8137.2011.03724.x

Wimmer, M.A. and Eichert, T. (2013). Review: mechanisms for boron deficiency-mediated changes in plant water relations. *Plant Science* 203–204: 25–32. http://dx.doi.org/10.1016/j.plantsci.2012.12.012.

Yau, S.K. and Ryan, J. (2008). Boron toxicity tolerance in crops: a viable alternative to soil amelioration. *Crop Science* 48: 854. https://www.crops.org/publications/cs/abstracts/48/3/854

Yuan, L., Shu, S., Sun, J. et al. (2012). Effects of 24-epibrassinolide on the photosynthetic characteristics, antioxidant system, and chloroplast ultrastructure in *Cucumis sativus* L. under $Ca(NO_3)_2$ stress. *Photosynthesis Research* 112: 205–214.

Yuan, L., Zhu, S., Shu, S. et al. (2015). Regulation of 2,4-epibrassinolide on mineral nutrient uptake and ion distribution in $Ca(NO_3)_2$ stressed cucumber plants. *Journal of Plant Physiology* 188: 29–36. https://linkinghub.elsevier.com/retrieve/pii/S0176161715001595

Zancani, M., Peresson, C., Biroccio, A. et al. (2004). Evidence for the presence of ferritin in plant mitochondria. *European Journal of Biochemistry* 271: 3657–3664.

Zhang, M., Zhai, Z., Tian, X. et al. (2008). Brassinolide alleviated the adverse effect of water deficits on photosynthesis and the antioxidant of soybean (*Glycine max* L.). *Plant Growth Regulation* 56: 257–264. http://link.springer.com/10.1007/s10725-008-9305-4

Zhang, X., Zhang, D., Sun, W. et al. (2019). The adaptive mechanism of plants to iron deficiency via iron uptake, transport, and homeostasis. *International Journal of Molecular Sciences* 20: 2424.

Zhang, B., Zhang, C., Tang, R. et al. (2022). Two magnesium transporters in the chloroplast inner envelope essential for thylakoid biogenesis in *Arabidopsis*. *New Phytologist* 236: 464–478. https://nph.onlinelibrary.wiley.com/doi/abs/10.1111/nph.18349 (accessed 22 November 2022).

Zhou, T., Hua, Y., and Xu, F. (2017). Involvement of reactive oxygen species and Ca^{2+} in the differential responses to low-boron in rapeseed genotypes. *Plant and Soil* 419: 219–236.

6

Role of Brassinosteroids on Postharvest Physiology of Plants

Naveed Mushtaq[1,2,3], Muhammad Ahsan Altaf[1,2,3], and Zhiwei Wang[1,2,3]*

[1]*Key Laboratory for Quality Regulation of Tropical Horticultural Crops of Hainan Province, School of Breeding and Multiplication (Sanya Institute of Breeding and Multiplication), Center of Nanfan and High-Efficiency Tropical Agriculture, Hainan University, Sanya, China*
[2]*Key Laboratory for Quality Regulation of Tropical Horticultural Crops of Hainan Province, School of Tropical Agriculture and Forestry, Hainan University, Danzhou, China*
[3]*Hainan Yazhou Bay Seed Laboratory, Sanya, China*

*Corresponding author: ahsanaltaf8812@gmail.com

6.1 Introduction

Brassinosteroids (BRs) have been termed the "New Class of plant hormones" (Clouse and Sasse 1998; Khripach et al. 1998), "polyhydroxylated steroidal plant hormone" (Fariduddin et al. 2014), as a novel and unique class of plant hormones. BRs were found in all plant kingdoms and have also been documented in various plant sections. Nevertheless, their amount varies depending on the part of the plant. Compared with mature tissues, the content of BR is higher in young, growing tissues like the pollen, and premature seeds are the largest sources of BR, while shoots and leaves typically have lower levels (Krishna et al. 2017).

BRs' structural resemblance to animal steroid hormones is evident, as they can control the expression of numerous genes, complex metabolic pathways, and the predetermined progression of morphogenesis. According to some studies, BR is non-toxic and safe for the environment. It can exist in its free form or conjugated with sugars or fatty acids (Ali 2017). As a result, they are frequently utilized to increase the yields of horticultural crops. BR deficiency impacts cell elongation and division, resulting in abnormal growth (dwarfism) and development (sterility, altered photomorphogenesis, etc.) in the lack of light (Hafeez et al. 2021). Other processes, such as the formation of ovules, anthers, and pollen, leaf erectness, seed size, regulation of organ boundaries, flowering, and fruit development, all depend on BR. Furthermore, BR is engaged in a variety of processes related to plant growth and development, such as photomorphogenesis, seed germination, flowering, stomata production, and stem elongation, in addition to light (Wang et al. 2018). BR can moreover enhance the resistance to biotic and abiotic stresses, like salt, temperature, chilling, and drought (Li et al. 2018a; Basit et al. 2021). BR mediates the adaptive responses in adverse conditions

Brassinosteroids and Abiotic Stress: Recent Developments on Physiological and Molecular Aspects,
First Edition. Edited by Anket Sharma, et al.
Published 2026 by John Wiley & Sons Ltd.

(Hafeez et al. 2021), initiates the antioxidant defense system (Zou et al. 2018), induces the osmoprotectants' accumulation, and renders the genes that respond to stress via the signaling cascade (Ye et al. 2017).

6.2 Detection of Brassinosteroids in Various Plant Species

BRs were first identified in *Brassica napus* pollen due to their capacity to stimulate growth (Mitchell et al. 1970). It has been found that BRs stimulate the division and elongation of plant cells. BRs were known as "brassins," and the most active BR is brassinolide (BL), which was discovered in 1979 (Grove et al. 1979). The BR-insensitive 1 (BRI1), a receptor kinase that initiates an intracellular signaling cascade in response to extracellular BR sensing, was the most important discovery (Li and Chory 1997). Numerous chemically distinct BRs have been found in the plant kingdom since the discovery of BL, including terrestrial plants and green algae, indicating that BRs developed early in the evolution of plants.

6.3 Occurrence of Brassinosteroids in Plants

All plant parts, including leaves, stems, roots, flowers, pollen, anthers, seeds, and organs, have been found to contain BRs. BR is generally spread in all developing parts of plants. However, fruit, pollen, and seed have been found to contain noticeably significantly higher amounts (Li and He 2020). Young tissues often have a greater quantity of BR (1–100 ng g^{-1} fresh weight) than mature tissues (0.01–0.1 ng g^{-1} fresh weight) (Hayat and Ahmad 2003). Due to their extensive distribution in plants and increased biological activity, BL and CS are the most significant BRs. Nevertheless, because of its commercial availability, the 24-epibrassinolide (EBR) is the most often used BR for conducting and exploring the physiological aspects of the phytohormones in plants. Following the discovery of BL, approximately 69 BRs were found in 64 plant species, comprising 6 gymnosperms, 3 algae, 53 angiosperms, 1 pteridophyte, 1 bryophyte, 2 lycophytes, a moss, and 13 fern species (Yokota et al. 2017). Regarding the concentration, the 6-deoxotyphasterol was found to be 6400-fold higher compared to BL in the pollen of *Cupressus arizonica*. In terms of their biological actions in plants, only 52 BRs have been reported (Hussain et al. 2020), and the most abundant BRs found in plants are CS, BL, TY, 6-deoxoCS, TE, and 28-norCS (Bajguz and Tretyn 2003; Hayat and Ahmad 2003). The seeds of unripe beans contained the most incredible variety of BRs, with 2 conjugated and 25 free available forms (Hayat and Ahmad 2003). In shoots, seeds, and pollen of several legumes, fewer BR members have been discovered; their amount ranged from 0.006 to 6280 ng g^{-1} fresh weight (Bajguz and Tretyn 2003).

6.4 Role of Brassinosteroids in the Growth and Development of Plants

BRs, steroid growth hormones, have a variety of functions in the growth and development of plants (Fang et al. 2019; Lin 2020) (Figure 6.1). BRs initiate and are involved in the regulation of several growth, developing and physiological crucial functions in plants,

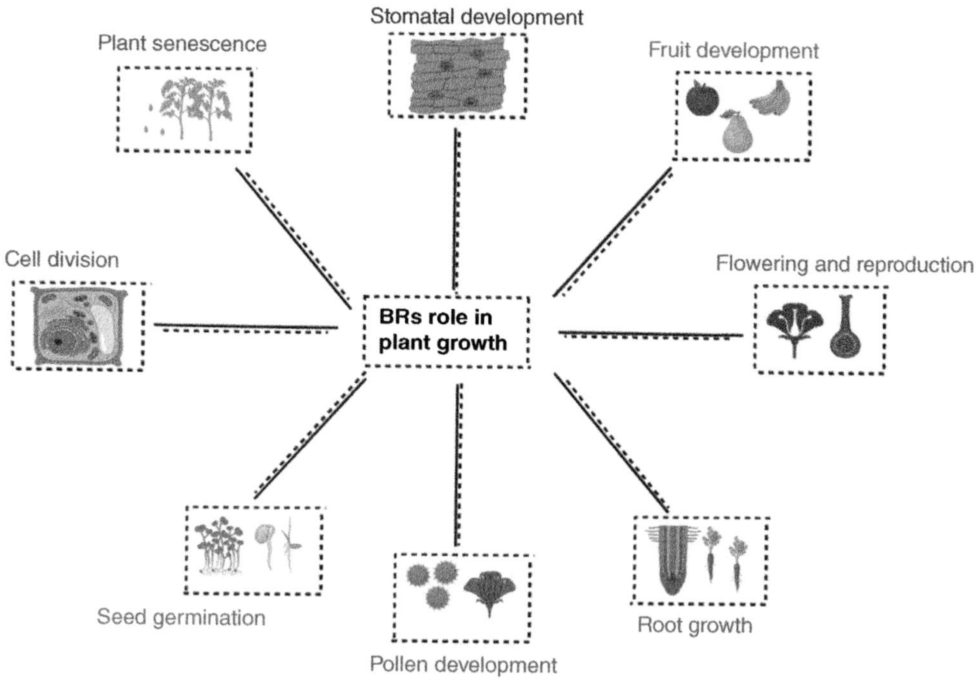

Figure 6.1 The role of BRs in plant growth, development, and several physiological aspects. *Source*: BioRender.

mainly the maintenance of the cellular stem, cell division process, the development of the vascular organs, elongation of diverse cells, and the floral growth process (Singh and Savaldi-Goldstein 2015; Lv et al. 2018). Furthermore, the BRs are also involved in different functions such as hypocotyl part elongation, root system growth (Hacham et al. 2011), shoot development (Bajguz et al. 2020), stomata functioning (Khan et al. 2013), pollen tube progression, germination, tracheary element separation (Yamamoto et al. 2001), xylem development (Khan et al. 2013), photomorphogenesis, and plant reproduction process (Ahammed et al. 2020), and aging process (Yin et al. 2002). Apart from this, the BRs could also stimulate cell progression in seed germination (Zadvornova et al. 2005), control cell cycle division (González-García et al. 2011), and regulate leaf cells' proliferation (Nakaya et al. 2002). Under various environmental conditions, the BRs also control the stress responses, the stomata movement, and the opening (Lin et al. 2020; Ackerman-Lavert and Savaldi-Goldstein 2020). Besides, the BR also plays a crucial role in regulating male and female parts' fertility in the plant (Makarevitch et al. 2012; Hartwig et al. 2011).

BR also plays a part and function in the development and the growth of the stigma, plant architecture system, chilling tolerance, proton movement, tiller growth, and leaf size development (Jaillais and Vert 2012; Zhu et al. 2015c). Besides, the exogenous supplementation of BR or changes in the signaling and biosynthesis could also enhance plant yields and production (Jaiswal et al. 2020).

Additionally, BRs regulate a number of genes in plants, and one research discovered that BRs induced the transcript level of the WRKY54, WRKY46, and WRKY70 genes, which were revealed to have positive effects on BR-induced plant development in *A. thaliana* (Chen et al. 2017). In a different study, SDG8 has been associated with BR-associated gene transcript level, and the knockout mutant of the SDG8 exhibited lessened growth of plants along with weakened BR responses in *A. thaliana*. The BR initiates seed formation and has a role in seed growth, size, and number via transcriptionally changing the genes and related pathways that trigger the development and formation of the ovule and seeds (Huang et al. 2013; Jiang et al. 2013). Besides, BRs can also affect the relationship of root nitrogen foraging in *A. thaliana* plants due to a slight nitrogen deficiency. The BR-associated BSK3 gene played a function in the development and growth of the primary root in slight nitrogen deficiency (Jia et al. 2019).

BR also plays a crucial role in the rice grain yield. Importantly, the architecture (Fujioka and Yokota 2003) and the transcript levels of the D11 and BRD1 are involved in the biosynthesis pathway of BR, and overall affect the growth and size of plants (Tanabe et al. 2005). BRs also play a role and impact the N fertilization process, which is important in the spikelet growth and also induces the spikelet growth process and induces the activities of the key antioxidant system AOS and energy metabolism during the development (Zhang et al. 2019). In other plants like maize, the inadequate BR biosynthesis system affects reproduction and impacts pollen failure and the growth of the anther (Hartwig et al. 2011).

The length, vigor, fresh, and dry shoot and root weights were all significantly improved as lucerne seeds were primed with BL (Zhang et al. 2007). The grape berry's ripening was enhanced by the application of BR (Symons et al. 2006). It was found that applying EBR significantly increased the sucrose synthase and soluble sugar content of berries (Xu et al. 2015). Furthermore, the deterioration of Jujube fruit was lessened after treatment with BR, most likely due to its capacity to delay senescence and promote disease resistance (Zhu et al. 2010). Furthermore, it has been demonstrated that in potato tubers, BRs stimulate the growth of the apical meristem (Meudt et al. 1983), accelerating the rate of cell division in Petunia hybrida's isolated protoplasts (Oh 2003). The inhibition of flowering in *Pharbitis nil* plants by CS and BL treatments, together with the inductive photoperiod, suggests that BR functions in concert with environmental cues to assure an effective reproductive transition process (Kesy et al. 2003). Different transcription factors regulate plant growth and development described in Table 6.1.

6.5 Brassinosteroids Function in the Seed and Yield of Plants

BRs can take part in physiological processes in response to stress by regulating plant development and enhancing plant performance through interactions with plant growth regulators or other plant hormones (Trevisan et al. 2020; Ahammed et al. 2015). Many developmental processes, such as seed formation (Jiang et al. 2013), pollen development (Ye et al. 2010), and flowering time (Domagalska et al. 2010), are impacted by the restriction of BR signaling. The BR deficit in plants caused dwarfism (Li et al. 1996), affected stomata development (Gudesblat et al. 2012), decreased male fertility, smaller hypocotyls, petioles, and internodes, and delayed the flowering process (Ye et al. 2010). Apart from this, BR-deficient plants also exhibit a compact plant structure due to the decreased lamina

Table 6.1 Different transcription factors regulates plant growth and development.

Crop/plant	Gene	Description of gene	Role in growth	References
Arabidopsis plants	*CESA*	CESA gene involved in cellulose	Initiation of the cellulose	Xie et al. (2011)
Arabidopsis plants	*CYCD3;1*	Cell division process	Cell cycle development	González-García et al. (2011)
Arabidopsis plants	*SDG8*	Arabidopsis contains 43 SDG	BR-regulated gene transcript level	Wang et al. (2014)
Arabidopsis plants	*WRKY46, WRKY54,* and *WRKY70*	WRKY family TFs composed of more than 69 members in Arabidopsis	Plant development	Chen et al. (2017)
Arabidopsis plants	*BZR1, BIMs*	BR-activated transcription factor and BIMs;bHLH transcription factor	Inhibiting the flowering	Li et al. (2018b)
Arabidopsis plants	*ATGs* genes	Autophagy-associated genes	Initiation autophagosomes	Wang et al. (2019)
Arabidopsis plants	*HMGR*	Mevalonate (MVA) pathway	Promote anthocyanin, coloration, and sugar level	Zheng et al. (2020)

inclination. Nevertheless, lack of BR can result in reduced germination of seeds (Zhu et al. 2013), improper stomatal distribution, and reductions in grain size, seed fertility, and tiller number (Tong and Chu 2018). BR-deficient mutants are frequently referred to as late flowering due to the slowed growth process (Zhao and Li 2012).

6.6 The involvement of Brassinosteroids in the Postharvest Physiology of Plants

The impact of BR on the postharvest is presented below, but the research on this domain is limited compared to the studies on the preharvest (Nolan et al. 2019). We have explored the research that has focused on the supplementation of BR or its inhibitor (brassinazole, BRZ) in postharvest plants, including fruits and vegetables.

6.6.1 Brassinosteroids and Ripening of Fruits

BRs have a crucial function in the ripening process in a variety of fruits, and applications of HBL and EBR have shown enhanced lycopene percentage in tomato plants. The increased ethylene biosynthesis was linked to improved tomato ripening, which in turn boosted the tomato ripening overall (Vardhini and Rao 2002). The application of BRs causes an increase in the BR endogenous production via upregulating the GhDWF4 gene, which in turn speeds up the ripening process of tomato fruits (Ye et al. 2015). Another research showed that the application of BRs increased the endogenous content and induced the expression of the FaBRI1 gene, which in turn sped up the ripening and coloring of strawberry fruits (Chai et al. 2013). In another study, an exogenous EBR treatment was found to upregulate the expression of the FaBRI1 receptor and the signaling pathway associated with the genes FaBRZ1 and FaBIN2 at the pink stage. In addition, EBR application changed the phenylpropanoid pathway, which improved the anthocyanin accumulation in strawberry fruits (Ayub et al. 2018).

It was discovered that an increase in the endogenous content of BRs caused the ripening of grape berries. The ripening of grape berries has been shown to be significantly affected by the upregulation of genes that encode for the production of BRs, such as DWARF1 and the enzyme BR-6-oxidase (Symons et al. 2006). The application of EBR in mango fruits under high temperatures led to an increased respiration rate and ethylene production, and improved softening and swift color development (Zaharah and Singh 2012). Tomato fruits subjected to 3–4 μmol L^{-1} BL exhibited more genes associated with lycopene production, such as LeGLK2 and LePSY1, and the ripening process improved compared to the control (Zhu et al. 2015b). The application of BRs causes an increase in their endogenous synthesis via upregulating the GhDWF4 gene transcript level, which in turn speeds up the ripening process of tomato fruits (Ye et al. 2015). In another study, it was observed that the application of BR increased their endogenous level and significantly upregulated the expression of the FaBRI1 gene, which in turn caused the ripening and development of red coloration in strawberry fruits to occur promptly (Chai et al. 2013). The treatment of EBR also involves regulating the transcript level of the FaBRI1 receptor and the critical signaling pathway associated with the dynamic essential genes like FaBIN2 and FaBRZ1 involved in the pink stage of the strawberry fruit. Moreover, EBR also affected the phenylpropanoid pathway, which improved the anthocyanin accumulation in strawberry fruits (Ayub et al. 2018).

6.6.2 The Role of Brassinosteroids in Chilling Injury

In horticultural plants, during cold storage, chilling injury (CI) is one of the significant postharvest physiological disorders affecting a wide range of plants. Most importantly, the CI usually varies with the maturity stage, and the variety of the plants and quality loss decreases the postharvest storage period and overall adversely disturbs the stability of cell membranes (Wang et al. 2012a; Gao et al. 2015; Li et al. 2018b). In the horticultural plants, the BR plays a significant role against the freezing resistance, and it was perceived that the higher endogenous BR contents are directly related to the enhanced cold stress and the cold storage resistance in the cucumber plants (Xia et al. 2011). The studies showed that the supplementation of 15 μmol L^{-1} BL to green bell pepper fruits improved the freezing temperature stress at 3 °C via the decrease of the calyx discoloration (Wang et al. 2012a). Another research showed that bamboo shoots treated with 0.5 μmol L^{-1} BL increased the level of the P5CS and OAT enzymes and decreased the activity of the PDH enzyme, which in turn led to a decrease in the reduction of the freezing temperature injury at the low temperature of 1 °C (Liu et al. 2016b). The supplementation of BRs caused noticeably improved low-temperature injury in the orange fruits at 3 °C and also reduced the oxidative injury in the lime fruits at the low temperature of about 5 °C (Ghorbani and Pakkish 2014). In loquat fruit, the application of the BR significantly affects the stability of cell walls and protects membrane damage (Chen et al. 2022). The application of 40 μmol L^{-1} EBR to banana fruits improved their fluorescence (Fv/Fm) and decreased electrolyte leakage. Additionally, there was a significant decrease in CI in banana fruits subjected to EBR at 8 °C due to the overexpression of 50 distinct proteins linked to energy biosynthesis and the upregulation of proteins associated with methionine production. The eggplants were treated with 10 μmol L^{-1} EBR, which enhanced the membrane stability via the reduction of the leakage in the plants and reduced the CI symptoms at the lower temperature about the 1 °C conditions (Gao et al. 2015).

6.6.3 The Role of Brassinosteroids in Color Metabolism

In horticultural plants, considerably, the coloring of the fruit is an important feature that is directly related to the quality and has a crucial role in purchasing in the marketplaces; the loss of color eventually decreases the visual feature and impacts the market perspective of the plants. The foliar application of 0.5 mg L^{-1} of EBR solution hindered the chlorophyll loss percentage and decreased the yellowing with a noticeable increase in the quality of the daylily flowers (Yao et al. 2017). The vacuum infiltration of 8–10 μmol L^{-1} BL lessened the chlorophyll degradation process and displayed green color because of repressed yellowing in Persian lime fruits. A similar application also exhibited a higher hue angle and lessened the chroma in lime plants (Tavallali 2018). Similarly, a 2 μmol L^{-1} EBR spray application decreased the yellowing, displayed increased chlorophyll concentrations, and showed an increased process of chlorophyll fluorescence. The reduction in chlorophyll breakdown and increase in chlorophyll fluorescence ultimately led to a notable decrease in the yellowing of broccoli florets (Cai et al. 2019). Functions of BRs as postharvest treatment in plants (Table 6.2).

6.6.4 The Impact of Brassinosteroids on the Postharvest Browning

One of the most common unfavorable conditions that affect many fruits and vegetables is postharvest browning (Ali et al. 2016, 2019). In addition to lowering the product's visual quality, browning also affects consumers' purchasing decisions and commercial potential. The anti-browning properties of BRs have been documented. In eggplant, the 10 μmol L^{-1} EBR treatment postponed pulp browning for 15 days at 1 °C and decreased the rise in PPO and POD activities (Gao et al. 2015). For 16 days, while being stored at 4 °C, the application of 3 μL BL pointedly decreased the browning of white button mushrooms. Similarly, in lotus root slices, 80 nmol L^{-1} EBR treatment effectively inhibited enzymatic browning by suppressing quinone synthesis, POD, PPO activity, and phenol oxidation process (Gao et al. 2017). Similarly, 5 μL HBL exposure to medlar fruits significantly reduced skin browning for 60 days at 0 °C (Ekinci et al. 2019). Similarly, 5 μL HBL application in the medlar fruits significantly decreased the skin browning process for about 55 days at the low temperature of 0 °C (Ekinci et al. 2019).

Table 6.2 The effect of BRs on the postharvest and treatment.

BR	Concentration	Plants	Solution	Effect	Application	Articles
BL	10	Mango	Water solution	Less freezing damage, less conductivity, higher FA, improve membrane integrity	Immersed	Li et al. (2012)
BL	4×10^5	Strawberry	Water solution	Boost ripening	Fruit injection	Chai et al. (2013)
EBR	10	Eggplant	Water solution	Less freezing damage, higher membrane stability, fewer phenols	Immersion	Gao et al. (2015)
BL	15	Green bell Pepper	Water solution	Less freezing damage, less MDA level	Immersion	Wang et al. (2012b)
EBR	1	Papaya	Ethanol solution (0.5% v/v)	Improve the antioxidant system	Vacuum infiltration process	Mazorra et al. (2013)
EBR	15	Peach	Water solution	Lower MDA levels and H_2O_2 Levels	Immersion	Gao et al. (2016)
BL	1.50 ppm	Grape	Water solution	Less freezing injury, less MDA, less membrane damage	Immersion	Pakkish et al. (2019)
BL	10	Asparagus	Water solution	Decrease lignin contents, increase chlorophyll content, increase the phenol level	Immersion (10 min)	Wu and Yang et al. (2016)
BL	5	Jujube	Water solution	Less deterioration, Improve the antioxidant system	Immersion 4–5 min	Zhu et al. (2010)
BL	5	Tomato	Water solution	Increase lycopene, decrease ascorbic acid	Immersion	Zhu et al. (2015c)

EBR	15	Loquat	Water solution	Less lignin contents, less cellulose content	Immersion	Chen et al. (2022)
EBR	2	Broccoli	Ethanol solution	Higher APX and PAL, less yellowing	Spray method	Fang et al. (2021)
EBR	2–3	Apple	Water solution	Less volatiles, improve antioxidant	Immersion method	Liu et al. (2022)
EBR	5	Kiwifruit	Water solution	Increase ascorbic acid, decrease O_2^{-1}	Immersion method	Wang et al. (2020)
28-HBL	5	Medlar	Water solution	Reduce freezing injury, reduce MDA contents	Immersion method	Duan et al. (2022)
EBR	5–15	Pomegranate	Water solution	Increase the proline level, decrease PDH activity	Immersion method	Islam et al. (2022)
BL	80	Banana	Water solution	Reduce freezing damage, reduce proline method	Immersion method 3 min	Zhang et al. (2022)
BRZ	0.30–2.44	Grape	Ethanol (0.1% v/v)	Higher soluble sugar contents, increase PAL level	Spray method	Xi et al. (2013)
EBR	0.1	Zucchini	DMSO (4–5 mg mL^{-1})	Decrease MDA contents, induce POD activity	Spray method	Massolo et al. (2022)
BL	3 or 6	Tomato	Water solution	Higher proline Level, fewer PLD activities	Immersion method	Aghdam and Mohammadkhani (2014)
BL	3 or 6	Tomato	Water solution	Decrease freezing injury, increase phenol content	Immersion method	Aghdam et al. (2012)

6.6.5 The Effect of Brassinosteroids on the Ethylene Biosynthesis and Respiration Rate

Respiration and ethylene are critical as their induction level mainly causes short storage life and rapid senescence process of the produce (Razzaq et al. 2014). Some appropriate hormone treatments are prerequisites to lessen the induction of respiration and the ethylene process during the postharvest storage of fruits. Nevertheless, the function and the studies regarding the effect of BR application during the postharvest storage, and mainly on the ethylene production and respiration rate, are contradictory. The EBR application considerably improved both the respiration rate and ethylene production in mango plant fruits in ambient storage (Zaharah and Singh 2012). Similarly, as the broccoli florets were subjected to the EBR (3 μmol L^{-1}), which induced the transcript level of genes associated with ACC synthase, like the BoACO3 and BoACS4, and overall increased the production of ethylene in the postharvest storage process (Zhu et al. 2015a), on the other side, a study showed a contrary result that the foliar application of the 0.5 mg L^{-1} of EBR solution displayed a lessened respiration rate in the daylily flower (Yao et al. 2017). Consequently, an additional inclusive investigation is still necessary to figure out the effect of BRs on different plant varieties and during the different post-storage processes.

6.6.6 Brassinosteroids and Lipid Peroxidation

In plants, the conservation process of the higher concentrations of unsaturated fatty acids (USFA) such as linoleic and linolenic acid is very significant; the decrease in the amount of USFA in membranes ultimately leads to enhanced membrane damage, fluidity, and lipid peroxidation (Aghdam et al. 2016). Enzymes such as lipoxygenase (LOX) and phospholipase D (PLD) produce irreversible damage to membranes and enhanced lipid peroxidation, which ultimately leads to CI in the postharvest storage process at low temperatures (Aghdam et al. 2016). It has been discovered that BRs are effective in suppressing membrane permeability and lipid peroxidation. Tomato fruits subjected to BRs during the postharvest process presented a significant reduction in lipid peroxidation, which overall preserved the membrane damage and inhibited the activities of the enzymes LOX and PLD (Aghdam and Mohammadkhani 2014). In the same manner, BR supplementation lessened the electrolyte leakage process and repressed the lipid peroxidation in the green bell peppers (Aghdam and Mohammadkhani 2014), eggplant (Gao et al. 2015), peach (Gao et al. 2016), daylily flower buds (Yao et al. 2017), and lotus root slices (Gao et al. 2017).

6.6.7 Brassinosteroids and Antioxidant Activities During the Postharvest

The antioxidant defense protection enzymes such as ascorbate peroxidase (APX), catalase (CAT), peroxidase (POD), and glutathione reductase (GR) and superoxide dismutase (SOD) are really essential to lessen the harmful damage caused due to oxidative stress in the postharvest storage of fruits (Ali et al. 2018). Studies have shown that the green bell peppers treated with 15 μmol L^{-1} BL improved the antioxidant defense enzyme activities (Wang et al. 2012b). The increased phenolic biosynthesis was additionally caused by the enhanced PAL activity in table grapes during postharvest storage (Liu et al. 2016a). Similar outcomes were found in green asparagus, where treatment with both carboxymethyl chitosan coating and BL dramatically increased the activity of the enzymes CAT, POD, SOD, and PAL (Wu and Yang 2016). The increased CAT, SOD, and APX enzyme activities were also found in the white button mushrooms treated with 3 μmol L^{-1} BL (Ding et al. 2016). Similarly, the postharvest application of 0.5 mg L^{-1} EBR demonstrated increased activities of APX, SOD, CAT, and POD in daylily, which ultimately led to decreased accumulation of H_2O_2 and $O2\bullet^-$ levels and directed an extended life period (Yao et al. 2017).

6.6.8 Brassinosteroids Effect on Sugar Metabolism and Energy Regulation

Sugar metabolism is involved in the senescence initiation process in certain plants, and the amount of glucose and fructose is induced promptly in the senescence, notably by the suppression of starch content. The exact mechanism was noticed in the fruit plants, like the prompt induction in the glucose and fructose content and decrease in the starch content, which was perceived in the kiwifruits during ripening. It has been discovered that applying the BRs in the postharvest can effectively postpone the induction of the sugar concentration. Kiwifruit treated with 5 μmol L^{-1} EBR showed delayed starch breakdown and decreased activity of enzymes such as acid invertase, sucrose phosphate synthase, hexokinase, and fructokinase. The enhanced mentioned enzymes' inhibited activity consequently suppressed the induction of glucose, sucrose, and fructose (Lu et al. 2019). Consequently, the prompt induction in the sugar content might be delayed with the supplementation of the BRs, and this probable impact needs to be studied in other plants.

Adequate energy accessibility is critical to sustain normal functioning and to postpone the senescence process in the postharvest shelf life of horticultural plants. The availability of appropriate extracellular ATP similarly delays the senescence process and lessens the oxidative stress damage in the postharvest storage of plants (Aghdam et al. 2018). Appropriate energy availability is similarly crucial for the maintenance of membrane stability in the postharvest. It has been detected that increased membrane stability is directly associated with a sufficient amount of ATP and energy (Jin et al. 2013). In bamboo shoots, the application of 0.5 μmol L^{-1} BL caused enhanced activities of succinate dehydrogenase (SDH), cytochrome C oxidase (CCO), H+-ATPase, and Ca^{+2}-ATPase (Liu et al. 2016b). Besides, it increased the amount of adenosine monophosphate (AMP) and decreased the decline in ATP, ADP, and energy charge during the 40–42 days' storage period. Additionally, it was noted that higher energy availability improved membrane integrity and, as a result, enhanced CI tolerance in bamboo shoots (Liu et al. 2016b). BL was examined in the energy regulation process, and it successfully retained higher energy. Therefore, research on other BRs, like EBR, brassinazole, and HBL, needs to be conducted on horticultural plants.

6.6.9 Brassinosteroids and Postharvest Fruit Softening

One of the main postharvest limitations that causes fruits to have a short storage life potential is rapid softening. It has been discovered that the breakdown and softening of cell walls are associated with several enzymes. Like the degradation of cell walls is catalyzed by several enzymes together with polygalacturonase (PG), pectin esterase (PE), β-galactosidase (β-gal), and pectate lyase (PL) (Razzaq et al. 2013; He et al. 2018). Subsequently, the reduction in the values of these critical enzymes is essential for the management of the fruit softening process, and the BRs have been perceived as proper in hindering the activities of the cell wall-degrading enzymes. Postharvest application of 10–15 μmol L^{-1} EBR and 5–7 μmol L^{-1} brassinazole considerably repressed the transcript level of the DkEGase1, DkPL1, DkPE2, and DkPG1 genes that consecutively lessened the activities of PL, EGase, PG, and β-gal.

Furthermore, a similar application likewise caused the increased pectin content, cellulose, and acid-soluble pectin in persimmon fruits. The reduced activities of PL, EGase, PG and β-gal and overall, the higher pectin contents ultimately caused significantly suppressed the softening of persimmon fruits in the low-temperature storage (He et al. 2018).

6.6.10 Brassinosteroids' Effect on the Postharvest Decay

The postharvest decay process is directly associated with the reduction in the quality and quantity of produce caused by bacterial and fungal diseases. On the other hand, fungal diseases are more significant since they cause significant economic loss in the value of the commodities (Aghdam et al. 2016). Fruits and vegetables are affected by postharvest decay because of latent conditions that develop in the field or wounds caused during gathering and management within supply systems (Hussain et al. 2015). The management of the postharvest decay process of vegetables and fruits is essential to minimize the damage to the produce. The application of EBR about 0.5 or 0.8 mg L^{-1} induced the PAL and POD activities that afterward inhibited the *Botrytis cinerea* persuaded disease of gray mold in the grapes in postharvest storage and increased the shelf life (Liu et al. 2016a). Likewise, during the postharvest, the application of 5 μmol L^{-1} BRs considerably induced the PAL and PPO contents, which significantly repressed the *Penicillium expansum* and prompted mold rot in the jujube fruits (Zhu et al. 2010).

6.6.11 BRs and the Fruit Cracking

Fruit cracking is one of the main issues with fruits like pomegranates, litchi, lemons, cherries, etc., and the abiotic variables, including temperature, water deficit or irregular precipitation, and nutritional deficiencies like calcium, zinc, and boron, among others, are the leading causes of it. Research has shown that the effect of BL application on fruits like litchi and persimmon fruit plays a vital role in increasing marketability by lessening the fruit's cracking. The BL application enhanced the activities of key enzymes like polygalacturonase (PG) and pectin methylesterase (PME), protopectin, and water-soluble pectin content. Apart from this, it also enhanced the calcium content percentage in the pericarp of the fruit and minimized the cracking of the fruit considerably (Peng et al. 2004).

6.7 Conclusion

The BRs are significant indeed occurring phytohormones involved in a series of physiological processes in plants, mainly the induction of development and growth, vascular development, cell division, elongation, yield, germination process, ripening, and enhanced tolerance against several environmental stresses, including abiotic and biotic. The studies regarding the impact of BRs show that it alleviates the CI effect softening process, inhibits the enzymatic browning, induces the sugar and energy metabolism during postharvest storage, and improves the shelf life and life span of fruits and vegetables. Apart from this, the supplementation of BRs activates the activities of critical enzymes like PAL and PPO and significantly minimizes the decay in the postharvest storage process. Several studies suggested that the application of BRs activates ethylene accumulation and increases the ripening process; however, some other research presented that BRs repress ethylene and delay ripening. Nevertheless, there is a gap and a need to focus on determining the impact of BRs on ethylene biosynthesis and respiration process. Secondly, it is important to explore effects and the involvement of BRs, especially on the metabolism of phenolics and aroma volatiles. There is also a need to conduct further studies regarding the involvement and crosstalk of BRs with other hormones such as jasmonic acid, abscisic acid, auxins, and GAs and their interaction and impact on the postharvest storage of fruits and vegetables and notably the numerous concentration and the overall effect of the several methods of application in the BRs also need to focused.

References

Ackerman-Lavert, M. and Savaldi-Goldstein, S. (2020). Growth models from a brassinosteroid perspective. *Current Opinion in Plant Biology* 53: 90–97.

Aghdam, M.S., Asghari, M., Farmani, B. et al. (2012). Impact of postharvest brassinosteroids treatment on PAL activity in tomato fruit in response to chilling stress. *Scientia Horticulturae* 144: 116–120.

Aghdam, M.S., Babalar, M., and Sarcheshmeh, M.A.A. (2016). Impact of brassinosteroids on postharvest physiology of fruits and vegetables. In: *Eco-friendly Technology for Postharvest Produce Quality*, 203–218. Academic Press.

Aghdam, M.S., Jannatizadeh, A., Luo, Z. et al. (2018). Ensuring sufficient intracellular ATP supplying and friendly extracellular ATP signaling attenuates stresses, delays senescence and maintains quality in horticultural crops during postharvest life. *Trends in Food Science & Technology* 76: 67–81.

Aghdam, M.S. and Mohammadkhani, N. (2014). Enhancement of chilling stress tolerance of tomato fruit by postharvest brassinolide treatment. *Food and Bioprocess Technology* 7: 909–914.

Ahammed, G.J., Li, X., Liu, A. et al. (2020). Brassinosteroids in plant tolerance to abiotic stress. *Journal of Plant Growth Regulation* 39 (4): 1451–1464.

Ahammed, G.J., Li, X., Xia, X.J. et al. (2015). Enhanced photosynthetic capacity and antioxidant potential mediate brassinosteriod-induced phenanthrene stress tolerance in tomato. *Environmental Pollution* 201: 58–66.

Ali, B. (2017). Practical applications of brassinosteroids in horticulture—some field perspectives. *Scientia Horticulturae* 225: 15–21.

Ali, S., Khan, A.S., Anjum, M.A. et al. (2019). Aloe vera gel coating delays post-cut surface browning and maintains quality of cold stored lotus (*Nelumbo nucifera* Gaertn.) root slices. *Scientia Horticulturae* 256: 108612.

Ali, S., Khan, A.S., Malik, A.U. et al. (2016). Effect of controlled atmosphere storage on pericarp browning, bioactive compounds and antioxidant enzymes of litchi fruits. *Food Chemistry* 206: 18–29.

Ali, S., Khan, A.S., Malik, A.U. et al. (2018). Pre-storage methionine treatment inhibits postharvest enzymatic browning of cold stored 'Gola' litchi fruit. *Postharvest Biology and Technology* 140: 100–106.

Ayub, R.A., Reis, L., Bosetto, L. et al. (2018). Brassinosteroid plays a role on pink stage for receptor and transcription factors involved in strawberry fruit ripening. *Plant Growth Regulation* 84: 159–167.

Bajguz, A., Chmur, M., and Gruszka, D. et al. (2020). Comprehensive overview of the brassinosteroid biosynthesis pathways: substrates, products, inhibitors, and connections. *Frontiers in Plant Science* 11: 535077.

Bajguz, A. and Tretyn, A. (2003). The chemical characteristic and distribution of brassinosteroids in plants. *Phytochemistry* 62(7): 1027–1046.

Basit, F., Liu, J., An, J. et al. (2021). Brassinosteroids as a multidimensional regulator of plant physiological and molecular responses under various environmental stresses. *Environmental Science and Pollution Research* 28 (33): 44768–44779. https://doi.org/10.1007/s11356-021-15087-8.

Cai, J.H., Luo, F., Zhao, Y.B. et al. (2019). 24-Epibrassinolide treatment regulates broccoli yellowing during shelf life. *Postharvest Biology and Technology* 154: 87–95.

Chai, Y.M., Zhang, Q., Tian, L. et al. (2013). Brassinosteroid is involved in strawberry fruit ripening. *Plant Growth Regulation* 69: 63–69.

Chen, G., Hou, Y., Zheng, Y. et al. (2022). 2, 4-epibrassinolide enhance chilling tolerance of loquat fruit by regulating cell wall and membrane fatty acid metabolism. *Scientia Horticulturae* 295: 110813.

Chen, J., Nolan, T.M., Ye, H. et al. (2017). Arabidopsis WRKY46, WRKY54, and WRKY70 transcription factors are involved in brassinosteroid-regulated plant growth and drought responses. *The Plant Cell* 29 (6): 1425–1439.

Clouse, S.D. and Sasse, J.M. (1998). Brassinosteroids: Essential regulators of plant growth and development. *Annual Review of Plant Biology* 49 (1): 427–451.

Ding, Y., Zhu, Z., Zhao, J. et al. (2016). Effects of postharvest brassinolide treatment on the metabolism of white button mushroom (*Agaricus bisporus*) in relation to development of browning during storage. *Food and Bioprocess Technology* 9: 1327–1334.

Domagalska, M.A., Sarnowska, E., Nagy, F. et al. (2010). Genetic analyses of interactions among gibberellin, abscisic acid, and brassinosteroids in the control of flowering time in *Arabidopsis thaliana*. *PLoS One* 5 (11): e14012.

Duan, W., Mekontso, F.N., Li, W. et al. (2022). Alleviation of postharvest rib-edge darkening and chilling injury of carambola fruit by brassinolide under low temperature storage. *Scientia Horticulturae* 299: 111015.

Ekinci, N., Gökbayrak, Z., Çavuşoğlu, Ş. et al. (2019). Influence of postharvest application of 28-homobrassinolide on storage quality of medlar fruit. *Erwerbs-Obstbau* 61 (2): 113–118.

Fang, H., Zhou, Q., Cheng, S. et al. (2021). 24-epibrassinolide alleviates postharvest yellowing of broccoli via improving its antioxidant capacity. *Food Chemistry* 365: 130529.

Fang, P., Yan, M., Chi, C. et al. (2019). Brassinosteroids act as a positive regulator of photoprotection in response to chilling stress. *Plant Physiology* 180 (4): 2061–2076.

Fariduddin, Q., Yusuf, M., Ahmad, I. et al. (2014). Brassinosteroids and their role in response of plants to abiotic stresses. *Biologia Plantarum* 58: 9–17.

Fujioka, S. and Yokota, T. (2003). Biosynthesis and metabolism of brassinosteroids. *Annual Review of Plant Biology* 54 (1): 137–164.

Gao, H., Chai, H., Cheng, N. et al. (2017). Effects of 24-epibrassinolide on enzymatic browning and antioxidant activity of fresh-cut lotus root slices. *Food Chemistry* 217: 45–51.

Gao, H., Kang, L., Liu, Q. et al. (2015). Effect of 24-epibrassinolide treatment on the metabolism of eggplant fruits in relation to development of pulp browning under chilling stress. *Journal of Food Science and Technology* 52: 3394–3401.

Gao, H., Zhang, Z., Lv, X. et al. (2016). Effect of 24-epibrassinolide on chilling injury of peach fruit in relation to phenolic and proline metabolisms. *Postharvest Biology and Technology* 111: 390–397.

Ghorbani, B. and Pakkish, Z. (2014). Brassinosteroid enhances cold stress tolerance of Washington navel orange (*Citrus sinensis* L.) fruit by regulating antioxidant enzymes during storage. *Agriculturae Conspectus Scientificus* 79 (2): 109–114.

González-García, M.P., Vilarrasa-Blasi, J., Zhiponova, M. et al. (2011). Brassinosteroids control meristem size by promoting cell cycle progression in *Arabidopsis* roots. *Development* 138 (5): 849–859.

Grove, M.D., Spencer, G.F., Rohwedder, W.K. et al. (1979). Brassinolide, a plant growth-promoting steroid isolated from *Brassica napus* pollen. *Nature* 281 (5728): 216–217.

Gudesblat, G.E., Schneider-Pizoń, J., Betti, C. et al. (2012). SPEECHLESS integrates brassinosteroid and stomata signalling pathways. *Nature Cell Biology* 14 (5): 548–554.

Hacham, Y., Holland, N., Butterfield, C. et al. (2011). Brassinosteroid perception in the epidermis controls root meristem size. *Development* 138 (5): 839–848.

Hafeez, M.B., Zahra, N., Zahra, K. et al. (2021). Brassinosteroids: molecular and physiological responses in plant growth and abiotic stresses. *Plant Stress* 2: 100029. https://doi.org/10.1016/j.stress.2021.100029.

Hartwig, T., Chuck, G.S., Fujioka, S. et al. (2011). Brassinosteroid control of sex determination in maize. *Proceedings of the National Academy of Sciences of the United States of America* 108 (49): 19814–19819.

Hayat, S. and Ahmad, A. (2003). *Brassinosteroids: Bioactivity and Crop Productivity*. Springer Science & Business Media.

He, Y., Li, J., Ban, Q. et al. (2018). Role of brassinosteroids in persimmon (*Diospyros kaki* L.) fruit ripening. *Journal of Agricultural and Food Chemistry* 66 (11): 2637–2644.

Huang, H.Y., Jiang, W.B., Hu, Y.W. e al. (2013). BR signal influences *Arabidopsis* ovule and seed number through regulating related genes expression by BZR1. *Molecular Plant* 6 (2): 456–469.

Hussain, M.A., Fahad, S., Sharif, R. et al. (2020). Multifunctional role of brassinosteroid and its analogues in plants. *Plant Growth Regulation* 92: 141–156.

Hussain, M., Hamid, M.I., and Ghazanfar, M.U. (2015). Salicylic acid induced resistance in fruits to combat against postharvest pathogens: a review. *Archives of Phytopathology and Plant Protection* 48 (1): 34–42.

Islam, M., Ali, S., Nawaz, A. et al. (2022). Postharvest 24-epibrassinolide treatment alleviates pomegranate fruit chilling injury by regulating proline metabolism and antioxidant activities. *Postharvest Biology and Technology* 188: 111906.

Jaillais, Y. and Vert, G. (2012). Brassinosteroids, gibberellins and light-mediated signalling are the three-way controls of plant sprouting. *Nature Cell Biology* 14 (8): 788–790.

Jaiswal, S., Båga, M., and Chibbar, R.N. (2020). Brassinosteroid receptor mutation influences starch granule size distribution in barley grains. *Plant Physiology and Biochemistry* 154: 369–378.

Jia, Z., Giehl, R.F., Meyer, R.C. et al. (2019). Natural variation of BSK3 tunes brassinosteroid signaling to regulate root foraging under low nitrogen. *Nature Communications* 10 (1): 2378.

Jiang, W.B., Huang, H.Y., Hu, Y.W. et al. (2013). Brassinosteroid regulates seed size and shape in *Arabidopsis*. *Plant Physiology* 162 (4): 1965–1977.

Jin, P., Zhu, H., Wang, J. et al. (2013). Effect of methyl jasmonate on energy metabolism in peach fruit during chilling stress. *Journal of the Science of Food and Agriculture* 93 (8): 1827–1832.

Khan, M., Rozhon, W., Bigeard, J. et al. (2013). Brassinosteroid-regulated GSK3/Shaggy-like kinases phosphorylate mitogen-activated protein (MAP) kinase kinases, which control stomata development in *Arabidopsis thaliana*. *Journal of Biological Chemistry* 288 (11): 7519–7527.

Khripach, V.A., Zhabinskii, V.N., and de Groot, A.E. (1998). *Brassinosteroids: A New Class of Plant hormones*. Academic Press.

Krishna, P., Prasad, B.D., and Rahman, T. (2017). Brassinosteroid action in plant abiotic stress tolerance. *Methods in Molecular Biology* 1564: 193–202. https://doi.org/10.1007/978-1-4939- 6813-8_16.

Li, B., Zhang, C., Cao, B. et al. (2012). Brassinolide enhances cold stress tolerance of fruit by regulating plasma membrane proteins and lipids. *Amino Acids* 43: 2469–2480.

Li, J. and Chory, J. (1997). A putative leucine-rich repeat receptor kinase involved in brassinosteroid signal transduction. *Cell* 90 (5): 929–938.

Li, J., Nagpal, P., Vitart, V. et al. (1996). A role for brassinosteroids in light-dependent development of Arabidopsis. *Science* 272 (5260): 398–401.

Li, Q., Lü, J., Yu, J. et al. (2018a). The brassinosteroid-regulated transcription factors BZR1/BES1 function as a coordinator in multisignal-regulated plant growth. *Biochimica et Biophysica Acta (BBA)-Gene Regulatory Mechanisms* 1861 (6): 561–571. https://doi.org/10.1016/j.bbagrm.2018.04.003.

Li, Z. and He, Y. (2020). Roles of brassinosteroids in plant reproduction. *International Journal of Molecular Sciences* 21 (3): 872.

Li, Z., Ou, Y., Zhang, Z. et al. (2018b). Brassinosteroid signaling recruits histone 3 lysine-27 demethylation activity to *FLOWERING LOCUS C* chromatin to inhibit the floral transition in *Arabidopsis*. *Molecular Plant* 11 (9): 1135–1146.

Lin, W.H. (2020). Designed manipulation of the brassinosteroid signal to enhance crop yield. *Frontiers in Plant Science* 11: 538293.

Liu, Q., Xi, Z., Gao, J. et al. (2016a). Effects of exogenous 24-epibrassinolide to control grey mould and maintain postharvest quality of table grapes. *International Journal of Food Science & Technology* 51 (5): 1236–1243.

Liu, X., Li, D., Li, Y. et al. (2022). Brassinosteroids are involved in volatile compounds biosynthesis related to MdBZR1 in 'Ruixue' (Malus× domestica Borkh.) fruit. *Postharvest Biology and Technology* 189: 111931.

Liu, Z., Li, L., Luo, Z. et al. (2016b). Effect of brassinolide on energy status and proline metabolism in postharvest bamboo shoot during chilling stress. *Postharvest Biology and Technology* 111: 240–246.

Lu, Z., Wang, X., Cao, M. et al. (2019). Effect of 24-epibrassinolide on sugar metabolism and delaying postharvest senescence of kiwifruit during ambient storage. *Scientia Horticulturae* 253: 1–7.

Lv, B., Tian, H., Zhang, F. et al. (2018). Brassinosteroids regulate root growth by controlling reactive oxygen species homeostasis and dual effect on ethylene synthesis in *Arabidopsis. PLoS Genetics* 14 (1): e1007144.

Makarevitch, I., Thompson, A., Muehlbauer, G.J. et al. (2012). Brd1 gene in maize encodes a brassinosteroid C-6 oxidase. *PLoS One* 7 (1): e30798.

Massolo, J.F., Sánchez, R., Zaro, M.J. et al. (2022). Low-dose prestorage 24-epibrassinolide spray enhances postharvest chilling tolerance in zucchini squash (*Cucurbita pepo* L.) by eliciting peroxidase and phenolic antioxidants. *Journal of Food Processing and Preservation* 46 (5): e16576.

Mazorra, L.M., Oliveira, M.G., Souza, A.F. et al. (2013). Involvement of brassinosteroids and ethylene in the control of mitochondrial electron transport chain in postharvest papaya fruit. *Theoretical and Experimental Plant Physiology* 25, 203–212.

Meudt, W.J., Thompson, M.J., and Bennett, H.W. (1983). Investigations on the mechanism of the brassinosteroid response. III. Techniques for potential enhancement of crop production [Barley, bean]. In: *Proceedings Annual Meeting Plant Growth Regulator Society of America.*

Mitchell, J.W., Mandava, N., Worley, J.F. et al. (1970). Brassins—A new family of plant hormones from rape pollen. *Nature* 225 (5237): 1065–1066.

Nakaya, M., Tsukaya, H., Murakami, N. et al. (2002). Brassinosteroids control the proliferation of leaf cells of *Arabidopsis thaliana. Plant and Cell Physiology* 43 (2): 239–244.

Nolan, T.M. Vukašinović, N., Liu, D. et al. (2019). Brassinosteroids: multidimensional regulators of plant growth, development, and stress responses. *The Plant Cell* 32 (2): 295–318. https://doi.org/10.1105/tpc.19.00335

Oh, M.H. (2003). Brassinosteroids accelerate the rate of cell division in isolated petal protoplasts of Petunia hybrida. *Journal of Plant Biotechnology* 5 (1): 63–67.

Pakkish, Z., Ghorbani, B., and Najafzadeh, R. et al. (2019). Fruit quality and shelf life improvement of grape cv. Rish Baba using brassinosteroid during cold storage. *Journal of Food Measurement and Characterization* 13: 967–975.

Peng, J., Tang, X., and Feng, H. (2004). Effects of brassinolide on the physiological properties of litchi pericarp (*Litchi chinensis* cv. nuomoci). *Scientia Horticulturae* 101 (4): 407–416.

Razzaq, K., Khan, A.S., Malik, A.U. et al. (2013). Ripening period influences fruit softening and antioxidative system of 'Samar Bahisht Chaunsa' mango. *Scientia Horticulturae* 160: 108–114.

Singh, A.P. and Savaldi-Goldstein, S. (2015). Growth control: brassinosteroid activity gets context. *Journal of Experimental Botany* 66 (4): 1123–1132.

Symons, G.M., Davies, C., Shavrukov, Y. et al. (2006). Grapes on steroids. Brassinosteroids are involved in grape berry ripening. *Plant Physiology* 140 (1): 150–158.

Tanabe, S., Ashikari, M., Fujioka, S. et al. (2005). A novel cytochrome P450 is implicated in brassinosteroid biosynthesis via the characterization of a rice dwarf mutant, dwarf11, with reduced seed length. *The Plant Cell* 17 (3): 776–790.

Tavallali, V. (2018). Vacuum infiltration of 24-epibrassinolide delays chlorophyll degradation and maintains quality of lime during cold storage. *Acta Scientiarum Polonorum. Hortorum Cultus* 17 (1): 35–48.

Tong, H. and Chu, C. (2018). Functional specificities of brassinosteroid and potential utilization for crop improvement. *Trends in Plant Science* 23 (11): 1016–1028.

Trevisan, S., Forestan, C., Brojanigo, S. et al. (2020). Brassinosteroid application affects the growth and gravitropic response of maize by regulating gene expression in the roots, shoots and leaves. *Plant Growth Regulation* 92 (1): 117–130.

Vardhini, B.V. and Rao, S.S.R. (2002). Acceleration of ripening of tomato pericarp discs by brassinosteroids. *Phytochemistry* 61 (7): 843–847.

Wang, Q., Ding, T., Gao, L. et al. (2012a). Effect of brassinolide on chilling injury of green bell pepper in storage. *Scientia Horticulturae* 144: 195–200.

Wang, Q., Yu, F., & and Xie, Q. (2020a). Balancing growth and adaptation to stress: Crosstalk between brassinosteroid and abscisic acid signaling. *Plant, Cell & Environment* 43 (10): 2325–2335.

Wang, W., Lu, X., Li, L. et al. (2018). Photoexcited CRYPTOCHROME1 interacts with dephosphorylated BES1 to regulate brassinosteroid signaling and photomorphogenesis in *Arabidopsis. Plant Cell* 30 (9): 1989–2005. https://doi.org/10.1105/tpc.17.00994.

Wang, X., Chen, J., Xie, Z. et al. (2014). Histone lysine methyltransferase SDG8 is involved in brassinosteroid-regulated gene expression in *Arabidopsis thaliana. Molecular Plant* 7 (8): 1303–1315.

Wang, Y., Cao, J. J., Wang, K. X. et al. (2019). BZR1 mediates brassinosteroid-induced autophagy and nitrogen starvation in tomato. *Plant Physiology* 179 (2): 671–685.

Wang, Z. Y., Bai, M. Y., Oh, E., & Zhu, J. Y et al. (2012b). Brassinosteroid signaling network and regulation of photomorphogenesis. *Annual Review of Genetics* 46: 701–724.

Wu, L. and Yang, H. (2016). Combined application of carboxymethyl chitosan coating and brassinolide maintains the postharvest quality and shelf life of green asparagus. *Journal of Food Processing and Preservation* 40 (2): 154–165.

Xi, Z.M., Zhang, Z.W., Huo, S.S. et al. (2013). Regulating the secondary metabolism in grape berry using exogenous 24-epibrassinolide for enhanced phenolics content and antioxidant capacity. *Food Chemistry* 141 (3): 3056–3065.

Xia, X.J., Zhou, Y.H., Ding, J. et al. (2011). Induction of systemic stress tolerance by brassinosteroid in *Cucumis sativus. New Phytologist* 191 (3): 706–720.

Xie, L., Yang, C., and Wang, X. (2011). Brassinosteroids can regulate cellulose biosynthesis by controlling the expression of CESA genes in *Arabidopsis. Journal of Experimental Botany* 62 (13): 4495–4506.

Xu, F., Xi, Z.M., Zhang, H. et al. (2015). Brassinosteroids are involved in controlling sugar unloading in *Vitis vinifera* 'Cabernet Sauvignon' berries during véraison. *Plant Physiology and Biochemistry* 94: 197–208.

Yamamoto, R., Fujioka, S., Demura, T. et al. (2001). Brassinosteroid levels increase drastically prior to morphogenesis of tracheary elements. *Plant Physiology* 125 (2): 556–563.

Yao, Y., Zhao, N., Xian, T. et al. (2017). Effect of 2, 4-epibrassinolide treatment on the postharvest quality and physiological metabolism of fresh daylily flower buds during storage. *Scientia Horticulturae* 226: 110–116.

Ye, H., Liu, S., Tang, B. et al. (2017). RD26 mediates crosstalk between drought and brassinosteroid signalling pathways. *Nature Communications* 8: 14573. https://doi.org/10.1038/ncomms14573.

Ye, Q., Zhu, W., Li, L. et al. (2010). Brassinosteroids control male fertility by regulating the expression of key genes involved in *Arabidopsis* anther and pollen development. *Proceedings of the National Academy of Sciences of the United States of America* 107 (13): 6100–6105.

Ye, S.E., Fang, L.I., Li, X.B. et al. (2015). Over-expression of GhDWF4 gene improved tomato fruit quality and accelerated fruit ripening. *Journal of Integrative Agriculture* 14 (10): 1980--1991.

Yin, Y., Wang, Z.Y., Mora-Garcia, S. et al. (2002). BES1 accumulates in the nucleus in response to brassinosteroids to regulate gene expression and promote stem elongation. *Cell* 109 (2): 181–191.

Yokota, T., Ohnishi, T., Shibata, K. et al. (2017). Occurrence of brassinosteroids in non-flowering land plants, liverwort, moss, lycophyte and fern. *Phytochemistry* 136: 46–55.

Zadvornova, Y.V., Alekseichuk, G.N., Laman, V.A. et al. (2005). Effect of brassinosteroids on activation of the cell cycle during germination of *Brassica oleracea* L. seeds. *Doklady Natsional'noi Akademii Nauk Belarusi* 49: 70–73.

Zaharah, S.S. and Singh, Z. (2012). Role of brassinosteroids in mango fruit ripening. *Acta Horticulturae* 934: 929–935.

Zhang, L., Cao, X., Wang, Z. et al. (2022). Brassinolide alleviated chilling injury of banana fruit by regulating unsaturated fatty acids and phenolic compounds. *Scientia Horticulturae* 297: 110922.

Zhang, S., Hu, J., Zhang, Y. et al. (2007). Seed priming with brassinolide improves lucerne (*Medicago sativa* L.) seed germination and seedling growth in relation to physiological changes under salinity stress. *Australian Journal of Agricultural Research* 58 (8): 811–815.

Zhang, W., Zhu, K., Wang, Z. et al. (2019). Brassinosteroids function in spikelet differentiation and degeneration in rice. *Journal of Integrative Plant Biology* 61 (8): 943–963.

Zhao, B. and Li, J. (2012). Regulation of brassinosteroid biosynthesis and inactivation F. *Journal of Integrative Plant Biology* 54 (10): 746–759.

Zheng, T., Dong, T., Haider, M.S. et al. (2020). Brassinosteroid regulates 3-hydroxy-3-methylglutaryl CoA reductase to promote grape fruit development. *Journal of Agricultural and Food Chemistry* 68 (43): 11987–11996.

Zhu, F., Yun, Z., Ma, Q. et al. (2015a). Effects of exogenous 24-epibrassinolide treatment on postharvest quality and resistance of Satsuma mandarin (*Citrus unshiu*). *Postharvest Biology and Technology* 100: 8–15.

Zhu, J.Y., Sae-Seaw, J., and Wang, Z.Y. (2013). Brassinosteroid signalling. *Development* 140 (8): 1615–1620.

Zhu, T., Tan, W.R., Deng, X.G. et al. (2015b). Effects of brassinosteroids on quality attributes and ethylene synthesis in postharvest tomato fruit. *Postharvest Biology and Technology* 100: 196–204.

Zhu, X., Liang, W., Cui, X., Chen. et al. (2015c). Brassinosteroids promote development of rice pollen grains and seeds by triggering expression of *Carbon Starved Anther*, a MYB domain protein. *The Plant Journal* 82 (4): 570–581.

Zhu, Z., Zhang, Z., Qin, G. et al. (2010). Effects of brassinosteroids on postharvest disease and senescence of jujube fruit in storage. *Postharvest Biology and Technology* 56 (1): 50–55.

Zou, L., Deng, X., Zhang, L. et al. (2018). Nitric oxide as a signaling molecule in brassinosteroid-mediated virus resistance to Cucumber mosaic virus in *Arabidopsis thaliana*. Physiologia Plantarum (København. *1948)* 163 (2): 196–210. https://doi.org/10.1111/ppl.12677.

7

Brassinosteroid-mediated Physiological and Molecular Mechanisms in Plants Under Heavy Metal Stress

*Deepak Kumar[1], Mohd Ali[1], Raman Tikoria[2], Parkirti[1], Roohi Sharma[1], Vikram[1], Rajesh Kumari Manhas[3], Harish Changotra[4], and Puja Ohri[1]**

[1]*Department of Zoology, Guru Nanak Dev University, Amritsar, Punjab, India*
[2]*Department of Zoology, Central University of Punjab, Ghudda, Bathinda, Punjab, India*
[3]*Department of Microbiology, Guru Nanak Dev University, Amritsar, Punjab, India*
[4]*Department of Biochemistry and Molecular Biology, Guru Nanak Dev University, Amritsar, Punjab, India*

*Corresponding author: ohri_puja_11@rediffmail.com

7.1 Introduction

Environmental stressors like drought, salinity, cold, heat, and heavy metals have a negative impact on plant growth and agricultural output. They may reduce the average yield of the majority of crop plants by more than 50% (Zulfiqar and Ashraf 2023). Globally, heavy metals are some of the most prominent abiotic factors that decrease crop quality and productivity. Heavy metal buildup in the environment is hazardous to soil fertility, plants, animals, and humans due to its extremely poisonous, long-term persistence, non-degradability, mutagenic, carcinogenic, and teratogenic properties. Genetic makeup, mental health, and the central nervous system of living organisms can be negatively impacted by inhaling and consuming heavy metals (Okoro et al. 2020; Abd Elnabi et al. 2023). Those who have direct contact with heavy metals may be more susceptible to cancer. Heavy metals also have a deleterious impact on the ecological dynamics of the rhizosphere niche. They negatively impact long-term food production, plant development, photosynthesis, and biomass of plants (Ghuge et al. 2023). Chromium, cadmium, mercury, nickel, lead, and copper are some of the heavy metals that are unpleasant for plants (Islam and Sandhi 2023; Kumar and Ohri 2023). These metals are accountable for the damage to agro-biological systems and decreased profitability. Heavy metals have the potential to cause nutritional deficiencies and imbalances in plants by interfering with the transport and absorption of essential metals from their usual binding sites, as well as by disrupting the structure of nucleic acids and enzyme function. They negatively affect plant physiology and cause stress to the plants (Kaur et al. 2022). Various defensive mechanisms are available to plants to counteract the detrimental effects of metal stress. The two primary mechanisms involved in the detoxification of metals are (i) exclusion methods, which stop metal ions from entering the plant

Brassinosteroids and Abiotic Stress: Recent Developments on Physiological and Molecular Aspects,
First Edition. Edited by Anket Sharma, et al.
Published 2026 by John Wiley & Sons Ltd.

system and (ii) built-in tolerance mechanisms, which involve the synthesis of different secondary metabolites and metal-chelates that may bind to metal ions and lessen their toxicity (Ali et al. 2023). In addition to their antioxidant defense system against heavy metal-induced oxidative stress, plants initiate a series of signaling molecules that control metabolic fluxes. In these signaling networks, phytohormones play a crucial role as their actions connect stress perception to a range of transcriptional responses that support the plant's ability to adapt to a changing and unfavorable environment (Rahman et al. 2023).

Brassinosteroids (BRs) are a class of phytohormones that play an important role in plant development, growth, and stress response by regulating a variety of physiological, biochemical, and molecular activities (Sharma et al. 2022a). While interacting with other phytohormones including auxins, jasmonates, and abscisic acid as well as other plant signaling pathways, BRs regulate a wide range of physiological and developmental activities (Li et al. 2021). BRs play a critical role in improving plant performance and resistance to various environmental stresses like drought, salinity, extreme temperature conditions, and metal toxicity by increasing productivity, photosynthesis, consolidating the antioxidant defense system and detoxification ability, and also by modifying the activity of stress-associated genes (Manghwar et al. 2022). It has been discovered that BRs and related signaling components interact extensively with both harmful heavy metals and nutritious components in the soil microenvironment (Kapoor et al. 2022). Enhancement in morphological elements like germination, shoot and root length, yield, and biomass is the stimulating properties of BRs under heavy metal stress (Kour et al. 2021). BRs enhance the permeability of cell membranes and have an affinity for membrane-bound proteins. Plants treated with BRs under heavy metal stress activate their antioxidant defense system and reduce the quantity of harmful substances in them (Yusuf et al. 2022). BRs also improve the activity of antioxidant enzymes to scavenge reactive oxygen species (ROS) and prevent metal stress-related lipid damage caused by ROS. When BRs are administered to plants in a certain amount, overall metal-induced toxicity is reduced (Zhong et al. 2020; Yaqoob et al. 2022). This chapter highlights the uptake and toxicity of heavy metals to plants and BR-mediated morphological and biochemical changes in plants under heavy metal stress.

7.2 Uptake and Translocation of Heavy Metals

In plants, heavy metals are absorbed via roots. Specialized membrane transport proteins are present in root cells that facilitate the absorption of heavy metals. Different kinds of intercellular and interorganellar transporters are present in plants for the uptake of heavy metals like Iron Regulated Transporter 1 (IRT), Natural Resistance Associated Macrophage Protein 1 (NRAMP), Zinc Regulated Transporters (ZRT), Heavy Metal Tolerance (HMT), P1B-type ATPase, and cation diffusion transporters. Heavy metals either actively or passively enter into the plant and are translocated via xylem or phloem to the whole plant. The transporters located in root plasma membrane help take up the heavy metals from the environment (Khan et al. 2021, 2023a). It has been observed that Cd was translocated with the help of *OsLCT1* translocators located in plasma membrane, and knockdown of this transporter reduced the Cd accumulation in rice plant (Songmei et al. 2019). Similarly, selenium

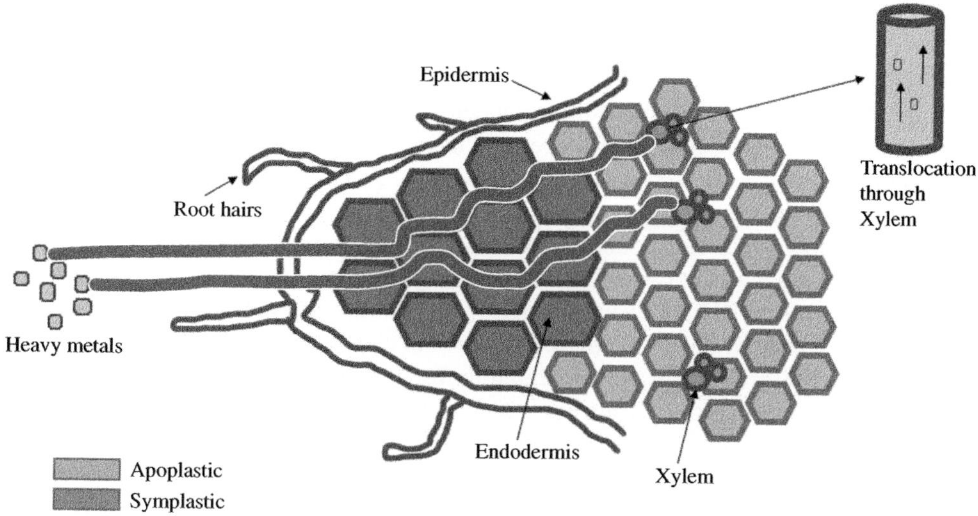

Figure 7.1 Pathway of heavy metal uptake and translocation in plants.

is taken up by Sulphate Transporters (SULTRs) while for selenite SeO_3^{2-} the phosphate transporters and aquaporins are present in plants. Further, selenate (SeO_4^{2-}) is taken up by roots in an active manner and accumulates in roots while SeO_3^{2-} is not normally accumulated in plant roots (Asgher et al. 2023). Mostly metals are immobile in their ionic form, but among all heavy metals, Zn and Cd are more mobile comparatively, so plants uptake these heavy metals readily. Therefore, heavy metals must be mobilized in the soil solution for their easy uptake by plant roots. Metals have strong affinity toward matter present in soil and soil particles itself, but they have low solubility in water. Sometimes plants also release some root exudates that increase heavy metals' bioavailability by changing their rhizosphere pH and increasing the solubility of heavy metals (Pasricha et al. 2021). Figure 7.1 gives a general illustration of heavy metal uptake and translocation in plants.

Heavy metals are absorbed at root surfaces either via crossing plasma membrane or through intracellular spaces. The metal ions are then transported from root apoplast to xylem system of roots, from where they are translocated to the aerial parts of plants. Some metal ions are insoluble so are not able to move freely as they form precipitates of phosphates, carbonates, and sulfates. As a result, the metal ions get stuck in intracellular and extracellular spaces. Thus, for metals to move and reach xylem by crossing the cell wall, they have to move symplastically. Through root symplasm, the metal ions reach the stele and enter into the xylem, and from there, they are sent to other parts of the plant. In xylem, the membrane proteins transport these metals via active transportation, which is an energy-dependent process regulated by some specific channels or carriers like coupled proteins, H^+ ion-coupled proteins, organic acid, or amino acid (Kajala et al. 2019; Sharma et al. 2022b). They regulate the specific movement of specific metals across the cell membrane and their translocation in the plants. These common metal transporters are broadly divided into families such as metal transporter proteins (MTPs), naturally resistant associated macrophage proteins (NRAMPs), heavy metal ATPases (HMAs), and ZRT–IRT-like proteins (ZIPs), and other families including iron-regulated transporters (IRTs) and heavy metal transporters like arsenic transporters. MTPs apart from translocating heavy metals like Zn and Ni toward extracellular space or internal compartments also regulate plant homeostasis. NRAMPs are associated with transport of heavy metal ions such as Mn^{2+}, Co^{2+}, Fe^{2+}, Cd^{2+}, and Cu^{2+}. These are present in the plasma membrane and tonoplast. Sometimes, organic acids in higher concentrations in roots also help in binding heavy metals for uptake and translocation. HMA's family involves transportation of heavy metals such as Cd, Co, Pb, and Zn. It also plays a role in plant homeostasis and heavy metal tolerance. ZIP family uptakes many cations such as Zn, Mn, and Fe, and helps shoot to root transportation and their accumulation in plants. Some free amino acids, such as histidine and nicotinamide, are also found to form efficient transporters of metal ions by forming stable complexes (Kajala et al. 2019; Kaur et al. 2021; Kosakivska et al. 2021; Pasricha et al. 2021).

7.3 Heavy Metal Toxicity to Plants

Heavy metal toxicity is one of the serious hazards for crops. Heavy metals such as Cu, Ni, Co, Hg, As, and Pb, in a concentration-dependent manner directly alter plant performance. Heavy metal toxicity induces alterations in morphology, anatomy, and physiology of plants. These alterations involve disruption of chlorophyll pigments, impaired cell development, disturbed hormonal pathways, induced oxidative stress, disturbed mineral uptake, chlorosis, disrupted nutrient uptake activity, decreased root growth, decreased leaf number, decreased plant biomass, impaired photosynthesis and transpiration rate, ultimately leading to plant death (Riyazuddin et al., 2021; Khan et al. 2023a). Various studies have revealed the ill effects of heavy metals on plants. In a study on the effect of four metals, Ni, Cd, Cu, and Zn, in their sub-lethal concentrations on maize plant, it was found that different parts of the plant had different degrees of heavy metal accumulation. The older leaves had lower amount of heavy metal accumulation as compared to the roots. In addition, reduced biomass, reduced photosynthesis, and induced oxidative stress were other effects of heavy metal toxicity on plants (AbdElgawad et al. 2020). Similarly, exogenous application of Cu and Pb was found to increase lipid peroxidation and hydrogen peroxide levels. It also led to declined growth and pigment degradation and lowered mineral nutrition, decreased photosynthetic rate, and enhanced membrane integrity disruption (Giannakoula et al. 2021). In another study, Khanna et al. (2019) reported that treatment of tomato seedlings with Cd resulted in reduced shoot and root length, decreased content of photosynthetic pigments, and modulated expression of various metal transporter genes leading to reduction of metal tolerance index in treated tomato seedlings. Further, reduced plant height, chlorophyll *a* and chlorophyll *b* content, and increased oxidative stress and electrolyte leakage were reported in Al-stressed tomato plants (Ghorbani et al. 2023). Reduced plant length and weight, content of chlorophyll *a, b,* total chlorophyll and carotenoid, stomatal conductance, photosynthetic rate along with elevated levels of malondialdehyde (MDA) and hydrogen peroxide (H_2O_2) were reported on application of 260 μM Cr in *Cicer arietinum.* Treated plants also exhibited increased activity of enzymatic antioxidant and levels of non-enzymatic antioxidants, including ascorbic acid and glutathione (El-Ballat et al. 2023). Thus, heavy metal accumulation shows toxic effects on plants like reduced nutrient uptake, reduced yield, sluggish growth, chlorosis, root deterioration, etc., which eventually results in death of plants. Table 7.1 lists some studies from the last five years depicting the negative effects of heavy metal toxicity in different plants.

Table 7.1 Effect of heavy metals on plants.

Heavy metal	Plant	Effect	References
As	Bamboo plants	Enhanced reactive oxygen species (ROS) production and decreased photosynthesis	Emamverdian et al. (2023)
Cd, Cr, and Pb	Maize	Lowered nutrient availability and diminished plant growth	Saleem et al. (2023)
Cu, Zn, Cd, and Ni	Mung bean	Inhibited phosphate solubilization ability and enhanced indole-3-acetic acid concentration	Thomas and Archana (2023)
Al	Barley	Reduced plant growth, reduced photosynthetic pigment formation, elevated lipoxygenase activity, and increased ROS species and MDA contents	Dawood et al. (2022)
Zn, Pb, Ni, Cu, Cr, Co, and Cd	Summer squash	Reduced dry biomass of plant and accumulation of heavy metals in plant	Ibrahim et al. (2022)
Pb, Hg, and Cd	Pak choi	Inhibited growth and development of plant, imbalanced nutrient availability, and induced oxidative stress	Zhu et al. (2022)
Cd	Rice	Increased Cd toxicity, increased leaf senescence, and induced oxidative stress	Jiang et al. (2021)
Cd	Barley	Decreased root growth, increased cell death, and increased ROS generation	Demecsová et al. (2020)
Cd and Pb	Dove tree	Increased chlorophyll degradation	Yang et al. (2020)
Cd and Zn	Tobacco	Inhibited growth, decreased chlorophyll and carotenoid content, and plant biomass	Zhang et al. (2020)
Pb, Zn, Cd, and Cu	Sorghum	Degraded photosynthetic pigments, declined growth	El-Meihy et al. (2019)
Pb, Zn, and Cd	Wheat	Decreased root length, germination rate, and antioxidant activity	Shafiq et al. (2019)
Pb and Cd	Sweet basil and Garden cress	Younger plants were more susceptible to heavy metal accumulation than older plants	Souri et al. (2019)
PB and Cd	Wheat	Affected growth, weakened antioxidant defense system, and altered physicochemical attributes	Taie et al. (2019)
Cd and Zn	Rice	Decreased chlorophyll content, increased H_2O_2, MDA, and increased heavy metal concentration in plants	Tian et al. (2019)

7.4 Biosynthesis of BRs

Biosynthesis of BRs occurs through complex network pathways and is mostly controlled by transcriptional regulation of BR biosynthetic genes and have been discovered in *Arabidopsis*, rice, and tomato (Vriet et al. 2013; Liu et al. 2023a). The BR-biosynthetic pathway was first established by feeding cultured cells of *Catharanthus roseus* and *Arabidopsis* seedlings with deuterium-labeled substrates, and then gas chromatography-mass spectrometry (GC-MS) was used to identify various metabolites and reaction sequences involved in the biosynthetic pathway (Noguchi et al. 2000; Parada et al. 2022). Figure 7.2 gives a general illustration of biosynthesis of BRs. Campesterol (CR), a type of phytosterol, serves as a precursor for the formation of brassinolide (BL), the most active BR. Initially, it was believed that early C-6 and late C-6 oxidation routes were the two parallel processes for the synthesis of BRs (Fujioka et al. 1998). These pathways state that CR is first changed into campestanol (CN), followed by castasterone (CS), which ultimately leads to the synthesis of BL. During this process, CN is first transformed into 6-oxocampestanol in the early C-6 oxidation pathway, followed by cathasterone, teasterone, 3-dehydroteasterone, typhasterol, and finally into CS. In the late C-6 oxidation process, CN first undergoes hydroxylation at C-22 to produce 6-deoxocathasterone, which is subsequently transformed into

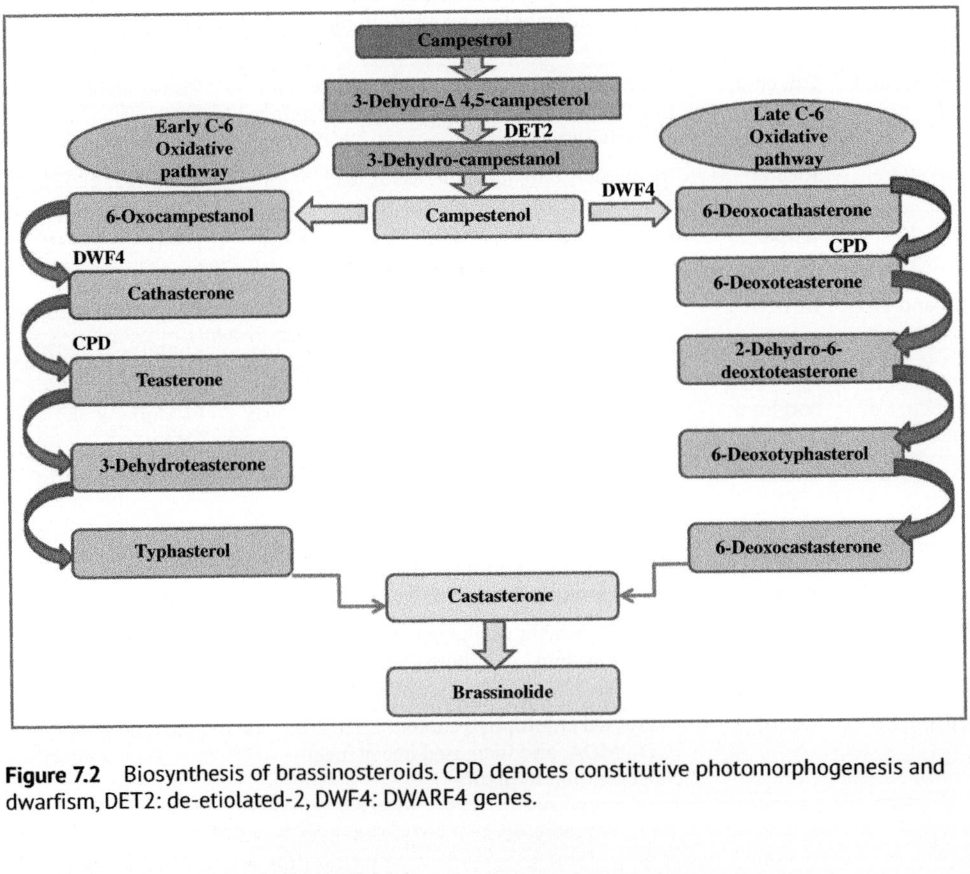

Figure 7.2 Biosynthesis of brassinosteroids. CPD denotes constitutive photomorphogenesis and dwarfism, DET2: de-etiolated-2, DWF4: DWARF4 genes.

the same intermediates as in the early C-6 oxidation pathway, but in C-6 de-oxy forms. These two pathways eventually meet at CS, which is then transformed into BL (Zhou et al. 2013). Another branching method, called the CN-independent pathway, which is an early C-22 oxidation branch, has been reported (Fujioka et al. 2002). Recently, a simpler approach employing C-23 hydroxylation that turns CR into 6-deoxytyphasterol has also been described (Ohnishi et al. 2006). The genes, including photomorphogenesis and dwarfism (*CPD*), de-etiolated-2 (*DET2*), and DWARF4 (*DWF4*), all have a role in BR-biosynthesis (Bhanu 2019). It is possible to control the constitutive expression of these genes in order to control the endogenous levels of BR in plants. Monocots such as rice and maize lack the CYP85A2 enzyme necessary for C-6 oxidation, suggesting that castasterone is the final product of BR production in monocots (Kim et al. 2008). In order to boost the biological activity of BR in rice, it has been demonstrated that an alternative pathway for the synthesis of functionally less active C-29-BR exists in addition to castasterone.

7.5 BRs in Plant Growth and Development

Plant growth and development are influenced by BRs in a variety of ways (Fang et al. 2019). The biosynthetic and signaling pathways of BRs have been widely explored in model plants, which have substantially increased our understanding of BR regulatory mechanisms, notably in several biological processes related to plant growth and development (Planas-Riverola et al. 2019). BR-induced growth enhancement was once assumed to be a result of cell elongation. However, a role for BRs in cell division was later discovered. It is now well documented that, in addition to cell elongation and cell division, BRs govern xylem differentiation, photomorphogenesis, plant reproduction, and responses to abiotic and biotic stressors (Nolan et al. 2020).

Although BR production is assumed to take place solely in the endoplasmic reticulum, BR sensing occurs at the cell surface through the plasma membrane-localized receptors BRASSINOSTEROID INSENSITIVE 1 (BRI1) and similar homologs (Northey et al. 2016). Following BR recognition by BRI1 and co-receptors, BR signals are conveyed via a well-known signaling cascade to downstream transcription factors BR1-ETHYL METHANE SULFONATE SUPPRESSOR1 (BES1) and BRASSINAZOLE-RESISTANT1 (BZR1), which eventually influence transcription of BR-regulated genes (Nolan et al. 2020; Planas-Riverola et al. 2019). Low seed germination, dwarfism, delayed blooming and senescence, reduced male fertility, and de-etiolation in the dark are all consequences of BR shortage or its perception incapacity in plants (Clouse 2015). Overexpression of BR biosynthesis genes, on the other hand, raises endogenous BR levels, resulting in higher crop productivity and stress tolerance (Xia et al. 2018). For instance, overexpression of a gene encoding sterol C-22 hydroxylases enhances endogenous BR levels in rice, resulting in enhanced grain size and yield by up to 40% (Wu et al. 2008). BR has been shown in recent research on soybean (*Glycine max* L. Merrill) to prevent leaf senescence (Yin et al. 2019). A growing body of research implies that the ability of BR to regulate crucial agronomic features has the potential to transform agriculture's future (Tong and Chu 2018). Some studies from the last two years depicting the effect of BRs on plant growth and development have been summarized in Table 7.2.

Table 7.2 Role of brassinosteroids (BRs) in plant growth and development.

Plant species	BRs source	Effect	References
Zea mays	24-Epibrassinolide (EBL)	Improved plant growth in terms of increased root and shoot length and photosynthetic rate	Ahmad et al. (2023)
Z. mays	EBL	Improved germination percentage and ion accumulation	Alhammad et al. (2023)
Brassica juncea	EBL	Improved content of photosynthetic pigments and activity of antioxidants, and provided stress tolerance	Gupta and Seth (2023)
B. rapa	28-Homobrassinolide (HBL)	Improved germination percentage, shoot and root length, and plant fresh and dry weight	Khan et al. (2023b)
Salvia officinalis	HBL	Improved active substances in essential oil and quality and quantity of yield	Maghsoudi et al. (2023)
Triticum aestivum	EBL	Increased growth and yield characteristics under normal and stressed conditions	Raza et al. (2023)
Z. mays	EBL	Improved photosynthetic efficiency and modulated antioxidant activity leading to improved growth	Seleiman et al. (2023)
Z. mays	Brassinosteroid (BR)	Improved plant height and photosynthetic rate	Wang et al. (2023)
Pisum sativum	EBL	Increased plant length, fresh biomass, leaf area, chlorophyll content, and antioxidant activity	Yusuf et al. (2023)
Solanum tuberosum	EBL	Enhanced photosynthetic capacity and activity of antioxidants	Zheng et al. (2022)
S. quitoense	BR	Enhanced plant height, leaf area, photosynthetic rate, relative water content, and intrinsic water use efficiency during well-watered and drought conditions	Castañeda-Murillo et al. (2022)
Musa spp.	EBL	Restored photosynthetic activity and enhanced productivity during stress	Helaly et al. (2022)
Z. mays	HBL	Hampered leaf senescence, modulated endogenous hormonal levels, and improved grain weight and dry mass accumulation in grains	Hussain et al. (2022)
Z. mays	HBL	Improved photosynthesis, reduced oxidative stress, and increased plant yield	Sun et al. (2022)
Z. mays	BR	Promoted elongation of primary roots and early development of lateral roots, and increased stress-resistance factors, stress tolerance, and yield	Zhang et al. (2022a)

7.6 Role of BRs During Heavy Metal Stress

The survival of plants is threatened in a number of ways, but heavy metal stress is one of the main issues in agricultural research (Ali et al. 2023; Gupta and Shrestha 2023; Salam et al. 2023). However, phytohormones like the BRs mitigate the effect of heavy metal stress in plants by scavenging ROS and activating the antioxidant defense enzymes such as superoxide dismutase (SOD), peroxidase (POD), catalase (CAT), ascorbate peroxidase (APOX), glutathione reductase (GR), and glutathione-S-transferase (GST) (Basit et al. 2023). BRs increase crop yield and quality by providing tolerance to plants against various heavy metals. The bioactive BRs, 28-homobrassinolide (HBL) and 24-epibrassinolide (EBL), which help the plant in removing harmful metals, act as the crucial components of the aided phytoremediation system (Qureshi et al. 2023). By modifying cell membrane permeability and inducing a set of protective enzymes, BRs prevent the uptake of heavy metals (Zhang et al. 2023). Due to increased expression of ATPase, they induce the formation of stress-proteins by acting on anti-stress genes (Madhan et al. 2014). When BRs are applied exogenously, they temporarily produce H_2O_2, which activates mitogen-activated protein kinases and causes the synthesis of NADPH oxidase (Jiang et al. 2012), and overexpression of protective enzymes and stress-proteins that reduce metal stress (Yin et al. 2016). Table 7.3 highlights the role of BRs in modulating heavy metal-induced morphological and physiological changes in plants.

7.6.1 Morphology and Photosynthesis

The most prominent effect of heavy metal stress is on the plant morphology. Heavy metals are reported to reduce the growth of plants in terms of reduced shoot and root length and root and shoot weight. Further, photosynthetic pigments and gas exchange parameters are also reported to be reduced in the presence of heavy metals (Sharma et al. 2023). BRs have been reported to have a significant growth boosting effect in plants (Vardhini 2017). Because of their interaction with other plant hormones, BRs play a broad and critical function in controlling plant metabolism (Ohri et al. 2019). BRs minimize the harmful effects of metal toxicity on plant morphology and photosynthesis by modifying cell permeability and damage by activating defensive enzymes. *Glycine max* plants treated with EBL and stressed by high zinc supply exhibited modulation of ROS scavenging and protected the chloroplast membranes, with clear benefits to photosystem II efficiency and photosynthetic pigments (dos Santos et al. 2020). This positive regulation of photosynthesis by EBL could be attributed to EBL-induced stimulation of ribulose 1,5-bisphosphate carboxylase/oxygenase and its protective role against overexcitation of PSII (Fariduddin et al. 2015). EBL also increased the photochemical efficiency of PSII, gas exchange characteristics, and photosynthetic pigments, causing a reduction in the accumulation of oxidative compounds. According to the findings of another study, EBL could be employed as a possible growth regulator to increase plant growth and metabolic parameters of *Vigna radiata* under zinc stress (Mir et al. 2015). In a recent study, it was reported that application of EBL in *Capsicum annuum* mitigated the negative effects of heavy metals in terms of growth promotion and alleviation of chloroplast damage (Mumtaz et al. 2022). Similarly, seed priming with EBL enhanced the growth parameters and photosynthetic pigments of Cd and Pb-stressed maize plants.

Table 7.3 Role of brassinosteroids (BRs) in mitigating heavy metal stress.

BRs source	Heavy metal	Test crop	Mode of application	Effect reported	References
28-Homobrassinolide (HBL)	Ni	*Raphanus sativus*	Seed priming	Increased the radical scavenging, plant morphological attributes, and gas exchange characteristics	Ahmed and Sardar (2023)
HBL	Cd	*Vigna radiata*	Seed priming	Elevated the radical scavenging activity and improved plant growth	Ahmed et al. (2023)
HBL	Cr	*Trigonella corniculata*	Seedling treatment	Improved activities of SOD (108%), POD (249%), CAT (132%), and APOX (55%)	Aslam et al. (2023)
24-epibrassinolide (EBL)	Cr	*Glycine max*	Seedling treatment	Reduced the MDA by 23.2%, electrolyte leakage by 27.9%, H_2O_2 by 30.1%, superoxide radical by 26.5%, and methylglyoxal MG by 33.5%	Basit et al. (2023)
EBL	Cd	*Primula forbesii*	Seedling treatment	Decreased MDA (20.62%) and H_2O_2 content (46.85%)	Jia et al. (2023)
EBL	Zn	*V. radiata*	Exogenous	Improved antioxidative enzymes like CAT, APOX, PPO, and POD	Kumar et al. (2023b)
Brassinolide (BL)	Cd	*Oryza sativa*	Foliar spray	Dropped the MDA content in shoots by 14.96%	Li et al. (2023)
EBL	Zn	*Citrullus lanatus*	Spray	Decreased the Zn accumulation and content of H_2O_2 and MDA by 36.57% and 32.79%, respectively	Liu et al. (2023b)
EBL	Zn	*C. lanatus*	Seedling treatment	Boosted the SOD, POD, CAT, and APOX activities by 22.6, 47.3, 37.0, and 26.2%, respectively	Liu et al. (2023c)
EBL	Ni	*Solanum lycopersicum*	Exogenous treatment	Elevated ROS detoxifying enzymes SOD (44%), APOX (46%), and POD (35%)	Maia et al. (2023)
EBL	Cd	*Poa pratensis*	Seedling treatment	Increased the activity of antioxidants like SOD, CAT, and APOX to lower ROS along with upregulation of genes PpSOD, PpCAT, and PpGR by 2.3, 3.5, and 3.1 times, respectively	Niu et al. (2023)
BR	Zn	*Medicago sativa*	Seedling treatment	Reduced lipid peroxide levels while activated antioxidant enzymes such as SOD, POD, and CAT	Ren et al. (2023)

EBL	Pb	*T. foenum-graecum*	Foliar	Activated the antioxidative activities by APOX (9%), POD (24%), SOD (72%), and CAT (26%) along with upregulation of SOD and CAT genes	Sharma et al. (2023)
EBL	Cd	*B. juncea*	Foliar treatment	Decreased the lipid peroxidation by lowering the H_2O_2 by 39.6% and superoxide anion by 44.3%	Siddiqui et al. (2023)
BR	Hg	*Z. mays*	Seedling treatment	Enhanced antioxidative efficiency and associated transcription factors	Tang et al. (2023)
BR	Al	*O. sativa* (CY927 and YLY689)	Seed Priming	Decreased the MDA levels in roots and shoots of both cultivars	Basit et al. (2022b)
BR	Cr	*G. max*	Foliar application	Increased the levels of SOD, POD, CAT, and GR in roots and shoots of soybean cultivars	Basit et al. (2022a)
EBL	Cu and Cd	*Pleioblastus pygmaeus*	In vitro treatment	Improved plant growth by significantly reducing MDA and H_2O_2 content by 45 and 36%, respectively	Emamverdian et al. (2022)
EBL	Cd	*Vitis vinifera*	Seedling treatment	Decreased content of oxidative stress markers such as MDA and H_2O_2 by 18.34 and 27.15%	Li et al. (2022)
BR	Hg	*Z. mays*	Exogenous	Enhanced antioxidant enzymes to combat oxidative stress	Shao et al. (2022)
EBL	Mn	*Malus robusta*	Seedling treatment	Enhanced the activities of antioxidant enzymes	Zhang et al. (2022b)
BR	Cd	*B. nigra*	Glasshouse treatment	Reduced the MDA content	Barghi et al. (2021)
EBL	Cu	*S. lycopersicum*	Plantlets submerged in EBL solution for 2 hours	Increased leaf gas exchange parameters and improved photosynthetic efficiency	Faizan et al. (2021)
EBL	Pb	*O. sativa*	Exogenous	Declined oxidative stress on photosynthetic machinery due to surge in SOD (26%), CAT (24%), APOX (54%), and POD (63%)	Guedes et al. (2021)
EBL	Zn	*Sedum lineare*	Seedling treatment	Decreased MDA buildup in the roots of plants	Zhang and Liao (2021)

(Continued)

Table 7.3 (Continued)

BRs source	Heavy metal	Test crop	Mode of application	Effect reported	References
EBL	Zn	*G. max*		Improved photosystem II	dos Santos et al. (2020)
EBL	Cd	*Capsicum annuum*	Seedling treatment	Increased APOX and GR activity and other enzymes connected to the antioxidant defense system	Kaya et al. (2020)
EBL	Ni	*Eucalyptus urophylla*	Foliar spray	Elevated the antioxidant enzymes, thereby decreasing the amounts of superoxide radicals and H_2O_2	Ribeiro et al. (2020)
EBL	Cd	*Cucumis sativus*	Seed Priming	Diminished amounts of H_2O_2 and MDA content	Shah et al. (2020)
EBL	Pb	*B. juncea*	Seed Priming	Elevated the activity of SOD, CAT, APOX, and POX and reduced ROS production	Soares et al. (2020)
EBL	Cd	*Amaranthus hypochondriacus*	Soil treatment	Enhanced the plant's SOD and CAT activities	Xie et al. (2020)
EBL	Mn	*B. juncea*	Foliage spray	Increased photosynthetic rate	Hussain et al. (2019)
EBL	Zn	*Vitis vinifera*	Applied at three pre-bloom	Boosted photosynthetic rate and improved plant growth	Tadayon and Moafpourian (2019)
EBL	Pb	*B. juncea*	Seed treatment	Reduced metal content in roots and increased root length	Dalyan et al. (2018)
EBL and HBL	As and Cd	*O. sativa*	Seedling spray	Reduced As and Cd concentration in leaves and increased growth of rice plants	Xu et al. (2018)
HBL	Cd	*S. lycopersicum*	Foliar spray	Mitigated toxic effects of Cd on *Solanum* by enhancing the enzymes of photosystem II, carbohydrate and nitrogen assimilation	Singh and Prasad (2017)
EBL	Cr	*O. sativa*	Seed soaking for 12 hours	Improved the growth of seedlings	Sharma et al. (2016)

EBL	Mn	*B. juncea*	Sprayed on the seedlings	Upregulated antioxidant defense system and photosynthetic efficiency of seedlings	Fariduddin et al. (2015)
EBL	Zn	*V. radiata*	The seedlings were sprayed at the 15-day stage	Enhanced antioxidant activity	Mir et al. (2015)
EBL/HBL	Zn	*R. sativus* L.	The seedlings were sprayed at the 15-day stage	Increased chlorophyll content and growth of plants	Ramakrishna and Rao (2015)
EBL	Cu	*B. juncea*	Seed priming	Enhanced antioxidant enzyme activity and reduced copper toxicity	Poonam et al. (2014)
HBL	Pb	*T. foenum-graecum*	Foliar spray	Increased root and shoot length and improved foliage growth	Swamy et al. (2014)
EBL	Cu	*R. sativus*	Two-day-old radish seedlings were treated with 4 mL of MS medium solution containing EBL	Increased seedling growth	Choudhary et al. (2012)
EBL/HBL	Cd	*S. lycopersicum*	The plants were sprayed with at the 59-day stage	Increased content of photosynthetic pigments	Hayat et al. (2012)
EBL	Ni	*B. juncea*	Foliar spray	Enhanced the activity of antioxidant enzymes	Kanwar et al. (2012)
EBL	Ni	*V. radiata*	Seed soaking	Modulated activities of POD, CAT, and SOD	Yusuf et al. (2012)
EBL	Cd	*S. lycopersicum*	Plants were sprayed at the 59-day stage	Restored photosynthetic efficiency	Hasan et al. (2011)
EBL	Cd	*Phaseolus vulgaris*	15 days old seedlings were sprayed	Significantly increased the activities of antioxidative enzymes and the level of proline	Rady (2011)
EBL	Ni	*R. sativus* L.	Seed soaking	Activated antioxidant defense system	Sharma et al. (2011)

Treated plants exhibited increased plant length, dry biomass, ear diameter and length, and chlorophyll *a*, chlorophyll *b*, total chlorophyll, and carotenoid content (Shafi et al. 2023).

7.6.2 Oxidative Stress

Oxidative stress occurs when a cell's defense is overwhelmed by the excessive presence of ROS (Hasanuzzaman et al. 2020). When ROS levels surpass a critical threshold, they triggers lipid peroxidation, a process where unsaturated fatty acids within phospholipids undergo oxidative damage. This results in the formation of MDA and H_2O_2, both of which can harm cell membranes (Kumar et al. 2023a). In plants, oxidative stress often arises due to heavy metal exposure, which leads to an elevated production of ROS (Hossain et al. 2015). These ROS, including H_2O_2, superoxide radicals, MDA, hydroxide ions (OH), and methylglyoxal (MG), are the primary culprits behind oxidative damage to various cellular components, including proteins, lipids, nucleic acids, and membranes (Sachdev et al. 2021). This damage can disrupt normal cellular functions and contribute to various stress-related plant diseases and conditions. To summarize, oxidative stress in plants results from an imbalance between ROS production and the cell's antioxidant defense, leading to harmful effects on cellular components. Application of BRs has been reported to reduce the heavy metal-induced accumulation of ROS in plants. For instance, *Pleioblastus pygmaeus* (Bamboo) plant tolerance (41% enhanced) to Cu and Cd toxicity was promoted by co-application of EBL and titanium oxide nanoparticles by increasing antioxidant activity, photosynthetic capacity, and minimizing oxidative stress markers including MDA, H_2O_2, and superoxide radicals by 45, 36, and 62%, respectively (Emamverdian et al. 2022). Also, EBL mitigated the harmful effects of Cd on plants by boosting the Ascorbate-Glutathione (AsA-GSH) cycle, promoting the activity of antioxidant enzymes, and upregulating the expression of genes such as VvSOD, VvPOD, VvAPOX, VvGR, VvDHAR, and VvMDHAR. Interestingly, the protective effect of EBL against Cd-induced stress was more pronounced at a concentration of 0.05 mg L^{-1} compared to concentrations of 0.1 and 0.2 mg L^{-1} (Li et al. 2022). Similar results were observed in other studies where heavy metal-induced oxidative stress was alleviated by the application of BR and its sources in various plant species. For instance, Mn in *Zea mays* (Wang et al. 2009), Zn in *M. sativa* (Ren et al. 2023), As in *Brassica juncea* (Kanwar and Bhardwaj 2015), and Sb in *A. thaliana* (Wu et al. 2019).

7.6.3 Antioxidative Defense

Various environmental pressures can lead to the unreasonable production of ROS within cells, ultimately actuating antioxidant enzyme activity and, if left unchecked, leading to cell death (Basit et al. 2021; Dufil et al. 2021). When heavy metals amass in plants beyond a certain threshold, they yield free radicals and ROS, causing various harmful effects. Notwithstanding, plants have inherent mechanisms to counteract the damaging effects of ROS. These mechanisms involve protein antioxidants responsible for scavenging ROS. The antioxidant system in plants consists of three main enzymes: SOD, POD, and CAT, all of which work to neutralize ROS (Anwar et al. 2018). In addition to these enzymes, the plant's antioxidant system includes other enzymes such as GR, GST, DHAR (dehydroascorbate reductase), and MDHAR (monodehydroascorbate reductase) (Yaqoob et al. 2022). SOD primarily converts radical superoxide into hydrogen peroxide, which is further converted into oxygen and water by CAT and APOX. CAT plays a vital role in eliminating excess hydrogen peroxide from the cell. A study on plants subjected to Pb stress found that the application of EBL in *B. juncea* significantly enhanced the activities of these antioxidant enzymes. Notably, there was a gradual increase in SOD and CAT activity in plants under heavy metal stress (Soares et al. 2020). Likewise, scientists explored the effects of stress caused by zinc oxide nanoparticles in tomato and reported that EBL dissolved in half-strength MS growth medium ameliorated the zinc oxide nanoparticles-induced oxidative stress by enhancing antioxidant enzymes such as SOD, CAT, POD, and GST for quick ROS scavenging and thereby improving the growth of tomato seedlings (Li et al. 2015). Under BR treatment, the amount of lipid peroxides in *Medicago sativa* reduced and the levels of antioxidant enzymes such as SOD, POD, and CAT were found to be elevated. Additionally, the BR treatment reduced the transcription of *M. sativa ZRT–IRT-like proteins 1/3/5 (MsZIP1/3/5)*, the zinc uptake carriers in root cells. Also, BR raised the expression of genes associated with long-distance transport, vacuolar storage, zinc efflux and chelation, including *M. sativa heavy metal transporter 4 (MsHMA1), zinc-induced facilitator 1 (MsZIF1), metal tolerance protein 1 (MsMTP1), nicotianamine synthase 1 (MsNAS1)*, and *yellow stripe-like protein 1 (MsYSL1)* (Ren et al. 2023). In another study, the foliar application of BRs on *G. max* under Cr stress enhanced the antioxidant activities by raising the levels of SOD, CAT, POD, and GR (Basit et al. 2022a). Furthermore, the toxic effects of Pb on *Trigonella foenum-graecum* seedlings were reduced by EBL by upregulating the activation of antioxidant enzymes APOX, POD, SOD, CAT by 9, 24, 72 and 26%, respectively. The genes that controlled SOD and CAT were also found to be upregulated in EBL treated plants (Sharma et al. 2023).

7.7 Mechanism of Action of BRs

A number of receptors, coreceptors, and secondary messengers form an intricate part of BRs' signal transduction induced by stress. The general process is as outlined below:

When sensed by its coreceptor complex, BR acts as a signal transducer. Both BRASSINOSTEROID INSENSITIVE 1 (BR1) (a receptor kinase) and BR1-ASSOCIATED RECEPTOR KINASE 1 (BAK1) are found on the cell membrane. In the absence of BR, BR1 associates with BR1 KINASE INHIBITOR 1 (BK1), rendering BR1 inactive and inhibiting heterodimerization of BR1 with BAK1 (Clouse 2011). BRASSINOSTEROID INSENSITIVE 2 (BIN2; a glycogen synthase kinase) functions as a negative regulator in BR signaling by phosphorylating two important transcription factors (TFs), BRASSINAZOLE RESISTANT1 (BZR1) and BR1-ETHYL METHANE SULFONATE SUPPRESSOR 1 (BES1). This phosphorylation event promotes the binding of 14-3-3 proteins to the BZR1–BES1 complex, which helps the TF complex to remain in the cytoplasm and prevents it from entering the nucleus, restricting the process of BR-triggered gene transcription (Jaillais et al. 2011; Bhandari and Nailwal 2020).

The presence of BR modifies the BR1 receptor. Interaction of BR with BR1 receptor causes autophosphorylation at Tyr-211 and leads to activation of BR1 fragmentary kinase. This separates BK1 from BR1 and causes heterodimerization of BR1 with BAK1. These two molecules transphosphorylate each other, giving BR1 complete kinase activity (Wang and Chory 2006). Now that BR1 is fully activated, it initiates a signaling cascade and phosphorylates BR-SIGNALING KINASES (BSKs) and CONSTITUTIVE DIFFERENTIAL

GROWTH 1 (CDG1) at Ser-230 and Ser-234, respectively, to activate them. These activated molecules then bind to and phosphorylate BR1 SUPPRESSOR 1 (BSU1) and add a phosphate group to it. This phosphorylated BSU1 acts as a phosphatase and eventually dephosphorylates BIN2 (Kim et al. 2011). This process relieves BIN2-mediated suppression of BES1 and BZR1, resulting in the accumulation of dephosphorylated BZR1 and BES1 in the cellular environment. Such dephosphorylated BES1 and BZR1 molecules become permeable and enter the nucleus. Within the nucleus, these molecules either directly or indirectly regulate BR-targeted gene expression by interacting with other transcription factors such as GATA (TF with one or two highly conserved zinc-finger DNA binding domains), BES-INTERACTING MYC-LIKE 1 (BIM1, helix-loop-helix transcription factor), and *Arabidopsis thaliana*-INTERACTS-WITH-SPT6 (AtIWS, a protein that aids in the post-recruitment of RNA polymerase) (Saidi et al. 2012). According to recent studies, additional signaling molecules are also involved in the BR signaling pathway. PROTEIN PHOSPHATASE 2A (PP2A) acts as a positive controller, dephosphorylating BZR1 and BES1, whereas SUPPRESSOR OF BR1 (SBI1) deactivates BR1 by methylating PP2A (Tang et al. 2011). Figure 7.3 depicts possible signal transduction pathways of BR signaling.

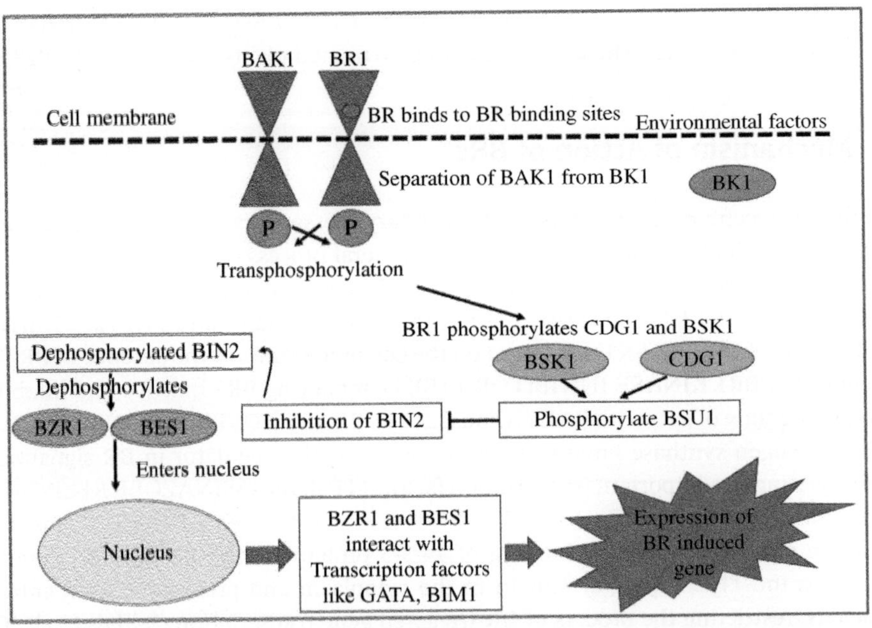

Figure 7.3 Signal cascade of BRs during abiotic stress response. BR1 denotes BRASSINOSTEROID-INSENSITIVE 1; BAK1: BR1-ASSOCIATED RECEPTOR KINASE 1; BK1: BR1 KINASE INHIBITOR 1; CDG1: CONSTITUTIVE DIFFERENTIAL GROWTH 1; BSK1: BR-SIGNALING KINASE 1; BSU: BR1 SUPPRESSOR 1; BIN2: BRASSINOSTEROID-INSENSITIVE 2 (a glycogen synthase kinase); BZR: BRASSINAZOLE-RESISTANT 1; BES1: BR1-ETHYL METHANE SULFONATE SUPPRESSOR 1; GATA: A transcription factor with one or two highly conserved zinc-finger DNA binding domains and BIM1: BES-INTERACTING MYC-LIKE 1 transcription factor.

7.8 Conclusion

The research unequivocally demonstrates that BRs are essential phytohormones that control signals to improve plants' ability to withstand heavy metal toxicity. As the primary controllers of the plant growth regulation network, BRs modulate important morphological and physiological functions like increased germination, plant length and weight, elevated levels of photosynthetic pigments, and decreased ROS accumulation. Some of the primary defense mechanisms used by BRs to lessen the detrimental impacts of heavy metals on plants include activation of the antioxidant system through increased production of APOX, CAT, SOD, POD, and overexpression of stress-related genes. A variety of BR treatment techniques, including foliar spraying, soaking in BR-containing solutions, and seed priming, have been used to assess the effects of exogenous BR administration in plants. Therefore, owing to the significance of BRs in crop improvement, more research is required to determine the essential regulatory components of the BR signaling pathways as well as the fundamental mechanisms of BR-modulated growth and developmental responses in substantial crop plants. This will enable the development of strategies that can maximize crop yield and improve plant performance during heavy metal stress.

References

Abd Elnabi, M.K., Elkaliny, N.E., Elyazied, M.M. et al. (2023). Toxicity of heavy metals and recent advances in their removal: a review. *Toxics* 11(7): 580.

AbdElgawad, H., Zinta, G., Hamed, B.A. et al. (2020). Maize roots and shoots show distinct profiles of oxidative stress and antioxidant defense under heavy metal toxicity. *Environmental Pollution* 258: 113705.

Ahmad, A., Tola, E., Alshahrani, T.S. et al. (2023). Enhancement of morphological and physiological performance of *Zea mays* L. under saline stress using ZnO nanoparticles and 24-epibrassinolide seed priming. *Agronomy* 13(3): 771.

Ahmed, S., Mudassar, S., Sardar, R. et al. (2023). 28-Homo-Brassinolide confers cadmium tolerance in *Vigna radiate* L. through modulating minerals uptake, antioxidant system and gas exchange attributes. *Journal of Plant Growth Regulation* 42: 1–15.

Ahmed, S. and Sardar, R. (2023). Improvement in growth and physiochemical attributes of *Raphanus sativus* L. through exogenous application of 28-homobrassinolide under nickel stress. *Scientia Horticulturae* 311: 111791.

Alhammad, B.A., Ahmad, A., Seleiman, M.F. et al. (2023). Seed priming with nanoparticles and 24-epibrassinolide improved seed germination and enzymatic performance of *Zea mays* L. in salt-stressed soil. *Plants* 12(4): 690.

Ali, M., Kumar, D., Tikoria, R. et al. (2023). Exploring the potential role of hydrogen sulfide and jasmonic acid in plants during heavy metal stress. *Nitric Oxide* 140–141: 1629.

Anwar, A., Liu, Y., Dong, R. et al. (2018). The physiological and molecular mechanism of brassinosteroid in response to stress: a review. *Biological Research* 51: 46.

Asgher, M., Rehaman, A., Islam, S.N.U. et al. (2023). Appraisal of functions and role of selenium in heavy metal stress adaptation in plants. *Agriculture* 13(5): 1083.

Aslam, M.A., Ahmed, S., Saleem, M. et al. (2023). Mitigation of chromium-induced phytotoxicity in 28-homobrassinolide treated *Trigonella corniculata* L. by modulation of oxidative biomarkers and antioxidant system. *Ecotoxicology and Environmental Safety* 263: 115354.

Barghi, A., Golipoori, A., Ghavidel, A. et al. (2021). Effect of plant growth promoting rhizobacteria, salicylic acid and brassinosteroid on physiological properties of black mustard in cadmium stress condition. *Journal of Plant Production Research* 28(1): 153–168.

Basit, F., Bhat, J.A., Hu, J. et al. (2022a). Brassinosteroid supplementation alleviates chromium toxicity in soybean (*Glycine max* L.) via reducing its translocation. *Plants* 11(17): 2292.

Basit, F., Liu, J., An, J. et al. (2021). Brassinosteroids as a multidimensional regulator of plant physiological and molecular responses under various environmental stresses. *Environmental Science and Pollution Research* 28: 44768–44779.

Basit, F., Liu, J., An, J. et al. (2022b). Seed priming with brassinosteroids alleviates aluminum toxicity in rice via improving antioxidant defense system and suppressing aluminum uptake. *Environmental Science and Pollution Research* 29: 1–15.

Basit, F., Tao, J., An, J. et al. (2023). Nitric oxide and brassinosteroids enhance chromium stress tolerance in *Glycine max* L. (Merr.) by modulating antioxidative defense and glyoxalase systems. *Environmental Science and Pollution Research* 30(18): 51638–51653.

Bhandari, S. and Nailwal, T.K. (2020). Role of brassinosteroids in mitigating abiotic stresses in plants. *Biologia* 75(12): 2203–2230.

Bhanu, A.N. (2019). Brassinosteroids: relevance in biological activities of plants and agriculture. *Journal of Plant Science and Research* 35: 1–15.

Castañeda-Murillo, C.C., Rojas-Ortiz, J.G., Sánchez-Reinoso, A.D. et al. (2022). Foliar brassinosteroid analogue (DI-31) sprays increase drought tolerance by improving plant growth and photosynthetic efficiency in lulo plants. *Heliyon* 8(2): e08977.

Choudhary, S.P., Oral, H.V., Bhardwaj, R. et al. (2012). Interaction of brassinosteroids and polyamines enhances copper stress tolerance in *Raphanus sativus*. *Journal of Experimental Botany* 63(15): 5659–5675.

Clouse, S.D. (2011). Brassinosteroid signal transduction: from receptor kinase activation to transcriptional networks regulating plant development. *The Plant Cell* 23(4): 1219–1230.

Clouse, S.D. (2015). A history of brassinosteroid research from 1970 through 2005: thirty-five years of phytochemistry, physiology, genes and mutants. *Journal of Plant Growth Regulation* 34: 828–844.

Dalyan, E., Yüzbaşıoğlu, E., and Akpınar, I. (2018). Effect of 24-epibrassinolide on antioxidative defence system against lead-induced oxidative stress in the roots of *Brassica juncea* L. seedlings. *Russian Journal of Plant Physiology* 65: 570–578.

Dawood, M.F., Tahjib-Ul-Arif, M., Sohag, A.A.M. et al. (2022). Fluoride mitigates aluminum-toxicity in barley: morpho-physiological responses and biochemical mechanisms. *BMC Plant Biology* 22(1): 1–17.

Demecsová, L., Zelinová, V., Liptáková, Ľ. et al. (2020). Indole-3-butyric acid priming reduced cadmium toxicity in barley root tip via NO generation and enhanced glutathione peroxidase activity. *Planta* 252: 1–16.

dos Santos, L.R., da Silva, B.R.S., Pedron, T. et al. (2020). 24-Epibrassinolide improves root anatomy and antioxidant enzymes in soybean plants subjected to zinc stress. *Journal of Soil Science and Plant Nutrition* 20: 105–124.

Dufil, G., Bernacka-Wojcik, I., Armada-Moreira, A. et al. (2021). Plant bioelectronics and biohybrids: the growing contribution of organic electronic and carbon-based materials. *Chemical Reviews* 122(4): 4847–4883.

El-Ballat, E.M., Elsilk, S.E., Ali, H.M. et al. (2023). Metal-resistant PGPR strain *Azospirillum brasilense* EMCC1454 enhances growth and chromium stress tolerance of chickpea (*Cicer arietinum* L.) by modulating redox potential, osmolytes, antioxidants and stress-related gene expression. *Plants* 12(11): 2110.

El-Meihy, R.M., Abou-Aly, H.E., Youssef, A.M. et al. (2019). Efficiency of heavy metals-tolerant plant growth promoting bacteria for alleviating heavy metals toxicity on sorghum. *Environmental and Experimental Botany* 162: 295–301.

Emamverdian, A., Ding, Y., Barker, J. et al. (2022). Co-application of 24-epibrassinolide and titanium oxide nanoparticles promotes *Pleioblastus pygmaeus* plant tolerance to Cu and Cd toxicity by increasing antioxidant activity and photosynthetic capacity and reducing heavy metal accumulation and translocation. *Antioxidants* 11(3): 451.

Emamverdian, A., Ghorbani, A., Li, Y. et al. (2023). Responsible mechanisms for the restriction of heavy metal toxicity in plants via the co-foliar spraying of nanoparticles. *Agronomy* 13(7): 1748.

Faizan, M., Bhat, J.A., Noureldeen, A. et al. (2021). Zinc oxide nanoparticles and 24-epibrassinolide alleviates Cu toxicity in tomato by regulating ROS scavenging, stomatal movement and photosynthesis. *Ecotoxicology and Environmental Safety* 218: 112293.

Fang, P., Yan, M., Chi, C. et al. (2019). Brassinosteroids act as a positive regulator of photoprotection in response to chilling stress. *Plant Physiology* 180(4): 2061–2076.

Fariduddin, Q., Ahmed, M., Mir, B.A. et al. (2015). 24-Epibrassinolide mitigates the adverse effects of manganese induced toxicity through improved antioxidant system and photosynthetic attributes in *Brassica juncea*. *Environmental Science and Pollution Research* 22: 11349–11359.

Fujioka, S., Noguchi, T., Yokota, T. et al. (1998). Brassinosteroids in *Arabidopsis thaliana*. *Phytochemistry* 48(4): 595–599.

Fujioka, S., Takatsuto, S., and Yoshida, S. (2002). An early C-22 oxidation branch in the brassinosteroid biosynthetic pathway. *Plant Physiology* 130(2): 930–939.

Ghorbani, A., Emamverdian, A., Pishkar, L. et al. (2023). Melatonin-mediated nitric oxide signaling enhances adaptation of tomato plants to aluminum stress. *South African Journal of Botany* 162: 443–450.

Ghuge, S.A., Nikalje, G.C., Kadam, U.S. et al. (2023). Comprehensive mechanisms of heavy metal toxicity in plants, detoxification and remediation. *Journal of Hazardous Materials* 450: 131039.

Giannakoula, A., Therios, I., and Chatzissavvidis, C. (2021). Effect of lead and copper on photosynthetic apparatus in citrus (*Citrus aurantium* L.) plants. The role of antioxidants in oxidative damage as a response to heavy metal stress. *Plants* 10(1): 155.

Guedes, F.R.C.M., Maia, C.F., da Silva, B.R.S. et al. (2021). Exogenous 24-epibrassinolide stimulates root protection and leaf antioxidant enzymes in lead stressed rice plants: central roles to minimize Pb content and oxidative stress. *Environmental Pollution* 280: 116992.

Gupta, B. and Shrestha, J. (2023). Abiotic stress adaptation and tolerance mechanisms in crop plants. *Frontiers in Plant Science*, 14: 1278895. https://doi.org/10.3389/fpls.2023.1278895.

Gupta, P. and Seth, C.S. (2023). 24-epibrassinolide regulates functional components of nitric oxide signalling and antioxidant defense pathways to alleviate salinity stress in *Brassica juncea* L. cv. Varuna. *Journal of Plant Growth Regulation* 42(7): 4207–4222.

Hasan, S.A., Hayat, S., and Ahmad, A. (2011). Brassinosteroids protect photosynthetic machinery against the cadmium induced oxidative stress in two tomato cultivars. *Chemosphere* 84(10): 1446–1451.

Hasanuzzaman, M., Bhuyan, M.B., Zulfiqar, F. et al. (2020). Reactive oxygen species and antioxidant defense in plants under abiotic stress: revisiting the crucial role of a universal defense regulator. *Antioxidants* 9(8): 681.

Hayat, S., Alyemeni, M.N., and Hasan, S.A. (2012). Foliar spray of brassinosteroid enhances yield and quality of *Solanum lycopersicum* under cadmium stress. *Saudi Journal of Biological Sciences* 19(3): 325–335.

Helaly, M.N., El-Hoseiny, H.M., Elsheery, N.I. et al. (2022). 5-Aminolevulinic acid and 24-epibrassinolide improve the drought stress resilience and productivity of banana plants. *Plants* 11(6): 743.

Hossain, M.A., Bhattacharjee, S., Armin, S.M. et al. (2015). Hydrogen peroxide priming modulates abiotic oxidative stress tolerance: insights from ROS detoxification and scavenging. *Frontiers in Plant Science* 6: 420.

Hussain, A., Nazir, F., and Fariduddin, Q. (2019). 24-Epibrassinolide and spermidine alleviate Mn stress via the modulation of root morphology, stomatal behavior, photosynthetic attributes and antioxidant defense in *Brassica juncea*. *Physiology and Molecular Biology of Plants* 25: 905–919.

Hussain, M., Wang, Z., Mo, Y. et al. (2022). Influence of exogenous 28-homobrassinolide optimized dosage and EDAH application on hormone status, grain filling and maize production. *Processes* 10(6): 1118.

Ibrahim, E.A., El-Sherbini, M.A., and Selim, E.M.M. (2022). Effects of biochar on soil properties, heavy metal availability and uptake and growth of summer squash grown in metal-contaminated soil. *Scientia Horticulturae* 301: 111097.

Islam, M. and Sandhi, A. (2023). Heavy metal and drought stress in plants: the role of microbes—a review. *Gesunde Pflanzen* 75(4): 695–708.

Jaillais, Y., Hothorn, M., Belkhadir, Y. et al. (2011). Tyrosine phosphorylation controls brassinosteroid receptor activation by triggering membrane release of its kinase inhibitor. *Genes and Development* 25(3): 232–237.

Jia, Y., Yin, X., Zhao, J. et al. (2023). Effects of 24-epibrassinolide, melatonin and their combined effect on cadmium tolerance in *Primula forbesii* Franch. *Ecotoxicology and Environmental Safety* 262: 115217.

Jiang, M., Dai, S., Wang, B. et al. (2021). Gold nanoparticles synthesized using melatonin suppress cadmium uptake and alleviate its toxicity in rice. *Environmental Science: Nano* 8(4): 1042–1056.

Jiang, Y. P., Cheng, F., Zhou, Y. H. et al. (2012). Cellular glutathione redox homeostasis plays an important role in the brassinosteroid-induced increase in CO_2 assimilation in *Cucumis sativus*. *New Phytologist* 194(4): 932–943.

Kajala, K., Walker, K.L., Mitchell, G.S. et al. (2019). Real-time whole-plant dynamics of heavy metal transport in *Arabidopsis halleri* and *Arabidopsis thaliana* by gamma-ray imaging. *Plant Direct* 3(4): e00131.

Kanwar, M.K. and Bhardwaj, R. (2015). Arsenic induced modulation of antioxidative defense system and brassinosteroids in *Brassica juncea* L. *Ecotoxicology and Environmental Safety* 115: 119–125.

Kanwar, M.K., Bhardwaj, R., Arora, P. et al. (2012). Plant steroid hormones produced under Ni stress are involved in the regulation of metal uptake and oxidative stress in *Brassica juncea* L. *Chemosphere* 86(1): 41–49.

Kapoor, D., Bhardwaj, S., Gautam, S. et al. (2022). Brassinosteroids in plant nutrition and heavy metal tolerance. In: *Brassinosteroids in Plant Developmental Biology and Stress Tolerance* (ed. G.J. Ahammed, A. Sharma, and J. Yu), 217–235. Academic Press.

Kaur, R., Das, S., Bansal, S. et al. (2021). Heavy metal stress in rice: uptake, transport, signaling and tolerance mechanisms. *Physiologia Plantarum* 173(1): 430–448.

Kaur, R., Sharma, N., Tikoria, R. et al. (2022). Insights into biosynthesis and signaling of cytokinins during plant growth, development and stress tolerance. In: *Auxins, Cytokinin and Gibberellins Signaling in Plants* (ed. T. Aftab), 153–187. Cham: Springer.

Kaya, C., Ashraf, M., Alyemeni, M.N. et al. (2020). The role of nitrate reductase in brassinosteroid-induced endogenous nitric oxide generation to improve cadmium stress tolerance of pepper plants by upregulating the ascorbate-glutathione cycle. *Ecotoxicology and Environmental Safety* 196: 110483.

Khan, I., Awan, S.A., Rizwan, M. et al. (2021). Effects of silicon on heavy metal uptake at the soil-plant interphase: a review. *Ecotoxicology and Environmental Safety* 222: 112510.

Khan, I.U., Qi, S.S., Gul, F. et al. (2023a). A green approach used for heavy metals 'phytoremediation' via invasive plant species to mitigate environmental pollution: a review. *Plants* 12(4): 725.

Khan, M., Ahmed, S., Yasin, N. A. et al. (2023b). 28-Homobrassinolide primed seed improved lead stress tolerance in *Brassica rapa* l. through modulation of physio-biochemical attributes and nutrient uptake. *Plants* 12(20): 3528.

Khanna, K., Jamwal, V.L., Gandhi, S.G. et al. (2019). Metal resistant PGPR lowered Cd uptake and expression of metal transporter genes with improved growth and photosynthetic pigments in *Lycopersicon esculentum* under metal toxicity. *Scientific Reports* 9(1): 5855.

Kim, B.K., Fujioka, S., Takatsuto, S. et al. (2008). Castasterone is a likely end product of brassinosteroid biosynthetic pathway in rice. *Biochemical and Biophysical Research Communications* 374(4): 614–619.

Kim, T.W., Guan, S., Burlingame, A.L. et al. (2011). The CDG1 kinase mediates brassinosteroid signal transduction from BRI1 receptor kinase to BSU1 phosphatase and GSK3-like kinase BIN2. *Molecular Cell* 43(4): 561–571.

Kosakivska, I.V., Babenko, L.M., Romanenko, K.O. et al. (2021). Molecular mechanisms of plant adaptive responses to heavy metals stress. *Cell Biology International* 45(2): 258–272.

Kour, J., Kohli, S.K., Khanna, K. et al. (2021). Brassinosteroid signaling, crosstalk and, physiological functions in plants under heavy metal stress. *Frontiers in Plant Science* 12: 608061.

Kumar, D., Manhas, R.K., and Ohri, P. (2023a). Deciphering the growth promoting and stress curtailing role of nitric oxide in *Meloidogyne incognita* infested *Solanum lycopersicum* seedlings. *Scientia Horticulturae* 319: 112147.

Kumar, D. and Ohri, P. (2023). Say "NO" to plant stresses: unravelling the role of nitric oxide under abiotic and biotic stress. *Nitric Oxide* 130: 36–57.

Kumar, N., Sharma, V., Kaur, G. et al. (2023b). Brassinosteroids as promoters of seedling growth and antioxidant activity under heavy metal zinc stress in mung bean (*Vigna radiata* L.). *Frontiers in Microbiology* 14: 1259103.

Li, B.B., Fu, Y.S., Li, X.X. et al. (2022). Brassinosteroids alleviate cadmium phytotoxicity by minimizing oxidative stress in grape seedlings: toward regulating the ascorbate-glutathione cycle. *Scientia Horticulturae* 299: 111002.

Li, B., Wang, S., You, X. et al. (2023). Effect of foliar spraying of gibberellins and brassinolide on cadmium accumulation in rice. *Toxics*, 11(4): 364.

Li, N., Guo, S.R., Shu, S. et al. (2015). Effects of exogenous 24-epibrassinolde on leaf morphology and photosynthetic characteristics of tomato seedlings under low light stress. *Ying Yong Sheng tai xue bao= The Journal of Applied Ecology* 26(3): 847–852.

Li, S., Zheng, H., Lin, L. et al. (2021). Roles of brassinosteroids in plant growth and abiotic stress response. *Plant Growth Regulation* 93: 29–38.

Liu, L., Chen, G., Li, S. et al. (2023a). A brassinosteroid transcriptional regulatory network participates in regulating fiber elongation in cotton. *Plant Physiology* 191(3): 1985–2000.

Liu, X., Zhu, Q., Liu, W. et al. (2023b). Exogenous brassinosteroid enhances zinc tolerance by activating the phenylpropanoid biosynthesis pathway in *Citrullus lanatus* L. *Plant Signaling and Behavior* 18(1): 2186640.

Liu, X., Zhu, Q., Liu, W. et al. (2023c). 24-Epibrassinolide confers zinc stress tolerance in watermelon seedlings through modulating antioxidative capacities and lignin accumulation. *PeerJ* 11: e15330.

Madhan, M., Mahesh, K., and Rao, S.S. (2014). Effect of 24-epibrassinolide on aluminium stress induced inhibition of seed germination and seedling growth of *Cajanus cajan* (L.) Millsp. *International Journal of Multidisciplinary and Current Research* 2: 286–290.

Maghsoudi, E., Abbaspour, H., Ghasemi Pirbalouti, A. et al. (2023). Influence of the foliar applications of paclobutrazol and 24-epibrassinolide on the quantitative and qualitative traits of sage (*Salvia officinalis* L.) volatile oil under different soil moisture conditions. *Journal of Plant Growth Regulation* 42: 5495–5506.

Maia, C.F., Pereira, Y.C., da Silva, B.R.S. et al. (2023). Exogenously applied 24-epibrassinolide favours stomatal performance, ROS detoxification and nutritional balance, alleviating oxidative damage against the photosynthetic apparatus in tomato leaves under nickel stress. *Journal of Plant Growth Regulation* 42(4): 2196–2211.

Manghwar, H., Hussain, A., Ali, Q. et al. (2022). Brassinosteroids (BRs) role in plant development and coping with different stresses. *International Journal of Molecular Sciences* 23(3): 1012.

Mir, B.A., Khan, T.A., and Fariduddin, Q. (2015). 24-epibrassinolide and spermidine modulate photosynthesis and antioxidant systems in *Vigna radiata* under salt and zinc stress. *International Journal of Advanced Research* 3(5): 592–608.

Mumtaz, M.A., Hao, Y., Mehmood, S. et al. (2022). Physiological and transcriptomic analysis provide molecular insight into 24-epibrassinolide mediated Cr (VI)-toxicity tolerance in pepper plants. *Environmental Pollution* 306: 119375.

Niu, K., Zhu, R., Wang, Y. et al. (2023). 24-epibrassinolide improves cadmium tolerance and lateral root growth associated with regulating endogenous auxin and ethylene in Kentucky bluegrass. *Ecotoxicology and Environmental Safety* 249: 114460.

Noguchi, T., Fujioka, S., Choe, S. et al. (2000). Biosynthetic pathways of brassinolide in *Arabidopsis*. *Plant Physiology* 124(1): 201–210.

Nolan, T.M., Vukašinović, N., Liu, D. et al. (2020). Brassinosteroids: multidimensional regulators of plant growth, development and stress responses. *The Plant Cell* 32(2): 295–318.

Northey, J.G., Liang, S., Jamshed, M. et al. (2016). Farnesylation mediates brassinosteroid biosynthesis to regulate abscisic acid responses. *Nature Plants* 2(8): 1–7.

Ohnishi, T., Szatmari, A.M., Watanabe, B. et al. (2006). C-23 hydroxylation by *Arabidopsis* CYP90C1 and CYP90D1 reveals a novel shortcut in brassinosteroid biosynthesis. *The Plant Cell* 18(11): 3275–3288.

Ohri, P., Bhardwaj, R., Kaur, R. et al. (2019). Emerging trends on crosstalk of BRS with other phytohormones. In; *Brassinosteroids: Plant Growth and Development* (ed. S. Hayat, M. Yusuf, R. Bhardwaj et al.), 425–441. Singapore: Springer.

Okoro, E.E., Okolie, A.G., Sanni, S.E. et al. (2020). Toxicology of heavy metals to subsurface lithofacies and drillers during drilling of hydrocarbon wells. *Scientific Reports* 10(1): 6152.

Parada, F., Oklestkova, J., and Arce-Johnson, P. (2022). Characterization of endogenous levels of brassinosteroids and related genes in grapevines. *International Journal of Molecular Sciences* 23(3): 1827.

Pasricha, S., Mathur, V., Garg, A. et al. (2021). Molecular mechanisms underlying heavy metal uptake, translocation and tolerance in hyperaccumulators-an analysis: heavy metal tolerance in hyperaccumulators. *Environmental Challenges* 4: 100197.

Planas-Riverola, A., Gupta, A., Betegón-Putze, I. et al. (2019). Brassinosteroid signaling in plant development and adaptation to stress. *Development* 146(5): dev151894.

Poonam, R.K., Bali, S., Singh, R. et al. (2014). Treatment of 24-EBL to *Brassica juncea* plants under Cu-metal stress lowers oxidative burst by activity antioxidative enzymes. *Journal of Stress Physiology & Biochemistry* 10: 315–327.

Qureshi, F.F., Rasheed, R., Hussain, I. et al. (2023). Signaling crosstalk between brassinosteriods and jasmonates in plant defense, growth and development. In: *Hormonal Cross-Talk, Plant Defense and Development* (ed. A. Husen and W. Zhang), 123–148. Academic Press.

Rady, M.M. (2011). Effect of 24-epibrassinolide on growth, yield, antioxidant system and cadmium content of bean (*Phaseolus vulgaris* L.) plants under salinity and cadmium stress. *Scientia Horticulturae* 129(2): 232–237.

Rahman, S.U., Li, Y., Hussain, S. et al. (2023). Role of phytohormones in heavy metal tolerance in plants: a review. *Ecological Indicators* 146: 109844.

Ramakrishna, B. and Rao, S.S.R. (2015). Foliar application of brassinosteroids alleviates adverse effects of zinc toxicity in radish (*Raphanus sativus* L.) plants. *Protoplasma* 252: 665–677.

Raza, M.A.S., Ibrahim, M.A., Ditta, A. et al. (2023). Exploring the recuperative potential of brassinosteroids and nano-biochar on growth, physiology and yield of wheat under drought stress. *Scientific Reports* 13(1): 15015.

Ren, Y., Li, X., Liang, J. et al. (2023). Brassinosteroids and gibberellic acid actively regulate the zinc detoxification mechanism of *Medicago sativa* L. seedlings. *BMC Plant Biology* 23: 75.

Ribeiro, A.T., de Oliveira, V.P., de Oliveira Barros Junior, U. et al. (2020). 24-Epibrassinolide mitigates nickel toxicity in young *Eucalyptus urophylla* ST Blake plants: nutritional, physiological, biochemical, anatomical and morphological responses. *Annals of Forest Science* 77: 1–19.

Riyazuddin, R., Nisha, N., Ejaz, B. et al. (2021). A comprehensive review on the heavy metal toxicity and sequestration in plants. *Biomolecules* 12(1): 43.

Sachdev, S., Ansari, S.A., Ansari, M.I. et al. (2021). Abiotic stress and reactive oxygen species: generation, signaling and defense mechanisms. *Antioxidants* 10(2): 277.

Saidi, Y., Hearn, T.J., and Coates, J.C. (2012). Function and evolution of 'green'GSK3/Shaggy-like kinases. *Trends in Plant Science* 17(1): 39–46.

Salam, A., Afridi, M.S., Khan, A.R. et al. (2023). Cobalt induced toxicity and tolerance in plants: insights from omics approaches. In: *Heavy Metal Toxicity and Tolerance in Plants: A Biological, Omics and Genetic Engineering Approach* (ed. M.A. Hossain, A.Z. Hossain, S. Bourgerie et al.), 207–229. John Wiley and Sons Ltd.

Saleem, A., Ur Rahim, H., Khan, U. et al. (2023). Organic materials amendments can improve NPK availability and maize growth by reducing heavy metals stress in calcareous soil. *International Journal of Environmental Science and Technology* 21: 1–14.

Seleiman, M.F., Ahmad, A., Tola, E. et al. (2023). Exogenous application of 24-epibrassinolide confers saline stress and improves photosynthetic capacity, antioxidant defense, mineral uptake and yield in maize. *Plants* 12(20): 3559.

Shafi, Z., Shahid, M., AlGarawi, A.M. et al. (2023). The exogenous application of 24-epibrassinolide (24-EBL) increases the Cd and Pb resilience in *Zea mays* (L.) by regulating the growth and physiological mechanism. *Applied Biochemistry and Biotechnology* 196: 1–25.

Shafiq, S., Zeb, Q., Ali, A. et al. (2019). Lead, cadmium and zinc phytotoxicity alter DNA methylation levels to confer heavy metal tolerance in wheat. *International Journal of Molecular Sciences* 20(19): 4676.

Shah, A.A., Ahmed, S., Abbas, M. et al. (2020). Seed priming with 3-epibrassinolide alleviates cadmium stress in Cucumis sativus through modulation of antioxidative system and gene expression. *Scientia Horticulturae* 265: 109203.

Shao, R., Tang, Y., Zhang, J. et al. (2022). Combined mercury and water deficit stress reduces photosynthetic efficiency but is alleviated by brassinosteroids. *Available at SSRN 4217601*.

Sharma, A., Kapoor, D., Gautam, S. et al. (2022b). Heavy metal induced regulation of plant biology: recent insights. *Physiologia Plantarum* 174(3): e13688.

Sharma, D., Bhardwaj, S., Raza, A. et al. (2023). Alleviatory effects of silicon and 24-epibrassinolide in modulation of growth, osmolytes, metabolites, antioxidant defense system and gene expression in lead-exposed fenugreek (*Trigonella foenum-graecum* L.) plants. *Agronomy* 13(7): 1884.

Sharma, I., Pati, P.K., and Bhardwaj, R. (2011). Effect of 24-epibrassinolide on oxidative stress markers induced by nickel-ion in *Raphanus sativus* L. *Acta Physiologiae Plantarum* 33: 1723–1735.

Sharma, N., Kour, S., Kumar, D. et al. (2022a). Role of brassinosteroids (BRs) in modulating antioxidative defense mechanism in plants growing under abiotic and biotic stress conditions. In: *Antioxidant Defense in Plants* (ed. T. Aftab and K.R. Hakeem), 325–367. Singapore: Springer.

Sharma, P., Kumar, A., and Bhardwaj, R. (2016). Plant steroidal hormone epibrassinolide regulate–heavy metal stress tolerance in *Oryza sativa* L. by modulating antioxidant defense expression. *Environmental and Experimental Botany* 122: 1–9.

Siddiqui, H., Ahmed, K.B.M., Alam, P. et al. (2023). 24-Epibrassinolide-mediated mitigation of Cd-induced toxicity in hyperaccumulator–*Brassica juncea*: influence on photosynthesis, cell death, redox and elemental status. *Journal of Plant Growth Regulation* 42(4): 2646–2661.

Singh, S. and Prasad, S.M. (2017). Effects of 28-homobrassinoloid on key physiological attributes of *Solanum lycopersicum* seedlings under cadmium stress: photosynthesis and nitrogen metabolism. *Plant Growth Regulation* 82: 161–173.

Soares, T.F.S.N., dos Santos Dias, D.C.F., Oliveira, A.M.S. et al. (2020). Exogenous brassinosteroids increase lead stress tolerance in seed germination and seedling growth of *Brassica juncea* L. *Ecotoxicology and Environmental Safety* 193: 110296.

Songmei, L., Jie, J., Yang, L. et al. (2019). Characterization and evaluation of OsLCT1 and OsNramp5 mutants generated through CRISPR/Cas9-mediated mutagenesis for breeding low Cd rice. *Rice Science* 26(2): 88–97.

Souri, M.K., Hatamian, M., and Tesfamariam, T. (2019). Plant growth stage influences heavy metal accumulation in leafy vegetables of garden cress and sweet basil. *Chemical and Biological Technologies in Agriculture* 6(1): 1–7.

Sun, S., Yao, X., Liu, X. et al. (2022). Brassinolide can improve drought tolerance of maize seedlings under drought stress: by inducing the photosynthetic performance, antioxidant capacity and ZmMYB gene expression of maize seedlings. *Journal of Soil Science and Plant Nutrition* 22(2): 2092–2104.

Swamy, K.N., Vardhini, B.V., Ramakrishna, B. et al. (2014). Role of 28-homobrassinolide on growth biochemical parameters of *Trigonella foenum-graecum* L. plants subjected to lead toxicity. *International Journal of Multidisciplinary and Current Research* 2: 317–321.

Tadayon, M.S. and Moafpourian, G. (2019). Effects of exogenous epi-brassinolid, zinc and boron foliar nutrition on fruit development and ripening of grape (*Vitis vinifera* L. clv.'Khalili'). *Scientia Horticulturae* 244: 94–101.

Taie, H.A., Seif El-Yazal, M.A., Ahmed, S.M. et al. (2019). Polyamines modulate growth, antioxidant activity and genomic DNA in heavy metal–stressed wheat plant. *Environmental Science and Pollution Research* 26: 22338–22350.

Tang, W., Yuan, M., Wang, R. et al. (2011). PP2A activates brassinosteroid-responsive gene expression and plant growth by dephosphorylating BZR1. *Nature Cell Biology* 13(2): 124–131.

Tang, Y., Zhang, J., Wang, L. et al. (2023). Water deficit aggravated the inhibition of photosynthetic performance of maize under mercury stress but is alleviated by brassinosteroids. *Journal of Hazardous Materials* 443: 130365.

Thomas, J. and Archana, G. (2023). Differential influence of heavy metals on plant growth promoting attributes of beneficial microbes and their ability to promote growth of *Vigna radiata* (mung bean). *Biocatalysis and Agricultural Biotechnology* 47: 102592.

Tian, S., Liang, S., Qiao, K. et al. (2019). Co-expression of multiple heavy metal transporters changes the translocation, accumulation and potential oxidative stress of Cd and Zn in rice (*Oryza sativa*). *Journal of Hazardous Materials* 380: 120853.

Tong, H. and Chu, C. (2018). Functional specificities of brassinosteroid and potential utilization for crop improvement. *Trends in Plant Science* 23(11): 1016–1028.

Vardhini, B.V. (2017). Modifications of morphological and anatomical characteristics of plants by application of brassinosteroids under various abiotic stress conditions-a review. *Plant Gene* 11: 70–89.

Vriet, C., Russinova, E., and Reuzeau, C. (2013). From squalene to brassinolide: the steroid metabolic and signaling pathways across the plant kingdom. *Molecular Plant* 6(6): 1738–1757.

Wang, H.H., Tao, F.E.N.G., Peng, X.X. et al. (2009). Ameliorative effects of brassinosteroid on excess manganese-induced oxidative stress in *Zea mays* L. leaves. *Agricultural Sciences in China* 8(9): 1063–1074.

Wang, X. and Chory, J. (2006). Brassinosteroids regulate dissociation of BKI1, a negative regulator of BRI1 signaling, from the plasma membrane. *Science* 313(5790): 1118–1122.

Wang, Y., Liu, Z., Xiemuxiding, A. et al. (2023). Fulvic acid, brassinolide and uniconazole mediated regulation of morphological and physiological traits in maize seedlings under water stress. *Journal of Plant Growth Regulation* 42(3): 1762–1774.

Wu, C., Li, F., Xu, H. et al. (2019). The potential role of brassinosteroids (BRs) in alleviating antimony (Sb) stress in *Arabidopsis thaliana*. *Plant Physiology and Biochemistry* 141: 51–59.

Wu, C.Y., Trieu, A., Radhakrishnan, P. et al. (2008). Brassinosteroids regulate grain filling in rice. *The Plant Cell* 20(8): 2130–2145.

Xia, X.J., Fang, P.P., Guo, X. et al. (2018). Brassinosteroid-mediated apoplastic H_2O_2–glutaredoxin 12/14 cascade regulates antioxidant capacity in response to chilling in tomato. *Plant, Cell and Environment* 41(5): 1052–1064.

Xie, Y., Luo, Y., Sheng, M. et al. (2020). 24-Epibrassinolide combined with heavy metal resistant bacteria enhancing phytoextraction of *Amaranthus hypochondriacus* L. in Cd-contaminated soil. *Journal of Hazardous Materials* 399: 123031.

Xu, B., Yu, J.Y., Xie, T. et al. (2018). Brassinosteroids and iron plaque affect arsenic and cadmium uptake by rice seedlings grown in hydroponic solution. *Biologia Plantarum* 62: 362–368.

Yang, Y., Zhang, L., Huang, X. et al. (2020). Response of photosynthesis to different concentrations of heavy metals in *Davidia involucrata*. *PLoS One* 15(3): e0228563.

Yaqoob, U., Jan, N., Raman, P.V. et al. (2022). Crosstalk between brassinosteroid signaling, ROS signaling and phenylpropanoid pathway during abiotic stress in plants: does it exist? *Plant Stress* 4: 100075.

Yin, W., Dong, N., Niu, M. et al. (2019). Brassinosteroid-regulated plant growth and development and gene expression in soybean. *The Crop Journal* 7(3): 411–418.

Yin, Y.L., Zhou, Y., Zhou, Y.H. et al. (2016). Interplay between mitogen-activated protein kinase and nitric oxide in brassinosteroid-induced pesticide metabolism in *Solanum lycopersicum*. *Journal of Hazardous Materials* 316: 221–231.

Yusuf, M., Fariduddin, Q., and Ahmad, A. (2012). 24-Epibrassinolide modulates growth, nodulation, antioxidant system and osmolyte in tolerant and sensitive varieties of *Vigna radiata* under different levels of nickel: a shotgun approach. *Plant Physiology and Biochemistry* 57: 143–153.

Yusuf, M., Khan, M.T.A., Faizan, M. et al. (2022). Role of brassinosteroids and its cross talk with other phytohormone in plant responses to heavy metal stress. In: *Brassinosteroids Signalling: Intervention with Phytohormones and Their Relationship in Plant Adaptation to Abiotic Stresses* (ed. M.T.A. Khan, M. Yusuf, F. Qazi et al.), 179–201. Singapore: Springer Singapore.

Yusuf, M., Saeed, T., Almenhali, H.A. et al. (2023). Melatonin improved efficiency of 24-epibrassinolide to counter the collective stress of drought and salt through osmoprotectant and antioxidant system in pea plants. *Scientia Horticulturae* 323: 112453.

Zhang, H., Xu, Z., Guo, K. et al. (2020). Toxic effects of heavy metal Cd and Zn on chlorophyll, carotenoid metabolism and photosynthetic function in tobacco leaves revealed by physiological and proteomics analysis. *Ecotoxicology and Environmental Safety* 202: 110856.

Zhang, H., Zhao, D., Tang, Z. et al. (2022a). Exogenous brassinosteroids promotes root growth, enhances stress tolerance and increases yield in maize. *Plant Signaling and Behavior* 17(1): 2095139.

Zhang, Q., Liang, M., Song, R. et al. (2022b). Brassinosteroids enhance resistance to manganese toxicity in *Malus robusta* Rehd. via modulating polyamines profile. *Journal of Plant Physiology* 277: 153808.

Zhang, Y. and Liao, H. (2021). Epibrassinolide improves the growth performance of Sedum lineare upon Zn stress through boosting antioxidative capacities. *PLoS One* 16(9): e0257172.

Zhang, Z., Chen, Z., Song, H. et al. (2023). From plant survival to thriving: exploring the miracle of brassinosteroids for boosting abiotic stress resilience in horticultural crops. *Frontiers in Plant Science* 14: 1218229. https://doi.org/10.3389/fpls.2023.1218229.

Zheng, H., Ma, J., Huang, W. et al. (2022). Physiological and comparative transcriptome analysis reveals the mechanism by which exogenous 24-epibrassinolide application enhances drought resistance in potato (*Solanum tuberosum* L.). *Antioxidants* 11(9): 1701.

Zhong, W., Xie, C., Hu, D. et al. (2020). Effect of 24-epibrassinolide on reactive oxygen species and antioxidative defense systems in tall fescue plants under lead stress. *Ecotoxicology and Environmental Safety* 187: 109831.

Zhou, Y.X., Guo, X.M., Lu, S.B. et al. (2013). Effect of water, nutrient and brassinolides on number of blossom, leaf nutrition and seed oil content of *Camellia oleifera*. *Zhiwu Yingyang Yu Feiliao Xuebao* 19: 387–395.

Zhu, Y., Dong, Y., Zhu, N. et al. (2022). Foliar application of biosynthetic nano-selenium alleviates the toxicity of Cd, Pb and Hg in *Brassica chinensis* by inhibiting heavy metal adsorption and improving antioxidant system in plant. *Ecotoxicology and Environmental Safety* 240: 113681.

Zulfiqar, F. and Ashraf, M. (2023). Proline alleviates abiotic stress induced oxidative stress in plants. *Journal of Plant Growth Regulation* 42(8): 4629–4651.

8

Brassinosteroid-mediated Physiological and Molecular Mechanisms in Plants Under Heat Stress

*Neetu Jagota, Swapnil Singh, Ravneet Kaur, Harleen Kaur, and Ashish Sharma**

Department of Botany and Environment Studies, DAV University, Jalandhar, India

*Corresponding author: ashish10210@davuniversity.org

8.1 Introduction

Due to abiotic factors, including salinity stress, water, wind, and temperature as well as global warming and changes in climate, worldwide crop productivity is declining extremely (Hassan et al. 2021). Consequently, improving the productivity of agriculture by novel methods to meet population needs that are always rising on a sustainable basis is challenging for all governments worldwide (Chávez-Dulanto et al. 2021). The main issues preventing sustainable crop production from fulfilling the rising food needs are abiotic factors like dryness by drought, salt, and heat stress (Hassan et al. 2022). Temperature's essential effect on the germination of seeds, crop growth, and final productivity cannot be overstated. However, the rapid rise in global temperature brought on by climate change poses a grave threat to the continuation of all Earthly life forms (Hassan et al. 2021). Heat stress (HS) is therefore the most significant abiotic stress among the other topics of conversation right now because of its adverse impacts on the expansion and output of growth in plants. Low and high-temperature extremes both harm the plant's growth. Temperatures might vary with exposure time beyond the threshold levels can slow down both plant growth and metabolism. Additionally, HS becomes even more damaging since it triggers the beginning of drought stress. HS negatively impacts the germination of seeds, photosynthesis, respiration, and accumulation of dry matter (Fahad et al. 2019). Additionally, HS denatures plant proteins, alters membrane permeability, and inactivates several enzymes, all of which ultimately lead to a decline in growth (Hassan et al. 2021). Additionally, HS causes the creation of ROS, which harms plants through oxidative stress (Hamza et al. 2023).

Around the world, several strategies are being employed to deal with heat stress. Exogenous administration of several plant growth regulators is regarded as a straightforward strategy that can provide noticeable effects. Mitchell et al. (1970) made the discovery of brassinosteroids (BRs), unique plant growth regulators, when they noticed that rapeseed

Brassinosteroids and Abiotic Stress: Recent Developments on Physiological and Molecular Aspects,
First Edition. Edited by Anket Sharma, et al.
Published 2026 by John Wiley & Sons Ltd.

(*Brassica napus* L.) pollen's crude lipid extract had the ability to stimulate growth. The well-known phytohormone known as BR is thought to be crucial for regulating a variety of processes in plants. During the different processes in plants for the plant life cycle, several cell types undergo differentiation of cells, photomorphogenesis in different plants, differentiation of xylem, proton pump activation, expression of genes, regulation of photosynthesis, and induction of the biosynthesis of ethylene (Nolan et al. 2020). Nevertheless, an inherited variation that alters the signaling and synthesis of BR hormone pathways results in reduced reproduction, delayed developmental stages, dwarfing in plants, and poor development of crop productivity (Planas-Riverola et al. 2019). In situations of short nutrients, BRs have recently been recognized as crucial root development regulators that increase root area by boosting root foraging (Jia et al. 2019). Abiotic and biotic stressors including the presence of toxic substances and biological contaminants like chemical fertilizers, as well as drought, excessive water retention, and low temperatures, are all controlled in part by BRs (Tanveer et al. 2019). There were numerous attempts to recommend the use of BR hormone as a regulator in the agricultural field for development and growth (Nolan et al. 2020).

The emergence of RESPIRATORY BURST OXIDASE HOMOLOG 1 (RBOH1), which encodes the oxidase, i.e. NADPH, in the apoplast, is what leads to the production of reactive oxygen species (Figure 8.1), is induced by biosynthesis genes by either overexpressing BRs or applying BRs exogenously (Jia et al. 2019). Through increased synthesis of protein, activation of transcription factors (TFs), which influence the transcription of responsive

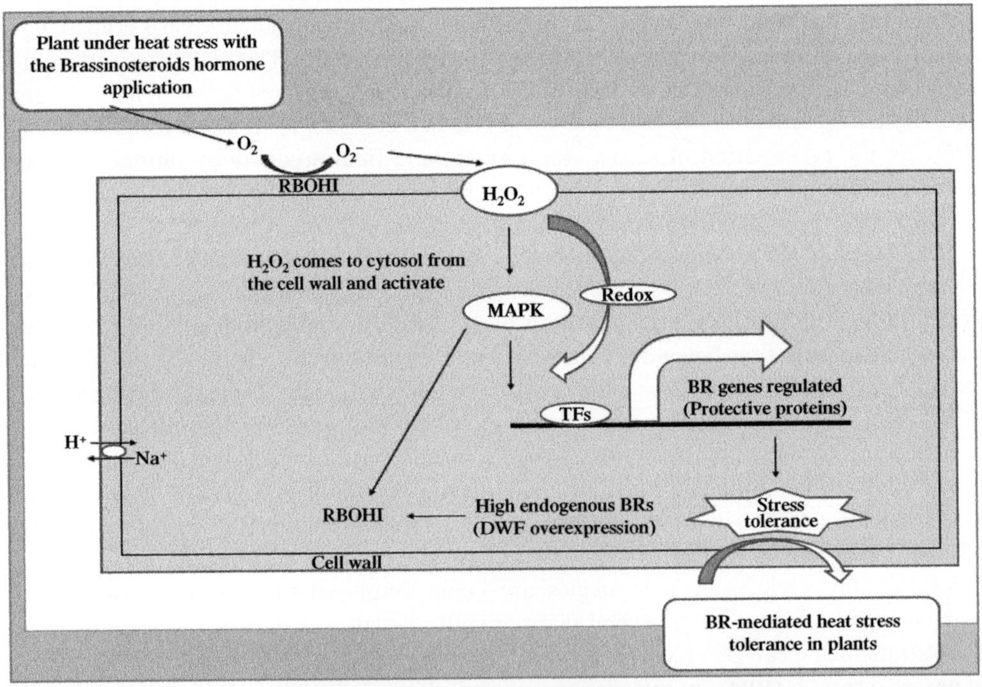

Figure 8.1 In the cell membrane, heat stress forms reactive oxygen species by the brassinosteroid hormone.

genes for stress and regulated BRs that increase the stress tolerance. Redox homeostasis is regulated by reactive oxygen species signals mediated by BRs (Ahammed et al. 2020).

Multiple parts of a plant's life depend on the action of BRs, which are steroid hormones with a polyhydroxylated nature according to their chemical classification (Nolan et al. 2020). The three main types of BRs that have been widely used in numerous studies recently are 24-epibrassinolide, brassinolide, and 28-homobrassinolide. In their structure according to the carbon number, BRs are commonly categorized as C-29, C-28, or C-27 BRs. (Vardhini 2019). So, in this chapter, we have spoken about how BRs mediate the numerous molecular and physiological pathways that cause plants to become heat-stress tolerant.

8.2 Hormones

A class of naturally occurring hormones called "plant hormones," small chemicals that have numerous functions in the kingdom of plants, including controlling development and growth in response to environmental or physical conditions (Rico-Jiménez et al. 2022). Less commonly referred to as "the big five," the five most significant plant hormones are IAA (auxin), Ck (cytokinins), GA (gibberellins), ABA (abscisic acid), and polyethylene (ethylene). These hormones regulate a wide range of biological functions; one of them frequently controls several distinct functions in plants. While Ck regulates the division of cells and aids in the development of the new organ, gibberellins (GAs) aid in the elongation of stems, abscisic acid regulates moisture, and the gas hormone ethene affects ripening, decaying, and growth promoted by the IAA hormone (Bhattacharya 2022). BRs, jasmonic acid (JA), salicylic acid (SA), and strigolactones (SLs) are a few more hormones whose role as plant hormones has also been established to date (Raza et al. 2022). Plants must detect environmental stress and react swiftly to environmental changes and various pressures because they are sessile organisms. Although Darwin had already postulated depends upon the finding of which coleoptile phototropic curving uses a distance difference between the sensor's location at the tip of the shoot and the bending location below the shoot (Park et al. 2017), the hormone concept wasn't actually coined until the beginning of the 20th century. The tip of the coleoptile, where the perception of light occurs, produces the plant hormone IAA, which is then delivered to the appropriate location in the shoot for action, which is what Darwin referred to as a "certain influence" (Park et al. 2017). This information has since been revealed (Christie and Murphy 2013). Transporters of hormones are required for the precise regulation of the development and use in plants. We now know that hormone-transporting proteins are necessary for the appropriate fusion of external factors and internal signaling in plants. Numerous IAA, ABA, Ck, GA, BRs, SLs, and SA are examples of plant hormones and be capable of long-distance transport (Anfang and Shani 2021). The xylem must be loaded with hormones before they can be released at the target cells in the transpiration stream. Similar loading and unloading procedures may be necessary for hormones that are delivered through the phloem.

Hormone synthesis, metabolism, transport, perception, and signal transduction all work together to direct hormone homeostasis, which in turn controls related plant functions. Hormone perception might occur close to or far from the location of synthesis. Therefore, hormones can be controlled in their distribution by active transporters, resulting in a variety of reactions. Although it also affects the hormones' intermediate and conjugated forms, this spatial modulation occurs in the bioactive hormone forms. The characterization of hormone transporters discovered in recent years has shown that hormone distribution and homeostasis are dynamically regulated (Anfang and Shani 2021). There are several families of hormone transporters, each with its own special traits. Additionally, the plant also translocates additional mobile signaling components such as short peptides and hormone-like compounds. BRs, another class of hormones with well-known growth-promoting properties, have undergone extensive biosynthesis and signaling research (Sadura et al. 2019). However, there hasn't been much information made available about BR transport so far. The widespread consensus is that BRs do not travel across or within organs. Due to the cellular synthesis and plasma membrane location, the receptor binding locations are open to the extracellular environment. Because BR signal transduction requires the transport of BRs from within the cell to the outside, even though these sterol compounds have membrane permeability, a plasma membrane exporter may still be necessary (Vaishnav and Chowdhury 2023). If protein transporters play a part in BR trafficking, further research might be beneficial. For understanding the intricate signals of plant hormones, in addition to signal transduction regulators, metabolism, and transport are also helpful.

8.3 Brassinosteroid Hormone History and Biosynthesis

8.3.1 History

BR hormones have been found as endogenous developmental phytohormones in plants that are crucial for sustaining plant's normal growth in both highly stressed and unstressed environments (Singh et al. 2018). Previous research discovered that the extraction of the pollen grains from the orchid plant encourages the extension of the cells as well as showed individual applying secretions of pollen grains causes dicot plant stems to bend. A "pollen research project" was started in the 1930s at the US Horticultural Station in Beltsville, Maryland, United States as a result of several more studies. The first article on the exploration of the naturally occurring steroid molecule brassinolide was published in 1941, in the Botanical Gazette, which was noticeably the start of extensive study on BRs (Mitchell and Whitehead 1941). He also showed in 1941 that the combination of lanolin and extraction of pollen grains from *Zea mays* significantly increased the extension of the internodes in *Phaseolus vulgaris* due to the extension of the cells; however, no response was exhibited in the etiolated stems; this was only observed in the seedlings which receiving the light that had robust internodes. Later in 1970, an original brassins hormone from the pollens extract of *Brassica napus* was found (Mitchell et al. 1970). When a brassin hormone was applied, bean second internodes exhibited similar organ enlargement to that seen by Mitchell and Whitehead (1941). At last, *brassinolide* (BL) was given the name by those who identified the chemical makeup of growth-promoting hormone for plants isolated from the pollen of *Brassica napus* (Grove et al. 1979). Following the identification of biologically active hormones brassins and brassinolide by Mandava et al. (1973), other related chemicals that resembled BL in various plant species were quickly found and isolated by Sakakibara and Mori (1982). Advanced, more chemicals in many numbers with basic resemblances to brassinolide were found, resulting in the creation of the BR hormone family (Khripach et al. 1998). Numerous plant taxa, including angiosperms, algae, bryophytes, and pteridophytes, have been identified to contain BRs (Stirk et al. 2014). Based on different side chain alkyl-substitution structures, the unaltered BRs are categorized into three categories: C29, C28, or C27 (Tang et al. 2016). In general, a number of plant parts have had BRs separated from them, including grains, flowers, leaves, seeds, roots, stems, anthers, or pollen, as well as from insects and crown galls. A larger concentration of BRs is seen in immature seeds and pollen, and young, growing leaves than in mature leaves (Ahammed et al. 2020).

8.3.2 Biosynthesis

Plants produce phytohormones, which are BRs that usually help in response to stress conditions like heat stress, depending on the different types of tissue growth and development (Nolan et al. 2020). With the discovery of biosynthesis of these BRs enzymes, there is evidence that the endoplasmic reticulum of *Arabidopsis* is a hormone-producing organ. (Northey et al. 2016). According to Lv and Li (2020), the biosynthesis process of BRs primarily depends on the production of brassinolide from campesterol, and the synthesis of BL is started by the common sterol precursor cycloartenol with a series of metabolic processes that is methylation, denaturation, and reduction (Belkhadir and Jaillais 2015). During the heat stress conditions, when temperature increases, it leads to the accumulation of BR-responsive TFs BZR1 and BES1 from the pathway of signaling through the PIF4, which involves the ABA-stimulated synthesis of BRs (Ibañez et al. 2018). The membrane-specific receptor for the phytohormone BR is BR1, which binds to the BR ligand directly to initiate the signal cascade that causes cell proliferation and triggers the transcription of the BR-responsive genetic factor (Planas-Riverola et al. 2019). The plasma membrane enzymes BRI1 and BAK1 recognize BRs and provide signals to the TFs BES1 and BZR1 to control gene expression by acting as BR regulators (Nolan et al. 2020). These coreceptors bind to the target location of BRs to initiate the signaling pathway. BRI1 and BAK1 can be phosphorylated and activated when BKI1 and BIR3 separate from the receptor complex. Sreeramulu et al. (2013) state that when BRs are phosphorylated, SIGNALING KINASES (BSKs) activate the BSU1 enzyme, an inhibitor of the phosphatase enzyme, which inactivates the BIN2. BZR1 and BES1 interact with other cofactors and TFs to decrease the expression of the gene for BR inhibition because PROTEIN PHOSPHATASE2A (PP2A) dephosphorylates BR-induced gene expression (Rehman et al. 2022). The abscisic acid hormone pathway was believed to serve as a signaling route that responds to environmental stress circumstances, possibly interacting with the BR hormone to regulate stress and maintain normal plant growth and development (Planas-Riverola et al. 2019). The BR hormone mediates adaptation to the stress environment by modifying the production of TFs or osmoprotectants in plants and stimulating the antioxidant enzyme activity (Ye et al. 2017; Fàbregas et al. 2018; Zou et al. 2018). BR signaling has a substantial effect on how plants adjust to various temperature stressors. Numerous studies have shown that BRs increase the ability of *Brassica napus* L. to withstand heat stress, and this increase in HSP synthesis was associated with it (Singh et al. 2018). BR hormone acts as a regulator for the metabolism of reactive oxygen species in the plant cell membrane under conditions of heat stress, according to Tanveer et al. (2019). The most recent research has demonstrated that the hormone of BRs dramatically increased CO_2 absorption capacity by increasing RuBisCO activity (Ahammed et al. 2020). The C3 cycle electron flow is maximized and the oxygen flow is minimized during temperature stress caused by the generation of reactive oxygen species (Tanveer et al. 2018; Nolan et al. 2020). Although the mechanisms by which the hormone BR regulates the activity of the PS system and growth of plants under high heat stress challenges are currently poorly understood, further studies at the molecular level are required to recognize the precise processes by which BRs serve as growth-promoting hormones in challenging conditions.

8.4 Physiological Mechanisms Mediated by BRs Under Heat Stress

Hormones are signaling molecules that plants utilize to mediate heat stress and other environmental stress conditions (Sadura and Janeczko 2018; Ahammed et al. 2020). More than 40 plant hormones are naturally produced in plants, collectively known as BRs are widely distributed across the plant world (Su et al. 2020). BR hormones are essential for numerous physiological processes in plants, including epinasty of the leaves, elongation of the stem, pollen tube development, gene expression, and other stress hormones biosynthesis, such as ethylene (Planas-Riverola et al. 2019). Several researchers have noted that BRs affect a plant's ability to withstand heat stress (Cukor et al. 2018). This chapter explains how BR has an impact on how plants behave in terms of their morphology, chloroplast function, and photosynthetic properties (Figure 8.2). Additionally, it has been emphasized how BR signaling activates the defense mechanism in plants counter to temperature stress and how BR and ABA interact to promote thermotolerance, with Table 8.1 providing significant insights into BRs roles in plant metabolism.

8.4.1 Grain Development, Seed Germination, and Plant Growth

On the physiological characteristics such as germination of seeds, growth of plants, and biomass of plants development of plants at high temperatures has a major negative impact

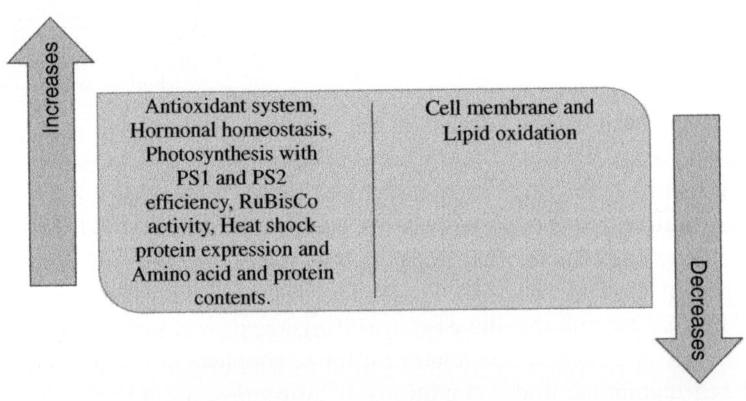

Figure 8.2 Brassinosteroids' action under heat stress.

Table 8.1 Effects of brassinosteroid hormones on plant species.

Plants name	Treatment of BRs with concentrations	Effect on plants	References
Brassica napus L. and *Lycopersicon esculentum* L.	24-epibrassinolide (EBL) with 1 μm	Increases the heat resistance	Dhaubhadel et al. (1999)
A. thaliana L.	2,4-EBL with 0, 10, and 100 nM	Increased basal heat resistance	Chen et al. (2022)
Lycopersicon esculentum L.	Epibrassinolide (EBL) with 1 μm	Increases the formation of chl. pigment through photosynthesis, germination of pollen grains, and total productivity	Mazorra et al. (2011)
Leymus chinensis L.	EBLEBL with 1 nM	Increase antioxidant enzyme activity and chl. content	Niu et al. (2016)
Cucumis melo L.	24-EBL with 0, 0.05, 0.1, 0.5, 1.0, and 1.5 mg dm^{-3}	Increased MDA, proline content, antioxidant enzyme activity, ameliorated photosynthetic pigment losses, maintained P_N and gas exchange, reduced root, shoot, and protein content	Zhang et al. (2013, 2014)
Oryza sativa L. and *Cucumis melo* L.	24-epibrassinolide with 1 nM	P_N, gas exchange, decreased internal CO_2 concentration, increased chlorophyll content, reduced MDA buildup, sugar content, root, and shoot mass, and a negative impact on PS II	Thussagunpanit et al. (2015)
Oryza sativa L.	BRs with 4.0 ppm	Protein content, carbohydrates, and male fertility all decreased, whereas proline content increased.	Fahad et al. (2016)
Solanum melongena L.	24-epibrassinolide with 0.1 μM	Reductions in plant fresh weight and height, MDA content, antioxidant enzyme activities, chlorophyll content, the ability of BRs to improve gas exchange, a decline in PS II efficiency, higher soluble sugar content, increased protein, and proline content	Wu et al. (2014)
Zea Mays L.	24-epibrassinolide with 1 μM	Decrease chlorophyll content	Singh et al. (2012)

(Continued)

Table 8.1 (Continued)

Plants name	Treatment of BRs with concentrations	Effect on plants	References
Hordeum vulgare L.	24-epibrassinolide with concentrations 0.25 and 0.005 mg dm^{-3}	Reduced absorption of energy and energy flux for electron transport were signs of irregular flow of energy in the photosystem	Janeczko et al. (2011)
Cucumis sativus L.	24-epibrassinolide	Increased photosystem relative electron transport rates	Xia et al. (2009)
Chlorella vulgaris L.	Brassinolide with 10 nM	Increased protein content	Bajguz (2009)
Brassica napus L.	24-epibrassinolide with 1 μM	Heat shock proteins that are expressed to encourage the accumulation of transcripts	Dhaubhadel et al. (1999, 2002)
Bromus diandrus L. and *B. rigidus* L.	24-epibrassinolide with 10 μM	Enhanced cell viability	Wilen et al. (1995)
Lycopersicon esculentum L.	24-epibrassinolide with 1 μM	Pollen bursting was reduced, and germination was boosted with a germination tube.	Hayat et al. (2010a)
Lycopersicon esculentum L.	EBL with 0.2 μM	Decreased photosynthesis and MDA concentration with increased enzyme antioxidant activity	Ogweno et al. (2008)
Cucumis melo L.	24-EBL with 1.0 mg L^{-1}	Increased chlorophyll content	Zhang et al. (2013)
Solanum melongena L.	24-epibrassinolide with 0.2–0.05 μM	Increased the activity of antioxidant enzymes	Wu et al. (2014)
Triticum aestivum L.	24-epibrssinolide with 0.01 μM	Increased antioxidant activity, chlorophyll content, and biomass with growth	Hussain et al. (2019)
Oryza sativa L.	αDHECD with 0.0001 μM	Increase the number of seeds, carbohydrates, and chlorophyll	Sonjaroon et al. (2018)
Lycopersicon esculentum L.	24-EBL with 1.0, 0.1, and 0.01 mg L^{-1}	Increased activity of antioxidant enzymes	Ogweno et al. (2008)
Lycopersicum esculentum L.	24-EBL 10 nm, 24-EPBL+ABA with 0, 30, 100, 200, 500, and 1000 nM + 0, 10, 20, 50, 100, and 200 uM, 24-epiBL + acetyl SA with 3 mM + 0.75 mM	Increased chlorophyll concentration and antioxidant enzyme activity	Mazorra et al. (2002), Khan et al. (2015), and Zhou et al. (2014)

Chlorella Vulgaris L.	24-EBL with 10 nM	Increased abscisic hormone and osmolyte buildup	Bajguz (2009)
Hordeum vulgare L.	24-EBL with 0.005 and 0.25 mg dm^{-3}	Increased chlorophyll content	Janeczko et al. (2011)
Brassica juncea L.	28-HBL with 10^{-9}, 10^{-6}, 10^{-8}, 10^{-10}, and 10^{-12}	MDA and proline content were reduced, whereas antioxidant activity, protein, and sugar content were enhanced	Geetika et al. (2014)
Cucumis melo L.	24-epibrassinolide with 1.5–0.05 mg dm^{-3}	Increased chlorophyll content	Zhang et al. (2014)
Ficus concinna L.	24-EBL with 0.25 μM	Increased antioxidant activity and content of chlorophyll	Jin et al. (2015)
Oryza sativa L.	24-EBL + DHECD with 1 nM	Increased photosystem activity	Thussagunpanit et al. (2015)
Leymus chinensis L.	BL with 1.0, 0.1, and 0.01 mg L^{-1}	Increased proline, protein, and antioxidant enzyme activity	Niu et al. (2016)
Zea mays L.	24-EBL with 1 μm	Increased activity of antioxidant enzymes	Yadava et al. (2016)
Triticum aestivum L.	24-EBL with 0.01 μM	Increased antioxidant activity, chlorophyll, and proline content	Hussain et al. (2019)
Triticum aestivum L.	24-EBL with 1 μM	Improved growth and biomass, together with increased chlorophyll and antioxidant activity	Kumari and Hemantaranjan (2018, 2019)
Brassica rapa L.	Brassinosteroids with 200 nM	Increased the activity of antioxidant enzymes	Lee et al. (2019)
Cucumis melo L.	24-EBL with 1.0 mg L^{-1}	Improvements in antioxidant defense, RuBisCO activity, and photosynthetic efficiency	Jiang et al. (2012)
Vigna radiata L.	28-HBL (homobrassinolide) with 0.1 mM	Enhanced efficiency of gas exchange, the carbonic anhydrase enzyme, chlorophyll, and carboxylation	Hayat et al. (2010b)
Festuca arundinacea L.	24-EBL with 0.1 mM	Protein, chlorophyll, and antioxidant activity increased, but MDA and reactive oxygen species decreased, increase heat stress tolerance by expression of heat shock protein	Alam et al. (2018)

(Farooq et al. 2017). However, the detrimental effect of heat stress on plants is greatly mitigated by the application of BR hormone (Li et al. 2020). For crops to be produced sustainably, seed germination and growth are crucial. During the germination phase, seedlings are extremely sensitive and are readily harmed by little temperature stress (Hassan et al. 2021). Application of the plant hormone BR increases the rate of germination and enhances seedling growth under increased temperature stress (Mahesh et al. 2013). For example, high heat stress on pine plant seeds manifests as an increase in seed germination (Cukor et al. 2018). The crucial genes, both GCR1 and GPA1, which are in charge of seed germination, are controlled by the implementation of BR hormone (Li et al. 2020). By enhancing the accumulation of nutrients and enzyme activity of antioxidants that promote the germination of seeds and development, the application of BRs reduces the impacts of temperature stress. Additionally, temperature-stressed plants responded to the BR application extremely quickly by accelerating the growth of shoots and roots (Yamagami et al. 2017). By enhancing nitrogen metabolism, the exogenous application of EBL enhanced plant nutrient absorption (Su et al. 2020). When BR hormone is given to plants in response to temperature stress, several enzymes involved in N_2 metabolism, such as GOGAT, nitrite, nitrate reductase, and glutamine and glutamate synthase, exhibit increased activity (Zhang et al. 2020). BR hormone was given to *Oryza sativa* at a concentration of 108 M DHECD, and the results suggest that the effect on plants is an increase in chlorophyll concentration and water content with a decrease in leaf wilting (Thussagunpanit et al. 2015). Rice plant also shows that when treated with BR, 7,8-dihydro-8α-20-hydroxyecdysone, and exposed to temperatures above optimum (40/30 °C day/night for 7 days) produced more biomass and had better leaf area.

8.4.2 Photosynthesis, ATP Production, and Chlorophyll

A crucial component for the development of chloroplasts and a fundamental property that indicates photosynthetic activity is chlorophyll. The functioning of the chloroplasts and mitochondria is hampered by high-temperature stress, which causes an excessive buildup of ROS, oxidative injury, as evidenced by the oxidation of proteins, damage to DNA, and lipid peroxidation (Yi et al. 2014). When temperatures are greater, plants' photosystems become inactive because of damage to the chloroplast organelle, which is caused by a reduction in the activity of proteins and enzymes and an imbalance in the redox reaction, which ultimately results in the death of the organelle (Li et al. 2018). The chloroplast is crucial for the plant's photosynthesis process, which initiates the cellular signaling process but is vulnerable to temperature stress (Chen et al. 2017). Nonetheless, it has been discovered that BR use improves the features used to assess plant photosynthetic performance under less-than-ideal growth conditions (Niu et al. 2016). In plants, treatment with BR hormone has a mechanism for tolerating heat in both heat-tolerant and sensitive genotypes (Zhang et al. 2013). Additionally, BRs regulate the chloroplast grana and thylakoids' structure and function, and their suboptimal levels interfere with those activities. Optimal BR levels are also required for better photosynthetic activity (Krumova et al. 2013). Additionally, BRs aid in the regulation of antioxidant enzymes and the stabilization of D1 protein, protecting the photosynthetic machinery under less-than-ideal growth conditions. Additionally, it aids in the recovery of PSI and PSII's photoinhibition caused by ROS (Siddiqui et al. 2018a, b).

8.4.3 Antioxidant Defense

When plants are exposed to temperature stress, an increase in reactive oxygen species disrupts their metabolism, resulting in oxidative stress in plants that initiates oxygenase activities by producing H_2O_2 (Emami et al. 2020; Bhattacharya 2022). Ahmad et al. (2018) published a study on how certain plants minimize heat stress by functioning as antioxidant enzymes. According to Emami et al. (2020), plants under heat stress exhibit low production, but when the hormone BRs are applied, the activity of the enzymatic antioxidant and nonenzymatic antioxidants (Wu et al. 2014) increases. In dwarf mutant plants, the hormone BR EBL at a dose of 1 mM enhanced thermotolerance, although only partially due to mutant allele-mediated regulation of enzyme antioxidant activity with reduced oxidative stress (Mazorra et al. 2011).

8.5 Molecular Mechanisms Mediated by Brassinosteroids Hormone Under High Temperature Stress

Temperature stress increases membrane fluidity, which in turn activates Ca^{2+} channels found on plasma membranes. This Ca^{2+} influx causes excessive ROS to be produced by activating NADPH oxidase (Esteras et al. 2021). Ca^{2+}-dependent CDPKs and CIPK proteins regulate the activity of NADPH oxidase by phosphorylation (Costa et al. 2017). The apoplastic signaling molecule is reactive oxygen species that is present during heat stress and affect the scavenging of these species by triggering the activity of antioxidant enzymes (Mansoor et al. 2022). BR hormone signaling receptors, specifically BRI1, plays a crucial part in the molecular mechanism of heat-stressed plants by activating the signal via receptors and attaching to the BR hormone to convey the signal and perform transphosphorylation with BAK1 (Kim and Russinova 2020). This signal is deactivated when BIN 2 is present in the cycle, which inhibits the TFs of the BR signaling pathway, i.e. BZR1 and BZR2 (Yin et al. 2018). As it regulates FERONIA (FER) homologs, BZR1 is an essential regulator of BR response for heat stress tolerance. Recent research by Yin et al. (2018) showed that FER homologs (FER2 and FER3) partially mediate ROS signaling that is dependent on RESPIRATORY BURST OXIDASE HOMOLOG1 (RBOH1) to regulate tomato's high-temperature stress responses. They also noted that the mutant of BRZ1 had less development and there was no response to 24-epibrassinolide, reduced RBOH1 initiation, H_2O_2 production in the apoplast, and a reduced ability to withstand temperature stress. However, BZR1 overexpression controlled the responses to heat stress, apoplastic H_2O_2 generation, and FER2/3 homologs. Reactive oxygen species and heat shock protein control are associated with tolerance in plants under temperature stress (Chen et al. 2022).

The hormonal BRs and ABA cross-talk caused by BR signaling, in particular BR, is essential for plants to respond to environmental challenges (Figure 8.3). High BR levels inhibit ABA production, but they also have the ability to cross-regulate and are an essential component of stress tolerance in high heat. Although ABA and BR are opposed, they positively cross-regulate. EBL and low ABA levels cause stomatal apparatus to close (Ha et al. 2018). For example, BRs increase the level of abscisic acid in *Chlorella vulgaris* when the plant is stressed by high temperatures. However, in mutants of the plant *Arabidopsis* lacking

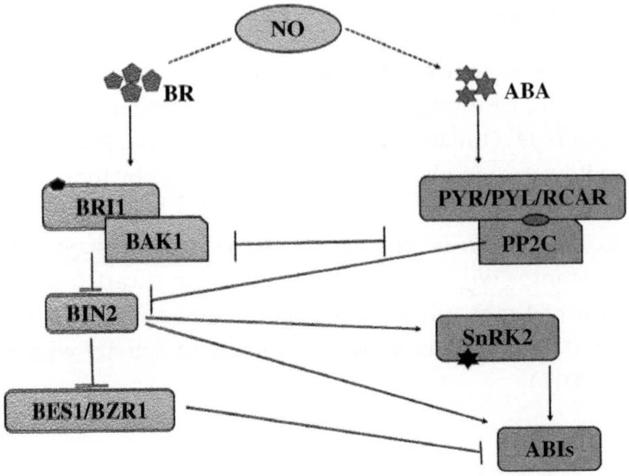

Figure 8.3 Crosstalk between BR and ABA occurs when stress reactions are controlled.

ABA that were subjected to heat stress and EBL treatment, HSPs accumulated more (Divi et al. 2010). In a 2014 study, Zhou et al. described how the hormones abscisic acid and BR interact under conditions of heat stress, providing molecular and genetic proof that H_2O_2 is generated. They found that following BR signal detection, RBOH1-NADPH oxidase stimulates H_2O_2 generation, and increases reactive oxygen species, starting a faster manufacture of ABA, which leads to higher H_2O_2 formation and a longer time for thermotolerance to develop. According to Setsungnern et al. (2020), *A. thaliana* L., the BES1 mutant is more vulnerable to photoinhibition and photooxidative stress than the wild type or another BR mutant. This is because, under heat stress, BR-mediated signaling increased photoinhibition and oxidative stress decreased the efficacy of photosystem, MDA. When exposed to heat stress, the BES1 mutant likewise displays a high level of ABA, and EBL administration can reverse the alterations caused by the BES1 mutant under heat stress.

8.5.1 Brassinosteroids Control HSP Accumulation and ROS Homeostasis to Reduce Heat Stress

By assessing certain antioxidant activities and MDA content, Mazorra et al. (2011) reported in their studies how the BR hormone affects plants under heat stress. Demonstrating thermotolerance is downstream of BR signaling and independent of BR homeostasis, the excessively produced and BR-deficient seedlings displayed identical tolerance to heat stress. Heat stress boosts HSP expression in *Hordeum vulgare* L. Under heat stress, BR-deficient or BR-signaling mutant plants produced fewer HSP group proteins than wild-type plants did (Sadura et al. 2020a). A fascinating interaction between the BR-induced HSP90 protein and two homologous TFs of the BRI1-EMS suppressor 1 (BES1), the BR signaling pathway, and brassinazole-resistant 1 (BZR1) (Samakovli et al. 2020) has been observed. According to Samakovli et al. (2020), heat shock protein 90 may have an impact on the BES1 protein's stability to promote BR-dependent gene expression. In vivo, protein synthesis assays revealed that greater protein synthesis in BR-treated plants is what causes the enhanced accumulation of HSPs (Li et al. 2021). Proton-pumping ATPase and aquaporins were two membrane proteins that were subject to heat-induced accumulation, and BR was implicated in controlling this process (Sadura et al. 2020b). By moving the TFs bZIP17 and bZIP28 from the endoplasmic reticulum membrane to the nucleus in *Arabidopsis*, temperature stress activates the endoplasmic reticulum chaperone and BR signaling activation of genes (Malini et al. 2020). According to He et al. (2005), brassinazole-resistant 1 is a crucial TFs of the BRs signaling pathway. BZR1 builds up in the nucleus after heat stress, where it controls the production of development-promoting genes (Ibañez et al. 2018).

8.5.2 Mitigating Harmful Effects of Heat Stress by BRs

According to Nolan et al. (2020), BRs have a moderating influence on plant growth and development that is based on physiological and molecular factors. BR sensing, which occurs at the cell surface via BR receptors, is considered to activate the main TF BZR1, which regulates the nuclear transcription of BR-responsive genes. This activation acts as an important cue to increase resistance to adverse environmental factors, including heat stress (Planas-Riverola et al. 2019). Some research suggests that in response to heat stress, BRs may increase the production of heat shock proteins, protecting proteins from irreparable heat damage that reduces denaturation and encourages protein folding after injury (Yang et al. 2021). BRs have been demonstrated to sustain the expression of genes encoding ascorbate-glutathione cycle enzymes such as CAT, SOD, ascorbate peroxidase, GR, and MDA reductase during heat stress (Yonca Surgun-Acar et al. 2019). According to reports, BRs play an important role in protecting the photosynthesis apparatus under HT stress by enhancing the production of chlorophyll content, the rate at which Rubisco carboxylates, photosystem activation, and the flow of energy in plants (Nawaz et al. 2017). Reduced HT stress may possibly be a result of BRs' interactions with other phytohormones. Notably, a study on barley seedlings found that BR-deficient and signaling mutants were extra resistant to extreme heat than wild varieties (Sadura et al. 2019). Under HT stress, six BZR genes have been decreasing in seedlings of maize plants, according to reports given by Manoli et al. (2018). It is assumed that BRs do not produce heat shock proteins to promote heat tolerance because *Arabidopsis thaliana* BR-deficient mutants expressed heat shock proteins at substantial amounts after heat stress. These findings suggest that the functions and processes of BRs in facilitating the negative effects of heat stress on plant development and growth significantly differ depending on the species of plant variety, developmental phase, and environmental conditions.

8.6 Conclusion

Due to climate change, high-temperature stress has become more frequent, which is one of the main reasons for decreased agricultural yield. Temperature (heat) stress affects the photosynthetic apparatus, causes reactive oxygen species to build up, and interferes with the antioxidant system in plants. The administration of phytohormones can aid in enhancing the response of plants to ecological challenges. In this sense, BR hormone and its associated complexes show a significant role in the control of several vital processes connected to the growth and development of plants in both optimal and unfavorable growth situations, and their deficiency or insensitivity may result in a number of Physiological and morphological problems in plants. Studies on BR hormone began in the 1940s, and in the early 1970s, a component (brassins) from the rapeseed pollen was discovered that had properties that encouraged plant development. The BR family of phytohormones has seen a large number of them identified and isolated over the last 50 years. These plant hormones are typically produced in the plant's endoplasmic reticulum cells, and a wide variety of genes and TFs regulate their manufacture. Preserving the exogenic distribution of BRs improves the morphological characteristics of the plant, such as germination of seeds, development of leaves, and development of grains, by improving the structure and function under heat stress, the photosynthetic system, and RuBisCO activity. Additionally, BRs aid in the upregulation of a number of genetic factors and TFs that elaborate in plants' responses to high temperatures. The BR typically exerts its effects close by, and the relationship between BRs and other plant hormones aids in the control of ROS homeostasis and signaling as well as the conferral of plant resistance to heat stress.

References

Ahammed, G.J., Li, X., Liu, A. et al. (2020). Brassinosteroids in plant tolerance to abiotic stress. *Journal of Plant Growth Regulation* 39: 1451–1464.

Ahmad, F., Singh, A., and Kamal, A. (2018). Crosstalk of brassinosteroids with other phytohormones under various abiotic stresses. *Journal of Applied Biology and Biotechnology* 6(1): 56–62.

Alam, M.N., Zhang, L., Yang, L. et al. (2018). Transcriptomic profiling of tall fescue in response to heat stress and improved thermotolerance by melatonin and 24-epibrassinolide. *BMC Genomics* 19: 1–14.

Anfang, M. and Shani, E. (2021). Transport mechanisms of plant hormones. *Current Opinion in Plant Biology* 63: 102055.

Bajguz, A. (2009). Brassinosteroid enhanced the level of abscisic acid in *Chlorella vulgaris* subjected to short-term heat stress. *Journal of Plant Physiology* 166(8): 882–886.

Belkhadir, Y. and Jaillais, Y. (2015). The molecular circuitry of brassinosteroid signaling. *New Phytologist* 206(2): 522–540.

Bhattacharya, A. (2022). Plant growth hormones in plants under low-temperature stress: a review. In: *Physiological Processes in Plants Under Low Temperature Stress*, 517–627. Singapore: Springer.

Chávez-Dulanto, P.N., Thiry, A.A., Glorio-Paulet, P. et al. (2021). Increasing the impact of science and technology to provide more people with healthier and safer food. *Food and Energy Security* 10(1): e259.

Chen, S.T., He, N.Y., Chen, J.H. et al. (2017). Identification of core subunits of photosystem II as action sites of HSP 21, which is activated by the GUN 5-mediated retrograde pathway in *Arabidopsis*. *The Plant Journal* 89(6): 1106–1118.

Chen, X., Xue, H., Zhu, L. et al. (2022). ERF49 mediates brassinosteroid regulation of heat stress tolerance in *Arabidopsis thaliana*. *BMC Biology* 20(1): 1–17.

Christie, J.M. and Murphy, A.S. (2013). Shoot phototropism in higher plants: new light through old concepts. *American Journal of Botany* 100(1): 35–46.

Costa, A., Luoni, L., Marrano, C.A. et al. (2017). Ca^{2+}-dependent phosphoregulation of the plasma membrane Ca^{2+}-ATPase ACA8 modulates stimulus-induced calcium signatures. *Journal of Experimental Botany* 68(12): 3215–3230.

Cukor, J., Rasakova, N.A.M., Linda, R. et al. (2018). Effects of brassinosteroid application on seed germination of Scots pine under standard and heat stress conditions. *Baltic Forestry* 24(1): 60–67.

Dhaubhadel, S., Browning, K.S., Gallie, D.R. et al. (2002). Brassinosteroid functions to protect the translational machinery and heat-shock protein synthesis following thermal stress. *The Plant Journal* 29(6): 681–691.

Dhaubhadel, S., Chaudhary, S., Dobinson, K.F. et al. (1999). Treatment with 24-epibrassinolide, a brassinosteroid, increases the basic thermotolerance of *Brassica napus* and tomato seedlings. *Plant Molecular Biology* 40: 333–342.

Divi, U.K., Rahman, T., and Krishna, P. (2010). Brassinosteroid-mediated stress tolerance in *Arabidopsis* shows interactions with abscisic acid, ethylene and salicylic acid pathways. *BMC Plant Biology* 10: 1–14.

Emami, N.K., Jung, U., Voy, B. et al. (2020). Radical response: effects of heat stress-induced oxidative stress on lipid metabolism in the avian liver. *Antioxidants* 10(1): 35.

Esteras, N., Kundel, F., Amodeo, G.F. et al. (2021). Insoluble tau aggregates induce neuronal death through modification of membrane ion conductance, activation of voltage-gated calcium channels and NADPH oxidase. *The FEBS Journal* 288(1): 127–141.

Fàbregas, N., Lozano-Elena, F., Blasco-Escámez, D. et al. (2018). Overexpression of the vascular brassinosteroid receptor BRL3 confers drought resistance without penalizing plant growth. *Nature Communications* 9(1): 4680.

Fahad, S., Adnan, M., Hassan, S. et al. (2019). Rice responses and tolerance to high temperature. In: *Advances in Rice Research for Abiotic Stress Tolerance*, 201–224. Woodhead Publishing.

Fahad, S., Hussain, S., Saud, S. et al. (2016). Exogenously applied plant growth regulators affect heat-stressed rice pollens. *Journal of Agronomy and Crop Science* 202(2): 139–150.

Farooq, M., Nadeem, F., Gogoi, N. et al. (2017). Heat stress in grain legumes during reproductive and grain-filling phases. *Crop and Pasture Science* 68(11): 985–1005.

Geetika, S., Harpreet, K., Renu, B. et al. (2014). Thermo-protective role of 28-homobrassinolide in *Brassica juncea* plants. *American Journal of Plant Sciences* 5(15): 48074.

Grove, M.D., Spencer, G.F., Rohwedder, W.K. et al. (1979). Brassinolide, a plant growth-promoting steroid isolated from *Brassica napus* pollen. *Nature* 281(5728): 216–217.

Ha, Y.M., Shang, Y., Yang, D. et al. (2018). Brassinosteroid reduces ABA accumulation leading to the inhibition of ABA-induced stomatal closure. *Biochemical and Biophysical Research Communications* 504(1): 143–148.

Hamza, A.A., Heeba, G.H., Hassanin, S.O. et al. (2023). Hibiscus-cisplatin combination treatment decreases liver toxicity in rats while increasing toxicity in lung cancer cells via oxidative stress-apoptosis pathway. *Biomedicine and Pharmacotherapy* 165: 115148.

Hassan, M.U., Aamer, M., Chattha, M.U. et al. (2022). Salicylic acid-mediated physiological and molecular mechanisms in plants under heat stress. In: *Managing Plant Stress Using*

Salicylic Acid: Physiological and Molecular Aspects (ed. A. Sharma, R. Bhardwaj, V. Kumar et al.), 163–182. John Wiley & Sons Ltd.

Hassan, M.U., Chattha, M.U., Khan, I. et al. (2021). Heat stress in cultivated plants: Nature, impact, mechanisms, and mitigation strategies—a review. *Plant Biosystems-An International Journal Dealing with all Aspects of Plant Biology* 155(2): 211–234.

Hayat, S., Hasan, S.A., Hayat, Q. et al. (2010a). Brassinosteroids protect *Lycopersicon esculentum* from cadmium toxicity applied as shotgun approach. *Protoplasma* 239: 3–14.

Hayat, S., Hasan, S.A., Yusuf, M. et al. (2010b). Effect of 28-homobrassinolide on photosynthesis, fluorescence and antioxidant system in the presence or absence of salinity and temperature in *Vigna radiata*. *Environmental and Experimental Botany* 69(2): 105–112.

He, J.X., Gendron, J.M., Sun, Y. et al. (2005). BZR1 is a transcriptional repressor with dual roles in brassinosteroid homeostasis and growth responses. *Science* 307(5715): 1634–1638.

Hussain, M., Khan, T.A., Yusuf, M. et al. (2019). Silicon-mediated role of 24-epibrassinolide in wheat under high-temperature stress. *Environmental Science and Pollution Research* 26: 17163–17172.

Ibañez, C., Delker, C., Martinez, C. et al. (2018). Brassinosteroids dominate hormonal regulation of plant thermomorphogenesis via BZR1. *Current Biology* 28(2): 303–310.

Janeczko, A., Oklešťková, J., Pociecha, E. et al. (2011). Physiological effects and transport of 24-epibrassinolide in heat-stressed barley. *Acta Physiologiae Plantarum* 33: 1249–1259.

Jia, Z., Giehl, R.F., Meyer, R.C. et al. (2019). Natural variation of BSK3 tunes brassinosteroid signaling to regulate root foraging under low nitrogen. *Nature Communications* 10(1): 2378.

Jiang, Y.P., Cheng, F., Zhou, Y.H. et al. (2012). Cellular glutathione redox homeostasis plays an important role in the brassinosteroid-induced increase in CO_2 assimilation in *Cucumis sativus*. *New Phytologist* 194(4): 932–943.

Jin, S.H., Li, X.Q., Wang, G.G. et al. (2015). Brassinosteroids alleviate high-temperature injury in *Ficus concinna* seedlings via maintaining higher antioxidant defence and glyoxalase systems. *AoB Plants* 7: plv009.

Khan, A.R., Hui, C.Z., Ghazanfar, B. et al. (2015). Acetyl salicylic acid and 24-epibrassinolide attenuate decline in photosynthesis, chlorophyll contents and membrane thermo-stability in tomato (*Lycopersicon esculentum* Mill.) under heat stress. *Pakistan Journal of Botany* 47(1): 63–70.

Khripach, V.A., Zhabinskii, V.N., and de Groot, A.E. (1998). *Brassinosteroids: A New Class of Plant Hormones*. Academic Press.

Kim, E.J. and Russinova, E. (2020). Brassinosteroid signalling. *Current Biology* 30(7): R294–R298.

Krumova, S., Zhiponova, M., Dankov, K. et al. (2013). Brassinosteroids regulate the thylakoid membrane architecture and the photosystem II function. *Journal of Photochemistry and Photobiology B: Biology* 126: 97–104.

Kumari, A. and Hemantaranjan, A. (2018). Morpho-physiological attributes of Wheat (*Triticum aestivum* L.) genotypes as influenced by brassinosteroids under heat stress. *Journal of Pharmacognosy and Phytochemistry* 7(6): 2111–2115.

Kumari, A. and Hemantaranjan, A. (2019). Mitigating effects of 24-epibrassinolide on heat stress damage by shifting biochemical and antioxidant defense mechanisms in wheat (*Triticum aestivum* L.) at pre-flowering stage and post-flowering stage. *Journal of Pharmacognosy and Phytochemistry* 8(1): 1157–1161.

Lee, H.J., Lee, J.H., Lee, S.G. et al. (2019). Foliar application of biostimulants affects physiological responses and improves heat stress tolerance in Kimchi cabbage. *Horticulture, Environment, and Biotechnology* 60(6): 841–851.

Li, X., Cai, C., Wang, Z. et al. (2018). Plastid translation elongation factor Tu is prone to heat-induced aggregation despite its critical role in plant heat tolerance. *Plant Physiology* 176(4): 3027–3045.

Li, N., Euring, D., Cha, J.Y. et al. (2021). Plant hormone-mediated regulation of heat tolerance in response to global climate change. *Frontiers in Plant Science* 11: 627969.

Li, J.G., Fan, M., Hua, W. et al. (2020). Brassinosteroid and hydrogen peroxide interdependently induce stomatal opening by promoting guard cell starch degradation. *The Plant Cell* 32(4): 984–999.

Lv, M. and Li, J. (2020). Molecular mechanisms of brassinosteroid-mediated responses to changing environments in *Arabidopsis*. *International Journal of Molecular Sciences* 21(8): 2737.

Mahesh, K., Balaraju, P., Ramakrishna, B. et al. (2013). Effect of brassinosteroids on germination and seedling growth of radish (*Raphanus sativus* L.) under PEG-6000 induced water stress. *American Journal of Plant Sciences* 4(12): 40442.

Malini, M.K., Lekshmy, V.S., Pal, M. et al. (2020). Unfolded protein response (UPR) mediated under heat stress in plants. *Plant Physiology Reports* 25: 569–582.

Mandava, N., Sidwell, B.A., Mitchell, J.W. et al. (1973). Production of brassins from rape pollen. A convenient preparatory method. *Industrial and Engineering Chemistry Product Research and Development* 12(2): 138–139.

Manoli, A., Trevisan, S., Quaggiotti, S. et al. (2018). Identification and characterization of the BZR transcription factor family and its expression in response to abiotic stresses in *Zea mays* L. *Plant Growth Regulation* 84: 423–436.

Mansoor, S., Ali Wani, O., Lone, J.K. et al. (2022). Reactive oxygen species in plants: from source to sink. *Antioxidants* 11(2): 225.

Mazorra, L.M., Holton, N., Bishop, G.J. et al. (2011). Heat shock response in tomato brassinosteroid mutants indicates that thermotolerance is independent of brassinosteroid homeostasis. *Plant Physiology and Biochemistry* 49(12): 1420–1428.

Mazorra, L.M., Nunez, M., Hechavarria, M. et al. (2002). Influence of brassinosteroids on antioxidant enzymes activity in tomato under different temperatures. *Biologia Plantarum* 45: 593–596.

Mitchell, J.W., Mandava, N., Worley, J.F. et al. (1970). Brassins—a new family of plant hormones from rape pollen. *Nature* 225(5237): 1065–1066.

Mitchell, J.W. and Whitehead, M.R. (1941). Responses of vegetative parts of plants following application of extract of pollen from *Zea mays*. *Botanical Gazette* 102(4): 770–791.

Nawaz, F., Naeem, M., Zulfiqar, B. et al. (2017). Understanding brassinosteroid-regulated mechanisms to improve stress tolerance in plants: a critical review. *Environmental Science and Pollution Research* 24: 15959–15975.

Niu, J.H., Ahmad Anjum, S., Wang, R. et al. (2016). Exogenous application of brassinolide can alter morphological and physiological traits of Leymus chinensis (Trin.) Tzvelev under room and high temperatures. *Chilean Journal of Agricultural Research* 76(1): 27–33.

Nolan, T.M., Vukašinović, N., Liu, D. et al. (2020). Brassinosteroids: multidimensional regulators of plant growth, development, and stress responses. *The Plant Cell* 32(2): 295–318.

Northey, J.G., Liang, S., Jamshed, M. et al. (2016). Farnesylation mediates brassinosteroid biosynthesis to regulate abscisic acid responses. *Nature Plants* 2(8): 1–7.

Ogweno, J.O., Song, X.S., Shi, K. et al. (2008). Brassinosteroids alleviate heat-induced inhibition of photosynthesis by increasing carboxylation efficiency and enhancing antioxidant systems in *Lycopersicon esculentum*. *Journal of Plant Growth Regulation* 27: 49–57.

Park, J., Lee, Y., Martinoia, E. et al. (2017). Plant hormone transporters: what we know and what we would like to know. *BMC Biology* 15: 1–15.

Planas-Riverola, A., Gupta, A., Betegón-Putze, I. et al. (2019). Brassinosteroid signaling in plant development and adaptation to stress. *Development* 146(5): dev151894.

Raza, A., Salehi, H., Rahman, M.A. et al. (2022). Plant hormones and neurotransmitter interactions mediate antioxidant defenses under induced oxidative stress in plants. *Frontiers in Plant Science* 13: 961872.

Rehman, A., Shahzad, B., Haider, F.U. et al. (2022). An introduction to brassinosteroids: history, biosynthesis, and chemical diversity. In: *Brassinosteroids in Plant Developmental Biology and Stress Tolerance* (ed. G.J. Ahammed, A. Sharma, and J. Yu), 1–14. Academic Press.

Rico-Jiménez, M., Roca, A., Krell, T. et al. (2022). A bacterial chemoreceptor that mediates chemotaxis to two different plant hormones. *Environmental Microbiology* 24(8): 3580–3597.

Sadura, I. and Janeczko, A. (2018). Physiological and molecular mechanisms of brassinosteroid-induced tolerance to high and low temperature in plants. *Biologia Plantarum* 62(4): 601–616.

Sadura, I., Libik-Konieczny, M., Jurczyk, B. et al. (2020a). HSP transcript and protein accumulation in brassinosteroid barley mutants acclimated to low and high temperatures. *International Journal of Molecular Sciences* 21(5): 1889.

Sadura, I., Libik-Konieczny, M., Jurczyk, B. et al. (2020b). Plasma membrane ATPase and the aquaporin HvPIP1 in barley brassinosteroid mutants acclimated to high and low temperature. *Journal of Plant Physiology* 244: 153090.

Sadura, I., Pociecha, E., Dziurka, M. et al. (2019). Mutations in the HvDWARF, HvCPD and HvBRI1 genes-involved in brassinosteroid biosynthesis/signalling: altered photosynthetic efficiency, hormonal homeostasis and tolerance to high/low temperatures in barley. *Journal of Plant Growth Regulation* 38: 1062–1081.

Sakakibara, M. and Mori, K. (1982). Facile synthesis of (22R, 23R)-homobrassinolide [homologue of brassinolide, having plant growth promoting activity, isolated from the pollen of rape]. In: *Agricultural and Biological Chemistry (Japan)*, 2769–2779. Japan: Agricultural and Biological Chemistry.

Samakovli, D., Roka, L., Plitsi, P.K. et al. (2020). Active BR signalling adjusts the subcellular localisation of BES1/HSP90 complex formation. *Plant Biology* 22(1): 129–133.

Setsungnern, A., Muñoz, P., Pérez-Llorca, M. et al. (2020). A defect in BRI1-EMS-SUPPRESSOR 1 (bes1)-mediated brassinosteroid signaling increases photoinhibition and photo-oxidative stress during heat stress in *Arabidopsis*. *Plant Science* 296: 110470.

Siddiqui, H., Ahmed, K.B.M., and Hayat, S. (2018b). Comparative effect of 28-homobrassinolide and 24-epibrassinolide on the performance of different components influencing the photosynthetic machinery in *Brassica juncea* L. *Plant Physiology and Biochemistry* 129: 198–212.

Siddiqui, H., Yusuf, M., Faraz, A. et al. (2018a). 24-Epibrassinolide supplemented with silicon enhances the photosynthetic efficiency of *Brassica juncea* under salt stress. *South African Journal of Botany* 118: 120–128.

Singh, A.P., Fridman, Y., Holland, N. et al. (2018). Interdependent nutrient availability and steroid hormone signals facilitate root growth plasticity. *Developmental Cell* 46(1): 59–72.

Singh, I., Kumar, U., Singh, S.K. et al. (2012). Physiological and biochemical effect of 24-epibrassinoslide on cold tolerance in maize seedlings. *Physiology and Molecular Biology of Plants* 18: 229–236.

Sonjaroon, W., Jutamanee, K., Khamsuk, O. et al. (2018). Impact of brassinosteroid mimic on photosynthesis, carbohydrate content and rice seed set at reproductive stage under heat stress. *Agriculture and Natural Resources* 52(3): 234–240.

Sreeramulu, S., Mostizky, Y., Sunitha, S. et al. (2013). BSK s are partially redundant positive regulators of brassinosteroid signaling in *Arabidopsis*. *The Plant Journal* 74(6): 905–919.

Stirk, W.A., Bálint, P., Tarkowská, D. et al. (2014). Effect of light on growth and endogenous hormones in *Chlorella minutissima* (Trebouxiophyceae). *Plant Physiology and Biochemistry* 79: 66–76.

Su, Q., Zheng, X., Tian, Y. et al. (2020). Exogenous brassinolide alleviates salt stress in *Malus hupehensis* Rehd. by regulating the transcription of NHX-Type Na+ (K+)/H+ antiporters. *Frontiers in Plant Science* 11: 38.

Tang, J., Han, Z., and Chai, J. (2016). Q&A: What are brassinosteroids and how do they act in plants? *BMC Biology* 14: 1–5.

Tanveer, M., Shahzad, B., Sharma, A. et al. (2018). 24-Epibrassinolide; an active brassinolide and its role in salt stress tolerance in plants: a review. *Plant Physiology and Biochemistry* 130: 69–79.

Tanveer, M., Shahzad, B., Sharma, A. et al. (2019). 24-Epibrassinolide application in plants: an implication for improving drought stress tolerance in plants. *Plant Physiology and Biochemistry* 135: 295–303.

Thussagunpanit, J., Jutamanee, K., Sonjaroon, W. et al. (2015). Effects of brassinosteroid and brassinosteroid mimic on photosynthetic efficiency and rice yield under heat stress. *Photosynthetica* 53: 312–320.

Vaishnav, D. and Chowdhury, P. (2023). Types and function of phytohormone and their role in stress. In: *Plant Abiotic Stress Responses and Tolerance Mechanisms* (ed. S. Hussain, T.H. Awan, E.A. Waraich et al.). IntechOpen.

Vardhini, B.V. (2019). Does application of brassinosteroids mitigate the temperature stress in plants. *International Journal of Earth Science and Geology* 1(2): 59–65.

Wilen, L.A., Wettlaufer, J.S., Elbaum, M. et al. (1995). Dispersion-force effects in interfacial premelting of ice. *Physical Review B* 52(16): 12426.

Wu, X., Yao, X., Chen, J. et al. (2014). Brassinosteroids protect photosynthesis and antioxidant system of eggplant seedlings from high-temperature stress. *Acta Physiologiae Plantarum* 36(2): 251–261.

Xia, X.J., Wang, Y.J., Zhou, Y.H. et al. (2009). Reactive oxygen species are involved in brassinosteroid-induced stress tolerance in cucumber. *Plant Physiology* 150(2): 801–814.

Yadava, P., Kaushal, J., Gautam, A. et al. (2016). Physiological and biochemical effects of 24-epibrassinolide on heat-stress adaptation in maize (*Zea mays* L.). *Natural Science* 8(4): 171–179.

Yamagami, A., Saito, C., Nakazawa, M. et al. (2017). Evolutionarily conserved BIL4 suppresses the degradation of brassinosteroid receptor BRI1 and regulates cell elongation. *Scientific Reports* 7(1): 5739.

Yang, J., Miao, W., and Chen, J. (2021). Roles of jasmonates and brassinosteroids in rice responses to high temperature stress–a review. *The Crop Journal* 9(5): 977–985.

Ye, H., Liu, S., Tang, B. et al. (2017). RD26 mediates crosstalk between drought and brassinosteroid signalling pathways. *Nature Communications* 8(1): 14573.

Yi, D., Alvim Kamei, C.L., Cools, T. et al. (2014). The *Arabidopsis* Siamese-related cyclin-dependent kinase inhibitors SMR5 and SMR7 regulate the DNA damage checkpoint in response to reactive oxygen species. *The Plant Cell* 26(1): 296–309.

Yin, Y., Qin, K., Song, X. et al. (2018). BZR1 transcription factor regulates heat stress tolerance through FERONIA receptor-like kinase-mediated reactive oxygen species signaling in tomato. *Plant and Cell Physiology* 59(11): 2239–2254.

Yonca Surgun-Acar, F.Z.N. (2019). 24-Epibrassinolide promotes arsenic tolerance in *Arabidopsis thaliana* L. by altering stress responses at biochemical and molecular level. *Journal of Plant Physiology* 238: 12–19.

Zhang, J., Li, S., Cai, Q. et al. (2020). Exogenous diethyl aminoethyl hexanoate ameliorates low temperature stress by improving nitrogen metabolism in maize seedlings. *PLoS One* 15(4): e0232294.

Zhang, Y.P., He, J., Yang, S.J. et al. (2014). Exogenous 24-epibrassinolide ameliorates high temperature-induced inhibition of growth and photosynthesis in *Cucumis melo*. *Biologia Plantarum* 58: 311–318.

Zhang, Y.P., Zhu, X.H., Ding, H.D. et al. (2013). Foliar application of 24-epibrassinolide alleviates high-temperature-induced inhibition of photosynthesis in seedlings of two melon cultivars. *Photosynthetica* 51: 341–349.

Zhou, J., Wang, J., Li, X. et al. (2014). H_2O_2 mediates the crosstalk of brassinosteroid and abscisic acid in tomato responses to heat and oxidative stresses. *Journal of Experimental Botany* 65(15): 4371–4383.

Zou, L.J., Deng, X.G., Zhang, L.E. et al. (2018). Nitric oxide as a signaling molecule in brassinosteroid-mediated virus resistance to cucumber mosaic virus in *Arabidopsis thaliana*. *Physiologia Plantarum* 163(2): 196–210.

9

Brassinosteroids Biology in Plants Under Salt Stress: Physiological and Molecular Aspects

Muhammad Ahsan Altaf[1] *and Riaz Ahmad*[2]*

[1]*Key Laboratory for Quality Regulation of Tropical Horticultural Crops of Hainan Province, Sanya Nanfan Research Institute, Hainan University, Sanya, China*
[2]*Department of Horticulture, The University of Agriculture, Dera Ismail Khan, Pakistan*

*Corresponding author: riazahmadbzu@gmail.com

9.1 Introduction

Salinity stress drastically reducing plant productivity occurs due to fluctuations in climate change (Jouyban 2012). Sodium chloride (NaCl), calcium chloride ($CaCl_2$), and sodium phosphate (Na_2SO_4) are causes of salinity stress in plants and soil (Nawaz et al. 2010). Generally, plants retain a low level of Na^+ and a high level of K^+ under optimum conditions (Naz et al. 2022; Muhammad et al. 2024). Irregularities in the ionic balance system due to induced salt stress cause a plant to become more sensitive to Na^+ ions, which in turn causes ion stress, damages the structure of the plant's cell membrane, and modifies the normal levels of physiological activities (Ahmad et al. 2013; Ondrasek et al. 2022). Osmotic stress and water deficit conditions are caused by a high concentration of Na^+ in the soil, which lowers the soil water potential and hinders the roots' ability to absorb water due to the buildup of soluble salt (Rasool et al. 2013; Naz et al. 2024). Moreover, plant production and quality are greatly decreased due to Na^+ absorption through the roots (Isayenkov 2012; Parihar et al. 2015). Salinity toxicity significantly hampers plant physiology and metabolism (Figure 9.1).

Na^+ absorption is linked with the availability of Na^+ within the root biosphere. Additionally, abscisic acid (ABA) buildup may result from osmotic stress. The cell membrane, which has the ability to control the ion balance within the cell, is an important protective component of plant cells (Park et al. 2016). Plants that are stressed by salt release a lot of reactive oxygen species (ROS), which can cause lipid peroxidation, protein denaturation, denaturation of cell membranes, and possibly death or damage to cells (Yokoi et al. 2002). Plant photosynthesis is altered by elevated salt concentrations, leading to the breakdown of the chloroplast structure, a decrease in the amount of pigment and the activity of photosynthetic enzymes, and ultimately, a suppression of plant development (Serrano and Rodriguez-Navarro 2001). The impacts of salt on plants in these soils are

Brassinosteroids and Abiotic Stress: Recent Developments on Physiological and Molecular Aspects,
First Edition. Edited by Anket Sharma, et al.
Published 2026 by John Wiley & Sons Ltd.

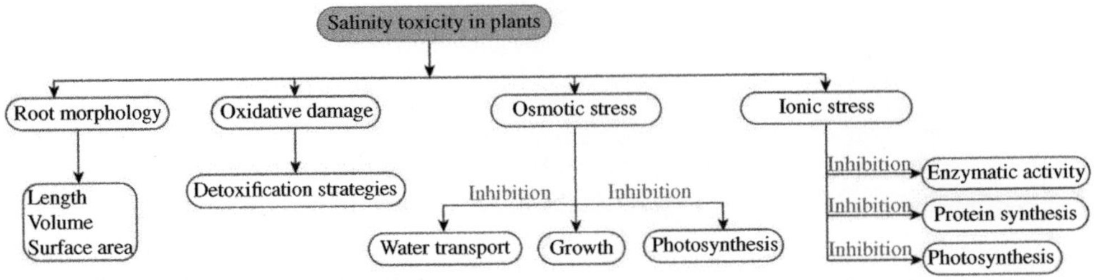

Figure 9.1 Salinity toxicity in plants.

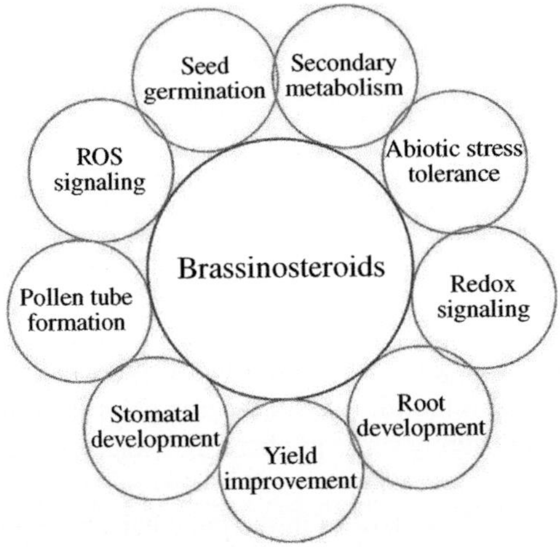

Figure 9.2 BRs regulate many plant physiological functions and enhance abiotic stress tolerance.

minimized, along with their associated implications, such as the avoidance of secondary issues, through the exogenous supplementation and endogenous activation of brassinosteroids (BRs) (Batool et al. 2014; Yang and Guo 2018).

The application of BRs increased the vegetative growth and the reproductive phase of plants even under optimum and adverse growing conditions. It had excellent potential to improve plant immune system to survive under adverse conditions. BRs supplementation is effective during seed priming, soil applications, and foliar sprays. It can be used to improve plant height, plant biomass, photosynthesis, elimination of toxic substances, anatomical changes, and photochemical effectiveness of crops under normal and adverse environments (Ashraf et al. 2010; Irfan et al. 2017; Hussain et al. 2023). BRs regulate various plant physiological functions (Figure 9.2).

Adverse effects of salts and BRs impact on plants based on cultivars' genetic makeup, stress intensity, BRs concentration, application mode, and numerous climatic factors. BRs contribute in the maturity phases of plants by alteration in several physiological changes. Osmotic injuries occur from higher salinity levels, which can be regulated with BRs supplementation. The plant defense system was improved with the supplementation of BRs. The salt-tolerant cultivars had excellent plant defense system as compared to salt-sensitive cultivars. Expression of salt-tolerant genes was found to be greater in tolerant plants as compared to sensitive plants (Ali et al. 2008; Hegazi et al. 2017; Manghwar et al. 2022). Different management practices such as application of various phytohormones, organic amendments, biochar use, seaweed application, and molecular implications are necessary to explore for higher seed germination, seedling growth, and various developmental processes occurring within the plant cells (Sun et al. 2015). Moreover, the improvements in the plant defense system are necessary for higher-quality production of crops. This study examines the use of phytohormones in horticultural crops to lessen the adverse effects of salt toxicity. Among phytohormones, BRs are found to be very helpful for the mitigation of adverse effects of salt toxicity. Moreover, this chapter explores the molecular and physico-chemical programming in plants with strong defense mechanisms against the deleterious effects of climatic extremes (Dong et al. 2017; Tanveer et al. 2018).

9.2 BRs Involvement in the Enhancement of Crop Growth and Yield of Crops Under Salt Stress

Seed germination and seedling growth are important parameters for healthy crop production. Seeds are very delicate during germination and can be easily damaged by climatic stress (Irfan et al. 2018). BRs enhance germination rate and ultimately increase seedling growth under stress conditions (Ghoname et al. 2023). BRs application regulates those activities and maintains proper seedling growth under stress conditions (Shahbaz and Ashraf 2007). Crop yield, growth, and development are all negatively impacted by salt toxicity (Abbasi et al. 2016). BRs have been demonstrated to lessen the harmful effects of salinity in several crops such as pepper, cucumber, common bean, and black locust (Siddiqui et al. 2017).

The efficacy of the treatment to increase the tolerance of eggplants to salty conditions is demonstrated by the lowered levels of Na^+ and Cl^-, increased levels of K^+ and Ca^{2+}, and improved antioxidant enzyme activities. Similarly, saline-grown cucumber plants can have their amounts of NO_3 and NH_4 lowered with different levels of BRs supplementation. Nevertheless, foliar application of BRs to rapeseed at 30 and 45 days after sowing may effectively alleviate the adverse consequences of salt stress (Hussain et al. 2019). The enhanced BRs-induced resistance of cucumber plants to the salty condition is associated with higher levels of photosynthesis, nitrogen efficiency, and total polyamines (Ahammed et al. 2015). BRs even at low levels had the potential to boost plant development, while greater concentrations showed adverse effects on growth and growth-related parameters (Mir et al. 2015). After applying BRs, plants responded very quickly (Hossain et al. 2022), growing more shoots and roots (Zhang et al. 2023).

Numerous earlier studies have shown that cell elongation and expansion during plant growth are the cause of enhanced BRs levels (Alhaithloul et al. 2020; Wang et al. 2023). Plants overexpressing the BR receptor *BRI1* had a significant increase in the size of epidermal and mesophyll cells (Ghorbel and Brini 2023), while several BRs' loss-of-function mutants exhibited a significant drop in cell size, leaf epidermal, and mesophyll cell quantity (Dawood 2022). Photosynthesis, respiration, transpiration, and guttation are all carried out by the specialized organs, such as leaves. Many different physiological and developmental processes, including those that take place in leaves, are regulated by BRs.

Table 9.1 The impact of BRs supplementation on the growth and yield of salt-affected plants.

Crop name	BRs impact	Reference
Eggplant	Improves seed germination and seedling growth	Wu et al. (2014); Gisbert et al. (2009)
Tomato	Improves flower and fruit formation under excessive salt levels	Ahammed et al. (2012)
Pepper	Modifications in the root's biosphere and root morphology	Samira et al. (2012)
Radish	Eliminate toxic substances produced due to excess salts	Ramakrishna and Rao (2015)
Tomato	Improves the stomatal conductance for optimum gaseous exchange	Júnior et al. (2022)
Radish	Involves the formation of photosynthetic pigments and apparatus	Choudhary et al. (2012)
Apple	Repairing of membrane injury due to stress exposure	Kumari and Thakur (2019)
Broccoli	Boosts the plant defense system during stress exposure	Cai et al. (2019)
Grapes	Balance the absorption and assimilation of minerals and nutrients	El-Banna et al. (2022)
Peppers	Ion homeostasis regulations under salt exposure	Yang et al. (2019)
Mustard	Higher accumulation of several metabolites to enhance tolerance	Wani et al. (2019)
Maize	Improved fruit size, yield, overall quality, and shelf life	Rattan et al. (2014)

According to a recent study, using BRs significantly enhanced the growth of cucumber leaves, such as leaf area, leaf size, number of leaves, and leaf biomass (Arshad et al. 2023). The main components for plant growth under stress are photosynthesis, chlorophyll content, and leaf area, all of which were significantly improved by BRs application, according to earlier research (Fariduddin et al. 2014; Basit et al. 2021; Hafeez et al. 2021). BRs supplementation improves plant growth and yield of salt-affected plants by adjusting the defense system (Table 9.1).

9.3 BRs Regulate Photosynthetic Pigments Formation and Countenance Under Salt Stress

BRs have the capability to balance the synthesis of photosynthetic pigments under salt stress. Since salt stress negatively affects both the pigment system and the photosynthetic machinery, plants that are exposed to salt toxicity typically experience a loss in photosynthetic performance (Zahra et al. 2022). A decrease in chlorophyll content under salt stress may result from either a hampered or an increased breakdown of chlorophyll molecules

(Siddiqui et al. 2018). Earlier findings stated that sunflowers lose chlorophyll concentration due to decreased synthesis of chlorophyll or salinity-induced chlorophyll degradation. The chlorophyll synthesis is inhibited when 5-aminolaevulinic acid, a precursor of chlorophyll biosynthesis, is not present in sufficient amounts (Sousa et al. 2021). However, higher chlorophyll *a* and *b* improve the plant photosynthetic process according to Fariduddin et al. (2013). The oxidation of photosynthetic pigments is brought on by elevated ROS concentrations, which therefore lower the chlorophyll formation. The chlorophyll content may possibly be decreased with increased chlorophyllase activity (Parmoon et al. 2018). Regulating the transcription and translational pathways that result in the synthesis of photosynthetic pigment is essential to treating the symptoms of stress conditions. The chlorophyll breakdown is also slowed down by BRs treatments. Bioregulatory substances (BRSs) normalize chlorophyll content and reduce ROS to assist plants withstand the negative effects of salt stress (Ashraf et al. 2010). This increases the efficiency of photosynthesis. The amount of chlorophyll in pea plants was improved through BRs (Oliveira et al. 2019). According to He et al. (2022), plants treated with BRs had better transcription and translation of the enzymes involved in the synthesis of chlorophyll, as well as greater light capture efficiency. When treated with BRs under salinity, plants exhibit a considerable increase in chlorophyll content, including chlorophyll a and b (Al-Taey et al. 2024).

BRs are responsible for upregulating the expression of genes involved in the production of enzymes that make chlorophyll, thereby increasing the quantity of chlorophyll found in mustard. BRs have the ability to prevent the breakdown of related proteins and chlorophyll, which are mostly found in the light-harvesting complex. Consequently, there is an increase in the level of chlorophyll (Kumudini and Patil 2019). To be clear about the exact process, more research must yet be done by researchers. By enhancing the activity of RuBisCo and the other essential Calvin cycle enzymes, BRs raise the photosynthetic rate of plants (Siddiqi and Husen 2021). BRs also enhance CO_2 uptake in order to further increase stomatal conductance (Sirhindi et al. 2018). By encouraging synthesis and activity, application of BRs under salt stress raises gaseous exchange parameters, according to Shahzad et al. (2019). It improves the stability and repair of *D1* protein, which increases photosynthetic efficiency in plants. The detrimental effects of high-energy innervation in the reaction core are shielded from the plant by BRs. Antioxidant activities were enhanced under BRs supplementation. Membrane rupturing is regulated with supplementation of BRs (Chmur and Bajguz 2021).

Photosynthetic process improved with the elimination of membrane-rupturing activities. By facilitating CO_2 uptake by plants and enhancing RuBisCo activity, BRs increase the net photosynthetic rate (Jakhar et al. 2022). By accelerating the process of ROS scavenging, carotenoids are thought to be beneficial in protecting the photosynthetic system of plants that are typically grown in harsh environments (Ashraf et al. 2010). Moreover, under stressful circumstances, recovery of the carotenoid contents following the exogenous application of BRs. Plant leaves that have been treated with BRs have higher levels of carotenoids (Oliveira et al. 2019). This may be because BRs alter ethylene signaling and biosynthesis (Khan et al. 2022). In salt-stressed maize plants, elevated levels of carotenoids were observed upon BRs spray (Vardhini 2012). Under abiotic stress circumstances, plants' ability to photosynthesize generally decreases because the structure of the chloroplast and thylakoid electron transport, two essential components of the photosynthetic mechanism,

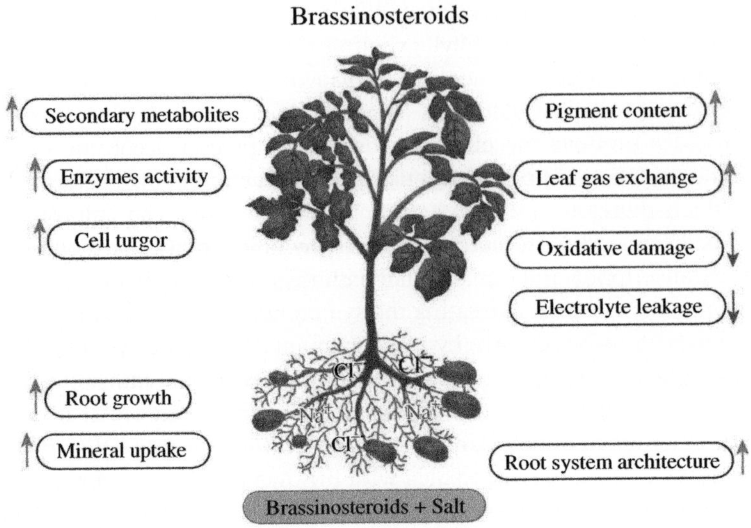

Figure 9.3 BRs application enhanced the antioxidant defense system and reduced oxidative stress.

are disturbed (Figure 9.3). Furthermore, a confirmed decrease in the activity of the major Calvin cycle enzymes and carboxylation efficiency may be caused by the inactivation of the Rubisco enzyme (Chaudhuri et al. 2022). Under conditions of salt stress, ions such as Na^+ and Cl^- accumulate more. By boosting chlorophyll concentrations, chlorophyll fluorescence features as well as by enhancing the activity of many enzymes in the chloroplast, BRs enhance photosynthesis in cucumbers (Nithila and Sivakumar 2019). The stimulation of stomatal and non-stomatal abilities under the impact of BRs may be the cause of the improvement in photosynthesis.

9.4 BRs Mitigate the Toxic Substances Within Plant Cells Generated from Salt Stress

Under salt stress, plant components produce more ROS, which eventually lead to oxidative stress through processes including lipid peroxidation and electrolyte leakage. Free radicals are toxic for cell membranes injury under stress conditions. Oxidative stress conditions are enhanced due to free radicals under salinity stress. Different investigations have established the effect of BRs in limiting oxidative damage and lowering the generation of ROS during salt stress (He et al. 2022). Research has demonstrated that the use of BRs can lessen the detrimental effects of oxidative stress by slowing down the processes of lipid peroxidation and electrolyte leakage (Arif et al. 2020; Liu et al. 2023). This results from the exogenous application of BRs under salt stress, which reduces ROS within the organelles. After exogenous application of BRs, Abd-Allah et al. (2018) reported a substantial decrease in the concentrations of toxic substances. According to Rattan et al. (2022), applying BRs topically to plants under salt stress suppressed the generation of malondialdehyde (MDA), which further reduced lipid peroxidation. Under salinity, exogenous application of BRs decreased non-enzymatic compounds in peppermint (Fathi et al. 2020), common beans (Flores et al. 2019), wheat (Shahbaz et al. 2008), pea (Jager et al. 2005), maize (Holá et al. 2010; Xing et al. 2022), and rice seedlings (Sharma et al. 2013; Larré et al. 2014). Indicators of oxidative stress include electrolyte leakage and membrane permeability, both of which tend to rise in plants under salt stress (Das et al. 2013). Under salt stress, electrolyte leakage was significantly reduced by the exogenous application of BRs (Özdemir et al. 2004). According to Alyemeni et al. (2013), electrolyte leakage in cucumber under salt stress was reduced by foliar application of BRs. Similarly, common bean plants cultivated under salt stress showed enhanced membrane stability index (MSI) and decreased electrolyte leakage upon administration of BRs, indicating the beneficial effect of BRs in plants under salt stress (Quamruzzaman et al. 2021). In carrot, strawberry, and peppermint plants, foliar application of BRs enhanced membrane permeability and decreased membrane damage (Ahanger et al. 2020). Alam et al. (2019) observed enhanced morphometric parameters and a considerable decrease in electrolyte loss in maize following the use of BRs. Following the exogenous application of BRs, several crops showed a notable improvement in MSI (Vighi et al. 2017). The improvements in MSI result in improved growth and higher yield.

Plants subjected to oxidative stress from salt stress, which triggers the activation of the plant's antioxidative defense system, which includes both enzymatic and non-enzymatic antioxidants (Noreen et al. 2021). ROS overgeneration was higher in plants because of oxidative stress due to adverse conditions resulting from excess of salts (Zia-ur-Rehman et al. 2023). Plants' antioxidant defense system lessens the damaging effects of oxidative burst by scavenging dangerous ROS (Shahzad et al. 2019). Plants' ability to withstand salt stress is enhanced by increased antioxidant enzyme activity (Hasanuzzaman et al. 2021). It is commonly recognized that the ascorbate-glutathione cycle improves ROS scavenging (Ranjan et al. 2023). When plants are exposed to harsh environments, exogenous administration of BRs activates their antioxidative defense system, which reduces oxidative stress and its effects (Sachdev et al. 2021). According to several studies, the administration of BRs activates several enzymatic and nonenzymatic antioxidants by acting as a signaling chemical and influencing the expression of antioxidant genes (Mishra et al. 2023). According to Tahjib-Ul-Arif et al. (2018), BRs controlled the expression of the peroxidase-encoding genes *ATP2* and *ATP24a* in *Arabidopsis*. The increased activity of antioxidant enzymes was crucial for scavenging ROS. Furthermore, the supplementation of BRs resulted in tolerance that was mediated by the upregulation of regulatory genes, including respiratory burst oxidase homologue (*RBOH*), mitogen-activated protein kinase 1 (*MAPK1*), and mitogen-activated protein kinase 3 (*MAPK3*). BRs-induced stress tolerance was associated with additional genes related to defense, antioxidant responses, and raised MDA levels as a result of increased oxidative harms.

A lower redox state under abiotic stress is the consequence of BRs' promotion of glutathione S-transferase and glutathione reductase activities as well as their ability to raise the level of reduced glutathione without changing the content of oxidized glutathione (Hasanuzzaman and Fujita 2022). Due to an increase in glutathione content, foliar application with BRs alone resulted in a considerable increase in the total pool of glutathione (Gupta and Huang 2014). When BRs were applied to stressed seedlings, it prevented glutathione and ascorbate from oxidizing, keeping redox molecules in their reduced forms. Therefore,

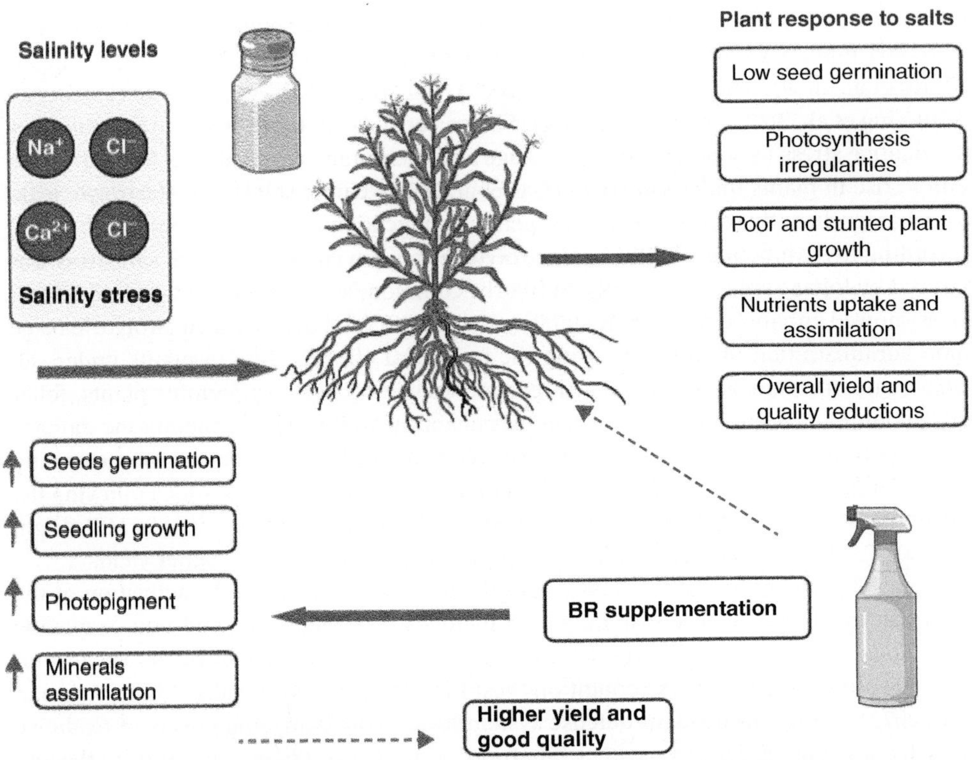

Figure 9.4 BRs enhanced salinity stress tolerance in plants.

as an antioxidant defense mechanism, the external application of BRs modifies the ascorbic acid and glutathione redox state to counteract oxidative stress through the ascorbic acid and glutathione redox cycle and glutathione metabolism (Mahmood et al. 2010). BRs have the potential to mitigate the adverse effects of salinity stress in plants (Figure 9.4).

9.5 Molecular Significance in Salt Stress Tolerance Mechanism

The role of ubiquitin-conjugating enzymes in BR-induced salt stress resistance has been established. It acts as a functional component of the endoplasmic reticulum-associated protein degradation (ERAD) system that influences the accumulation of *BRI1* receptor in cells. It also directs the ERAD pathway toward BR-enhanced osmotic adjustment in *Arabidopsis* (Sairam and Tyagi 2004). Moreover, BRs have been connected to DNA methylation regulation, which is essential for salt tolerance. The fact that seed priming with BRs increases overall methylation and improves salt tolerance levels to a role for BRs in epigenetic modification under salinity stress (Gupta et al. 2018). The essential genes for seed germination, G-protein α-subunit 1 (*GPA1*) and G-protein coupled receptor 1 (*GCR1*), are under the direction of BRs. *GCR1* is a potential G protein-coupled cell surface receptor, while *GPA1* is a subunit of the heterotrimeric G protein (Zhang et al. 2023). Early seed germination and seedling growth are regulated by *GPA1* and *Pirin1* interactions (Kapoor et al. 2021). A crucial initiator of the BRs signal pathway, BR-insensitive-long hypocotyl 4 (*BIL4*) interacts with the receiver BRs receptor *BRI1* and dynamically regulates cell elongation. When compared to wild-type plants, *BIL4-OX* mutants exhibited larger and taller cells, whereas *BIL4-RNAi* mutants showed a decrease in these characteristics (Bundó et al. 2022). Plant cell differentiation and enhanced induced stress tolerance were significantly influenced by the BRs signaling pathways (Gruszka 2013; Hao et al. 2013; Wei et al. 2015). A recent investigation identified the cucumber mutant known as super compact-2, which exhibited the characteristic phenotype of BRs biosynthesis deficiency. These mutants exhibited decreased endogenous BRs accumulation, abnormalities in cell elongation and vascular development, reduced leaf area, dark green and wrinkled leaves (Divi and Krishna 2009; Xia et al. 2015; Nolan et al. 2017). So, genome mapping, sequencing, transformation, and editing strategies are effective for the development of higher-yielding cultivars with greater level of tolerance against adverse climatic conditions.

9.6 Conclusion

The intricate defense system that plants develop to protect from salt stress is brought about by BRs. This system works through a variety of regulatory levels. Plants may, therefore, be less vulnerable to salt stress, which is bad for them. BRs maintain the ability of plants to withstand salt stress by controlling the ionic equilibrium, increasing the antioxidant response, and promoting osmoregulation. Over time, plants in salt-affected areas can immobilize nutrients and transform salt into organic compounds, protecting agriculture and the food supply from salt buildup. Acquiring a thorough understanding of the intricate mechanism via which BRs communicate salt stress response pathways will take time, as will producing crop cultivars that are predictable and salt-tolerant to climatic shifts such as increased soil salinization brought on by climate change. To effectively utilize these features in agriculture, molecular difference mechanisms underlying BR-induced salt stress resistance must be uncovered. Finally, the use of BRs has the potential to contribute to the development of a sustainable food supply for the expanding global population, particularly in terms of raising agricultural yields and increasing salinity tolerance.

References

Abbasi, H., Jamil, M., Haq, A. et al. (2016). Salt stress manifestation on plants, mechanism of salt tolerance and potassium role in alleviating it: a review. *Zemdirbyste-Agriculture* 103 (2): 229–238.

Abd-Allah, E.F., Alqarawi, A.A., Hashem, A. et al. (2018). Regulatory roles of 24-epibrassinolide in tolerance of *Acacia gerrardii* Benth to salt stress. *Bioengineered* 9 (1): 61–71.

Ahammed, G., Xia, X.J., Li, X. et al. (2015). Role of brassinosteroid in plant adaptation to abiotic stresses and its interplay with other hormones. *Current Protein and Peptide Science* 16 (5): 462–473.

Ahammed, G.J., Zhang, S., Shi, K. et al. (2012). Brassinosteroid improves seed germination and early development of tomato seedling under phenanthrene stress. *Plant Growth Regulation* 68: 87–96.

Ahanger, M.A., Mir, R.A., Alyemeni, M.N. et al. (2020). Combined effects of brassinosteroid and kinetin mitigates salinity stress in tomato through the modulation of antioxidant and osmolyte metabolism. *Plant Physiology and Biochemistry* 147: 31–42.

Ahmad, P., Azooz, M.M., and Prasad, M.N.V. (2013). *Salt Stress in Plants*. New York: Springer.

Alam, P., Albalawi, T.H., Altalayan, F.H. et al. (2019). 24-Epibrassinolide (EBR) confers tolerance against NaCl stress in soybean plants by up-regulating antioxidant system, ascorbate-glutathione cycle, and glyoxalase system. *Biomolecules* 9 (11): 640.

Alhaithloul, H.A.S., Abu-Elsaoud, A.M., and Soliman, M.H. (2020). Abiotic stress tolerance in crop plants: role of phytohormones. In: *Abiotic Stress in Plants* (ed. C. Wu, D. Wang, S. Fahad et al.), 233–256. IntechOpen.

Ali, Q., Athar, H.U.R., and Ashraf, M., (2008). Modulation of growth, photosynthetic capacity and water relations in salt stressed wheat plants by exogenously applied 24-epibrassinolide. *Plant Growth Regulation* 56: 107–116.

Al-Taey, D.K., Al-Musawi, Z.J., Kadium, S.M.A. et al. (2024). Brassinolides' function and involvement in salt stress response: a review. *IOP Conference Series: Earth and Environmental Science* 1371 (4): 042032.

Alyemeni, M.N., Hayat, S., Wijaya, L. et al. (2013). Foliar application of 28-homobrassinolide mitigates salinity stress by increasing the efficiency of photosynthesis in *Brassica juncea*. *Acta Botanica Brasilica* 27: 502–505.

Arif, Y., Singh, P., Siddiqui, H. et al. (2020). Salinity induced physiological and biochemical changes in plants: an omic approach towards salt stress tolerance. *Plant Physiology and Biochemistry* 156: 64–77.

Arshad, A., Mushtaq, N., Sajjad, M. et al. (2023). Role of exogenous phytohormones in mitigating stress in plants. In: *Phytohormones and Stress Responsive Secondary Metabolites* (ed. M. Ozturk, R.A. Bhat, M. Ashraf et al.), 111–131. Academic Press.

Ashraf, M., Akram, N.A., Arteca, R.N. et al. (2010). The physiological, biochemical and molecular roles of brassinosteroids and salicylic acid in plant processes and salt tolerance. *Critical Reviews in Plant Sciences* 29 (3): 162–190.

Basit, F., Liu, J., An, J. et al. (2021). Brassinosteroids as a multidimensional regulator of plant physiological and molecular responses under various environmental stresses. *Environmental Science and Pollution Research* 28: 44768–44779.

Batool, N., Shahzad, A., Ilyas, N. et al. (2014). Plants and salt stress. *International Journal of Agriculture and Crop Sciences* 7 (9): 582.

Bundó, M., Martín-Cardoso, H., Pesenti, M. et al. (2022). Integrative approach for precise genotyping and transcriptomics of salt tolerant introgression rice lines. *Frontiers in Plant Science* 12: 797141.

Cai, J.H., Luo, F., Zhao, Y.B. et al. (2019). 24-Epibrassinolide treatment regulates broccoli yellowing during shelf life. *Postharvest Biology and Technology* 154: 87–95.

Chaudhuri, A., Halder, K., Abdin, M.Z. et al. (2022). Abiotic stress tolerance in plants: brassinosteroids navigate competently. *International Journal of Molecular Sciences* 23 (23): 14577.

Chmur, M. and Bajguz, A., (2021). Brassinolide enhances the level of brassinosteroids, protein, pigments, and monosaccharides in *Wolffia arrhiza* treated with brassinazole. *Plants* 10 (7): 1311.

Choudhary, S.P., Kanwar, M., Bhardwaj, R. et al. (2012). Chromium stress mitigation by polyamine-brassinosteroid application involves phytohormonal and physiological strategies in *Raphanus sativus* L. *PLoS One* 7 (3): e33210.

Das, T.A.N.I.A., Shukla, Y.M., Poonia, T.C. et al. (2013). Effects of brassinolide on physiological characteristics of rice (*Oryza sativa* L.) with different salinity levels. *Annals of Biology* 29: 228–231.

Dawood, M.G. (2022). Protective role of brassinosteroids in plants during abiotic stress. In: *Jasmonates and Brassinosteroids in Plants* (ed. R. Akula and G. Sirhindi), 75–88. CRC Press.

Divi, U.K. and Krishna, P. (2009). Brassinosteroid: a biotechnological target for enhancing crop yield and stress tolerance. *New Biotechnology* 26: 131–136.

Dong, Y.J., Wang, W.W., Hu, G.Q. et al. (2017). Role of exogenous 24-epibrassinolide in enhancing the salt tolerance of wheat seedlings. *Journal of Soil Science and Plant Nutrition* 17 (3): 554–569.

El-Banna, M.F., Al-Huqail, A.A., Farouk, S. et al. (2022). Morpho-physiological and anatomical alterations of salt-affected Thompson seedless grapevine (*Vitis vinifera* L.) to brassinolide spraying. *Horticulturae* 8 (7): 568.

Fariduddin, Q., Mir, B.A., Yusuf, M. et al. (2013). Comparative roles of brassinosteroids and polyamines in salt stress tolerance. *Acta Physiologiae Plantarum* 35: 2037–2053.

Fariduddin, Q., Yusuf, M., Ahmad, I. et al. (2014). Brassinosteroids and their role in response of plants to abiotic stresses. *Biologia Plantarum* 58: 9–17.

Fathi, A., Oveysi, M., Nasri, M. et al. (2020). Physiological responses of peppermint (*Mentha piperita* L.) to plant growth regulators and salinity stress. *Iranian Journal of Plant Physiology* 11 (1): 3499–3508.

Flores, R.A., da Cunha, P.P., Damin, V. et al. (2019). Physiological quality and grain production of '*Phaseolus vulgaris*' (cv. BRS Perola) using boron (B) application under irrigation system. *Australian Journal of Crop Science* 13 (4): 520–528.

Ghoname, A.A., AbdelMotlb, N.A., Abdel-Al, F.S. et al. (2023). Brassinosteroids or proline can alleviate yield inhibition under salt stress via modulating physio-biochemical activities and antioxidant systems in snap bean. *The Journal of Horticultural Science and Biotechnology* 98 (4): 526–539.

Ghorbel, M. and Brini, F. (2023). Role of brassinosteroids in regulating physiological and molecular aspects of plants under abiotic stress. In: *The Role of Growth Regulators and Phytohormones in Overcoming Environmental Stress* (ed. A. Sharma, S. Pandey, R. Bhardwaj et al.), 197–233. Academic Press.

Gisbert, C., Prohens, J., and Nuez, F. (2009). Treatments for improving seed germination in eggplant and related species. In: *V International Symposium on Seed, Transplant and Stand Establishment of Horticultural Crops 898* (ed. J.A. Pascual and F. Pérez Alfocea), 45–51. ISHS.

Gruszka, D. (2013). The brassinosteroid signaling pathway-new key players and interconnections with other signaling networks crucial for plant development and stress tolerance. *International Journal of Molecular Sciences* 14 (5): 8740–8774.

Gupta, S.K., Goyal, M.R., and Singh, A. (2018). Biochemical, physiological, and molecular approaches for improving salt tolerance in crop plants: a review. *Engineering Practices for Management of Soil Salinity* (ed. S.K. Gupta, M.R. Goyal, and A. Singh), 201–250. Apple Academic Press.

Gupta, B. and Huang, B. (2014). Mechanism of salinity tolerance in plants: physiological, biochemical, and molecular characterization. *International Journal of Genomics* 2014 (1): 701596.

Hafeez, M.B., Zahra, N., Zahra, K. et al. (2021). Brassinosteroids: molecular and physiological responses in plant growth and abiotic stresses. *Plant Stress* 2: 100029.

Hao, J., Yin, Y., and Fei, S.Z. (2013). Brassinosteroid signaling network: implications on yield and stress tolerance. *Plant Cell Reports* 32: 1017–1030.

Hasanuzzaman, M. and Fujita, M. (2022). Plant responses and tolerance to salt stress: physiological and molecular interventions. *International Journal of Molecular Sciences* 23 (9): 4810.

Hasanuzzaman, M., Raihan, M.R.H., Masud, A.A.C. et al. (2021). Regulation of reactive oxygen species and antioxidant defense in plants under salinity. *International Journal of Molecular Sciences* 22 (17): 9326.

He, X., Wan, Z., Jin, N. et al. (2022). Enhancement of cucumber resistance under salt stress by 2, 4-epibrassinolide lactones. *Frontiers in Plant Science* 13: 1023178.

Hegazi, A.M., El-Shraiy, A.M., and Ghoname, A.A. (2017). Mitigation of salt stress negative effects on sweet pepper using arbuscular mycorrhizal fungi (AMF), *Bacillus megaterium* and brassinosteroids (BRs). *Gesunde Pflanzen* 69 (2): 91–102.

Holá, D., Rothová, O., Kočová, M. et al. (2010). The effect of brassinosteroids on the morphology, development and yield of field-grown maize. *Plant Growth Regulation* 61: 29–43.

Hossain, A., Pamanick, B., Venugopalan, V.K. et al. (2022). Emerging roles of plant growth regulators for plants adaptation to abiotic stress–induced oxidative stress. In: *Emerging Plant Growth Regulators in Agriculture* (ed. M. Naeem and T. Aftab), 1–72. Academic Press.

Hussain, S., Nanda, S., Ashraf, M. et al. (2023). Interplay impact of exogenous application of abscisic acid (ABA) and brassinosteroids (BRs) in rice growth, physiology, and resistance under sodium chloride stress. *Life* 13 (2): 498.

Hussain, S., Shaukat, M., Ashraf, M. et al. (2019). Salinity stress in arid and semi-arid climates: effects and management in field crops. *Climate Change and Agriculture* 13: 201–226.

Irfan, M., Alam, J., Ahmad, I. et al. (2017). Effects of exogenous and foliar applications of Brassinosteroid (BRs) and salt stress on the growth, yield and physiological parameters of *Lycopersicon esculentum* (Mill.). *Plant Science Today* 4 (3): 88–101.

Irfan, M., Alam, J., Jan, G. et al. (2018). Influence of brassinosteroid (BRs) in roots and foliar spray against salinity on physiological parameters and micro nutrients upon tomato (*Lycopersicon esculentum* Mill.). *Specialty Journal of Biological Sciences* 4 (2): 18–27.

Isayenkov, S.V. (2012). Physiological and molecular aspects of salt stress in plants. *Cytology and Genetics* 46 (5): 302–318.

Jager, C.E., Symons, G.M., Ross, J.J. et al. (2005). The brassinosteroid growth response in pea is not mediated by changes in gibberellin content. *Planta* 221: 141–148.

Jakhar, S., Rani, K., and Singh, S. (2022). Brassinosteroids: a wonder growth regulator to alleviate abiotic stresses in plants. In: *Plant Stress: Challenges and Management in the New Decade* (ed. S. Roy, P. Mathur, A.P. Chakraborty et al.), 97–110. Cham: Springer International Publishing.

Jouyban, Z. (2012). The effects of salt stress on plant growth. *Technical Journal of Engineering and Applied Sciences* 2 (1): 7–10.

Júnior, M.S.D.O., Andrade, J.R.D., Nascimento, R.D. et al. (2022). Brassinosteroid application increases tomato tolerance to salinity by changing the effects of stress on membrane integrity and gas exchange. *Acta Scientiarum. Agronomy* 44: e55647.

Kapoor, B., Kumar, P., Sharma, R. et al. (2021). Regulatory interactions in phytohormone stress signaling implying plants resistance and resilience mechanisms. *Journal of Plant Biochemistry and Biotechnology* 2021: 1–16.

Khan, E.A., Upadhyay, T.K., Prajapat, R.K. et al. (2022). Revisiting brassinosteroids signaling in plants: current advances and challenges. *Brassinosteroids in Plant Developmental Biology and Stress Tolerance* (ed. G.J. Ahammed, A. Sharma, and J. Yu), 15–41. Elsevier.

Kumari, S. and Thakur, A. (2019). The effects of water stress and brassinosteroid on apple varieties. *International Journal of Economic Plants* 6 (1): 1–6.

Kumudini, B.S. and Patil, S.V. (2019). Role of plant hormones in improving photosynthesis. *Photosynthesis, Productivity and Environmental Stress* (ed. P. Ahmad, M.A. Ahanger, M.N. Alyemeni et al.), 215–240. John Wiley & Sons, Inc.

Larré, C.F., Marini, P., Moraes, C.L. et al. (2014). Influence of the 24-epibrassinolide on tolerance to salt stress in rice seedlings. *Universidade Estadual de Londrina* 35: 67–75.

Liu, F., Yang, J., Mu, H. et al. (2023). Effects of brassinolide on growth, photosynthetic rate and antioxidant enzyme activity of ornamental gourd under salt stress. *Russian Journal of Plant Physiology* 70 (6): 137.

Mahmood, Q., Ahmad, R., Kwak, S.S. et al. (2010). Ascorbate and glutathione: protectors of plants in oxidative stress. *Ascorbate-Glutathione Pathway and Stress Tolerance in Plants* (ed. N. Anjum, M.T. Chan, and S. Umar), 209–229. Dordrecht: Springer.

Manghwar, H., Hussain, A., Ali, Q. et al. (2022). Brassinosteroids (BRs) role in plant development and coping with different stresses. *International Journal of Molecular Sciences* 23 (3): 1012.

Mir, B.A., Khan, T.A., and Fariduddin, Q. (2015). 24-epibrassinolide and spermidine modulate photosynthesis and antioxidant systems in *Vigna radiata* under salt and zinc stress. *International Journal of Advanced Research* 3 (5): 592–608.

Mishra, N., Jiang, C., Chen, L. et al. (2023). Achieving abiotic stress tolerance in plants through antioxidative defense mechanisms. *Frontiers in Plant Science* 14: 1110622.

Muhammad, H.M.D., Naz, S., Ahmad, R. et al. (2024). Melatonin triggers salinity tolerance in pansy (*Viola tricolor*) by regulation of defense system. *Journal of King Saud University-Science* 36: 103286.

Nawaz, K., Hussain, K., Majeed, A. et al. (2010). Fatality of salt stress to plants: morphological, physiological and biochemical aspects. *African Journal of Biotechnology* 9 (34): 5475–5480.

Naz, S., Muhammad, H.M.D., Ali, S. et al. (2024). Reprogramming of salt stress under the influence of melatonin. In: *Melatonin in Plants: A Pleiotropic Molecule for Abiotic Stresses and Pathogen Infection* (ed. R. Kumar, M.A. Altaf, M.K. Lal et al.), 45–58. Singapore: Springer Nature Singapore.

Naz, S., Mushtaq, A., Ali, S. et al. (2022). Foliar application of ascorbic acid enhances growth and yield of lettuce (*Lactuca sativa*) under saline conditions by improving antioxidant defence mechanism. *Functional Plant Biology* 51: FP22139.

Nithila, S. and Sivakumar, R. (2019). Salt stress in crop plants: physiological mechanisms and management. In: *Plant Abiotic Stresses Physiological Mechanisms Tools and Regulation* (ed. A. Hemantaranjan), 121–138. Scientific Publishers.

Nolan, T., Chen, J., and Yin, Y. (2017). Cross-talk of brassinosteroid signaling in controlling growth and stress responses. *Biochemical Journal* 474 (16): 2641–2661.

Noreen, S., Akhter, M.S., Noman, A. et al. (2021). Future perspectives of oxidative stress and antioxidant defence system in plants. *Developing Climate-Resilient Crops* (ed. S. Fahad, O. Sonmez, S. Saud et al.), 53–84. Boca Raton: CRC Press.

Oliveira, D.V.P., Lima, M.D.R., da Silva, B.R.S. et al. (2019). Brassinosteroids confer tolerance to salt stress in *Eucalyptus urophylla* plants enhancing homeostasis, antioxidant metabolism and leaf anatomy. *Journal of Plant Growth Regulation* 38: 557–573.

Ondrasek, G., Rathod, S., Manohara, K.K. et al. (2022). Salt stress in plants and mitigation approaches. *Plants* 11 (6): 717.

Özdemir, F., Bor, M., Demiral, T. et al. (2004) Effects of 24-epibrassinolide on seed germination, seedling growth, lipid peroxidation, proline content and antioxidative system of rice (*Oryza sativa* L.) under salinity stress. *Plant Growth Regulation* 42: 203–211.

Parihar, P., Singh, S., Singh, R. et al. (2015). Effect of salinity stress on plants and its tolerance strategies: a review. *Environmental Science and Pollution Research* 22: 4056–4075.

Park, H.J., Kim, W.Y., and Yun, D.J. (2016). A new insight of salt stress signaling in plant. *Molecules and Cells* 39 (6): 447–459.

Parmoon, G., Ebadi, A., Jahanbakhsh, S. et al. (2018). Effect of exogenous application of several plant growth regulators on photosynthetic pigments of fennel plants. *Notulae Scientia Biologicae* 10 (4): 508–515.

Quamruzzaman, M., Manik, S.N., Shabala, S. et al. (2021). Improving performance of salt-grown crops by exogenous application of plant growth regulators. *Biomolecules* 11 (6): 788.

Ramakrishna, B. and Rao, S.S.R. (2015). Foliar application of brassinosteroids alleviates adverse effects of zinc toxicity in radish (*Raphanus sativus* L.) plants. *Protoplasma* 252: 665–677.

Ranjan, S., Mangal, V., Prakash, A. et al. (2023). Melatonin-induced tolerance against salt stress. In: *Advancement of Melatonin Research in Plants* (ed. A. Roychoudhury), 186–201. Boca Raton: CRC Press.

Rasool, S., Hameed, A., Azooz, M.M. et al. (2013). Salt stress: causes, types and responses of plants. *Ecophysiology and Responses of Plants Under Salt Stress* (ed. P. Ahmad, M. Azooz, and M. Prasad), 1–24. New York, N.Y.: Springer.

Rattan, A., Kapoor, D., Kapoor, N. et al. (2014). Application of brassinosteroids reverses the inhibitory effect of salt stress on growth and photosynthetic activity of *Zea mays* plants. *International Journal of Theoretical and Applied Sciences* 6 (2): 13.

Rattan, A., Kapoor, D., Kapoor, N. et al. (2022). Involvement of brassinosteroids in plant response to salt stress. In: *Brassinosteroids in Plant Developmental Biology and Stress Tolerance* (ed. G.J. Ahammed, A. Sharma, and J. Yu), 237–253. Elsevier.

Sachdev, S., Ansari, S.A., Ansari, M.I. et al. (2021). Abiotic stress and reactive oxygen species: generation, signaling, and defense mechanisms. *Antioxidants* 10: 277.

Sairam, R.K. and Tyagi, A. (2004). Physiology and molecular biology of salinity stress tolerance in plants. *Current Science* 86(3): 407–421.

Samira, I.M.H., Bouthaina, D.M. et al. (2012). 24-Epibrassinolide ameliorates the adverse effect of salt stress (NaCl) on pepper (*Capsicum annuum* L.). *Journal of Stress Physiology & Biochemistry* 8 (1): 232–240.

Serrano, R. and Rodriguez-Navarro, A. (2001). Ion homeostasis during salt stress in plants. *Current Opinion in Cell Biology* 13 (4): 399–404.

Shahbaz, M. and Ashraf, M. (2007). Influence of exogenous application of brassinosteroid on growth and mineral nutrients of wheat (*Triticum aestivum* L.) under saline conditions. *Pakistan Journal of Botany* 39 (2): 513.

Shahbaz, M., Ashraf, M., and Athar, H.U.R. (2008). Does exogenous application of 24-epibrassinolide ameliorate salt induced growth inhibition in wheat (*Triticum aestivum* L.)? *Plant Growth Regulation* 55: 51–64.

Shahzad, B., Fahad, S., Tanveer, M. et al. (2019). Plant responses and tolerance to salt stress. In: *Approaches for Enhancing Abiotic Stress Tolerance in Plants* (ed. M. Hasanuzzaman, K. Nahar, M. Fujita et al.), 61–78. CRC Press.

Sharma, I., Ching, E., Saini, S. et al. (2013). Exogenous application of brassinosteroid offers tolerance to salinity by altering stress responses in rice variety Pusa Basmati-1. *Plant Physiology and Biochemistry* 69: 17–26.

Siddiqi, K.S. and Husen, A. (2021). Significance of brassinosteroids and their derivatives in the development and protection of plants under abiotic stress. *Biologia* 76 (10): 2837–2857.

Siddiqui, H., Hayat, S., and Bajguz, A. (2018). Regulation of photosynthesis by brassinosteroids in plants. *Acta Physiologiae Plantarum* 40: 1–15.

Siddiqui, M.N., Mostofa, M.G., Akter, M.M. et al. (2017). Impact of salt-induced toxicity on growth and yield-potential of local wheat cultivars: oxidative stress and ion toxicity are among the major determinants of salt-tolerant capacity. *Chemosphere* 187: 385–394.

Sirhindi, G., Bhardwaj, R., Kumar, M. et al. (2018). Physiological roles of brassinosteroids in conferring temperature and salt stress tolerance in plants. *Metabolic Adaptations in Plants During Abiotic Stress* (ed. A. Ramakrishna and S.S. Gill), 339–358. Boca Raton: CRC Press.

Sousa, V.Q., Messias, W.F.S., Pereira, Y.C. et al. (2021). Pretreatment with 24-epibrassinolide synergistically protects root structures and chloroplastic pigments and upregulates antioxidant enzymes and biomass in Na+-stressed tomato plants. *Journal of Plant Growth Regulation* 41: 1–17.

Sun, S., An, M., Han, L. et al. (2015). Foliar application of 24-epibrassinolide improved salt stress tolerance of perennial ryegrass. *HortScience* 50 (10): 1518–1523.

Tahjib-Ul-Arif, M., Siddiqui, M.N., Sohag, A.A.M. et al. (2018). Salicylic acid-mediated enhancement of photosynthesis attributes and antioxidant capacity contributes to yield improvement of maize plants under salt stress. *Journal of Plant Growth Regulation* 37: 1318–1330.

Tanveer, M., Shahzad, B., Sharma, A. (2018). 24-Epibrassinolide; an active brassinolide and its role in salt stress tolerance in plants: a review. *Plant Physiology and Biochemistry* 130: 69–79.

Vardhini, B.V. (2012). Mitigation of water stress and saline stress by brassinosteroids. In: *Brassinosteroids: Practical Applications in Agriculture, Forestry and Human Health* (ed. A.B. Pereira-Netto), 16–25. United States of America: Bentham Science Publishers.

Vighi, I.L., Benitez, L.C., Amaral, M.N. et al. (2017). Functional characterization of the antioxidant enzymes in rice plants exposed to salinity stress. *Biologia Plantarum* 61 (3): 540–550.

Wang, X., Chai, J., Liu, W. et al. (2023). Promotion of Ca^{2+} accumulation in roots by exogenous brassinosteroids as a key mechanism for their enhancement of plant salt tolerance: a meta-analysis and systematic review. *International Journal of Molecular Sciences* 24 (22): 16123.

Wani, A.S., Ahmad, A., Hayat, S. et al. (2019). Epibrassinolide and proline alleviate the photosynthetic and yield inhibition under salt stress by acting on antioxidant system in mustard. *Plant Physiology and Biochemistry* 135: 385–394.

Wei, L.J., Deng, X.G., Zhu, T. et al. (2015). Ethylene is involved in brassinosteroids induced alternative respiratory pathway in cucumber (*Cucumis sativus* L.) seedlings response to abiotic stress. *Frontiers in Plant Science* 6: 982.

Wu, X., Yao, X., Chen, J. et al. (2014). Brassinosteroids protect photosynthesis and antioxidant system of eggplant seedlings from high-temperature stress. *Acta Physiologiae Plantarum* 36 (2): 251–261.

Wu, W., Zhang, Q., Ervin, E.H. et al. (2017). Physiological mechanism of enhancing salt stress tolerance of perennial ryegrass by 24-epibrassinolide. *Frontiers in Plant Science* 8: 1017.

Xia, X.J., Zhou, Y.H., Shi, K. et al. (2015). Interplay between reactive oxygen species and hormones in the control of plant development and stress tolerance. *Journal of Experimental Botany* 66 (10): 2839–2856.

Xing, J., Wang, Y., Yao, Q. et al. (2022). Brassinosteroids modulate nitrogen physiological response and promote nitrogen uptake in maize (*Zea mays* L.). *The Crop Journal* 10 (1): 166–176.

Yang, Y. and Guo, Y. (2018). Unraveling salt stress signaling in plants. *Journal of Integrative Plant Biology* 60 (9): 796–804.

Yang, P., Wang, Y., Li, J. et al. (2019). Effects of brassinosteroids on photosynthetic performance and nitrogen metabolism in pepper seedlings under chilling stress. *Agronomy* 9 (12): 839.

Yokoi, S., Bressan, R.A., and Hasegawa, P.M. (2002). Salt stress tolerance of plants. *JIRCAS Working Report* 23 (1): 25–33.

Zahra, N., Al Hinai, M.S., Hafeez, M.B. et al. (2022). Regulation of photosynthesis under salt stress and associated tolerance mechanisms. *Plant Physiology and Biochemistry* 178: 55–69.

Zhang, Z., Chen, Z., Song, H. et al. (2023). From plant survival to thriving: exploring the miracle of brassinosteroids for boosting abiotic stress resilience in horticultural crops. *Frontiers in Plant Science* 14: 1218229.

Zia-ur-Rehman, M., Anayatullah, S., Irfan, E. et al. (2023). Nanoparticles assisted regulation of oxidative stress and antioxidant enzyme system in plants under salt stress: a review. *Chemosphere* 314: 137649.

10

Brassinosteroids Biology in Plants Under Pesticide Stress: Physiological and Molecular Aspects

Palak Bakshi[1,2], *Pooja Sharma*[2], *Sukhmeen Kaur Kohli*[3], *Dhriti Kapoor*[4], *Nahida Rehman Mir*[1], *Nitika Kapoor*[1], *and Renu Bhardwaj*[2,5,*]

[1]*Department of Botany, School of Biosciences, Baba Ghulam Shah Badshah University, Rajouri, Jammu and Kashmir, India*
[2]*Department of Botanical and Environmental Sciences, Guru Nanak Dev University, Amritsar, Punjab, India*
[3]*Department of Earth and Climate Sciences, Indian Institute of Science Education and Research (IISER) Tirupati, Tirupati, Andhra Pradesh, India*
[4]*School of Biological and Environmental Sciences, Shoolini University, Shimla, Himachal Pradesh, India*
[5]*University Centre for Research & Development, Chandigarh University, Mohali, Punjab, India*

*Corresponding author: renubhardwaj82@gmail.com

10.1 Introduction

The availability of land is continuously diminishing due to rapid industrialization and globalization, leading to a critical challenge for world leaders to address hunger needs (Raj et al. 2023). Farmers lose between 35% and 40% of their crops annually because of diseases, pests, and other biotic and abiotic variables (Saha et al. 2021). Traditionally, farmers would use a variety of techniques to reduce pest infestation, such as sprinkling salts or red pepper, or applying tobacco extract (Durán-Lara et al. 2020).

Furthermore, a significant rise in food production was made possible starting in the 1940s by the expanded use of synthetic crop protection agents. Additionally, the output of pesticides rose at a pace of roughly 11% per year globally, rising from 0.2 million tons in the 1950s to over 5 million tons by 2000 (Carvalho 2017). Despite the active increase in the annual global use of pesticides, only 1% effectively controls insect pests on target plants (Tudi et al. 2021). Pesticides negatively affect photosynthesis, photophosphorylation, protein synthesis, and membrane lipid integrity in many crops (Mehrian et al. 2023). Approximately 4.15 million tons of pesticides are employed annually worldwide, with people using them unaware of their negative effects. Excessive application of pesticides has been found to be associated with hormonal imbalances, and their accumulation within cells can contribute to the exacerbation of metabolic syndrome and obesity (Mehrian et al. 2023; Alavanja et al. 2013).

Plant hormones serve an important and effective weapon for plants to complete their life cycle (Kumari et al. 2023). The plant steroid hormone, brassinosteroids (BRs), regulates abiotic and biotic stress responses, as well as critical processes, including cell elongation, cell division,

Brassinosteroids and Abiotic Stress: Recent Developments on Physiological and Molecular Aspects,
First Edition. Edited by Anket Sharma, et al.
Published 2026 by John Wiley & Sons Ltd.

photomorphogenesis, xylem differentiation, and reproduction (Nolan et al. 2020). Significant progress has been made in understanding the signal transduction pathways from cell surface receptors to the nucleus. This has led to the significant alteration of hundreds of genes in response to BRs, resulting in a wide range of biological responses (Clouse 2011; Nolan et al. 2017).

BRs lower the detrimental effects of abiotic stress by reducing the negative effects of ROS. It was reported that BRs could enhance the capability of plants to tolerate pesticide stress and reduce pesticide residues in plants (Dehghanian et al. 2021). BRs enhance the growth, photosynthesis, and yield (Nolan et al. 2020), and upregulate the expression of carbon stabilization and RuBisCO's activity to bolster CO_2 perception in plants (Sharma et al. 2019). The exogenous application of BRs resulted in enhancing the metabolism of several pesticides by improving the action of enzymes like glutathione reductase (GR), glutathione S-transferase (GST), and peroxidase (POD), hence reducing the pesticide residue in the plant (Ahammed et al. 2013, 2017). Histone alterations are among the epigenetic processes involved in BR-regulated gene regulation. His-3 Lys-27 di- and trimethylation (H3K27me2/H3K27me3) is one such mark linked to the suppression of gene expression. BES1 interacts with RELATIVE OF EARLY FLOWER6 (REF6) and EARLY FLOWER6, which work together to eliminate repressive H3K27me2/H3K27me3 marks and act as positive regulators in the BR pathway (Yu et al. 2008).

10.1.1 Global Status of Pesticides

The term "pesticide" first came into use in 1908 to refer to chemical substances that were used to eliminate pests that were harming the health of different crops, such as nematicides, fungicides, weedicides, and insecticides (Raj and Kumar 2022; Raj et al. 2021; Rakshit et al. 2015). However, in the 1940s, synthetic pesticides like DDT, chlordane, aldrin, and dieldrin were developed because of the limitations of these traditional techniques, which included a lack of selectivity, a narrow action spectrum, and a high application rate (Zhang 2018; Rakshit et al. 2015). The global loss of crops like wheat, rice, maize, potato, and soybean due to pests and pathogens ranges between 10–28%, 25–41%, 20–41%, 8–21%, and 11–32%, respectively (Savary et al. 2019). Thus, to protect the crops and to increase their productivity and profitability, the use of pesticides has increased by about 46% between 1996 and 2006 (Nayak and Solanki 2021; WHO 2019). Annually, approximately 2 million tons of pesticides are used worldwide, with China at the top of the list, followed by the United States and Argentina (Nayak and Solanki 2021) (Figure 10.1).

The pattern of pesticide usage in India is insecticides–herbicides–fungicides+bactericides–other-pesticides while globally the pattern is herbicides–fungicides+bactericides–insecticides–other-pesticides has been documented. India currently holds the position of the fourth largest producer of pesticides, having an expected market of Rs. 316 billion by 2024, which was Rs. 214 billion in 2019 (TAAS et al. 2020). In India, the most widely used pesticide is chlorpyrifos, commonly used fungicide is zinc, and zinc phosphate is the widely used rodenticide due to their increased usage of 1431 MT in 2019–2020 3878 MT in 2019–2020 and 200 MT between 2014 and 2020, respectively, by the Government of India (2020). There are advantages and possible hazards associated with using pesticides. There is a drawback to these advantages in the form of possible harmful effects from the germination of plants (Parween et al. 2016) like impaired root length, decreased weight gain, high injury rate, increased MDA level, and chromosomal aberration (Acar and Singh 2023). Due to the growing environmental concern and adverse effects on humans and plants, a shift from chemical pesticides to bio-pesticides is needed.

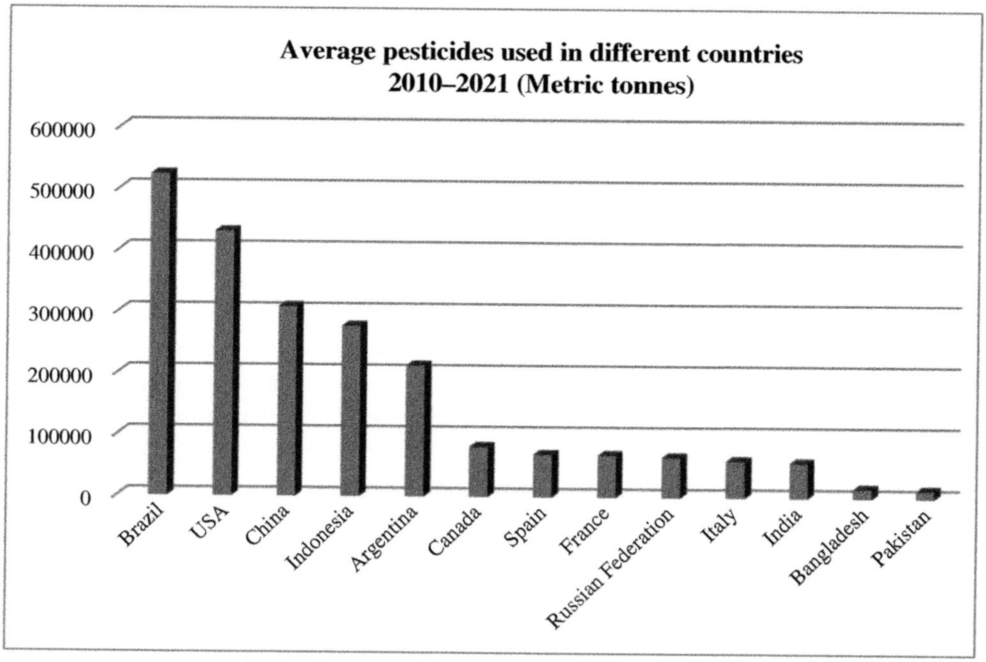

Figure 10.1 Consumption of different pesticides in different countries. *Source*: Adapted from Food and Agriculture Data, http://www.fao.org/faostat/ en/#data/RP/visualize/, last accessed on April 26, 2025.

To help and shape future strategies that favor agroecosystem compatibility under pesticide-stressed conditions, the present review intends to highlight the significance of BRs' implications and its genetic approachability.

10.2 Brassinosteroid (Nature, Signaling, and Its Regulation Mechanism)

10.2.1 Nature

BRs, a plant steroidal hormone, have a critical participation in modulating growth and development in plants, further regulating a wide array of biological phenomena in plants, including cell division, expansion, and elongation, photomorphogenesis, differentiation of xylem tissue, reproduction, and response to abiotic and biotic stress cues (Nolan et al. 2020). BRs are categorized into C27, C28, and C29 groups based on diverse alkyl-substitution patterns within their side chains. Generally, the presence of an active BR requires a transfused A/B ring system, characterized by two hydroxyl groups in ring A and a 6-ketone or 7-oxa-6-ketone system in ring B. The chemical composition of brassinolide, specifically (22R,23R,24S)-2α,3α,22,23-tetrahydroxy-24-methyl-B-homo-7-oxa-5α-cholestan-6-one, adheres to these specific criteria (Tang et al. 2016). Some BRs lack substitutes at C-23 or C-24, and others have no substitute at both the positions. BRs can also conjugate with glucose and fatty acids (Wendeborn et al. 2017; Zullo and Bajguz 2019). Approximately 40–50% of BRs belong to the C-28 category. The diverse cyclic group and side chain patterns predominantly result in the conversion of BRs into active analogs like 24-epibrassinolide (EBL) and 28-homobrassinolide (HBL) (Kang and Guo 2011). BRs including analogs like typhasterol, brassinolides, and castasterone are broadly distributed in specialized plant structures like pollens, flowers, seeds, shoots, roots, stems, and leaves. Notably, the highest BR concentration, i.e. 100 ng per fresh weight, is observed in pollen grains and seeds, contrasting with 0.01 and 0.1 ng per fresh weight in shoots and leaves, respectively (Kour et al. 2021).

Over 69 BRs and its derivatives, including conjugates, metabolites, and analogs, have been identified from various plants such as *Castanea crenata* and *Catharanthus roseus* (Liu et al. 2017). They are distributed across plant families and decreasing from angiosperm to gymnosperm, pteridophytes, bryophytes, and algae (Bajguz and Tretyn 2003). True BRs are formed from 22α and 23α-dihydroxylation, while others are considered BR precursors (Zullo and Kohout 2004). Natural BRs include those with 23-epoxy groups, 23-glycoside groups, 23-ester groups, 3-oxygenated-5α-cholestane-23α diols, and alkyl derivatives (Zullo and Bajguz 2019). Biological activity is influenced by the side chains of rings, with 22α and 23α-dihydroxy BRs showing activity as 28-EBL in the rice laminal inclination assay (Watanabe et al. 2001). Structural variability in ring A spans Δ2, 3-unsaturated to conjugated BRs, and functionality increases from 3β-hydroxy to 3α-hydroxy rings (Liu et al. 2017). This correlates with enhanced oxidation states in the biosynthetic pathway from 6-deoxo and 6α-hydroxy forms toward 6-oxo and 7-oxa-lactone states (Vriet et al. 2013). The decline in functionality from 2α, 3α toward 2β, 3β is attributed to the importance of the α oriented hydroxyl group at C-2 (Zou et al. 2020). Moreover, BRs are classified based on B-rings, including 6-oxo-7-oxalactonic BRs, 6-oxo or 6-keto BRs, 6α-hydroxy BRs, and 6-deoxo BRs (Zullo and Bajguz 2019).

10.2.2 Signaling and Its Regulation Mechanism

BRs-linked metabolic pathways contribute to successive signaling networks. Thus, a thorough understanding of BRs' equilibrium in plants is crucial for comprehensive insight into BR signaling cascades (Chung and Choe 2013). Studies conducted over decades have unraveled the intricate BR-associated pathways, exposing a complex signaling network pivotal for plant growth and development (Anwar et al. 2018). The BR-triggered signaling cascade comprises three main phases, including (i) BR reception and activation of BRI1 receptor kinases, (ii) deactivation of BIN2 inhibitors, kinases, and phosphatases, and (iii) role of transcription factors (TFs), including BES1 and BZR1 (Clouse 2011).

10.2.2.1 BR Reception and BRI1 Receptor Kinases Activation

BRs initiate signaling by binding to brassinosteroid-insensitive 1 (BRI1) receptor kinase on the cell surfaces, characterized by a leucine-enriched repeat receptor kinase structure. BRI1 associates with the co-receptor, somatic embryogenesis receptor kinase (SERK), forming an active complex that triggers downstream signals involving various kinases and phosphatases. This leads to the activation of transcription factors and modulation of gene expression (Belkhadir et al. 2012). Exogenous BRs, upon binding to BRI1, induce an association with BRI1-associated receptor 1 (BAK1), causing dissociation of BRI1 kinase inhibitor 1 (BKI1), a membrane negative regulator of BR signals, which interacts with BRI1 to prevent co-receptor interference (Jiang et al. 2015). BRI1 functions as a versatile kinase, phosphorylating Ser, Thr, and Tyr, with Tyr phosphorylation being crucial for specific BR signaling aspects (Oh et al. 2009). As a ligand-independent plasma membrane homodimer, BRI1 initiates and stabilizes

BR binding and signaling (Albrecht et al. 2012). BAK1 (SERK3) phosphorylates BRI1 and BAK1, while SERK4 (BKK1) interacts with BRI1 in a BR-dependent manner (Wang et al. 2009; He et al. 2007). FLS2 enhances BAK1's plant defense function, and SERK forms a significant heterodimer with BRI1, augmenting BR signaling during embryogenesis. Wang et al. (2009) investigated the phosphorylation and oligomerization of BRI1 and BAK1 in response to exogenous BR. They found that active BRI1, not BAK1 kinase, is responsible for BR-dependent association. Expression of BAK1-GFP in a bri-1-1 null mutant background resulted in significantly decreased phosphorylation of BAK1-GFP. In vitro studies revealed BAK1 positively stimulates BRI1 activity, and mutual transphosphorylation occurs at specific residues. LC-MS/MS assays, biochemical analysis, and functional characterization led to a new sequential transphosphorylation model for BRI1/BAK1 interaction, suggesting similarities to mammalian TGF-b and RTKs receptor kinases. Autophosphorylation of Tyr residues in BRI1, similar to BAK1, regulates the in vivo expression of a subset of BAK1 (Oh et al. 2010).

10.2.2.2 Deactivation of BIN2 Inhibitors, Kinases, and Phosphatases

Elevated BR signaling deactivates BIN2, similar to mammalian GSK3, crucial in Arabidopsis BR signal transduction (Chung and Choe 2013; Rozhon et al. 2019), leading to acclimation of transcription factors BES1/BZR2 and BZR1 (Yin et al. 2005; He et al. 2005). Recent Arabidopsis studies unveil the association between BRI1 and BAK1 activation (Kim and Wang 2010). Wang and Chory (2006) identify BKI1 as a negative BR signaling modulator, binding to and deactivating BRI1 in the absence of BAK1, hindering subsequent BAK1 binding. Exogenous BR treatment induces the detachment of BKI1 from the outer cell surface, inhibiting the BR signaling networks. In vivo and in vitro demonstrations revealed BRI1 and BKI1 associations, leading to BKI1-CD. Recent findings highlight BKI1's interaction with BRI1 through a conserved C-terminal 20-residue segment, suppressing BRI1-BAK1 association with a peptide around this site (Jaillais et al. 2011). A Lys-Arg-enriched domain in BKI1 directs it to the plasma membrane, where BR-dependent phosphorylation at Tyr-211 releases BKI1. Overexpression of a BKI1-Y21F mutant in Arabidopsis results in dwarfed plants, indicating that BR phosphorylation at Tyr-211 enables BRI1 to dissociate from the membrane. Furthermore, facilitating its association with BAK1 and initiation of BR signaling (Lemmon and Schlessinger 2010; Lim and Pawson 2010). Proteomic analysis identified BSKs as direct targets of BRI1, augmenting BR signaling. BSKs, including BSK1 and BSK3, independently engage with BRI1. BRI1 phosphorylates BSK1 at Ser230, facilitating its release from the receptor complex. Phosphorylated BSK1 then associates with BSU1, initiating BSU1-mediated dephosphorylation of phosphor-Tyr200 in BIN2, suppressing the BR signaling cascade (Tang et al. 2008; Kim et al. 2009).

10.2.2.3 Role of TFs Like BES1 and BZR1

Brassinozole-resistant 1 (BZR1) and BRI1-EMS suppressor-1 (BES1/BZR2) are primary transcription factors activated upon BR binding to BRI1, regulating genes in diverse physiological processes (Peres et al. 2019). The GSK3 kinase BIN2 negatively regulates BZR1 and BES1 in the absence of BR signaling, phosphorylating and sequestering them in the cytoplasm. In the presence of BR, BRI1 associates with BAK1, initiating a phosphorylation cascade involving BSKs and CDG1. This further stimulates BSU1 phosphatases, leading to BIN2 dephosphorylation, thus positively influencing BZR1 and BES1 by preventing their degradation. Phosphatases 2A, in the absence of BIN2, dephosphorylate BZR1 and BES1 in the cytoplasm allowing their translocation to the nucleus for modulation of BR-mediated gene expression (Chung and Choe 2013). In the intricate landscape of the BR-signaling pathway, the transcription factors BES1/BZR1 exhibit a versatile nature, functioning both as inducer and repressor. BES1, through its binding to CANNTG sequences, specifically the E-box, enhances the expression of genes. Conversely, BZR1 employs its interaction with the CGTGT/CG sequence, denoted as BRRE, to suppress gene expression (He et al. 2005). This dynamic duo plays a pivotal role in orchestrating the intricate gene regulation mechanisms triggered by BR. One of the down players in this regulatory symphony is MYB30, a target of BES1; MYB30 not only serves as an effector but also reinforces BR-induced signals by forming a functional interaction with BES1. This collaborative effort further amplifies the intricate network of gene expression controlled by BR signaling. MYBL2, another regulator in BR signaling, undergoes phosphorylation and stabilization, facilitated by BIN2 kinase activity.

The interplay between these transcription factors and their downstream targets showcases the nuanced and finely tuned nature of the BR-signaling network contributing toward plant growth and development (Ye et al. 2012). BR repression increases BRRE expression while BR-induced genes upregulate E-boxes (Yu et al. 2011). The E-box interaction with BES1, PIL6, GLK1, and GLK2 contributes to transcriptional networks triggering BR-induced gene expression. BZR1 forms a heterodimer with PIF4, recognizing the promoter elements CACGTG (G-box) (Oh et al. 2012). The interaction between BZR1 and PIFs is indispensable for normal hypocotyl growth, as evidenced by studies on mutants of PIF4 and its homologs (pif1, pif3, pif4, and pif5), revealing diminished plant growth in darkness and hindrance of BR-induced growth. There are approximately 2000 shared target genes between BZR1 and PIF4 in plants, encompassing diverse PREs associated with cell elongation (Guo et al. 2013). Within the GA-signaling pathway, the absence of GA results in the stable nuclear localization of DELLA protein, impeding genome binding with PIFs and BZR. Activation of the GA signal induces DELLA protein degradation, activating PIF and BZR to target specific genes (Liu et al. 2020). Pioneering investigations have unveiled BES1's role in the chloroplast inhibition through the involvement of GLK1 and GLK2, inducing structural changes in chloroplast such as enlarged plastoglobules (Yu et al. 2011).

Advancement in understanding the BR-signaling pathway and biosynthesis has been substantial, yet the activation mechanism of BRI1 kinase through BR-induced BRI1-BAK1 heterodimerization remains unclear. Resolving this requires detailed exploration of the full-length complex structure. Ongoing identification of new components in BRs signaling and their integration with other networks is vital for comprehension.

10.3 Impact of Pesticides on the Biology of Plants and Its Genetic Composition

Application of pesticide causes phytotoxicity, which is evident in the form of chlorosis, necrosis, twisting of leaves, stunting, and burns (Sharma et al. 2018). It has been speculated that pesticide application adversely affects plant growth and development (Shahzad et al. 2018). It leads to the generation of reactive oxygen species, causing oxidative stress to plants (Sharma et al. 2018). Reactive oxygen species react with proteins, lipids, nucleic acids, and degrade chlorophyll pigments, which inhibits photosynthetic efficiency of plants (Sharma et al. 2019; Sharma 2015). The effect of pesticides on different plant species has been mentioned in Table 10.1.

Table 10.1 Impact of pesticides on biology and genetic composition of different plants.

S. no.	Pesticide used	Concentration applied	Plant	Effect of pesticide on biology and its genetic composition	References
1.	Thiamethoxam	0.1, 0.5, and 0.9 μm a.i. L^{-1}	*Solanum tuberosum*	Enhanced expression of superoxide dismutase, peroxidase, catalase, acetylcholinesterase, carboxylesterase, and glutathione-*S*-transferase was observed.	Khurshid et al. (2024)
2.	Imidacloprid	50 mg L^{-1} 100 mg L^{-1}	Strawberry	Lowered fruit quality was observed in the mature strawberry fruits.	Yue et al. (2024)
3.	Chlorpyrifos	1000 mg L^{-1}	*Oryza sativa*	Reduction in root length, plant height, biomass, and vigour index was noticed.	Saengsanga and Phakratok (2023)
4.	Imidacloprid, Diazinon, Mancozeb	200 μg mL^{-1}	*Solanum lycopersicum* L.	Decrease in root length, shoot length, and biomass was observed.	Hatamleh et al. (2022)
5.	Imidacloprid	5.0×10^{-3} to 5.0×10^{-2} mol L^{-1}	*Phaseolus vulgaris*	Germination rate, seed vigor index, chlorophyll *a*, *b*, and total chlorophyll content were reduced.	Ajermoun et al. (2022)
6.	Chlorpyrifos	500 mg L^{-1}	*Brassica juncea*	Reduction in germination potential and morphological attributes were observed. Increased levels of superoxide anion (O_2^-), hydrogen peroxide (H_2O_2), MDA, and electrolyte leakage (EL) were noticed. Enhanced the expression of defense enzymes.	Bakshi et al. (2021)
7.	Endosulfan	1 mg kg^{-1}	*Phaseolus leptostachyus*	Root length, leaf fresh weight, and germination percentage were observed.	Landa-Faz et al. (2021)

(Continued)

Table 10.1 (Continued)

S. no.	Pesticide used	Concentration applied	Plant	Effect of pesticide on biology and its genetic composition	References
8.	Paraquat and Glyphosate	5.4 g L^{-1}	*Vicia faba* *Phaseolus vulgaris* *Sorghum bicolor*	Germination rate and photosynthetic pigments were reduced.	Maldani et al. (2021)
9.	Thiamethoxam	–	*Zea mays* hybrid zp341	Decrease in photosynthetic characteristics and pigment content was altered.	Todorenko et al. (2021)
10.	Glyphosate	30 mg kg^{-1}	*Solanum lycopersicum*	Reduction in germination potential, hypocotyl, and radicle length in seedlings was noticed. Root length, shoot length, and fresh and dry weight of roots were also reduced	Khan et al. (2020)
11.	Chlorpyrifos and Pyriproxyfen	–	*Triticum aestivum*	Roots number was decreased.	Munir et al. (2019)
12.	Thiram	6.6 mM	*Solanum lycoperscium* L.	Reduction in chl *a*, *b*, carotenoids, and an increase in H_2O_2 content and MDA content were observed	Yüzbaşıoğlu and Dalyan (2019)
13.	Nicosulfuron	100 μM	*Zea mays*	Increase in H_2O_2 and MDA. Elevated activities of GST and GR.	Liu et al. (2019)
14.	Chlorothalonil	1500 g hm^{-2}	*Vitis vinifera*	MDA content was increased. Activities of SOD, APOX, CAT, POD, and GR were increased.	Wang et al. (2017)
15.	Imazapic	400–800 μg (a.i.) kg^{-1}	*Zea mays*	Root length, shoot length, dry biomass mass, Chl *a*, *b*, and carotenoid content were reduced.	Su et al. (2017)
16.	Atrazine	10 mg kg^{-1}	*Pennisetum americanum*	Augmented SOD, POD, CAT, and APOX have been recorded.	Jiang et al. (2016)

10.4 Modulation of Plant Biology in the Presence of BR Under Pesticide Stress

Chemical pesticides are commonly used to control pests in plants, but improper use can cause ecological imbalances (Barański et al. 2014). The photosynthetic system and plant growth are inhibited by the pesticides used in plants, which can systemically alter the physicochemical properties by initiating necrosis (Xia et al. 2009). The adverse effects of pesticides are significantly alleviated by BRs application in plants through various detoxification mechanisms (Figure 10.2).

10.4.1 Morphological Attributes

Excessive use of pesticides can have negative effects on the growth of plants. In particular, it can cause the roots to become damaged or "flubbed," which can lead to uncertain growth patterns (Rajmohan et al. 2020). Sharma et al. (2016b) studied the effects of 24-EBL on *B. juncea* seedlings under imidacloprid (IMI) toxicity and noticed that 24-EBL application improved the growth of the seedlings, BR's potential to regulate cellulose biosynthesis, cell division, and cell elongation might be the reason for the EBL-mediated seedling growth (Gonzalez-Garcia et al. 2012; Hacham et al. 2011; Sharma et al. 2016b). Exogenous application of EBL and BR mimetics, i.e. bikinin and brazide, significantly mitigated the adverse effects of nicosulfuron (NSF) by increasing the growth parameters such as plant height, root length, and biomass of maize. This increase in growth might be due to an increase in net photosynthetic rate, accelerated degradation of NSF residues, and upregulated expression of detoxification-related genes (Liu et al. 2019).

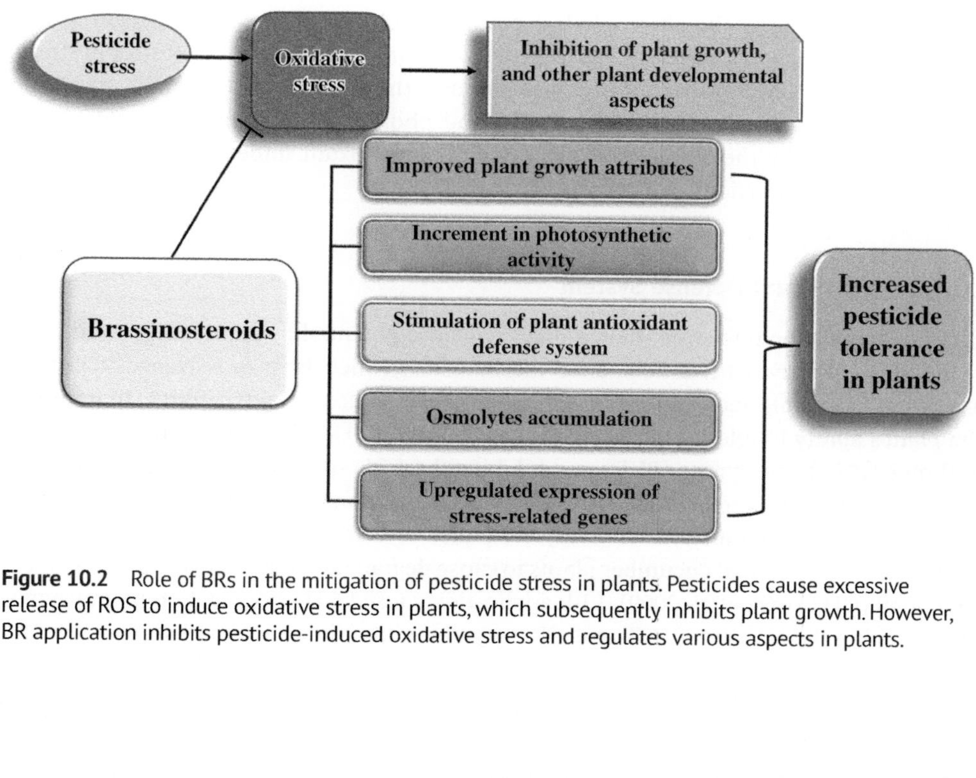

Figure 10.2 Role of BRs in the mitigation of pesticide stress in plants. Pesticides cause excessive release of ROS to induce oxidative stress in plants, which subsequently inhibits plant growth. However, BR application inhibits pesticide-induced oxidative stress and regulates various aspects in plants.

10.4.2 Photosynthetic Activity

It's amazing to see how BRs play a crucial role in regulating various components of photosynthetic machinery in plants. By influencing photochemistry, stomatal conductance, and enzymes of the Calvin cycle, BRs can enhance photosynthetic carbon fixation and promote plant growth and development. It's encouraging to note that BR-mediated regulation operates constitutively, ensuring the consistent promotion of net photosynthetic rate. Such mechanisms can help plants thrive and contribute to the overall development of the plant (Siddiqui et al. 2019). Exogenous application of 24-EBR improved *B. juncea* tolerance to IMI toxicity by increasing the content of photosynthetic pigments including carotenoid, xanthophyll, and anthocyanin (Sharma et al. 2016b). Bakshi et al. (2023) unraveled the ameliorative potential of 24-EBL in combination with plant growth-promoting rhizobacteria (PGPR) in *B. juncea* under chlorpyrifos (CP) stress. The total chlorophyll, carotenoids, xanthophyll, anthocyanin, and flavonoid contents were observed to be increased in EBL and *Burkholderia gladioli*-treated *B. juncea* seedlings under CP stress. Moreover, chlorophyll, carotenoids, and total phenols showed maximum fluorescence as shown by autofluorescence imaging. Pre-sowing seed treatment with EBR increased the level of photosynthetic pigments, i.e. chlorophyll, carotenoids, anthocyanins, and xanthophylls, and gas exchange attributes such as photosynthetic rate, stomatal conductance, inter-cellular CO_2, and transpiration rate of *B. juncea* plants which were grown in IMI-amended soil (Sharma et al. 2016a). This increase in chlorophyll content could be due to the fact that chlorophyll breakdown is reduced and transcription and translation processes during the chlorophyll biosynthesis are upregulated by BRs (Honnerová et al. 2010). Polychlorinated biphenyl (PCB)-induced photo-inhibition was significantly alleviated by the application of EBL through increasing the chlorophyll content, photosynthetic capacity, Fv/Fm, ΦPSII, and q_P (Ahammed et al. 2013). The application of brassinolide increased the chlorophyll *a* and chlorophyll *b* contents by activating the transcription of photosynthetic genes in *Arabidopsis thaliana* under pyraclostrobin stress (An et al. 2023). Chlorophyll content, net photosynthetic rate, PS II effective quantum yield, electron transport rate of PS I and II, and photochemical quantum yield of PSI were augmented by the application of brassinolide in foxtail millet (*Setaria italica* L.) under sigma Broad toxicity (Yuan et al. 2017).

10.4.3 Antioxidant Defense System

BRs can actually help to strengthen a plant's defense potential by stimulating both enzymatic and nonenzymatic antioxidative mechanisms, which help to scavenge ROS that are generated. Additionally, BRs can activate pesticidal detoxifying transcripts to improve the plant's ability to tolerate stress (Bakshi et al. 2022). Oxidative stress caused by phenanthrene (PHE) and pyrene (PYR) was inhibited by EBL application by augmenting the activity of guaiacol peroxidase, CAT, APX, and GR in tomato, and the xenobiotic detoxification system (Ahammed et al. 2012). Activities of GST, POD, and GR enzymes were increased in EBL-treated cucumber plants to cause degradation and detoxification of chlorothalonil (CHT) (Xia et al. 2009). EBL application caused CHT degradation in grapevine by improving the activity of GST and GR enzymes, glutathione, accumulation of soluble proteins, and free proline content (Wang et al. 2017). Activities of antioxidative enzymes, i.e. POX, CAT, GR, and GST enzymes, were increased in EBL-treated wheat plants under glyphosate stress (Shopova et al. 2021). Exogenous application of EBL escalated the activity of GST enzyme, accumulation of total glutathione and increased the ratio of GSJ/GSSG under CHT stress in tomato cultivars (Hou et al. 2018).

Yin et al. (2016) showed that BR-induced GSH biosynthesis and enhancement of CHT metabolism were modulated by nitric oxide (NO). Castasterone application alleviated the IMI-induced oxidative stress by stimulating the activities of antioxidative enzymes such as SOD, CAT, GR, POD, DHAR, and GST in mustard seedlings (Sharma et al. 2019). EBL application to maize plants maintained membrane integrity by reducing the lipid peroxidation and also upsurged the activities of various antioxidative enzymes such as SOD, CAT, POD, APX, GR, and GST enzymes under diazinon (DZ) toxicity (Mehrian et al. 2023). BR-mediated increase in the activity of antioxidant enzymes could be attributed to the potential up-regulation in the gene expression of these enzymes, leading to changes in their synthesis. Such changes could be beneficial for maintaining cellular health and combating oxidative stress (Cui et al. 2011). The fact that many antioxidant enzymes rely on calcium ions for their activity highlights the importance of calcium signaling in maintaining plant health. Furthermore, the activation of calcium/calmodulin-dependent protein kinases by BRs presents an exciting opportunity to enhance the activity of antioxidant pathways (Kretynin et al. 2021). Contents of H_2O_2 and $O_2^{\cdot-}$ were decreased, whereas activities of SOD, CAT, GR, GST, POD, and the content of non-enzymatic antioxidants i.e. GSH in EBR-treated *B. juncea* seedlings under IMI toxicity (Sharma et al. 2017). The ameliorative potential of BRs in mediating pesticide stress has been mentioned in Table 10.2.

10.4.4 Molecular Level

It is interesting to note that BR has been found to upregulate many genes related to metabolism, hormonal biosynthesis, cytoskeleton proteins, molecular chaperones, and redox metabolism. These genes are closely associated with various plant developmental processes (Hafeez et al. 2021). Expression of genes encoding key enzymes such as phenylalanine ammonia lyase, chalcone synthase, chlorophyllase, succinate dehydrogenase, succinyl Co-A ligase, citrate synthase, fumarate hydratase, malate synthase, and phytoene synthase was improved by 24-EBL application in *B. juncea* seedlings under IMI stress (Sharma et al. 2016b). Exogenous application of 28-HBL increased the gene expression of *Cu/Zn-SOD, Fe-SOD, Mn-SOD, APX, CAT*, and *GR* in Pusa Basmati-1 under CP and IMI stress (Sharma et al. 2015). Expression of genes encoding cytochrome P450, *GSH1-2, GR, GST1-7, ABC1-4* (ABC transporters), and *RBOH1* (*Respiratory burst oxidase homolog1*, encoding NADPH oxidase) was increased by the EBR application in tomato leaves under CHT toxicity (Zhou et al. 2015). Exogenous application of EBR improved the IMI tolerance in *B. juncea* seedlings by increasing the gene expressions of antioxidative genes (*SOD, CAT, GR, POD*),NADH-ubiquinone oxidoreductase, carboxylesterase, glutathione synthase, glutathione transporter-1, and cytochrome P450 monooxygenase (Sharma et al. 2017). The three-phased pesticide detoxification system is enhanced by BRs, which includes activation of the absorbed pesticides by P450 monooxygenase, peroxidase, and carboxylesterase

Table 10.2 Impact of brassinosteroids on plants under pesticide stress.

S. no.	BRs used	Concentration of BR applied	Pesticide used	Concentration	Plant used	Effect of BRs	References
1.	24-epibrassinolide (EBL)	0.01 μM	Diazinon (DZ), Imidacloprid (IMI), and Chlorpyrifos (CPS)	DZ (60% EC), IMI (35% EC), and CPS (40.8% EC)	*Zea mays*	Decreases DZ, IMI, and CPS residues, the toxic effects of these pesticides are reduced and transformed into less toxic forms	Mehrian et al. (2024)
2.	24-EBL	10 μM, 0.1 μM 0.01 μM	Diazinon	60% EC	*Z. mays*	Reduces the harmful effect of DZ on photosynthesis, improves CO_2 assimilation, and increases stomatal conductivity	Mehrian et al. (2023)
3.	Brassinolide	0.015%	Glufosinate-ammonium	88.8%	*Brassica napus*	Improved the resistance of seeds to glufosinate, also the CAT, SOD activity as well as the MDA content was significantly increased	Gan et al. (2022)
4.	Brassinolide	0.1–1.0 mg L^{-1}	Mesosulfuron-methyl + Iodosulfuron-methyl-sodium	2.81+0.56 g a.i./ha	*Setaria italica*	Increased leaf area, fresh mass, net photosynthesis rate, stomatal conductance, plant height, total chlorophyll and carotenoid content thus enhance the growth and photosynthesis resistance	Ning et al. (2022)

5.	24-EBL	1 μM	Glyphosate	0.5 mM	*Triticum aestivum* L.	Priming of wheat plants with EBL can partially mitigate the negative effect of oxidative stress and improve shoot growth	Shopova et al. (2021)
6.	24-EBL	100 nM	Chloropyrifos	500 mg L^{-1}	*Brassica juncea* L. (var. RLC-3)	24-EBL in combination with PGPRs reduces the content of ROS, MDA, and enhances the level of nitric oxide	Bakshi et al. (2021)
7.	Castasterone	100 nM	Imidacloprid	0, 150, 200 and 250 mg L^{-1}	*B. juncea* L. (var. RLC-1)	Castasterone helped in the recovery of seedlings' growth, reduced the production of ROS, and activated the expression of insecticide detoxification genes under imidacloprid stress	Sharma et al. (2019)
8.	24-EBL	0.2 μM	Chlorothalionil	11.2 mM	*Solanum lycopersicum* L.	The interaction between GRX and transcription factors is mediated by BRs induce pesticide metabolism, which helps in the expression of genes for pesticide detoxification	Hou et al. (2019)

(Continued)

Table 10.2 (Continued)

S. no.	BRs used	Concentration of BR applied	Pesticide used	Concentration	Plant used	Effect of BRs	References
9.	24-EBL	0.2 μM	Chlorothalonil	11.2 mM	*S. lycopersicum* L.	Metabolism of pesticide by brassinosteroid is mediated by the interaction between glutaredoxin (S25) and transcription factors, which prompts the upregulation of genes responsible for pesticide detoxification	Hou et al. (2019)
10.	24-EBL	0.2 μM	Chlorothalonil	11.2 mM	*S. lycopersicum* L.	Negative effects of chlorothalonil are significantly reduced due to expression of respiratory burst oxidase homolog1	Hou et al. (2018)
11.	24-EBL	0.2 μM	Chlorothalonil	11.2 mM	*S. lycopersicum* L.	GRXS16, a specific type of glutaredoxin, plays a pivotal role in the process of pesticide metabolism induced by brassinosteroid	Hou et al. (2018)

12.	24-EBL	100 nM	Imidacloprid	0, 150, 200 and 250 mg L^{-1}	*B. juncea* L.	Decreases the content of ROS, enhances the activity of antioxidant enzymes, decreases the expression of the RBO gene while increases the expression of SOD, CAT, POD, and GST genes	Sharma et al. (2017)
13.	24-EBL	100 nM	Chloropyrifos	10 μM	*Cucumis sativus* L.	Alleviates the inhibitory effects of chlorpyrifos on both shoot and root growth	Ahammed et al. (2017)
14.	24-EBL	0.10, 0.21, and 0.42 μM	Chlorothalonil	600 times diluted	*Vitis vinifera* L.	Accelerates chlorothalonil-degrading genes and glutathione reductase transcription	Wang et al. (2017)
15.	24-EBL	0.1, 1, and 100 nM	Imidacloprid	250, 300, and 350 mg kg^{-1}	*B. juncea* L.	Imidacloprid residue within tissue reduced, increased the activities of antioxidant enzymes	Sharma et al. (2016a)
16.	24-EBL	100 nM	Imidachloprid	150, 200, 250 mg L^{-1}	*B. juncea* L.	EBR-treated seeds increase the plant's dry weight and length of seedlings	Sharma et al. (2016b)

enzymes in the first phase. In the second phase, conjugates of activated components with glutathione and glucose by GST and UGT, respectively, are formed. In the third phase, compounds produced due to pesticide breakdown are stored in vacuoles or in the apoplast (Mehrian et al. 2023).

10.5 Conclusion and Future Prospects

Plant growth and agricultural productivity can be significantly reduced due to a variety of biotic and abiotic stressors. Plants have therefore evolved several defense mechanisms against these stressors, such as the synthesis of many phytohormones. The hormones known as BRs control a wide range of physiological and developmental processes. BRs are essential for several important antioxidant processes in plants, including the control and enhancement of plant resistance to diverse stressors. Because BRs are involved in a variety of physiological and developmental processes in plants, plant scientists have been interested in studying them over the past few decades. It is imperative to conduct an extensive analysis of the role of BRs in plant growth, development, and defense against various stressors. By taking proactive measures to identify and manage them, we can improve plant health and increase yields, ensuring a sustainable future for agriculture.

References

Acar, A. and Singh, D. (2023). Monitoring genotoxic, biochemical and morphotoxic potential of penoxsulam and the protective role of European blueberry (*Vaccinium myrtillus* L.) extract. *Scientific Reports* 13 (1): 6787.

Ahammed, G.J., Choudhary, S.P., Chen, S. et al. (2013). Role of brassinosteroids in alleviation of phenanthrene–cadmium co-contamination-induced photosynthetic inhibition and oxidative stress in tomato. *Journal of Experimental Botany* 64 (1): 199–213.

Ahammed, G.J., He, B.B., Qian, X.J. et al. (2017). 24-Epibrassinolide alleviates organic pollutants-retarded root elongation by promoting redox homeostasis and secondary metabolism in *Cucumis sativus* L. *Environmental Pollution* 229: 922–931.

Ahammed, G.J., Yuan, H.L., Ogweno, J.O. et al. (2012). Brassinosteroid alleviates phenanthrene and pyrene phytotoxicity by increasing detoxification activity and photosynthesis in tomato. *Chemosphere* 86 (5): 546–555.

Ajermoun, N., Aghris, S., Ettadili, F. et al. (2022). Phytotoxic effect of the insecticide imidacloprid in Phaseolus vulgaris L. plant and evaluation of its bioaccumulation and translocation by electrochemical methods. *Environmental Research* 214: 113794.

Alavanja, M.C., Ross, M.K., and Bonner, M.R. (2013). Increased cancer burden among pesticide applicators and others due to pesticide exposure. *CA: A Cancer Journal for Clinicians* 63 (2): 120–142.

Albrecht, C., Boutrot, F., Segonzac, C. et al. (2012). Brassinosteroids inhibit pathogen-associated molecular pattern–triggered immune signaling independent of the receptor kinase BAK1. *Proceedings of the National Academy of Sciences of the United States of America* 109 (1): 303–308.

An, Y.Q., Qin, Z.T., Li, D.D. et al. (2023). The combined formulation of brassinolide and pyraclostrobin increases biomass and seed yield by improving photosynthetic capacity in *Arabidopsis thaliana. Frontiers in Plant Science* 14: 1138563.

Anwar, A., Liu, Y., Dong, R. et al. (2018). The physiological and molecular mechanism of brassinosteroid in response to stress: a review. *Biological Research* 51: 46.

Bajguz, A. and Tretyn, A. (2003). The chemical structures and occurrence of brassinosteroids in plants. In: *Brassinosteroids: Bioactivity and Crop Productivity* (ed. S. Hayat and A. Ahmad), 1–44. Dordrecht: Kluwer Academic Publishers.

Bakshi, P., Bali, S., Sharma, P. et al. (2022). Mechanism associated with brassinosteroids-mediated detoxification of pesticides in plants. In *Brassinosteroids Signalling: Intervention with Phytohormones and Their Relationship in Plant Adaptation to Abiotic Stresses* (pp. 203–221). Singapore: Springer Singapore.

Bakshi, P., Chouhan, R., Sharma, P. et al. (2021). Amelioration of chlorpyrifos-induced toxicity in *Brassica juncea* L. by combination of 24-epibrassinolide and plant-growth-promoting rhizobacteria. *Biomolecules* 11 (6): 877.

Bakshi, P., Sharma, P., Chouhan, R. et al. (2023). Interactive effect of 24-epibrassinolide and plant growth promoting rhizobacteria inoculation restores photosynthetic attributes in *Brassica juncea* L. under chlorpyrifos toxicity. *Environmental Pollution* 320: 120760.

Barański, M., Średnicka-Tober, D., Volakakis, N. et al. (2014). Higher antioxidant and lower cadmium concentrations and lower incidence of pesticide residues in organically grown crops: a systematic literature review and meta-analyses. *British Journal of Nutrition* 112 (5): 794–811.

Belkhadir, Y., Jaillais, Y., Epple, P. et al. (2012). Brassinosteroids modulate the efficiency of plant immune responses to microbe-associated molecular patterns. *Proceedings of the National Academy of Sciences of the United States of America* 109 (1): 297–302.

Carvalho, F.P. (2017). Pesticides, environment, and food safety. *Food and Energy Security* 6 (2): 48–60.

Chung, Y. and Choe, S. (2013). The regulation of brassinosteroid biosynthesis in *Arabidopsis. Critical Reviews in Plant Sciences* 32 (6): 396–410.

Clouse, S.D. (2011). Brassinosteroids. *The Arabidopsis Book* 9: e0151.

Cui, J.X., Zhou, Y.H., Ding, J.G. et al. (2011). Role of nitric oxide in hydrogen peroxide-dependent induction of abiotic stress tolerance by brassinosteroids in cucumber. *Plant, Cell & Environment* 34(2): 347–358.

Dehghanian, Z., Bandehagh, A., Habibi, K. et al. (2021). Impact of abiotic stress on plant brassinosteroids. In: *Climate Change and the Microbiome: Sustenance of the Ecosphere* (ed. D.K. Choudhary, A. Mishra, A. Varma), 279–298. Cham: Springer.

Durán-Lara, E.F., Valderrama, A., and Marican, A. (2020). Natural organic compounds for application in organic farming. *Agriculture* 10 (2): 41.

Gan, Q., Luan, M., Hu, M. et al. (2022). Functional study of CYP90A1 and ALDH3F1 gene obtained by transcriptome sequencing analysis of *Brassica napus* seedlings treated with brassinolide. *Frontiers in Plant Science* 13: 1040511.

Gonzalez-Garcia, M. C., Maltoni, M. et al. (2012). Global fit to three neutrino mixing: critical look at present precision. *Journal of High Energy Physics 2012* (12): 1–24.

Guo, H., Li, L., Aluru, M. et al. (2013). Mechanisms and networks for brassinosteroid regulated gene expression. *Current Opinion in Plant Biology* 16 (5): 545–553.

Hacham, Y., Holland, N., Butterfield, C. et al. (2011). Brassinosteroid perception in the epidermis controls root meristem size. *Development* 138 (5): 839–848.

Hafeez, M. B., Zahra, N., Zahra, K. et al. (2021). Brassinosteroids: Molecular and physiological responses in plant growth and abiotic stresses. *Plant Stress* 2: 100029.

Hatamleh, A.A., Danish, M., Al-Dosary, M.A. et al. (2022). Physiological and oxidative stress responses of *Solanum lycopersicum* (L.) (tomato) when exposed to different chemical pesticides. *RSC Advances* 12 (12): 7237–7252.

He, J.X., Gendron, J.M., Sun, Y. et al. (2005). BZR1 is a transcriptional repressor with dual roles in brassinosteroid homeostasis and growth responses. *Science* 307 (5715): 1634–1638.

He, K., Gou, X., Yuan, T. et al. (2007). BAK1 and BKK1 regulate brassinosteroid-dependent growth and brassinosteroid-independent cell-death pathways. *Current Biology* 17 (13): 1109–1115.

Honnerová, J., Rothová, O., Holá, D. et al. (2010). The exogenous application of brassinosteroids to *Zea mays* (L.) stressed by long-term chilling does not affect the activities of photosystem 1 or 2. *Journal of Plant Growth Regulation* 29(4): 500–505.

Hou, J., Sun, Q., Li, J. et al. (2019). Glutaredoxin S25 and its interacting TGACG motif-binding factor TGA2 mediate brassinosteroid-induced chlorothalonil metabolism in tomato plants. *Environmental Pollution* 255: 113256.

Hou, J., Zhang, Q., Zhou, Y. et al. (2018). Glutaredoxin GRXS16 mediates brassinosteroid-induced apoplastic H_2O_2 production to promote pesticide metabolism in tomato. *Environmental Pollution* 240: 227–234.

Jaillais, Y., Hothorn, M., Belkhadir, Y. et al. (2011). Tyrosine phosphorylation controls brassinosteroid receptor activation by triggering membrane release of its kinase inhibitor. *Genes & Development* 25 (3): 232–237.

Jiang, J., Zhang, C., and Wang, X. (2015). A recently evolved isoform of the transcription factor BES1 promotes brassinosteroid signaling and development in *Arabidopsis thaliana. The Plant Cell* 27 (2): 361–374.

Jiang, Z., Ma, B., Erinle, K.O. et al. (2016). Enzymatic antioxidant defense in resistant plant: *Pennisetum americanum* (L.) K. Schum during long-term atrazine exposure. *Pesticide Biochemistry and Physiology* 133: 59–66.

Kang, Y.Y. and Guo, S.R. (2011). Role of brassinosteroids on horticultural crops. *Brassinosteroids: A Class of Plant Hormone* 269–288.

Khan, S., Zhou, J.L., Ren, L. et al. (2020). Effects of glyphosate on germination, photosynthesis and chloroplast morphology in tomato. *Chemosphere* 258: 127350.

Khurshid, A., Inayat, R., Basit, A. et al. (2024). Effects of thiamethoxam on physiological and molecular responses to potato plant (*Solanum tuberosum*), green peach aphid (*Myzus persicae*), and parasitoid (*Aphidius gifuensis*). *Pest Management Science* 80(6): 3000–3009.

Kim, T.W., Guan, S., Sun, Y. et al. (2009). Brassinosteroid signal transduction from cell-surface receptor kinases to nuclear transcription factors. *Nature Cell Biology* 11 (10): 1254–1260.

Kim, T.W. and Wang, Z.Y. (2010). Brassinosteroid signal transduction from receptor kinases to transcription factors. *Annual Review of Plant Biology* 61: 681–704.

Kour, J., Kohli, S.K., Khanna, K. et al. (2021). Brassinosteroid signaling, crosstalk and, physiological functions in plants under heavy metal stress. *Frontiers in Plant Science* 12: 608061.

Kretynin, S.V., Kolesnikov, Y.S., Derevyanchuk, M.V. et al. (2021). Brassinosteroids application induces phosphatidic acid production and modify antioxidant enzymes activity in tobacco in calcium-dependent manner. *Steroids* 168: 108444.

Kumari, S., Nazir, F., Maheshwari, C. et al. (2023). Plant hormones and secondary metabolites under environmental stresses: shedding light on defense molecules. *Plant Physiology and Biochemistry* 206: 108238.

Landa-Faz, A., González-Orenga, S., Boscaiu, M. et al. (2021). Effect of the pesticide endosulfan and two different biostimulants on the stress responses of *Phaseolus leptostachyus* plants grown in a saline soil. *Agronomy* 11(6): 1208.

Lemmon, M.A. and Schlessinger, J. (2010). Cell signaling by receptor tyrosine kinases. *Cell* 141 (7): 1117–1134.

Lim, W.A. and Pawson, T. (2010). Phosphotyrosine signaling: evolving a new cellular communication system. *Cell* 142 (5): 661–667.

Liu, J., Chen, J., Zheng, X. et al. (2017). GW5 acts in the brassinosteroid signalling pathway to regulate grain width and weight in rice. *Nature Plants* 3 (5): 1–7.

Liu, S., He, Y., Tian, H. et al. (2019). Application of brassinosteroid mimetics improves growth and tolerance of maize to nicosulfuron toxicity. *Journal of Plant Growth Regulation* 38 (2): 701–712.

Liu, X., Yang, H., Wang, Y. et al. (2020). Comparative transcriptomic analysis to identify brassinosteroid response genes. *Plant Physiology* 184 (2): 1072–1082.

Maldani, M., Aliyat, F.Z., Cappello, S. et al. (2021). Effect of glyphosate and paraquat on seed germination, amino acids, photosynthetic pigments and plant morphology of *Vicia faba, Phaseolus vulgaris* and *Sorghum bicolor. Environmental Sustainability* 4(4): 723–733.

Mehrian, S.K., Karimi, N., and Rahmani, F. (2023). 24-Epibrassinolide alleviates diazinon oxidative damage by escalating activities of antioxidant defense systems in maize plants. *Scientific Reports* 13 (1): 19631.

Mehrian, S.K., Karimi, N., and Rahmani, F. (2024). Detrimental impacts of concomitant application of cadmium and pesticides are ameliorated by 24-epibrassinolide through alteration in oxidative status and CYP genes expression in *Zea mays* L. *Rhizosphere* 29: 100872.

Munir, I., Bano, A., and Faisal, M. (2019). Impact of phosphate solubilizing bacteria on wheat (Triticum aestivum) in the presence of pesticides. *Brazilian Journal of Biology* 79 (1): 29–37.

Nayak, P. and Solanki, H. (2021). Pesticides and Indian agriculture—A review. *International Journal of Research – Granthaalayah* 9 (5): 250–263.

Ning, N., Huang, L., Wang, B. et al. (2022). Seed pre-soaking with brassinolide enhances photosynthetic resistance of foxtail millet seedlings under mesosulfuron-methyl+ iodosulfuron-methyl-sodium stress. *Chilean Journal of Agricultural Research* 82 (2): 266–275.

Nolan, T., Chen, J., and Yin, Y. (2017). Cross-talk of Brassinosteroid signaling in controlling growth and stress responses. *Biochemical Journal* 474(16): 2641–2661.

Nolan, T.M., Vukašinović, N., Liu, D. et al. (2020). Brassinosteroids: multidimensional regulators of plant growth, development, and stress responses. *The Plant Cell* 32 (2): 295–318.

Oh, E., Zhu, J.Y., and Wang, Z.Y., (2012). Interaction between BZR1 and PIF4 integrates brassinosteroid and environmental responses. *Nature Cell Biology*, 14 (8): pp. 802–809.

Oh, M.H., Clouse, S.D., and Huber, S.C. (2009). Tyrosine phosphorylation in brassinosteroid signaling. *Plant Signaling & Behavior* 4 (12): 1182–1185.

Oh, M.H., Wang, X., Wu, X. et al. (2010). Autophosphorylation of Tyr-610 in the receptor kinase BAK1 plays a role in brassinosteroid signaling and basal defense gene expression. *Proceedings of the National Academy of Sciences of the United States of America* 107 (41): 17827–17832.

Parween, T., Jan, S., Mahmooduzzafar, S. et al. (2016). Selective effect of pesticides on plant—A review. *Critical Reviews in Food Science and Nutrition* 56 (1): 160–179.

Peres, A.L.G., Soares, J.S., Tavares, R.G. et al. (2019). Brassinosteroids, the sixth class of phytohormones: A molecular view from the discovery to hormonal interactions in plant development and stress adaptation. *International Journal of Molecular Sciences* 20 (2): 331.

Raj, A., Dubey, A., Malla, M.A. et al. (2023). Pesticide pestilence: Global scenario and recent advances in detection and degradation methods. *Journal of Environmental Management* 338: 117680.

Raj, A. and Kumar, A. (2022). Recent advances in assessment methods and mechanism of microbe-mediated chlorpyrifos remediation. *Environmental Research* 214: 114011.

Raj, A., Kumar, A., and Dames, J.F. (2021). Tapping the role of microbial biosurfactants in pesticide remediation: an eco-friendly approach for environmental sustainability. *Frontiers in Microbiology* 12: 791723.

Rajmohan, K.S., Chandrasekaran, R., and Varjani, S. (2020). A review on occurrence of pesticides in environment and current technologies for their remediation and management. *Indian Journal of Microbiology* 60 (2): 125–138.

Rakshit, A., Raha, P., and De, N. (2015). *Manures, Fertilizers and Pesticides: Theory and Applications*, 266. CBS Publishers & Distributors.

Rozhon, W., Akter, S., Fernandez, A. et al. (2019). Inhibitors of brassinosteroid biosynthesis and signal transduction. *Molecules* 24 (23): 4372.

Saha, C., Chakraborty, B., Basu, K. et al. (2021). *Lectures on Pharmaceutical Chemistry and Pesticide Chemistry – Towards Skill Enhancement*. Techno World.

Savary, S., Willocquet, L., Pethybridge, S.J. et al. (2019). The global burden of pathogens and pests on major food crops. *Nature Ecology & Evolution* 3 (3): 430–439.

Shahzad, B., Tanveer, M., Che, Z. et al. (2018). Role of 24-epibrassinolide (EBL) in mediating heavy metal and pesticide induced oxidative stress in plants: a review. *Ecotoxicology and Environmental Safety* 147: 935–944.

Sharma, A., Kumar, V., Kumar, R. et al. (2018). Brassinosteroid-mediated pesticide detoxification in plants: a mini-review. *Cogent Food & Agriculture* 4(1): 1436212.

Sharma, A., Kumar, V., Singh, R. et al. (2016a). Effect of seed pre-soaking with 24-epibrassinolide on growth and photosynthetic parameters of *Brassica juncea* L. in imidacloprid soil. *Ecotoxicology and Environmental Safety* 133: 195–201.

Sharma, A., Kumar, V., Thukral, A.K. et al. (2019). Responses of plants to pesticide toxicity: an overview. *Planta Daninha* 37: e019184291.

Sharma, A., Thakur, S., Kumar, V. et al. (2016b). Pre-sowing seed treatment with 24-epibrassinolide ameliorates pesticide stress in *Brassica juncea* L. through the modulation of stress markers. *Frontiers in Plant Science* 7: 226602.

Sharma, A., Thakur, S., Kumar, V. et al. (2017). 24-epibrassinolide stimulates imidacloprid detoxification by modulating the gene expression of *Brassica juncea* L. *BMC Plant Biology* 17: 1–10.

Sharma, A., Yuan, H., Kumar, V. et al. (2019). Castasterone attenuates insecticide induced phytotoxicity in mustard. *Ecotoxicology and Environmental Safety* 179: 50–61.

Sharma, D.R. (2015). Use of pesticides and its residue on vegetable crops in Nepal. *Journal of Agriculture and Environment* 16: 33–42.

Sharma, I., Bhardwaj, R., and Pati, P.K. (2015). Exogenous application of 28-homobrassinolide modulates the dynamics of salt and pesticides induced stress responses in an elite rice variety Pusa Basmati-1. *Journal of Plant Growth Regulation* 34 (3): 509–518.

Shopova, E., Katerova, Z., Brankova, L. et al. (2021). Modulation of physiological stress response of *Triticum aestivum* L. to glyphosate by brassinosteroid application. *Life* 11 (11): 1156.

Siddiqui, H., Sami, F., Faizan, M. et al. (2019). Brassinosteroid mediated regulation of photosynthesis in plants. In *Brassinosteroids: Plant Growth and Development* (pp. 185–217). Singapore: Springer Singapore.

Su, W.C., Sun, L.L., Wu, R.H. et al. (2017). Effect of imazapic residues on photosynthetic traits and chlorophyll fluorescence of maize seedlings. *Photosynthetica* 55 (2): 294–300.

TAAS (Trust for Advancement of Agricultural Sciences), SPS (Society of Pesticide Science), IPS (Indian Phytopathological Society), and ESI (Entomological Society of India) (2020). *Stakeholders Dialogue on Current Challenges and Way Forward for Pesticides Management – A Road Map*. Indian Phytopathological Society.

Tang, J., Han, Z., and Chai, J. (2016). Q&A: What are brassinosteroids and how do they act in plants? *BMC Biology* 14: 1–5.

Tang, X., Alavi, S., and Herald, T.J., (2008). Effects of plasticizers on the structure and properties of starch–clay nanocomposite films. *Carbohydrate Polymers*, 74 (3): pp. 552–558.

Todorenko, D.A., Hao, J., Slatinskaya, O.V. et al. (2021). Effect of thiamethoxam on photosynthetic pigments and primary photosynthetic reactions in two maize genotypes (Zea mays). *Functional Plant Biology* 48 (10): 994–1004.

Tudi, M., Daniel Ruan, H., Wang, L. et al. (2021). Agriculture development, pesticide application and its impact on the environment. *International Journal of Environmental Research and Public Health* 18 (3): 1112.

Vriet, C., Russinova, E., and Reuzeau, C. (2013). From squalene to brassinolide: the steroid metabolic and signaling pathways across the plant kingdom. *Molecular Plant* 6 (6): 1738–1757.

Wang, H., Zhu, Y., Fujioka, S. et al. (2009). Regulation of *Arabidopsis* brassinosteroid signaling by atypical basic helix-loop-helix proteins. *The Plant Cell* 21 (12): 3781–3791.

Wang, X. and Chory, J. (2006). Brassinosteroids regulate dissociation of BKI1, a negative regulator of BRI1 signaling, from the plasma membrane. *Science* 313 (5790): 1118–1122.

Wang, Z., Jiang, Y., Peng, X. et al. (2017). Exogenous 24-epibrassinolide regulates antioxidant and pesticide detoxification systems in grapevine after chlorothalonil treatment. *Plant Growth Regulation* 81: 455–466.

Watanabe, T., Noguchi, T., Yokota, T. et al. (2001). Synthesis and biological activity of 26-norbrassinolide, 26-norcastasterone and 26-nor-6-deoxocastasterone. *Phytochemistry* 58 (2): 343–349.

Wendeborn, S., Lachia, M., Jung, P.M. et al. (2017). Biological activity of brassinosteroids–direct comparison of known and new analogs in planta. *Helvetica Chimica Acta* 100 (2).

WHO (2019). *Global Situation of Pesticide Management in Agriculture and Public Health: Report of a 2018 WHO-FAO Survey*. World Health Organization.

Xia, X. J., Zhang, Y., Wu, J. X. et al. (2009). Brassinosteroids promote metabolism of pesticides in cucumber. *Journal of Agricultural and Food Chemistry* 57(18): 8406–8413.

Yadav, R., Singh, S., and Singh, A.N. (2022). Biopesticides: current status and future prospects. *Proceedings of the International Academy of Ecology and Environmental Sciences* 12 (3): 211.

Ye, H., Li, L., Guo, H. et al. (2012). MYBL2 is a substrate of GSK3-like kinase BIN2 and acts as a corepressor of BES1 in brassinosteroid signaling pathway in *Arabidopsis. Proceedings of the National Academy of Sciences of the United States of America* 109 (49): 20142–20147.

Yin, Y., Vafeados, D., Tao, Y. et al. (2005). A new class of transcription factors mediates brassinosteroid-regulated gene expression in *Arabidopsis. Cell* 120 (2): 249–259.

Yin, Y.L., Zhou, Y., Zhou, Y.H. et al. (2016). Interplay between mitogen-activated protein kinase and nitric oxide in brassinosteroid-induced pesticide metabolism in *Solanum lycopersicum. Journal of Hazardous Materials* 316: 221–231.

Yu, X., Li, L., Li, L. et al. (2008). Modulation of brassinosteroid-regulated gene expression by Jumonji domain-containing proteins ELF6 and REF6 in Arabidopsis. *Proceedings of the National Academy of Sciences* 105(21): 7618–7623.

Yu, X., Li, L., Zola, J. et al. (2011). A brassinosteroid transcriptional network revealed by genome-wide identification of BESI target genes in *Arabidopsis thaliana. The Plant Journal* 65 (4): 634–646.

Yuan, X. Y., Zhang, L. G., Huang, L. et al. (2017). Spraying Brassinolide improves Sigma Broad tolerance in foxtail millet (*Setaria italica* L.) through modulation of antioxidant activity and photosynthetic capacity. *Scientific Reports* 7(1): 11232.

Yue, N., Zhang, C., Li, S. et al. (2024). Imidacloprid triggered changes in strawberry fruits on edible quality and phenolic profiles by applied at two growth stages. *Food Research International* 179: 114031.

Yüzbaşıoğlu, E. and Dalyan, E. (2019). Salicylic acid alleviates thiram toxicity by modulating antioxidant enzyme capacity and pesticide detoxification systems in the tomato (*Solanum lycopersicum* Mill.). *Plant Physiology and Biochemistry* 135: 322–330.

Zhang, W. (2018). Global pesticide use: Profile, trend, cost/benefit and more. *Proceedings of the International Academy of Ecology and Environmental Sciences*, 8 (1): 1.

Zhou, Y., Xia, X., Yu, G. et al. (2015). Brassinosteroids play a critical role in the regulation of pesticide metabolism in crop plants. *Scientific Reports* 5 (1): 9018.

Zou, L., Qu, M., Zeng, L. et al. (2020). The molecular basis of the interaction between brassinosteroid induced and phosphorous deficiency induced leaf inclination in rice. *Plant Growth Regulation* 91: 263–276.

Zullo, M.A.T. and Bajguz, A. (2019). The brassinosteroids family–structural diversity of natural compounds and their precursors. In: *Brassinosteroids: Plant Growth and Development* (ed. S. Hayat, M. Yusuf, R. Bhardwaj et al.), 1–44. Singapore: Springer Singapore.

Zullo, M.A.T. and Kohout, L. (2004). Semisystematic nomenclature of brassinosteroids. *Plant Growth Regulation* 42, 15–28.

11

Brassinosteroids Biology in Plants Under Drought Stress: Physiological and Molecular Aspects

*Muhammad Ahsan Altaf[1,2], Ravinder Kumar[3], Milan Kumar Lal[4], Rahul Kumar Tiwari[5], Hafiza Muniba Din Muhammad[6], and Riaz Ahmad[7]**

[1]*Key Laboratory for Quality Regulation of Tropical Horticultural Crops of Hainan Province, School of Breeding and Multiplication (Sanya Institute of Breeding and Multiplication), Center of Nanfan and High-Efficiency Tropical Agriculture, Hainan University, Sanya, China*
[2]*Key Laboratory for Quality Regulation of Tropical Horticultural Crops of Hainan Province, School of Tropical Agriculture and Forestry, Hainan University, Danzhou, China*
[3]*ICAR-Indian Agricultural Research Institute, New Delhi, India*
[4]*Division of Crop Physiology and Biochemistry, ICAR-National Rice Research Institute, Cuttack, Odisha, India*
[5]*Division of Crop Protection, Indian Institute of Sugarcane Research, Raibareli Road, P.O. Dilkusha, Lucknow, India*
[6]*Department of Horticulture, Bahauddin Zakariya University, Multan, Pakistan*
[7]*Department of Horticulture, The University of Agriculture, Dera Ismail Khan, Pakistan*

*Corresponding author: riazahmadbzu@gmail.com

11.1 Introduction

The rapidly changing environment introduces a broad spectrum of abiotic stressors, and any fluctuations in these stresses may disrupt the normal physiological and biochemical processes (Vardhini 2017). Among all factors affecting crop productivity in agriculture, drought is a significant threat to crop yield sustainability (Gill and Tuteja 2010). There can be significant productivity reductions in crops when they are exposed to environmental stresses that alter their morphology, physiology, and molecular characteristics (Altaf et al. 2023). To improve cultural practices, we must gain a comprehensive understanding of how environmental variables interact with physiological mechanisms (Tiwari et al. 2022). The environmental variables cold, heat, metals, salinity, and drought have a tremendous impact on crops (Devi et al. 2022). Globally, agricultural yield is reduced by around 50% due to environmental stressors.

Crop production worldwide is adversely affected by drought. When plants suffer from drought, their growth is impaired, water relations are disturbed, and their water use efficiency (WUE) is reduced (Heikal et al. 2023; Kaur and Asthir 2017). When a plant is exposed to excessive water stress, it can suffer from decline in photosynthesis, metabolic disturbance, and eventually declined plant growth, yield, and quality (Jaleel et al. 2009). Drought hinders plant growth by decreasing WUE, damaging root structure, hampering

Brassinosteroids and Abiotic Stress: Recent Developments on Physiological and Molecular Aspects, First Edition. Edited by Anket Sharma, et al.
Published 2026 by John Wiley & Sons Ltd.

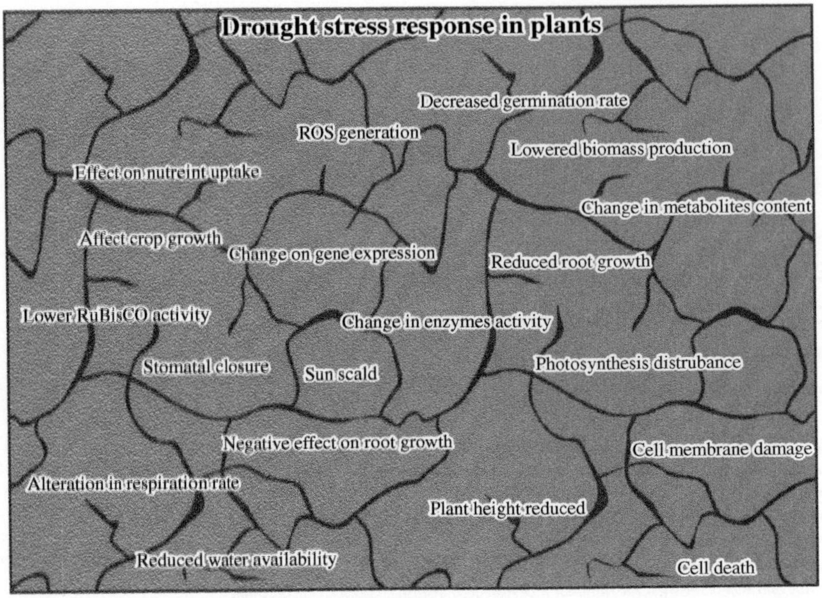

Figure 11.1 Drought stress response in plants.

the photosynthetic system, disturbed mineral nutrient uptake, altered antioxidant enzymes activity, unbalanced redox homeostasis, changes in secondary metabolites production in plants (Yang et al. 2021). Drought stress is a significant constraint during the early stages of plant development and establishment (Ozturk et al. 2021). Drought stress effectively decreases the growth and development of tomato (Jangid and Dwivedi 2016), rice (Farooq et al. 2009a), strawberry (Thokchom and Hazarika 2024), and maize (Bruce et al. 2002). Drought stress response in plants is presented in Figure 11.1.

Brassinosteroids (BRs) are a novel class of phytohormones that are becoming more and more prominent because of their diverse functions in plant growth and development (Fridman and Savaldi-Goldstein 2013). BRs modulate seed germination, photosynthetic functions, cell division, cell elongation, lateral root development, antioxidant machinery, balanced ROS homeostasis, maintained cellular membrane integrity, seedling growth, secondary metabolite production, and plant defense system (Yang et al. 2011; Fariduddin et al. 2014; Ahanger et al. 2018). BRs also regulate growth and development of plants under abiotic stress environment. BRs significantly enhance abiotic stress tolerance in plants (Bhandari and Nailwal 2020).

Recently, the use of BRs has attracted significant interest as a developing approach for improving agricultural output under both stress and control environments. This strategy shows promise in overcoming stress-related physiological disruptions (Ahammed et al. 2015; Hafeez et al. 2021). Numerous studies have been carried out on the effects of BRs in mitigating stress challenges (Nawaz et al. 2017; Li et al. 2021). The mechanisms and effects of this novel plant hormone in combating stress at morphological, physiological and molecular levels are not yet fully understood. Understanding BRs is complicated due to their strong interaction with other signaling pathways connected to stress reduction

(Yaqoob et al. 2022). This chapter addresses a few potential morphological, physiological, and molecular processes that BRs may regulate in response to drought stress. The information acquired from the literature may be used to understand and develop strategies to mitigate drought stress, particularly for highly significant agricultural crops.

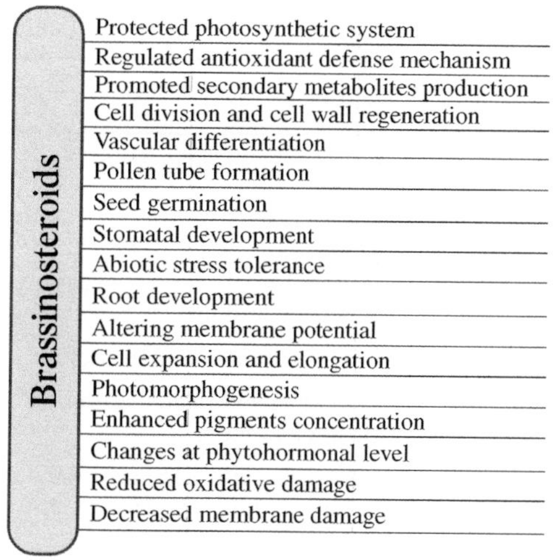

Figure 11.2 Physiological functions of Brassinosteroids in plants.

11.2 Brassinosteroids Functions in Plants

Plant steroidal molecules, namely BRs, have a wide range of functions related to plant growth, metabolism, signaling, and defense against various biotic and abiotic stressors. Seed germination, seedling growth, photosynthetic apparatus, secondary metabolite production, proline accumulation, WUE, vascular differentiation, oxidative damage, and antioxidant machinery are all regulated by BRs application. BRs play a crucial role in the regulation of plant growth, yield, and quality. Figure 11.2 explains some of the key functions of BRs in plants.

11.2.1 Brassinosteroids Enhanced Growth and Development Under Drought Stress

BRs have been investigated closely for their impact on plant metabolism. BRs application significantly regulated different physiological functions such as cell development, improved yield traits, xylem differentiation, biomass production, synthesis of proteins, and antioxidant enzyme activity (Nawaz et al. 2017; Manghwar et al. 2022). In addition, the effects on gene expression and enzyme activity play a part in cell enlargement, as well as increased metabolite production, nitrogen fixation, and nodulation (Nolan et al. 2020). BRs significantly influence plant morphology, which is directly related to crop yields. BRs application enhanced drought stress tolerance in soybean (Chen et al. 2021), melon (Liu et al. 2021), strawberry (Thokchom and Hazarika 2024), and wheat (Hayes 2019). The presence of BRs in plants is often modest, even at ng levels. Furthermore, the levels of endogenous BRs differ across plant tissues, with young developing tissues having a higher levels of BRs compared to mature tissues (Fariduddin et al. 2014; Jangid and Dwivedi 2016). Water stress treatment significantly reduced pepper growth and caused oxidative damage. In contrast, BRs application improved pepper yield-related traits and reduced cellular membrane damage (Kaya et al. 2019). Furthermore, Khan et al. (2021) reported that BRs greatly improved plant growth in both ideal and drought stress environments. BRs enhanced the growth of plants by increasing their spike length, fresh and dry biomass production, and

Table 11.1 Role of BRs in enhanced growth and development.

Species	Function	Reference
Strawberry	Improved yield-related parameters	Ali et al. (2022)
Tomato	Enhanced growth and yield traits, photosynthetic apparatus, and increased root morphological traits	Nazir et al. (2019)
Rice	Increased WUE, promoted gas exchange parameters, and enhanced drought stress tolerance	Farooq et al. (2009b)
Strawberry	Improved quality and yield traits of strawberry fruit	Furio et al. (2022)
Soybean	Increased yield-related attributes and enhanced drought stress tolerance	Phan et al. (2024)
Grape	Reduced oxidative damage, maintained cellular membrane damage, and regulated antioxidant machinery	Chen et al. (2019)
pepper	Enhanced drought stress tolerance, regulate secondary metabolites production, and proline uptake	Khamsuk et al. (2018)
Okra	Reduced oxidative damage, enhanced antioxidant defense mechanism, and better plant growth	Mahmood et al. (2024)
Potato	Increased sugar content, proline uptake, and lowered EL level	Zheng et al. (2022)
Maize	Increased secondary metabolites production, lowered oxidative stress level, and increased antioxidant defense mechanism	Kaya et al. (2018)

root length. EBR application efficiently improved physiological attributes such as biomass yield, shoot length, and leaf area of cucumber plants. In addition, EBR application improved stomatal conductance and leaf photosynthesis of cucumber (Yang et al. 2019). BRs BR application remarkably increased yield-related characteristics such as root weight, shoot weight, leaf area, and shoot length of strawberry. In addition, BRs treatment efficiently promoted vegetative growth, flowering, and yield index (Ali et al. 2022). Recent reports suggest that BRs application positively regulates the growth and development of many plant species under drought stress environment (Phan et al. 2024). The potential role of BRs is described in Table 11.1.

11.2.2 Brassinosteroids Improved Photosynthesis-related Parameters

Plant productivity is increased by photosynthesis. Leaf photosynthetic efficiency was significantly reduced by drought stress (Altaf et al. 2024). BRs supplementation potentially improved the leaf photosynthetic efficiency in many species under drought stress (Yu et al. 2004; Siddiqui et al. 2018). Furthermore, BRs have a crucial role in maintaining the integrity of the thylakoid membrane and managing the activity of the chlorophyllase enzyme, which in turn controls the chlorophyll molecules (Xia et al. 2009). Li et al. (2013) reported that BRs in cucumber modulate the plant defense system by influencing the transcription of defense-related genes and mitigate the challenges caused by drought. Seedlings treated with BRs exhibited an elevated quantum yield of PSII and increased activities of Rubisco, along with the upregulation of their large and small subunit genes, resulting in an enhanced rate of carbon dioxide absorption (Xia et al. 2009). Chlorophyll-synthesis enzymes are activated or induced by BRs to prevent the loss of pigment molecules. BRs potentially improved the leaf photosynthetic efficiency, maintained photosystem reaction, and increased gas exchange parameters in plants (Siddiqui et al. 2018). Drought stress treatment significantly decreased rice biomass production. However, the application of exogenous BRs enhanced many physiological parameters such as photosynthetic assimilation rate, WUE, leaf water status, secondary metabolite production, and antioxidant enzyme activity. Additionally, the application of BRs reduced the oxidative damage by decreasing the overproduction of MDA and H_2O_2 (Farooq et al. 2009b). BRs treatment effectively mitigated drought-induced damage by increasing pigment concentration, photosynthetic assimilation rate, and antioxidant defense mechanism in *Brassica rapa* L. (Ahmad Lone et al. 2022). The application of BR significantly improved the drought resistance and photosynthetic activity of tall fescue via a synergistic effect (Chen et al. 2018). BRs application protected leaf photosynthetic capacity in soybean under drought stress environment (Pereira et al. 2019). Sun et al. (2022) observed that BRs application efficiently enhanced the leaf gas exchange elements, chlorophyll fluorescence characteristics, photosynthetic pigment content, and upregulated the related gene expression in maize seedlings. Current findings suggested that BRs protected plant photosynthetic system in many plants (Fariduddin et al. 2014).

11.2.3 Brassinosteroids Improved Root Growth and Mineral Nutrient Uptake

The root structure of plants is modified so that nutrients are more efficiently utilized and absorbed by the plant, resulting in improved plant growth (Fageria and Moreira 2011). The roots are the primary organs that are susceptible to damage caused by soil stress (Karlidag et al. 2007). Drought stress hampered the mineral nutrient uptake and damaged root structure (Altaf et al. 2022). BRs application remarkably improved mineral homeostasis, reduced metal accumulation, and increased inorganic ions in canola leaves (Liu et al. 2014). Furthermore, Shu et al. (2016) reported that tomato plants treated with BR demonstrated increased N metabolism as well as glutamate dehydrogenase, glutamate synthase, and glutamine synthase activities. In another study, BR supplementation played a crucial role in the development and maintenance of mineral nutrient balance in soybean seedlings (Cheng et al. 2021). BRs significantly enhanced the levels of vitamins such as AsA in the roots, demonstrating their capacity to enhance the nutritional value of radish storage roots, which are the edible portions of the plant (Vardhini et al. 2011). BRs supplementation primarily improved lateral root growth and cell division development in plants (Kartal et al. 2009). BRs have a crucial role not only in the elongation of root cells but also in several other aspects of root development, including the regulation of meristem size, synthesis of root hairs, initiation of lateral roots, formation of mycorrhiza, and flowering in legume species (Wei and Li 2016) (Figure 11.3). Mussig et al. (2003) reported that the growth of *Arabidopsis* roots is regulated by BRs.

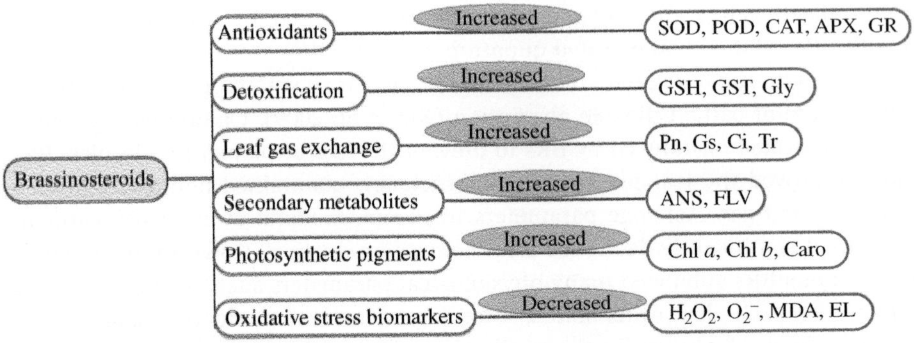

Figure 11.3 BRs promoted plant physiological and molecular functions in plants.

11.2.4 Brassinosteroids Reduced Oxidative Damage and Increased Antioxidant Activity

In a stressed environment, plants often produce a significant quantity of ROS, leading to the peroxidation of membrane lipids and oxidative damage (Altaf et al. 2022). In contrast, antioxidant enzymes are crucial in plants' defense mechanism against drought stress (Fariduddin et al. 2014). Under drought stress environment, antioxidant enzymes become more active and able to scavenge ROS more effectively (Mittler 2017). In a recent study, Kartal et al. (2009) observed that drought stress adversely affect the growth, photosynthesis, caused oxidative damage, and altered the antioxidant defense mechanism in tomato. In contrast, BRs application effectively decreased oxidative damage, maintained cellular membrane damage, and upregulated antioxidant defense mechanism in barley root. Antioxidant enzymes (SOD, POD, and CAT) significantly increased and oxidative stress biomarkers were decreased in maize under drought stress by the application of BR (Sun et al. 2022). Exogenous BRs supplementation significantly increased drought stress tolerance by increasing secondary metabolite production, protecting photosynthetic apparatus, and antioxidant machinery in *Brassica rapa* under drought stress (Ahmad Lone et al. 2022). BRs application balanced oxidative stress biomarkers and enhanced antioxidant enzyme activity in cucumber (Yang et al. 2019), wheat (Khan et al. 2021), pepper (Kaya et al. 2019), and okra (Mahmood et al. 2024). In addition, foliar application of BRs markedly improved soluble protein concentration, antioxidant enzyme activity, and yield-related characteristics in wheat under drought stress (Zeid et al. 2019). BRs treatment considerably decreases EL level, MDA concentration, and H_2O_2 accumulation in pepper under drought stress conditions. Furthermore, BR significantly increased the secondary metabolite production in pepper (Khamsuk et al. 2018). Applying EBL helped reduce the harmful impacts of drought and enhanced drought resistance by boosting SOD activity, fruit production, and other physiological functions. (Jangid and Dwivedi 2017). Foliar application of 24-EPBL significantly improved the antioxidant enzymes activity and modulated *WRKY* transcription factor in pepper under drought stress environment (Hu et al. 2015). BRs increased antioxidant enzymes (SOD and CAT) concentration and reduced EL in tomato under drought stress conditions (Behnamnia 2015) (Table 11.2).

Table 11.2 BRs regulated physiological and molecular functions in plants under drought stress.

Species	Response	Reference
Pepper	Enhanced secondary metabolites production, reduced oxidative damage, and enhanced drought stress tolerance	Khamsuk et al. (2018)
Wheat	Increased phytohormone production, leaf photosynthesis, and enhanced water use capacity	Hayes (2019)
Okra	Increase growth and biomass production, improved yield-related characteristics, upregulated antioxidant enzyme activity	Mahmood et al. (2024)
Rice	Enhanced antioxidant defense system, reduced oxidative damage, protected photosynthetic system	Farooq et al. (2009a)
Strawberry	Increased yield-related traits, improved photosynthesis, and shoot length	Ali et al. (2022)
Maize	Improved proline production, reduced oxidative damage, and altered antioxidant defense mechanism	Fariduddin et al. (2014)
Tomato	Enhanced nitrogen metabolism, increased drought stress tolerance	Shu et al. (2016)
Radish	Increased ascorbic acid concentration, improved nutritional status, and antioxidant defense system	Vardhini et al. (2011)
Soybean	Promoted plant morphology structure, increased biomass production, and yield-related characteristics	Chen et al. (2021)
Canola	Increased inorganic ion concentration, reduced oxidative damage, and promoted antioxidant defense system	Liu et al. (2014)

11.3 Conclusion and Future Prospects

The research clearly demonstrates that BRs play a crucial role as phytohormones in regulating signals that promote stress tolerance in plants. BRs play a crucial role in various physiological functions such as decreasing oxidative damage by decreasing EL, MDA production, enhancing secondary metabolite production, enzymatic and non-enzymatic antioxidant activity under stressful environment. Numerous approaches to BR treatment, including seed priming, foliar spray, and soaking in a BR solution, have been used to assess the impact of externally supplied BR on plants. Research suggests that BR-induced growth promotion is time- and dose-dependent, with factors like BR level, growth stage, species type, and application time affecting effectiveness.

Future studies should address crucial concerns such as the function of BR signaling in cellular processes such as cytoskeleton organization, vesicle trafficking, and cell wall integrity. What is the mechanism that modulates the amounts of BR in plant tissues and organs, especially when exposed to drought stress? What is the role of BRs in the interaction with other signaling pathways to impact plant metabolism? Further investigation is needed to address these unresolved inquiries, which would enhance our understanding of plant growth regulation. Additionally, it would provide insights into how the BR-signaling pathway may be manipulated by different stress elements to activate downstream defense pathways associated with stress.

References

Ahammed, G.J., Xia, X.J., Li, X. et al. (2015). Role of brassinosteroid in plant adaptation to abiotic stresses and its interplay with other hormones. *Current Protein and Peptide Science* 16: 462–473.

Ahanger, M.A., Ashraf, M., Bajguz, A. et al. (2018). Brassinosteroids regulate growth in plants under stressful environments and crosstalk with other potential phytohormones. *Journal of Plant Growth Regulation* 37: 1007–1024.

Ahmad Lone, W., Majeed, N., Yaqoob, U. et al. (2022). Exogenous brassinosteroid and jasmonic acid improve drought tolerance in *Brassica rapa* L. genotypes by modulating osmolytes, antioxidants and photosynthetic system. *Plant Cell Reports* 41: 603–617.

Ali, M.M., Anwar, R., Malik, A.U. et al. (2022). Plant growth and fruit quality response of strawberry is improved after exogenous application of 24-epibrassinolide. *Journal of Plant Growth Regulation* 41: 1786–1799.

Altaf, M.A., Shahid, R., Lal, P. et al. (2023). Current understanding of boosting power of salicylic acid for abiotic stress tolerance in horticultural crops. *South African Journal of Botany* 163: 285–293.

Altaf, M.A., Shahid, R., Naz, S. et al. (2024). Vanadium toxicity was alleviated by supplementation of silicon in tomato seedlings: upregulating antioxidative enzymes and glyoxalase system. *Plant Physiology and Biochemistry* 213: 108809.

Altaf, M.A., Shahid, R., Ren, M.X. et al. (2022). Melatonin improves drought stress tolerance of tomato by modulating plant growth, root architecture, photosynthesis, and antioxidant defense system. *Antioxidants* 11: 309.

Behnamnia, M. (2015). Protective roles of brassinolide on tomato seedlings under drought stress. *International Journal of Agriculture and Crop Sciences* 8: 552–559.

Bhandari, S. and Nailwal, T.K. (2020). Role of brassinosteroids in mitigating abiotic stresses in plants. *Biologia* 75: 2203–2230.

Bruce, W.B., Edmeades, G.O., and Barker, T.C. (2002). Molecular and physiological approaches to maize improvement for drought tolerance. *Journal of Experimental Botany* 53: 13–25.

Chen, Z.Y., Wang, Y.T., Pan, X.B. et al. (2019). Amelioration of cold-induced oxidative stress by exogenous 24-epibrassinolide treatment in grapevine seedlings: toward regulating the ascorbate–glutathione cycle. *Scientia Horticulturae* 244: 379–387.

Chen, Z., Wang, Z., Yang, Y. et al. (2018). Abscisic acid and brassinolide combined application synergistically enhances drought tolerance and photosynthesis of tall fescue under water stress. *Scientia Horticulturae* 228: 1–9.

Chen, L., Yang, H., Fang, Y. et al. (2021). Overexpression of GmMYB14 improves high-density yield and drought tolerance of soybean through regulating plant architecture mediated by the brassinosteroid pathway. *Plant Biotechnology Journal* 19: 702–716.

Cheng, L., Li, M., Min, W. et al. (2021). Optimal brassinosteroid levels are required for soybean growth and mineral nutrient homeostasis. *International Journal of Molecular Sciences* 22: 8400.

Devi, R., Behera, B., Raza, M.B. et al. (2022). An insight into microbes mediated heavy metal detoxification in plants: a review. *Journal of Soil Science and Plant Nutrition* 22: 914–936.

Fageria, N.K. and Moreira, A. (2011). The role of mineral nutrition on root growth of crop plants. *Advances in Agronomy* 110: 251–331.

Fariduddin, Q., Yusuf, M., Ahmad, I. et al. (2014). Brassinosteroids and their role in response of plants to abiotic stresses. *Biologia Plantarum* 58: 9–17.

Farooq, M., Wahid, A., and Basra, S.M.A. (2009b). Improving water relations and gas exchange with brassinosteroids in rice under drought stress. *Journal of Agronomy and Crop Science* 195: 262–269.

Farooq, M., Wahid, A., Lee, D.J. et al. (2009a). Advances in drought resistance of rice. *Critical Reviews in Plant Sciences* 28: 199–217.

Fridman, Y. and Savaldi-Goldstein, S. (2013). Brassinosteroids in growth control: how, when and where. *Plant Science* 209: 24–31.

Furio, R.N., Salazar, S.M., Mariotti-Martínez, J.A. et al. (2022). Brassinosteroid applications enhance the tolerance to abiotic stresses, production and quality of strawberry fruits. *Horticulturae* 8: 572.

Gill, S.S. and Tuteja, N. (2010). Polyamines and abiotic stress tolerance in plants. *Plant Signaling & Behavior* 5: 26–33.

Hafeez, M.B., Zahra, N., Zahra, K. et al. (2021). Brassinosteroids: molecular and physiological responses in plant growth and abiotic stresses. *Plant Stress* 2: 100029.

Hayes, S. (2019). BRacing for water stress: brassinosteroid signaling promotes drought survival in wheat. *Plant Physiology* 180: 18–19.

Heikal, Y.M., El-Esawi, M.A., El-Ballat, E.M. et al. (2023). Applications of nanoparticles for mitigating salinity and drought stress in plants: an overview on the physiological, biochemical and molecular genetic aspects. *New Zealand Journal of Crop and Horticultural Science* 51: 297–327.

Hu, W.H., Zan, X.H., Yan, X.H. et al. (2015). Effects of 24-epibrassinolide on antioxidant system and drought-tolerance relative genes in pepper seedlings under drought stress. *Scientia Horticulturae* 150: 232–237.

Jaleel, C.A., Manivannan, P., Wahid, A. et al. (2009). Drought stress in plants: a review on morphological characteristics and pigments composition. *International Journal of Agriculture and Biology* 11:100–105.

Jangid, K.K. and Dwivedi, P. (2016). Physiological responses of drought stress in tomato: a review. *International Journal of Agriculture, Environment and Biotechnology* 9: 53–61.

Jangid, K.K. and Dwivedi, P. (2017). Physiological and biochemical changes by nitric oxide and brassinosteroid in tomato (*Lycopersicon esculentum* Mill.) under drought stress. *Acta Physiologiae Plantarum* 39: 1–10.

Karlidag, H., Esitken, A., Turan, M. et al. (2007). Effects of root inoculation of plant growth promoting rhizobacteria (PGPR) on yield, growth and nutrient element contents of leaves of apple. *Scientia Horticulturae* 114: 16–20.

Kartal, G., Temel, A., Arican, E. et al. (2009). Effects of brassinosteroids on barley root growth, antioxidant system and cell division. *Plant Growth Regulation* 58: 261–267.

Kaur, G. and Asthir, B. (2017). Molecular responses to drought stress in plants. *Biologia Plantarum* 61: 201–209.

Kaya, C., Ashraf, M., Wijaya, L. (2019). The putative role of endogenous nitric oxide in brassinosteroid-induced antioxidant defence system in pepper (*Capsicum annuum* L.) plants under water stress. *Plant Physiology and Biochemistry* 143: 119–128.

Kaya, C., Aydemir, S., Akram, N.A. et al. (2018). Epibrassinolide application regulates some key physio-biochemical attributes as well as oxidative defense system in maize plants grown under saline stress. *Journal of Plant Growth Regulation* 37: 1244–1257.

Khamsuk, O., Sonjaroon, W., Suwanwong, S. et al. (2018). Effects of 24-epibrassinolide and the synthetic brassinosteroid mimic on chili pepper under drought. *Acta Physiologiae Plantarum* 40: 1–12.

Khan, I., Awan, S.A., Ikram, R. et al. (2021). Effects of 24-epibrassinolide on plant growth, antioxidants defense system, and endogenous hormones in two wheat varieties under drought stress. *Physiologia Plantarum* 172: 696–706.

Li, P., Chen, L., Zhou, Y. et al. (2013). Brassinosteroids-induced systemic stress tolerance was associated with increased transcripts of several defence-related genes in the phloem in *Cucumis sativus*. *PLoS One* 8: e66582.

Li, S., Zheng, H., Lin, L. et al. (2021). Roles of brassinosteroids in plant growth and abiotic stress response. *Plant Growth Regulation* 93: 29–38.

Liu, J., Gao, H., Wang, X. et al. (2014). Effects of 24-epibrassinolide on plant growth, osmotic regulation and ion homeostasis of salt-stressed canola. *Plant Biology* 16: 440–450.

Liu, W., Jiang, Y., Jin, Y. et al. (2021). Drought-induced ABA, H_2O_2 and JA positively regulate *CmCAD* genes and lignin synthesis in melon stems. *BMC Plant Biology* 21: 1–12.

Mahmood, K., Hussain, A., Ahmad, M. et al. (2024). Enhancing drought resilience in okra (*Abelmoschus esculentus*) through synergistic application of drought-tolerant rhizobacteria and brassinosteroids under drought stress. *Global NEST Journal* 26: 05730.

Manghwar, H., Hussain, A., Ali, Q. et al. (2022). Brassinosteroids (BRs) role in plant development and coping with different stresses. *International Journal of Molecular Sciences* 23: 1012.

Mittler, R. (2017). ROS are good. *Trends in Plant Science* 22: 11–19.

Mussig, C., Shin, G.H., and Altmann, T. (2003). Brassinosteroids promote root growth in *Arabidopsis*. *Plant Physiology* 133: 1261–1271.

Nawaz, F., Naeem, M., Zulfiqar, B. et al. (2017). Understanding brassinosteroid-regulated mechanisms to improve stress tolerance in plants: a critical review. *Environmental Science and Pollution Research* 24: 15959–15975.

Nazir, F., Hussain, A., and Fariduddin, Q. (2019). Interactive role of epibrassinolide and hydrogen peroxide in regulating stomatal physiology, root morphology, photosynthetic and growth traits in *Solanum lycopersicum* L. under nickel stress. *Environmental and Experimental Botany* 162: 479–495.

Nolan, T.M., Vukašinović, N., Liu, D. et al. (2020). Brassinosteroids: multidimensional regulators of plant growth, development, and stress responses. *The Plant Cell* 32: 295–318.

Ozturk, M., Turkyilmaz Unal, B., García-Caparrós, P. et al. (2021). Osmoregulation and its actions during the drought stress in plants. *Physiologia Plantarum* 172: 1321–1335.

Pereira, Y.C., Rodrigues, W.S., Lima, E.J.A. et al. (2019). Brassinosteroids increase electron transport and photosynthesis in soybean plants under water deficit. *Photosynthetica* 57: 181–191.

Phan, V.H., Le, T.T.H., Pham, D.M. et al. (2024). Effects of concentration and time of brassinosteroid treatment on growth and yield of soybean under drought stress conditions. *Plant Science Today* 11 (2): 612–619.

Shu, S., Tang, Y., Yuan, Y. et al. (2016). The role of 24-epibrassinolide in the regulation of photosynthetic characteristics and nitrogen metabolism of tomato seedlings under a combined low temperature and weak light stress. *Plant Physiology and Biochemistry* 107: 344–353.

Siddiqui, H., Hayat, S., and Bajguz, A. (2018). Regulation of photosynthesis by brassinosteroids in plants. *Acta Physiologiae Plantarum* 40: 59.

Sun, S., Yao, X., Liu, X. et al. (2022). Brassinolide can improve drought tolerance of maize seedlings under drought stress: by inducing the photosynthetic performance, antioxidant capacity and ZmMYB gene expression of maize seedlings. *Journal of Soil Science and Plant Nutrition* 22: 2092–2104.

Thokchom, A. and Hazarika, B.N. (2024). Morpho-physiological and biochemical changes under drought stress in strawberry: a review. *Agricultural Reviews* 45: 340.

Tiwari, R.K., Lal, M.K., Kumar, R. et al. (2022). Insight into melatonin-mediated response and signaling in the regulation of plant defense under biotic stress. *Plant Molecular Biology* 109: 385–399.

Vardhini, B.V. (2017). Modifications of morphological and anatomical characteristics of plants by application of brassinosteroids under various abiotic stress conditions-a review. *Plant Gene* 11: 70–89.

Vardhini, B.V., Sujatha, E., and Rao, S.S.R. (2011). Studies on the effect of brassinosteroids on the qualitative changes in the storage roots of radish. *Asian and Australasian Journal of Plant Science and Biotechnology* 5: 27–30.

Wei, Z. and Li, J. (2016). Brassinosteroids regulate root growth, development, and symbiosis. *Molecular Plant* 9: 86–100.

Xia, X.J., Huang, L.F., Zhou, Y.H. et al. (2009). Brassinosteroids promote photosynthesis and growth by enhancing activation of Rubisco and expression of photosynthetic genes in *Cucumis sativus. Planta* 230: 1185–1196.

Yang, P., Azher Nawaz, M., Li, F. et al. (2019). Brassinosteroids regulate antioxidant system and protect chloroplast ultrastructure of autotoxicity-stressed cucumber (*Cucumis sativus* L.) seedlings. *Agronomy* 9: 265.

Yang, X., Lu, M., Wang, Y. et al. (2021). Response mechanism of plants to drought stress. *Horticulturae* 7: 50.

Yang, C.J., Zhang, C., Lu, Y.N. et al. (2011). The mechanisms of brassinosteroids' action: from signal transduction to plant development. *Molecular Plant* 4: 588–600.

Yaqoob, U., Jan, N., Raman, P.V. et al. (2022). Crosstalk between brassinosteroid signaling, ROS signaling and phenylpropanoid pathway during abiotic stress in plants: does it exist? *Plant Stress* 4: 100075.

Yu, J.Q., Huang, L.F., Hu, W.H. et al. (2004). A role for brassinosteroids in the regulation of photosynthesis in *Cucumis sativus. Journal of Experimental Botany* 55: 1135–1143.

Zeid, F., El Shihy, O., and Ibrahim, F. (2019). Exogenous application of 24-epibrassinolide and spermine improves growth and yield of drought stressed wheat (*Triticum aestivum* L.). *Middle East Journal of Applied Sciences* 9: 1063–1080.

12

Illuminating Insight: Exploring Brassinosteroid Biology in Plants Amidst High Light and UV Exposure

Sukhmeen Kaur Kohli[1], Harsimran Kaur[2], Pooja Sharma[3], Priyanka Sharma[4], Manu Sharma[5], Shalini Dhiman[5], Nikita Kapoor[5], and Renu Bhardwaj[5,6*]*

[1]*Department of Earth and Climate Sciences (ECS) and Biology, Indian Institute of Science Education and Research (IISER) Tirupati, Tirupati, Andhra Pradesh, India*
[2]*Post Graduate Department of Botany, Khalsa College Amritsar, Punjab, India*
[3]*Department of Biotechnology, DAV College, Amritsar, Punjab, India*
[4]*School of Bioengineering Sciences and Research, MIT-Art Design and Technology University, Loni Kalbhor, Pune, Maharashtra, India*
[5]*Plant Stress Biology Lab, Department of Botanical and Environmental Sciences, Guru Nanak Dev University, Amritsar, Punjab, India*
[6]*University Centre for Research & Development, Chandigarh University, Mohali, Punjab, India*

*Corresponding authors: sukhmeenkohli@iisertirupati.ac.in, renubhardwaj82@gmail.com

12.1 Introduction

Plants face perpetual challenges due to the constantly shifting environmental conditions. Sunlight, a crucial environmental factor, not only serves as an energy source for photosynthesis but also conveys information about the time of day and the season. Plants experience continual fluctuations in light intensity and quality owing to weather variations (e.g. sunny or cloudy skies) and the changing sun's inclination throughout the year due to seasonal shifts. For plants beneath canopies, coping with these seasonal changes poses significant challenges. During spring, canopy closure diminishes light quantity, particularly affecting the red:far-red (R:FR) ratio and the total amount of blue light reaching understory plants (Casal 2013). Consequently, the plant's ability to perceive changes in light quality and quantity and mount appropriate responses becomes indispensable for survival in these dynamic and often adverse light environments. Distinct photoreceptors in plants specialize in the perception of specific segments of the solar light spectrum, as comprehensively discussed in the review by Paik and Huq (2019). In the model plant *Arabidopsis thaliana*, the phytochrome family, encompassing phyA to phyE, is dedicated to the perception of red to far-red (R/FR) light. Blue light, on the other hand, is sensed by cryptochromes (CRY1, CRY2, and CRY3), phototropins (PHOT1, PHOT2), and F-box-containing flavin-binding proteins (ZEITLUPE, FKF1/LKP2). Furthermore, the ultraviolet-B (UV-B) segment of the light spectrum is distinctly perceived by the UVR8 receptor. This intricate array of

Brassinosteroids and Abiotic Stress: Recent Developments on Physiological and Molecular Aspects, First Edition. Edited by Anket Sharma, et al.
Published 2026 by John Wiley & Sons Ltd.

photoreceptor specificity underscores the sophisticated mechanisms employed by plants to interpret and respond to the diverse light qualities encountered in their environment.

Elevated light levels pose a detriment to plants, precipitating oxidative stress that manifests in the form of photodamage and photoinhibition (Ganguly et al. 2018). The recurrent exposure to excess light, followed by subsequent phases of lower light intensities, commonly referred to as fluctuating light stress, elicits robust stress responses. Furthermore, the extension of the light period, known as photoperiod stress (Nitschke et al. 2016), and the UV segment of the light spectrum contribute significantly to stress, resulting in profound DNA damage (Hideg et al. 2013). These intricate light-induced stresses collectively underscore the inherent susceptibility of plants to fluctuations and imbalances in their radiant environment, highlighting the need for a nuanced understanding of the physiological responses to such stressors. Plants leverage sunlight not only for energy production and carbon compound synthesis but also as a vital source of environmental information. Sunlight serves as a signaling cue, prompting plants to initiate diverse systemic responses associated with photosynthesis, phototropism, photoperiodicity, and photomorphogenesis. The aberrant presence of UV radiation in the atmosphere can significantly influence these processes. Consequently, the repercussions of UV-induced damage on these crucial plant functions have been a subject of intensive study in recent decades (Robson et al. 2015). Photosynthesis, reliant on light, is inherently susceptible to UV radiation impact. Extensive literature reports detail damage to specific photosynthetic apparatus sites in green plants (Correa et al. 2023). Reactive oxygen species (ROS) inflict protein damage within light-harvesting complexes and electron carrier-concentrated protein complexes. ROS accumulation culminates in ribulose bisphosphate carboxylase/oxygenase degradation, a critical enzyme in photosynthesis. The nuanced interplay underscores the vulnerability of photosynthesis to UV radiation stressors (Nasibova et al. 2023). ROS inflict crucial damage by oxidizing fatty acids within membranes. This, coupled with peroxidation and photooxidation induced by UV light, compromises the vital integrity of thylakoid membranes in chloroplasts. Consequently, alterations in the organization of membrane-embedded photosynthetic complexes occur, leading to a reduction in their photosynthetic capacity. The intricate interplay between ROS and membrane components underscores the significant impact on the functional efficiency of the photosynthetic apparatus. Several investigations have delved into the transcriptional changes in plants exposed to both long- and short-wave UV-B radiation, with the aim of elucidating the cellular components that govern the UV response (Ulm and Jenkins 2015). The outcomes reveal that UV-B radiation elicits the activation of pathways associated with cell growth and morphogenesis (Ulm and Nagy 2005). Furthermore, the signals emanating from the UV-B response exhibit intercellular transmission, displaying distinctive organ-specific patterns (Casati et al. 2011). Genetic methodologies employed to scrutinize phenotypic responses to UV-B stress revolve around models incorporating increased tolerance or aberrant responses, such as perturbations in hypocotyl growth, upon exposure to UV-B irradiation (Ulm and Nagy 2005).

The photoprotective efficacy of various secondary metabolites, encompassing flavonoids (including anthocyanins), cinnamic acid derivatives, and xanthophylls, stems from the presence of conjugated double bonds characterized by delocalized π-electrons. This specific electronic configuration facilitates the absorption of both visible and UV radiation, leading to efficient electron and energy transfers. The prevalence of π-electron configuration is discernible in structures featuring closed-ring, short-side chain, and long-chain arrangements, as expounded by Demmig-Adams (2003). Alkaloids belonging to the quinolizidine type, exemplified by sparteine, have been identified in the epidermal cells of plants within the *Lupinus* genus. According to Ali et al. (2005), the observed storage pattern of various alkaloids aligns with a putative phytochemical role as antifeedant chemical defense compounds, and alternatively, may indicate an antioxidative function. The prospect of these alkaloids serving as UV light-filtering agents is deemed improbable, given their anticipated minimal absorption in the solar UV range (Bartwal et al. 2013). Liang et al. (2020) revealed that brassinosteroid (BR) signaling, mediated by BES1, inhibits UV-B stress responses in *Arabidopsis* and crops by suppressing flavonol biosynthesis. The study emphasizes BES1's dual role in promoting growth and inhibiting flavonoid biosynthesis, with UV-B stress down-regulating BES1 to prioritize energy allocation for flavonoid production and UV-B stress responses.

The chapter explores the intricate molecular and biochemical processes triggered by BR application in plants. This orchestrated response collectively enhances plant resilience, enabling them to endure and adapt to challenges posed by UV radiation stress. The study emphasizes the coordinated activation of various pathways and highlights the pivotal role of BR in modulating gene expression and defense mechanisms at the molecular level.

12.2 Light as a Plant Stressor

12.2.1 High Light Stress

Excessive light intensity beyond the energetic demand of photosynthesis eventually damages the photosynthetic apparatus, particularly the D1 protein of the reaction center of photosystem II (PSII) (Edelman and Mattoo 2008). D1 proteins are replaced during the repair cycle. However, if photodamage surpasses the D1 repair capacity, consequently resulting in photoinhibition occurs (Baena-González and Aro 2002). Plants avoid photodamage by several mechanisms, such as heliotropism, chloroplast avoidance response (Takahashi and Badger 2011). The photosynthetic pigments, when they absorb excess light, result in the generation of ROS and cause oxidative stress (Apel and Hirt 2004). ROS directly inhibit the D1 repair cycle (Takahashi and Murata 2008).

12.2.2 Fluctuating Light Stress

During the daytime, plants face fluctuations in light caused by various factors such as cloud movement, the shade in understory plants, or changes in sun elevation. Under fluctuating light, a change from light absorption to heat dissipation does not occur as in constant high light, due to slow non-photochemical quenching (NPQ) relaxation (Kono and Terashima 2014). Fluctuating light limits the flow of electrons on the acceptor side of Photosystem I (PSI) and causes PSI photoinhibition and an accretion of ROS, which disrupts the iron–sulfur centers of PSI (Sonoike 1996; Roeber et al. 2020).

12.2.3 Photoperiod Stress

Photoperiod stress is induced by alteration in particularly by prolonged light period, (Nitschke et al. 2016, 2017). An extended light period induced stress marker genes like *ZAT12* and *BAP1* and increased oxidative stress during the night, followed by a significant decrement of PSII maximum quantum efficiency (Fv/Fm), which ultimately leads to programmed cell death in the leaves. Photoperiod stress induces an oxidative burst is associated with simultaneously diminishing catalase activities and an increment in apoplastic peroxidase activity (Abuelsoud et al. 2020). Krasensky-Wrzaczek and Kangasjarvi (2018) gave a eustress-distress concept according to which shorter prolonged light period, are not considered as harmful and perceived as eustress (beneficial stress), whereas longer prolonged light periods are harmful and induce distress.

12.2.4 UV Light Stress

UV light is divided into three wavebands: UV-A (315–400 nm), UV-B (280–315 nm), and UV-C (200–280 nm). High doses of UV-B radiation are harmful and can cause programmed cell death, whereas low intensities are harmless "eustress" (Hideg et al. 2013). UV light damages photosynthetic machinery by degrading the D1 and D2 proteins of PSII (Takahashi et al. 2010). Furthermore, it can damage DNA by the formation of cyclobutene pyrimidine dimers, thereby inhibiting transcription and replication. It might also augment the production of ROS, which cause oxidative damage and oxidizes nucleic acids, lipids, and proteins in plant cells (D'Orazio et al. 2013; Demarsy et al. 2018). In addition, it inhibits Ribulose bisphosphate carboxylase oxygenase (Rubisco) activity and reduces chlorophyll content resulting in lowered photosynthetic activity in plants (Sztatelman et al. 2015; Chen et al. 2022a).

12.3 Detection of High Light and UV Radiation by Plants

Plants have the ability to sense the duration of light, intensity, and timing, but they are also able to sense the quality of light with photoreceptors, each having different absorption and action spectra (Meyer et al. 2021). Plants perceive UV radiation *via* a wide range of photoreceptors with a specific absorption spectrum. Cryptochromes (cry1 and cry2), phototropins (phot1 and phot2), and three zeitlupe proteins are attributed as UV-A/blue photoreceptors. UV RESISTANCE LOCUS 8 (UVR8) is ascribed to be a UV-B specific photoreceptor (Rizzini et al. 2011; Christie et al. 2015; Rai et al. 2021). UV-C specific photoreceptors are not known till date, but it is speculated that some photoreceptors like CRYs and UVR8 can be triggered by UV-C (Vanhaelewyn et al. 2020).

UVR8 photoreceptor is composed of a 150 kDa protein, and is present in an inactive homodimer in the cytoplasm, which is stabilized by cross-dimer salt bridges (Tossi et al. 2019). The UVR8 photoreceptor appears like a seven-bladed β-propeller. The tryptophan residues in the β-propeller core and at the dimer interface residues perceive UV-B light and do not involve a chromophore or specific cofactor (Christie et al. 2012; Wu et al. 2012). As the UV-B light is perceived by the UVR8 receptor, the cross-dimer salt bridges are disrupted and the inactive homodimer is dissociated into two active monomers, which are then translocated to the nucleus (Figure 12.1) (Christie et al. 2012; Wu et al. 2012). The activated monomeric form of UVR8 interacts with the COP1–SPA (CONSTITUTIVE PHOTOMORPHOGENIC 1; SUPPRESSOR OF PHYA-105) E3 ligase complex *via* highly conserved amino acid sequences present in the C terminus (Cloix et al. 2012; Camacho

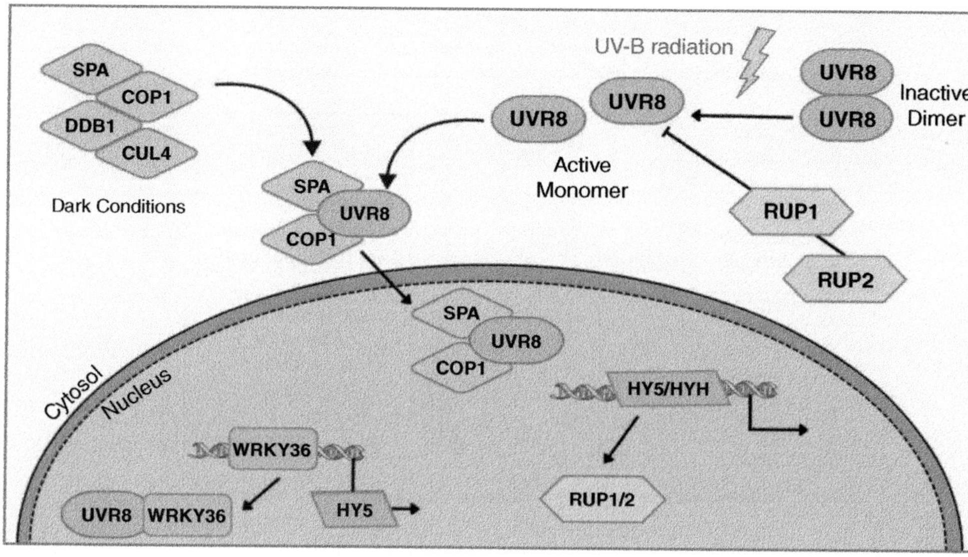

Figure 12.1 UVR8-mediated signaling.

et al. 2019). The COP1–SPA complex is located in the nucleus under dark conditions. Under light conditions, this complex is exported to the cytosol where the complex is disrupted and SPA is degraded (Saijo et al. 2003; Nguyen 2020). The COP1–SPA complex forms a ubiquitination complex by acting as an E3 ligation adaptor for CUL4–DDB1 (CULLIN4; DNA DAMAGE BINDING PROTEIN 1). The COP1–SPA complex alone or in association with CUL4–DDB1 destroys transcription factors like HY5 (ELONGATED HYPOCOTYL 5) and its homolog HYH (HY5 HOMOLOG), which are involved in photomorphogenesis (Huang et al. 2013; Zhu et al. 2015). However, under UV-B radiation, COP1–SPA is dissociated from CUL4–DDB1, which then interacts with monomeric UVR8. When COP1–SPA interacts with UVR8, UV-B light signaling is initiated, which stabilizes HY5 and HYH (Chen et al. 2010; Jenkins 2017). Further, UVR8 promotes the transcription of HY5 by interacting with the transcription factor WRKY36 (WRKY DNA-BINDING PROTEIN 36), which is a repressor of HY5 transcription. The conserved sequences in the C terminus of UVR8 responsible for COP1 binding are also requisite for the association of the negative regulators i.e. REPRESSOR OF UV-B PHOTOMORPHOGENESIS, viz. RUP1 and RUP2 of active monomer UVR8. The RUP proteins, located in the nucleus and cytosol, are responsible for the redimerization of UVR8 (Liang et al. 2018; Yang et al. 2018a, b). Figure 12.1 depicts UVR8-mediated signaling.

12.4 High Light and UV-induced Damage

The two main environmental parameters influencing plant growth and development are light intensity and UV radiation. Variations in their length, intensity, and range can cause morphological alterations, damage to photosystem, oxidative stress, cellular and DNA

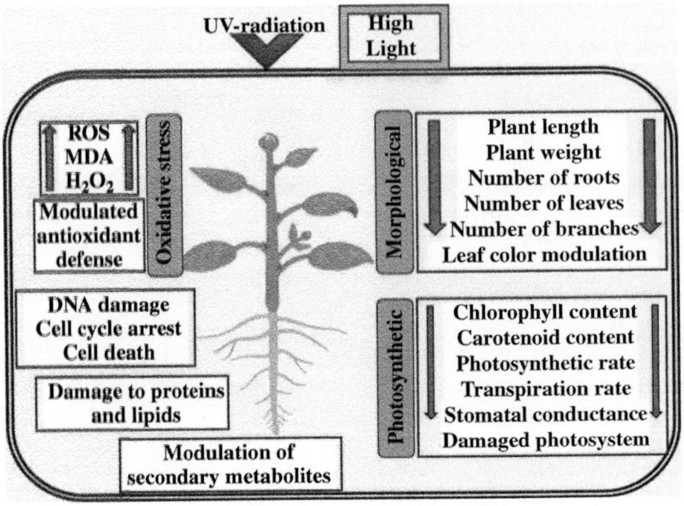

Figure 12.2 General illustration of high light and UV-induced damage.

damage, acclimatory reactions, and eventually plant death (Vanhaelewyn et al. 2020; Roeber et al. 2021). Plants are able to withstand adverse light and UV radiation because they have developed a number of effective defense mechanisms. The xanthophyll cycle, the photosynthetic electron transport chain, and the photorespiratory pathway are the main connections between these systems. Stressful situations result in high concentrations of ROS, which have negative impacts as well as signaling roles (Kumar et al. 2023; Sharma et al. 2023). Antioxidant enzymes and several low-molecular-weight antioxidants are produced in response to increased ROS production (Kumar and Ohri 2023). The amount of synthesized antioxidants varies depending on the severity and length of the stress. Low-molecular-weight antioxidants, including ascorbate, glutathione, and prenyllipids, tend to have lower concentrations during acute, brief light/UV stress, which is associated with an elevated need for ROS scavenging (Foyer and Noctor 2016). Studies on plant antioxidant responses indicate that a crucial part of the antioxidant network operates in chloroplasts, and its action shows a high level of interdependence. The antioxidant response also depends on plant stress tolerance and frequently varies for different species. High light and UV radiation induce damage at various levels (Figure 12.2), as explained in the subsequent sections.

12.4.1 Morphological Alterations

Plant sensitivity to high light and UV exposure might be determined by direct damage to cell structural and functional elements or by ineffective acclimatization process (Smith et al. 2000). UV radiation and high light may induce alterations in various morphological characteristics of plants including plant length and weight, number of branches and leaves, leaf coloration, leaf differentiation and senescence processes *via* modification of leaf structure (Szymańska et al. 2017; Verdaguer et al. 2017). Various studies have reported that duration and intensity of light and UV radiation can have positive or detrimental effects on crop plants' morphology. For instance, Yoshida et al. (2022) carried out two trial experiments to examine the impact of UV-B radiation three days before harvest. In the first trial, 60-day-old seedlings were exposed to UV-B irradiation at 0 (control), 6, and 10 W m^{-2} during a 16-hour light period. In the second trial, 45-day-old seedlings were exposed to UV-B irradiation at 0 (control), 4 W m^{-2} at continuous irradiation, and 6 W m^{-2} during 16-hour irradiation. They reported that in the first trial, in comparison to the control samples, the fresh weights of the above-ground and fourth leaves of plants subjected to UV-B irradiation of 6 and 10 W m^{-2} were remarkably reduced. When UV-B irradiation intensity got higher, the dry weights of the plant's above-ground and fourth-leaf samples, which were subjected to the radiation, dropped. In second trials, leaves turned green possibly due to the breakdown of stored anthocyanins, but there was no significant difference in the weight of treated and control plants (Yoshida et al. 2022). Another study evaluated whether the morphological parameters of Micro-Tom tomato plants throughout their 11-day treatment and three-day recovery period were affected by modest, non-stressful levels of UV-B radiation delivered above ground. Results revealed that root length and leaf number were slightly decreased on the 11th day of treatment (Mannucci et al. 2020). Similarly, UV-B radiation applied at 40 μW cm^{-2} in *Glycine max* was reported to reduce soybean sprout weight and length, possibly due to increased oxidative stress causing damage to the cell membrane's integrity (Jovanić et al. 2022). Additionally, application of high light (150 and 210 μmol m^{-2} s^{-1}) in strawberry was observed to decrease the number of roots and root length in contrast to the plants exposed to low light (90 μmol m^{-2} s^{-1}) (Zheng et al. 2019). Various reports of morphological alterations triggered by high light and UV radiation in different plants are summarized in Table 12.1.

12.4.2 Photosynthetic

Photosynthesis is an important, complex, and most delicate metabolic process, which makes it crucial to investigate how photosynthesis reacts to high light and UV exposure, as it is directly related to the generation of biomass and yield (Ferreyra et al. 2021). Despite the fact that high light and UV radiation have a variety of targets in plants, one of their imperative targets is the photosynthetic machinery, whose degradation considerably adds to the general harm triggered by high light and UV radiation Different laboratory growing conditions, plant growth stages, flow rate, the ratio of photosynthetically active radiation to UV-B radiation, dosages of light and UV radiation and how they interacts with other environmental stressors all affect how the photosynthetic pathways respond to UV radiation (Kataria et al. 2014; Gollan and Aro 2020). For instance, a study conducted by Jaiswal et al. (2023) determined how *Chlorophytum borivillianum* responds in terms of photosynthetic parameters to two different UV-B radiation doses i.e. ambient + low (3.2 kJ m^{-2} d^{-1}) and high (7.2 kJ m^{-2} d^{-1}) UV-B doses. Plants receiving low-dose UV-B showed a non-significant change, however, plants receiving high doses showed a greater reduction in photosynthetic rate. When exposed to high doses i.e. UV-B during the vegetative and reproductive stages, there was a significant decline in stomatal conductance as well. Similarly, another study reported that *Rhododendron chrysanthum* exposed to 2.3 W m^{-2} UV-B exhibited a 7.95% and 8.36% reduction in the maximum and effective quantum yield of PSII (Zhou et al. 2021). In another study it was reported that elevated UV-B

Table 12.1 Morphological and photosynthetic alterations in different plants, induced by high light or UV-radiation.

S. no.	Plant species	Stressor and intensity	Attribute affected	Mechanism of action	References
1	*Oryza sativa*	10 kJ UV-B radiation	Morphological	Decreased plant height, tiller number, leaf number, area and weight, root length and weight, shoot weight, and plant height, which might be due to decreased levels of growth hormones, damaged cell components, or decreased carbohydrate content.	Mathur et al. (2024)
2	*Triticum aestivum*	UV-B radiation (10.08 kJ m^{-2} d^{-1})	Morphological and photosynthetic	Reduced plant height, number of roots, root length, fresh weight, and chlorophyll content, which might be due to reduced germination index and germination potential.	Mu and Han (2024)
3	*Stevia rebaudiana*	UV-C radiation	Morphological	Inhibited growth resulted in curling, plant scorch, and reduced leaf blades.	Semenova et al. (2024)
4	*Glycine max*	UV-B radiation using an 8 W UV-B lamp	Morphological	Replenished plant length and weight (probably due to modulation of fatty acid composition).	Xu et al. (2024)
5	*Mentha spicata*	UV-radiation (43.2 and 86.4 kJ m^{-2})	Photosynthetic	Reduced chlorophyll a and b, which might be attributed to damage to the chloroplast and PSII degeneration.	Ali et al. (2023)
6	*Triticum aestivum* and triticale	High light (1200 μmol m^{-2} s^{-1} photons)	Photosynthetic	Reduced chlorophyll content and decreased intercellular CO_2, which might be attributed to damaged photosystem.	Chen et al. (2023)
7	*Panax notoginseng*	Photon exposure 29.8, 11.5, 9.6, 5.0, 3.6, 1.4, and 0.2% of full sunlight	Photosynthetic	High light-reduced chlorophyll a and b content and net photosynthetic rate, which might be attributed to reduced liquid phase diffusion of CO_2 in mesophyll cells and resulted in smaller and yellow leaves.	Cun et al. (2023)
8	Strawberry	High light (1800 μmol m^{-2} s^{-1} for 8 h)	Photosynthetic	Chlorophyll content decreased after 1–8 h and modulated gas exchange parameters, which might be due to degradation of chlorophyll.	Fu et al. (2023)

9	*Chlorophytum borivillianum*	eUV-B (7.2 and 3.2 kJ m^{-2} d^{-1})	Photosynthetic	Reduced content of chlorophyll a and b, total chlorophyll, carotenoid, and photosynthetic rate, which are attributed to reduced stomatal conductance.	Jaiswal et al. (2023)
10	*Solanum lycopersicum*	UV-B radiation for 15, 30, 45, and 60 min	Photosynthetic	Lowered content of chlorophyll a and b, total chlorophyll and carotenoid with increased dose of UV-B, which is attributed to reduced biosynthesis, or/and enhanced degradation of chlorophyll pigments.	Khare et al. (2023)
11	*Ocimum basilicum*	UV-B radiation (1.36 W m^{-2} for 2 h/day)	Morphological	Decrement in plant length and weight, which might be attributed to UV-B-mediated disruption of the cell cycle.	Santin et al. (2023)
12	*Solanum tuberosum*	UV-B radiation using 15 W UV-B lamp	Photosynthetic	High doses of UV-B reduced chlorophyll content, which may be due to stomata-limiting factors.	Wu et al. (2023)
13	*Iris tectorum*	UV-B radiation (210 and 252 μW cm^{-2})	Photosynthetic	Decreased levels of chlorophyll and carotenoid (which are attributed to damaged chloroplast ultrastructure) and reduced net photosynthetic and transpiration rate.	Chu et al. (2022)
14	*Glycine max*	UV-B radiation (40 μW cm^{-2})	Morphological	Reduced soybean sprout length and weight which might be attributed to increased oxidative stress, damaged the integrity of the cell membrane.	Jovanić et al. (2022)
15	*Solanum lycopersicum*	High red:far red light	Morphology	Lowered stem length and hyponasty, which might be attributed to the growth-inhibiting potential of high light.	Meijer et al. (2022)
16	*Psoralea corylifolia*	eUV-B (7.2 kJ m^{-2})	Morphological	Decline in above-ground and below-ground biomass, which might be attributed to the production of ROS and lower accumulation of photo-assimilates.	Pandey et al. (2022)
17	*Acanthopanax senticosus*	UV-B radiation (3.442 kJ m^{-2} d^{-1})	Morphological and Photosynthetic	Reduced plant height, stem diameter, petiole length, and gaseous exchange parameters, which might be the result of non-stomatal limiting factors.	Wu et al. (2022)

(Continued)

Table 12.1 (Continued)

S. no.	Plant species	Stressor and intensity	Attribute affected	Mechanism of action	References
18	*Perilla frutescens*	6 and 10 W cm^{-2} UV-B for 3 days (Experiment 1) and 4 and 6 W cm^{-2} UV-B (Experiment 2)	Morphological	Decreased weight in the first experiment and in the second experiment leaves turned green, probably due to decomposition of accumulated anthocyanins.	Yoshida et al. (2022)
19	*Arabidopsis thaliana*	UV-B (1 W m^{-2}) and High light (1000 μmol photons m^{-2} s^{-1})	Photosynthetic	Decline in transpiration rate and photosynthetic rate, which might be attributed to reduced photochemical activity of PSII	Kreslavski et al. (2021)
20	*Sinopodophyllum hexandrum*	UV-B radiation (107 μW cm^{-2})	Morphological and photosynthetic	Decline in the number of petioles, leaf area, stem diameter, root diameter, plant height, and content of chlorophyll a and b (due to downregulation of certain genes).	Lv et al. (2021)
21	*Solanum lycopersicum*	UV-B radiation (0.07642 kJ m^{-2}) for 5 min/day	Morphology	Reduced plant height, which might be attributed to interference at the cell cycle level.	Mariz-Ponte et al. (2021)
22	*Conyza blinii*	20, 40, and 80 μW cm^{-2} UV-B	Photosynthetic	Lowered chlorophyll a and b and carotenoid content.	Zhan et al. (2021)
23	*Rhododendron chrysanthum*	2.3 W m^{-2} UV-B	Photosynthetic	Maximum and effective quantum yield of PSII were lowered.	Zhou et al. (2021)
24	*Solanum lycopersicum*	High light (1000 μmol photons m^{-2} s^{-1}) for 6 h	Photosynthetic	Induced reversible PSII photoinhibition.	Bednarczyk et al. (2020)
25	*Arabidopsis thaliana*	High light (120 μmol photons m^{-2} s^{-1})	Photosynthetic	Decrease in carotenoid and total chlorophyll content, which might be attributed to the inhibition of chlorophyll biosynthesis.	Chen et al. (2020)
26	*Silene littorea*	UV light (90% and 25% of the UV-A and UV-B radiation transmitted)	Photosynthetic	Reduced photochemical efficiency, which might be attributed to photoinhibition of PSII by UV rays.	Del Valle et al. (2020)
27	*Lactuca sativa*	High light (500 μmol m^{-2} s^{-1} for 4 h)	Morphological	Reduced leaf area, which might be attributed to a response to high light stress by increasing thickness and reducing leaf area.	Shao et al. (2020)

28	*Solanum lycopersicum*	UV-B radiation (15 min/day equivalent to 1.19 kJ m^{-2})	Morphological and Photosynthetic	Root length and leaf number were slightly decreased after 11th day of treatment, while chlorophyll levels remained unchanged, which might be attributed to a decrease in de-epoxidation of the xanthophyll cycle.	Mannucci et al. (2020)
29	*Arabidopsis thaliana*	High light (500 μmol photons m^{-2} s^{-1})	Photosynthetic	Decreased chlorophyll and carotenoid content, which might be due to altered photosystem.	Velitchkova et al. (2020)
30	*Arabidopsis thaliana*	High light (6000 μmol m^{-2} s^{-1})	Photosynthetic	Induced damage to the photosystem and reduced stomatal aperture.	Balfagón et al. (2019)
31	*Fragaria* X *ananassa*	High light (150 and 210 μmol m^{-2} s^{-1})	Morphological	Decline in number of roots and length compared to low light (90 μmol m^{-2} s^{-1}) exposed plants, leaves color changed to purple-red, which might be attributed to altered pigment levels.	Zheng et al. (2019)
32	*Fagopyrum esculentum*	UV radiation	Morphological	Lowered plant height, which might be attributed to auxin sensitivity to UV-B.	Golob et al. (2018)
33	*Solanum lycopersicum*	High light (500 and 1000 μmol m^{-2} s^{-1})	Photosynthetic	Decrement in photosynthetic performance in terms of reduced net photosynthetic rate, stomatal conductance, and transpiration rate, which might be attributed to reduced activity of RuBisCO.	Lu et al. (2017)
34	*Prunella vulgaris*	UV-B (35 μW cm^{-2} nm^{-1})	Photosynthetic	Lowered chlorophyll content and carotenoid content, which might be attributed to the saturation of pathways that increase carotenoid induction.	Zhang et al. (2017)
35	*Solanum lycopersicum*	High light (800 μmol m^{-2} s^{-1})	Photosynthetic	High light for days reduced chlorophyll content, which might be due to inhibition of pigment biosynthetic enzymes.	Gerganova et al. (2016)

exposure of *Solanum lycopersicum* accompanied by lowered levels of carotenoid, total chlorophyll, and chlorophyll a and b, which were linked to either accelerated breakdown of chlorophyll pigments or deteriorates biomass (Khare et al. 2023). Additionally, chlorophyll fluorescence and the levels of photosynthetic pigments under high light were used to examine the photosynthetic properties of four distinct Triticeae crops, including wheat, triticale, barley, and highland barley. Results revealed that wheat and triticale plants exhibited reduced chlorophyll content and decreased intercellular CO_2 when exposed to the light intensity of 1200 μmol m^{-2} s^{-1} photons (Chen et al. 2023). Similarly, exposure of *Arabidopsis* to high light having intensity of 6000 μmol m^{-2} s^{-1} induced damage to the photosystem and reduced stomatal aperture (Balfagón et al. 2019). Various studies depicting the negative impact of high light and UV radiation on the photosynthetic parameters of plants are listed in Table 12.1.

12.4.3 Oxidative Stress Augmentation

UV radiation augment ROS production, which further induces oxidative stress in plants. ROS are produced by vital energy-generating activities in plants, including photosynthesis and respiration within the chloroplast, apoplast, mitochondria, peroxisomes, etc. ROS production is unavoidable in plants, and plants have several enzymatic and non-enzymatic mechanisms to scavenge excessive ROS, which maintains their balance. However, in the presence of UV radiation, this balance is disrupted. Plants produce more ROS under UV light stress due to metabolic activity disruption and elevated membrane-localized NADPH-oxidase activity that leads to oxidative stress (Hideg and Vass 1996; Kalbina and Strid 2006). These ROS are very reactive, and their overproduction can lead to cellular damage, programmed cell death (PCD), and more upsurge in ROS production (Raja et al. 2017; Mittler et al. 2022). ROS oxidize carbohydrates, lipids, nucleic acids, and proteins, followed by oxidative stress, which further causes molecular and cellular damage, ultimately leading to cell death (Hasanuzzaman et al. 2020). According to Tripathi et al. (2017), *Triticum aestivum* L. seedlings subjected to two levels of UV-B radiation *viz.* ambient (8.6 kJ m^{-2} d^{-1}) and enhanced (ambient + 2.8 kJ m^{-2} d^{-1}) had higher MDA electrolyte leakage (EL), increased concentrations of superoxide anion ($O_2^{-}\cdot$), hydrogen peroxide (H_2O_2), and malondialdehyde (MDA). In another study done by Yang et al. (2020), it was reported that levels of ROS, including $O_2^{-}\cdot$, H_2O_2, and MDA, were significantly increased in UV-C (0.284 mW cm^{-2} intensity) light-exposed *Glycine max* L. plants. Also, under dark conditions, 4 hours of UV-B radiation raises the levels of Electrolyte Leakage (EL), MDA, and O_2^{-}, contents in *Morus alba* seedlings as observed by Li et al. (2020). Exposure to excessive UV-B radiation also leads to the overproduction of ROS, which causes an increase in MDA content in plants (Tan et al. 2023).

Plants exposed to high light intensity also experience oxidative stress due to the induced generation of ROS, as reported by various studies. Vuleta et al. (2016) found that *Iris pumila* cultivated in full sunshine had 20% greater MDA concentration in contrast to plants that were grown in the shade. Similarly, Shengxin et al. (2016) also suggested that rapeseed (*Brassica napus* L. cv. Zhongshuang11) seedlings exposed to HL (550 ± 20 μmol photons m^{-2} s^{-1}, 16 hours) had higher levels of MDA, $O_2^{-}\cdot$, and H_2O_2.

Also, Zha and Liu (2019) categorized lettuce plants (*Lactuca sativa* L.) into three groups depending on the exposure of light intensity *viz.*, low light exposed lettuce plants (LL; 100 μmol m^{-2} s^{-1}), medium light exposed lettuce plants (ML; 200 μmol m^{-2} s^{-1}), and high light exposed lettuce plants (HL; 300 μmol m^{-2} s^{-1}). It was observed that H_2O_2 content rose at both ML and HL, but $O_2^{-}\cdot$ content only increased in the latter. MDA content followed a similar pattern to $O_2^{-}\cdot$ content. However, the H_2O_2 and $O_2^{-}\cdot$ contents were rather stable over the first 9 days of therapy under LL stress and showed an increasing trend on the 12th day same as HL lettuce leaves. Therefore, it is evident that increased ROS generation as a result of UV and high light exposure acts as an oxidative stressor.

12.4.4 DNA Damage and Imparities

DNA damage is caused by various agents, including chemicals, heavy metals, alkylating agents, crosslinking agents, etc. UV radiation and high light are two of the factors that induce DNA damage, like cross-links, adducts, and oxidative cleavages (Martin 2008). From the UV spectrum, UV-A has shown the least damaging effect because it is not absorbed by DNA, whereas other UV radiations like UV-B and UV-C are strongly absorbed by the nucleus, mitochondria, and chloroplasts causing mutations. Since DNA strongly absorbs UV-B radiation, it is the most critical target for UV-induced damage (Gill et al. 2015). Exposure to excessive UV-B radiation leads to the overproduction of ROS, which causes an increase in MDA content in plants (Tan et al. 2023). It is also detrimental to nucleotides and proteins. UV-B produces various major modifications in DNA, known as photoproducts. The most probable photoproducts to produce mutations are two different types of lesions that join nearby pyrimidine residues on the same strand by forming covalent bonds between adjacent pyrimidines. The principal lesions are cyclobutene pyrimidine dimers (CPDs), which account for roughly 75% of UV-B-induced DNA damage, and 6-4 photoproducts (6-4PPs) (Kciuk et al. 2020). Consequently, resulting in the inhibition of the transcription and replication process (Britt 2002). On the other hand, minor DNA damage comprises hydrated bases, oxidized bases, and single-strand breaks (Gill et al. 2015; Takahashi et al. 2011). CPDs and 6-4PPs cause DNA bending, whereas 6-4PPs produce DNA unwinding, which can result in growth retardation, death, or mutagenesis (Tuteja et al. 2009). Similarly, high light stress also induces DNA damage and imparities. While studying the impact of high light and high temperature on rice (*ls1*), Qiu et al. (2019) observed that there was more DNA damage in the *ls1* mutant under higher light stress conditions than under the low light condition. This DNA damage leads to overaccumulation of ROS, which was confirmed by measuring the contents of H_2O_2, O_2^{-}, MDA, and catalase (CAT) activity. The accumulation of ROS can exacerbate genomic damage. More severe DNA damage, together with excess ROS accumulation, causes oxidative stress, which can lead to chloroplast degradation and cell apoptosis, ultimately resulting in a local lesions phenotype. Thus, long exposure to UV radiation and high light induces oxidative stress in plants, thereby damaging essential biomolecules such as DNA, RNA, proteins, and membranes (Ferreyra et al. 2021). Table 12.2 enlists various studies depicting the impact of high light and UV radiation on oxidative stress.

Table 12.2 Enlists various studies depicting the impact of high light and UV radiation on oxidative stress.

S. No.	Plant species	Stressor and intensity	Mechanism of action	References
1	*Triticum aestivum*	UV light with 0–60 min of exposure time in increment of 10 min	An increase in peroxidase (POX) activity with 30 min exposure, and further exposure led to a significant decrease in enzyme activity	Sharlaeva and Chirkova (2021)
2	*Lactuca sativa*	High light (300 μmol m^{-2} s^{-1})	ROS production, lipid peroxidation, increase in MDA content	Zha and Liu (2019)
3	*Oryza sativa*	High light (~300 μmol photons m^{-2} s^{-1})	Higher accumulation of H_2O_2, decreased level of CAT and APX suggested a weakened antioxidant system	Hamdani et al. (2019)
4	*Triticum aestivum* Parabola-Stress Tolerant (ST) Radunia-Stress Sensitive (SS) *Avena sativa* Bingo (ST) Siwek (SS) *Hordeum vulgare* CAM/B1 (ST) Masesi (SS)	UV light UV-A and UV-B both have a dose of 0.48 J cm^{-2}	Generation of ROS ($O_2^{-\bullet}$, H_2O_2) and stable radicals (semiquinone, phenoxyl, and carbon-centered); lipid peroxidation takes place	Kurdziel et al. (2018)
5	*Rosa hybrida* cv. "*Avalanche*"	High light (1500 μmol m^{-2} s^{-1})	Ascorbate peroxidase (APOX) and catalase (CAT) activity decrease, Superoxide dismutase (SOD), H_2O_2, and malondialdehyde (MDA) content increased	Bayat et al. (2018)
6	*Phyllostachys edulis*	High light (1500 μmol m^{-2} s^{-1})	Increased level of ROS, MDA, antioxidant enzymes like SOD and CAT due to high light stress	Gu et al. (2017)
7	*Brassica napus*	High light (550 μmol photons m^{-2} s^{-1})	High content of $O_2^{\bullet-}$ and H_2O_2, and MDA	Shengxin et al. (2016)
8	*Iris pumice*	High light (1378 ± 44 μmol m^{-2} s^{-1})	Increased SOD, mitochondrial (MnSOD), APX activity only in exposed plant and CAT activity noticed was similar in both exposed and non-exposed plants	Vuleta et al. (2016)
9	*Arabidopsis thaliana*	High light (900 μmol m^{-2} s^{-1})	Accumulation of $O_2^{\bullet-}$ and H_2O_2, photo-bleaching of leaf tissues, and impairment in photosynthetic activity	Awad et al. (2015)
10	*Capsicum annuum* L.	UV light (UV-A) 6.1, (UV-B) 5.8, and (UV-C) 5.7 W m^{-2}	Activities of antioxidant enzymes like CAT, glutathione reductase (GR), glutathione peroxidase (GPx), and ASP were increased	Mahdavian et al. (2008)

12.5 Acclimation and Adaptation to UV and High Light in Plants

Light influences several developmental, physiological, and morphological processes in the life cycle of a plant. The exposure duration, intensity, and wavelength of light activate explicit signaling cascades and influence the expression of genes downstream, which further induce photomorphogenic responses (Mannucci et al. 2020). Besides influencing the development and growth, sunlight, in particular the UV part (10–400 nm), induces stress in plants (Müller-Xing et al. 2014).

12.5.1 UV Regulated Specialized Metabolites and Hormones

UV radiation modifies several crucial organic compounds in living organisms and alters plant's primary and secondary metabolism. The discovery of UVR8, a UV-B photoreceptor, helped in identifying the hormonal connection in response to light. Moreover, a considerable overlap between the UVR8 and phytochrome reactions suggests that a portion of UV-B responses triggered by UVR8 is under the control of PHYTOCHROME INTERACTING FACTOR (PIF) (Vanhaelewyn et al. 2016). COP1 is central to numerous photomorphogenic pathways that are regulated by many photoreceptors. In the presence of UV-B, the interaction between COP1 and UVR8 is not only crucial for UV-B signaling but is also vital for nuclear accretion of UVR8 in retortion to UV-B (Qian et al. 2016; Yin et al. 2016). As a defense mechanism to prevent UV-B-implicated damage, plants accumulate flavonoids, including flavonols, proanthocyanidins, and anthocyanins. UV-B persuades expression of genes involved in flavanol biosynthesis in *Arabidopsis* and its mutants lacking flavonoids express UV-B sensitivity (Clayton et al. 2018).

The UVR8 locus in *Chrysanthemum morifolium* (CmUVR8), expressed in leaves, flowers, and roots, regulates UV-B persuaded expression of four different genes related to flavonoid biosynthesis including *CHI (chalcone isomerase)*, *CHS (chalcone synthase)*, *FLS1 (flavonol synthase 1)*, and *F3H* (*Flavonoid 3-hydroxylase*) and the accumulation of flavonoids under UV-B stress (Yang et al. 2018a, b). Tocopherols, flavonoids, and polyamines accumulation in response to UV-induced stress in *Arabidopsis* was observed (Hectors et al. 2014). α-tocopherol accumulated rapidly and its levels remained elevated during UV exposure whereas polyamines sequestration was fast and transient implicating their short-term UV response. In contrast, flavanol sequestration took place in a sustained manner. These discrete secondary metabolite accumulation patterns confirmed the dynamic process of UV acclimation in *Arabidopsis*. The study also revealed that flavonol glycosides, including kaempferol and quercetin di- and triglycosides, were specifically rhamnosylated at the seventh position, suggesting 7-O-rhamnosylation as an imperative process in UV tolerance.

Hormones including auxins, gibberellins (GA), cytokinins, ethylene (ET), jasmonic acid (JA), abscisic acid (ABA), and salicylic acid (SA) are intricately involved in metabolic and morphological response regulation in plants. UV-B radiation influences the hormonal pathways, and its downstream effects on plants' morphological and defensive mechanisms have been largely investigated. A significant reduction in levels of indoleacetic acid and ET and augmentation in levels of SA have been observed in roots and leaves of Micro-Tom tomato plants in response to exposure to UV-B radiation without any marked changes in plant architecture, suggesting the role of hormones during UV-induced stress and acclimatization responses (Mannucci et al. 2020). Similarly, Mewis et al. (2012) reported augmentation in levels of flavonoids, including quercetin and kaempferol, in broccoli seedlings on exposure to UV-B. Besides, stimulation of glucosinolate production, especially 4-methylsulfinylbutyl glucosinolate and 4-methoxy-indol-3-ylmethyl glucosinolate, under the effect of UV-B treatment. Accretion of protective glucosinolate metabolites was further accompanied by amplified expression of genes related to JA and salicylate signaling defense cascades. A study in two different cultivars of *Pisum sativum* HUP-2 and HUDP-15 exposed to UV-B revealed an elevation in levels of SA in both cultivars, followed by an increase in levels of flavonoids. Higher buildup of SA in the HUP-2 cultivar is proposed to provide higher sensitivity toward UV-B, whereas higher levels of JA and antioxidant enzymes impart tolerance to HUDP-15 cultivar against the applied stress (Choudhary and Agrawal 2014). In another study, the levels of quercetin, rutin, and total flavonoids increased significantly during UV radiation treatment, along with a reduction in levels of IAA and ABA in soybean leaves. A positive correlation between ABA levels and seed yield and a negative correlation between the levels of IAA concentration and seed yield were observed (Mao et al. 2017).

12.5.2 Role of Brassinosteroids in Elevating Plant Tolerance to High Light and UV Exposure

Steroidal hormones, brassinosteroids (BRs), exhibit strong growth-stimulating properties and play an essential role in all stages of plant development. White light and BRs antagonistically control the change from etiolation in dark conditions to photomorphogenesis under light conditions. Transcriptomic data suggest that the effects of UV-B in plants are dependent upon BR signaling. In *Arabidopsis thaliana* UV-B regulates certain genes of BR pathways, including DWARF1 (DWF1), BES-INTERACTING MYC-LIKE PROTEIN (BIM1), BRASSINOSTEROID INSENSITIVE 1 SUPPRESSOR 1 (BRS1), and RELATED TO ABA-INSENSITIVE3/VIVIPAROUS1 (RAV1) (Hectors et al. 2007). BR signaling is responsible for UVR8-controlled hypocotyl inhibition, and it negatively affects plant tolerance to UV-B stress. BR signaling and BR-deficient mutants have transformed UV-B stress tolerance capacity, and BES1 (BRI1-EMS-SUPPRESSOR 1) acts downstream of BRI1 to impede UV-B stress (Liang et al. 2018). BR signaling obstructs UV-B stress reactions in *Arabidopsis* and several other crops by regulating flavonol biosynthesis. BES1, a vital transcription factor implicated in BR signaling, suppresses the expression of *MYB11, MYB12*, and *MYB111* (transcription factor genes), which further stimulate flavanol biosynthesis. BES1 binds directly to the *MYB* promoters in a BR-augmented manner to inhibit their expression, thus decreasing flavonol accumulation. However, broadband UV-B exposure downregulates the expression of *BES1*, thereby stimulating flavonol accumulation. Also, the buildup of flavonol is substantially higher in BR-deficient mutants in comparison to the wild-type (Liang et al. 2020).

12.6 Genetic Responses in Relation to Brassinosteroids

Hormones play a major role as regulators of growth and development in both plants and animals. Among these hormones, BRs, a phytohormone, exerts a wide array of effects on various processes including plant architecture, cell elongation and division, biomass production, photomorphogenesis, root development, and seed germination (Hafeez et al. 2021). Additionally, BRs are pivotal in mediating plant responses to diverse environmental stresses like heat, light, salt, drought, oxidative stress, heavy metal exposure, and pathogen attacks. Over the last 20 years, significant advancements have been attained in understanding the signal transduction pathways that transmit messages from cell surface receptors to the nucleus, where modulation of thousands of genes in response to BRs occurs. The intricate signaling cascades triggered by brassinosteroids and genetic responses to BRs elicit precise cellular and developmental reactions (Guo et al. 2013; Song et al. 2019; Nolan et al. 2020).

12.6.1 Signaling Pathways and Their Transcriptional Response

Light serves not solely as an energy provider but also as a crucial environmental cue governing plant growth and development. Factors such as light quality (spectral composition), quantity, and duration (photoperiod) influence the plant's morphological and architectural adaptation *via* processes like photomorphogenesis, phototropism, and avoidance responses (Lv and Li 2020). Heat stress represents a significant abiotic challenge, leading to a reduction in net photosynthetic rates through the interruption between PSII and photochemical activities. However, over time, plants have developed diverse adaptive mechanisms to mitigate the adverse effects of heat stress by maintaining membrane integrity, producing heat shock proteins (HSPs) and ROS-scavenging enzymes, and initiating molecular chaperone signaling. In higher plants, HSPs are significantly induced across various cells and organisms to promote thermotolerance (Falcioni et al. 2024). The earlier work suggests that HSPs during heat stress primarily function as molecular chaperones, aiding in the restoration of cellular protein homeostasis. For instance, HSP70 and HSP90 play crucial roles in protein folding, transport, oligomeric assembly, reactivation of denatured proteins, activating genes which are specific to binding motifs found in the promoters of heat shock genes, thus enhancing plant thermotolerance (Figure 12.3) (Wang et al. 2023). It was found that 21 HSFs encoded in the *Arabidopsis thaliana* genome grouped into three major classes (HSFA, HSFB, and HSFC), distinguished by the characteristics of their oligomerization domain (Park et al. 2015; Albertos et al. 2022; Yang et al. 2024).

Among the myriad plant regulators, brassinosteroids (BRs) govern various facets of plant development. Zhang et al. (2013) described that pre-treatment with brassinosteroids (BRs) in tomato plants mitigated heat-induced photosynthetic losses by strengthening antioxidant enzyme activity and modulating thermotolerance across various plant cultivars, irrespective of their heat sensitivity. Furthermore, BR signaling influences plant growth under high-temperature stress conditions. Multiple components of the BR signaling pathway have been identified in plants (Planas-Riverola et al. 2019). Earlier findings have revealed that the membrane steroid receptor BRI1 forms a binding complex with the BR ligand (BRI1-ASSOCIATED KINASE1 (BAK1). These interactions trigger a series of cytoplasmic signaling events, leading to the activation of gene expression associated with BR, thereby propelling various cellular processes (Planas-Riverola et al. 2019). Among these, BZR1 (Brassinazole Resistant 1) plays a crucial role as a positive regulator of BR signaling, serving dual functions in both activating BR responses and regulating the negative feedback loop of BR biosynthesis. BRs are perceived by the receptor-like kinase BRI1 and trigger a phosphorylation-dependent signaling cascade. This cascade modulates

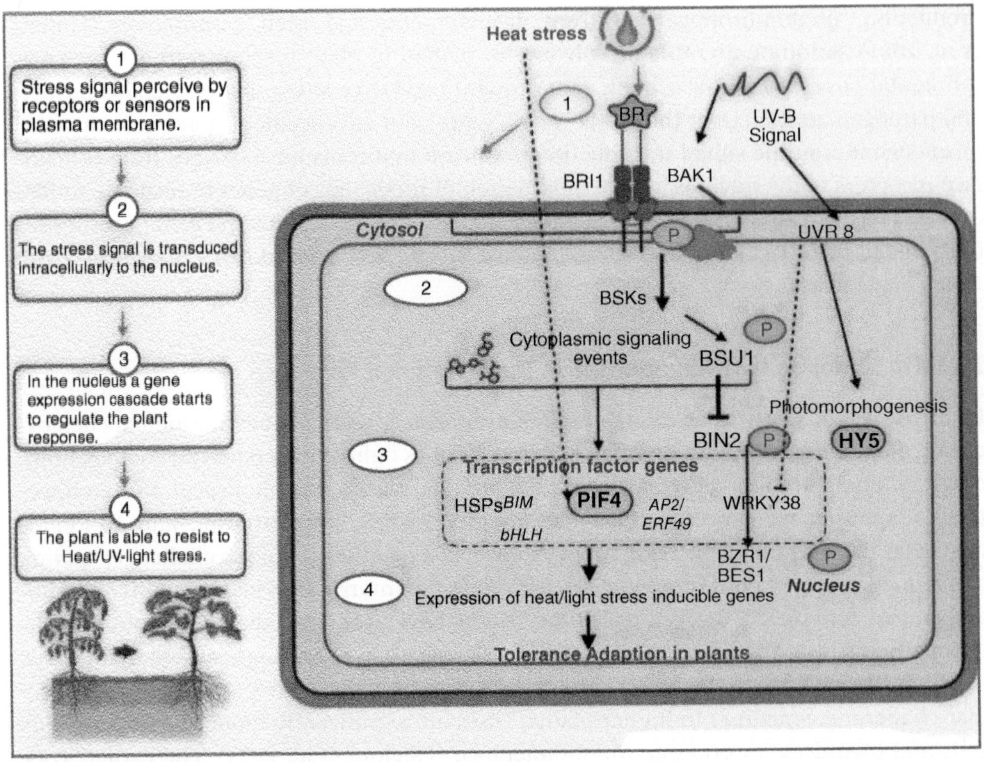

Figure 12.3 Multiple components of the brassinosteroid (BR) signaling cascade stimulate plant growth under high-temperature/light stress conditions.

the phosphorylation of transcription factors (TFs), thereby affecting their protein stability, nuclear localization, and DNA binding abilities (Belkhadir and Jaillais 2015). Transcription factors are regulated through proteins like BIM proteins, PIF4, and BEE 1–3, a BRs basic helix-loop-helix (bHLH) proteins (Bernardo-García et al. 2014; Albertos et al. 2022). Additionally, a small subfamily of six proteins with bHLH-like features, notably BES1 and BZR1, is also controlled by BR signaling. The phosphorylation status of BES1 and BZR1 is regulated by GSK3/shaggy-like kinases such as BR-INSENSITIVE 2 (BIN2) and phosphatases like BRI1 SUPPRESSOR 1 (BSU1). Similarly, it is also noted that BR signaling through BIN2 interacts with the autophagy pathway to direct plant growth and survival during stress and starvation (Figure 12.3). In the absence of BR, the kinase BIN2 phosphorylates and inactivates BES1 and BZR1, hindering their transcriptional regulatory functions (Han et al. 2023). Recent investigations have highlighted the significance of BR in enhancing thermotolerance in *Arabidopsis*, with BIN2 emerging as a critical regulator of heat stress responses. BES1, functioning as a transcriptional regulator, undergoes dephosphorylation to directly modulate the heat shock element, thereby regulating the expression of heat shock factors (HSFs) to counteract heat stress (Luo et al. 2022). Upon activation of BR signaling, there is an increase in the abundance of BES1 and BZR1 proteins and an enrichment of their non-phosphorylated active forms (Albertos et al. 2022).

BZR1 exerts its influence on BR responses by modulating the expression of its direct target genes (Nolan et al. 2020). It is noticed that the control of thermomorphogenesis by phytochrome interacting factor 4 (PIF4) relies on the activity of BZR1 (Ibanez et al. 2018). This study demonstrated that BZR1 is a temperature-dependent enhancer of PIF4, functioning as a primary coordinator of growth. Moreover, studies unveil a collaborative interplay between BZR1, ERF49, and PIF4, forming a heterodimer to jointly regulate gene expression and participate in the plant's response to elevated temperatures (Ibanez et al. 2018). Increased temperatures stimulate active accumulation of PIF4, favoring the formation of BES1–PIF heterodimers in nuclear protein complexes over BES1 homodimers. This diminishes the pool of available active BES1 homodimers, suppressing BR biosynthesis and impeding BR signaling (Jiroutová et al. 2019; Khan et al. 2023).

It is noticed that the ethylene responsive factors (ERF), members of the AP2/ERF multi-gene family, also emerge as one of the largest clusters of transcription factors, showing significant influence over various biological processes integration of various stress-responsive genes with crosstalk of BRs. Among these families, ERF49, also known as DREB2D, belongs to the DREB2 subfamily of AP2/ERF domain transcription factors (Chen et al. 2022b). Their observations emphasized the role of ERF49 in heat stress response, elucidating its negative regulatory role in plant thermotolerance. Furthermore, they revealed ERF49 capacity to modulate the expression of heat-inducible gene (HSFA2), suggesting its involvement in transcriptional machinery post-heat stress. Unravelling the intricate regulatory networks between ERF49, HSFA2, and downstream targets holds promise for a deeper understanding of plant thermotolerance mechanisms in response to heat stress.

A recent discovery unveils another interaction involving BES1 and BES-interacting Myc-like (BIM1) with the protein UVR8, a UV-B light photoreceptor mediating plant UV-B light response. In the absence of UV-B radiation, UVR8 proteins form homodimers, but exposure to UV-B triggers the accumulation of UVR8 monomers (Liang et al. 2018). These monomers directly interact with the COP1 E3 ubiquitin ligase to transmit the UV-B signal (Yin et al. 2016). UV-B light hampers plant growth by suppressing the BR response. This inhibition occurs through UVR8's physical interaction with active BES1 and BIM1 transcription factors, leading to the suppression of their activities. While the interaction between UVR8 and BES1/BIM1 is not dependent on UV-B, UV-B exposure enhances this interaction. Upon activation by UV-B, UVR8 localizes in the nucleus, where it hinders the DNA-binding abilities of TFs of BES1 and BIM1, thereby impeding BES1-mediated transcription. It is notable that the regulation of BES1 and BIM1 is a crucial network between the BR and light signaling pathways. BES1 and BIM1 function downstream of UVR8 to control the expression of genes associated with cell elongation (Figure 12.4) (Lin et al. 2021).

12.6.2 Alteration in Epigenetic Markers

Epigenetics refers to the heritable modulation of gene expression without altering the genetic sequence itself. This intricate process comprises various mechanisms like DNA methylation, histone modifications, chromatin remodeling, and regulation by non-coding RNAs (Figure 12.4) (Lv and Li 2020). Epigenetic regulation plays a crucial role in maintaining the balance of growth-promoting phytohormones like BR and GA. The regulation of gene expression by brassinosteroids (BRs) involves epigenetic mechanisms, notably histone

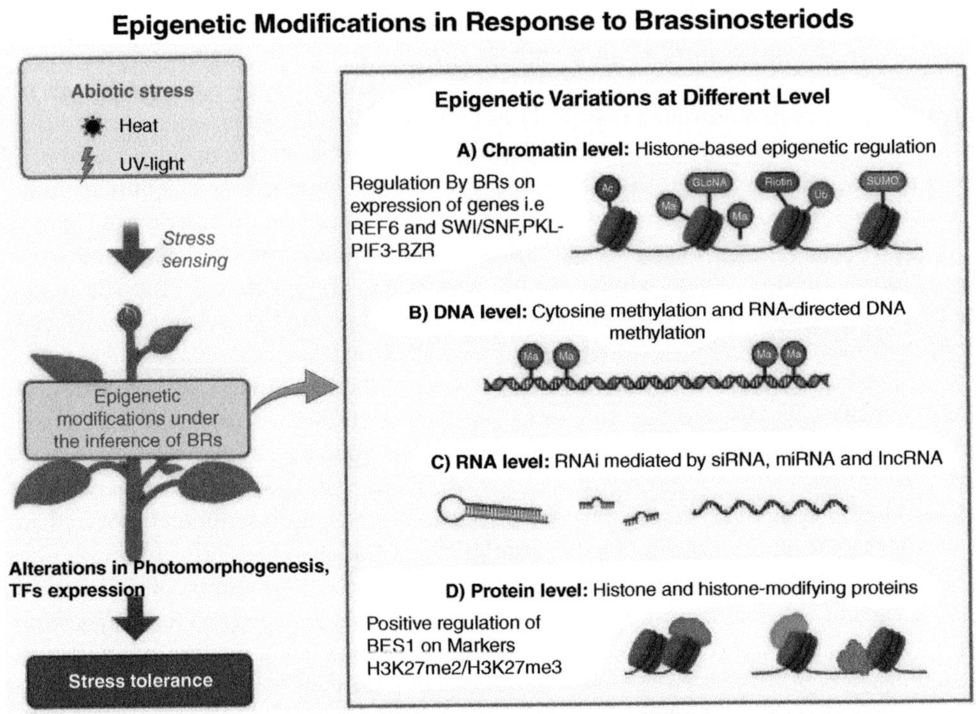

Figure 12.4 Alteration in epigenetic markers under the inference of brassinosteroids.

modifications (Jiang et al. 2023). One key modification associated with the suppression of gene expression is the di- and trimethylation of Histone 3 at Lysine 27 (H3K27me2/H3K27me3) (Lu et al. 2011). The studies indicate that BES1 interrelates with photomorphogenesis TFs like EARLY FLOWER6 and RELATIVE OF EARLY FLOWER6 (REF6), which act as positive regulators in the BR signaling pathway by eliminating suppressive markers *viz.* H3K27me2/H3K27me3 (Li et al. 2016). REF6 functions as an H3K27 demethylase, targeting the removal of methyl groups specifically from double- and triple-methylated residues of H3K27. The positive regulation of BES1 thereby activates gene expression and involves chromatin remodeling on account of REF6 and SWI/SNF chromatin remodeler interactions (Li et al. 2016). Similarly, another work also shows the involvement of BZR1 and PIF3 with the PICKLE, a chromatin remodeler, to alleviate H3K27me3 repression on target gene promoters to facilitate BR-induced gene expression (Zhang et al. 2014). These findings underscore the critical role of histone-modifying enzymes and chromatin-remodeling factors in relieving the repression of BR-regulated genes.

Studies by Wei et al. (2014) demonstrate that in rice, OsDCL3a directly influences genes responsible for BR and GA homeostasis. Furthermore, the BZR1/BES1 transcription factors also engage with the chromatin-remodeling factor PKL and form a complex of PKL-PIF3-BZR1. This complex promotes skotomorphogenesis by suppressing trimethylation of histone H3 Lys-27 on target gene promoters (Zhang et al. 2014). It is observed that BZR1 employs a histone demethylase to the intron of FLOWERING LOCUS C (FLC) and leads to decreased H3K27 trimethylation activity; however, it activates FLC expression to suppress flowering (Li et al. 2018). Consequently, PKL serves as a crucial turning point in light, BR, and GA signaling pathways, facilitating epigenetic regulation of plant development. BZR1, ARF6, and PIF4 transcription factors interact and share numerous target genes, enhancing their DNA-binding and transcriptional activities, thereby promoting plant growth synergistically. This cooperative interaction, along with the inhibitory effect imposed by DELLAs, is referred to as the BZR1-ARF6-PIF4/DELLA (BAP/D) modulation.

Epigenetic mechanisms also facilitate communication between BR and ABA (abscisic acid). The findings of Ryu et al. (2014) explained the repressive action of BES1 on a complex of TPL and HDA19. This is on account of negatively regulating ABI3 transcription by deacetylating histones on ABI3 chromatin. Furthermore, BZR1 enhances the expression of histone demethylase JMJ30, which removes H3K27me histone from the BZR1 locus in response to ABA that specifying epigenetic-mediated crosstalk between BR and ABA (Wu et al. 2020). Analysis of the *Arabidopsis* RNA-seq database (ARS) unveiled distinctive expression patterns of DNA METHYLTRANSFERASE 2, and MET2b genes responsible for encoding a DNA methyltransferase. Additionally, earlier findings showed ectopic expression of MET2b in seedlings subjected to brassinosteroid (BR) inhibitors, indicating a regulatory link between BR signaling and MET2b expression (Lu et al. 2022).

12.7 Conclusion and Outcome

Decades of research have made the BR pathway one of the scrutinized signaling cascades in plants. Genetic approaches and various investigations exhibit intricate crosstalk with hormones and stress responses, and their cell- and tissue-specific variations offer manipulable avenues for optimizing plant growth and stress responses. The extant data pertaining to the interplay between BR and UV radiation and high light stress are limited, despite suggestive indications of a functional association with components of the UV RESISTANT LOCUS 8 (UVR8) pathway. This association posits the potential involvement of BR in UV-mediated photomorphogenic processes. Crucial research is needed to understand the intricate aspects of BR biosynthesis, transport, and its role in plant growth under UV and high light stress conditions. Identifying BR transporters is vital for mitigating the adverse effects of UV stress and high light exposure. Strategic manipulation of the BR signaling pathway in crops is essential for enhancing plants' resilience in response to fluctuations in these environmental parameters.

References

Abuelsoud, W., Cortleven, A., and Schmülling, T. (2020). Photoperiod stress induces an oxidative burst-like response and is associated with increased apoplastic peroxidase and decreased catalase activities. *Journal of Plant Physiology* 253: 153252. https://doi.org/10.1016/j.jplph.2020.153252.

Albertos, P., Dundar, G., Schenk, P. et al. (2022). Transcription factor BES1 interacts with HSFA1 to promote heat stress resistance of plants. *EMBO Journal* 41 (3): e108664. https://doi.org/10.15252/embj.2021108664.

Ali, A., Santoro, P., Mori, J. et al. (2023). Effect of UV-B elicitation on spearmint's (*Mentha spicata* L.) morpho-physiological traits and secondary metabolites production. *Plant Growth Regulation* 104: 63–76. https://doi.org/10.1007/s10725-023-01028-7.

Ali, M.B., Hahn, E.J. and Paek, K.Y. (2005). Effects of light intensities on antioxidant enzymes and malondialdehyde content during short-term acclimatization on micropropagated Phalaenopsis plantlet. *Environmental and Experimental Botany 54(2)*: 109-120. https://doi.org/J.ENVEXPBOT.2004.06.005.

Apel, K. and Hirt, H. (2004). Reactive oxygen species: metabolism, oxidative stress, and signal transduction. *Annual Reviews of Plant Biology* 55: 373–399. https://doi.org/10.1146/annurev.arplant.55.031903.141701.

Awad, J., Stotz, H.U., Fekete, A. et al. (2015). 2-cysteine peroxiredoxins and thylakoid ascorbate peroxidase create a water-water cycle that is essential to protect the photosynthetic apparatus under high light stress conditions. *Plant Physiology* 167 (4): 1592–1603. https://doi.org/10.1104/pp.114.255356.

Baena-González, E. and Aro, E.M. (2002). Biogenesis, assembly and turnover of photosystem II units. *Philosophical Transactions of the Royal Society of London. Series B: Biological Sciences* 357 (1426): 1451–1460. https://doi.org/10.1098/rstb.2002.1141.

Balfagón, D., Sengupta, S., Gómez-Cadenas, A. et al. (2019). Jasmonic acid is required for plant acclimation to a combination of high light and heat stress. *Plant Physiology* 181 (4): 1668–1682. https://doi.org/10.1104/pp.19.00956.

Bartwal, A., Mall, R., Lohani, P. et al. (2013). Role of secondary metabolites and brassinosteroids in plant defense against environmental stresses. *Journal of Plant Growth Regulation* 32: 216–232. https://doi.org/10.1007/s00344-012-9272-x.

Bayat, L., Arab, M., Aliniaeifard, S. et al. (2018). Effects of growth under different light spectra on the subsequent high light tolerance in rose plants. *AoB Plants* 10 (5): ply052. https://doi.org/10.1093/aobpla/ply052.

Bednarczyk, D., Aviv-Sharon, E., Savidor, A. et al. (2020). Influence of short-term exposure to high light on photosynthesis and proteins involved in photo-protective processes in tomato leaves. *Environmental and Experimental Botany* 179: 104198. https://doi.org/10.1016/j.envexpbot.2020.104198.

Belkhadir, Y. and Jaillais, Y. (2015). The molecular circuitry of brassinosteroid signaling. *New Phytologist* 206 (2): 522–540. https://doi.org/10.1111/nph.13269.

Bernardo-García, S., de Lucas, M., Martínez, C. et al. (2014). BR-dependent phosphorylation modulates PIF4 transcriptional activity and shapes diurnal hypocotyl growth. *Genes and Development* 28 (15): 1681–1694. https://doi.org/10.1101/gad.243675.114.

Britt, A. (2002). Repair of damaged bases. *The Arabidopsis Book/American Society of Plant Biologists* 1: e0005. https://doi.org/10.1199/tab.0005.

Camacho, I.S., Theisen, A., Johannissen, L.O. et al. (2019). Native mass spectrometry reveals the conformational diversity of the UVR8 photoreceptor. *Proceedings of the National Academy of Sciences* 116 (4): 1116–1125. https://doi.org/10.1073/pnas.1813254116.

Casal, J.J. (2013). Photoreceptor signaling networks in plant responses to shade. *Annual Review of Plant Biology* 64: 403–427. https://doi.org/10.1146/annurev-arplant-050312-120221.

Casati, P., Morrow, D.J., Fernandes, J. et al. (2011). Rapid maize leaf and immature ear responses to UV-B radiation. *Frontiers in Plant Science* 2: 33. https://doi.org/10.3389/fpls.2011.00033.

Chen, H., Huang, X., Gusmaroli, G. et al. (2010). *Arabidopsis* CULLIN4-damaged DNA binding protein 1 interacts with CONSTITUTIVELY PHOTOMORPHOGENIC1-SUPPRESSOR OF PHYA complexes to regulate photomorphogenesis and flowering time. *The Plant Cell* 22 (1): 108–123. https://doi.org/10.1105/tpc.109.065490.

Chen, L.X., Mao, H.T., Lin, S. et al. (2023). Different photosynthetic response to high light in four triticeae crops. *International Journal of Molecular Sciences* 24 (2): 1569. https://doi.org/10.3390/ijms24021569.

Chen, X., Xue, H., Zhu, L. et al. (2022a). ERF49 mediates brassinosteroid regulation of heat stress tolerance in *Arabidopsis thaliana*. *BMC Biology* 20: 254. https://doi.org/10.1186/s12915-022-01455-4.

Chen, Y.E., Mao, H.T., Wu, N., Mohi Ud Din, A., Khan, A., Zhang, H.Y., and Yuan, S et al. (2020). Salicylic acid protects photosystem II by alleviating photoinhibition in *Arabidopsis thaliana* under high light. *International Journal of Molecular Sciences* 21 (4): 1229. https://doi.org/10.3390/ijms21041229.

Chen, Z., Dong, Y., and Huang, X. (2022b). Plant responses to UV-B radiation: Signaling, acclimation and stress tolerance. *Stress Biology* 2 (1): 51. https://doi.org/10.1007/s44154-022-00076-9.

Choudhary, K.K. and Agrawal, S.B. (2014). Cultivar specificity of tropical mung bean (*Vigna radiata* L.) to elevated ultraviolet-B: changes in antioxidative defense system, nitrogen metabolism and accumulation of jasmonic and salicylic acids. *Environmental and Experimental Botany* 99: 122–132. https://doi.org/10.1016/j.envexpbot.2013.11.006.

Christie, J.M., Arvai, A.S., Baxter, K.J. et al. (2012). Plant UVR8 photoreceptor senses UV-B by tryptophan-mediated disruption of cross-dimer salt bridges. *Science* 335 (6075): 1492–1496. https://doi.org/10.1126/science.1218091.

Christie, J.M., Blackwood, L., Petersen, J. et al. (2015). Plant flavoprotein photoreceptors. *Plant and Cell Physiology* 56: 401–413. https://doi.org/10.1093/pcp/pcu196.

Chu, R., Zhang, Q.H., and Wei, Y.Z. (2022). Effect of enhanced UV-B radiation on growth and photosynthetic physiology of *Iris tectorum* maxim. *Photosynthesis Research* 153 (3): 177–189. https://doi.org/10.1007/s11120-022-00933-8.

Clayton, W.A., Albert, N.W., Thrimawithana, A.H. et al. (2018). UVR8-mediated induction of flavonoid biosynthesis for UVB tolerance is conserved between the liverwort *Marchantia polymorpha* and flowering plants. *The Plant Journal* 96: 503–517. https://doi.org/10.1111/tpj.14044.

Cloix, C., Kaiserli, E., Heilmann, M. et al. (2012). C-terminal region of the UV-B photoreceptor UVR8 initiates signaling through interaction with the COP1 protein. *Proceedings of the National Academy of Sciences* 109 (40): 16366–16370. https://doi.org/10.1073/pnas.1210898109.

Correa, M.D.S.S., Saavedra, M.E.R.R., Parra, E.A.E. et al. (2023). Ultraviolet radiation and its effects on plants. In: *Abiotic Stress in Plants - Adaptations to Climate Change* (ed. M. Oliveira and A.A. Fernandes Silva). IntechOpen. https://doi.org/10.5772/intechopen.109474.

Cun, Z., Xu, X.Z., Zhang, J.Y. et al. (2023). Responses of photosystem to long-term light stress in a typically shade-tolerant species *Panax notoginseng*. *Frontiers in Plant Science* 13: 1095726. https://doi.org/10.3389/fpls.2022.1095726.

D'Orazio, J., Jarrett, S., Amaro-Ortiz, A. et al. (2013). UV radiation and the skin. *International Journal of Molecular Sciences* 14 (6): 12222–12248. https://doi.org/10.3390/ijms140612222.

Del Valle, J.C., Buide, M.L., Whittall, J.B. et al. (2020). UV radiation increases phenolic compound protection but decreases reproduction in *Silene littorea*. *PLoS One* 15 (6): e0231611. https://doi.org/10.1371/journal.pone.0231611.

Demarsy, E., Goldschmidt-Clermont, M., and Ulm, R. (2018). Coping with 'dark sides of the sun' through photoreceptor signaling. *Trends in Plant Science* 23 (3): 260–271. https://doi.org/10.1016/j.tplants.2017.11.007.

Demmig-Adams, B. (2003). Linking the xanthophyll cycle with thermal energy dissipation. *Photosynthesis Research* 76: 73–80. https://doi.org/10.1023/A:1024902927403.

Edelman, M. and Mattoo, A.K. (2008). D1-protein dynamics in photosystem II: the lingering enigma. *Photosynthesis Research* 98: 609–620. https://doi.org/10.1007/s11120-008-9342-x.

Falcioni, R., Chicati, M.L., de Oliveira, R.B. et al. (2024). Decreased photosynthetic efficiency in *Nicotiana tabacum* L. under transient heat stress. *Plants* 13 (3): 395. https://doi.org/10.3390/plants13030395.

Ferreyra, M.L.F., Serra, P., and Casati, P. (2021). Recent advances on the roles of flavonoids as plant protective molecules after UV and high light exposure. *Physiologia Plantarum* 173 (3): 736–749. https://doi.org/10.1111/ppl.13543.

Foyer, C.H. and Noctor, G. (2016). Stress-triggered redox signalling: what's in pROSpect? *Plant, Cell and Environment* 39 (5): 951–964. https://doi.org/10.1111/pce.12621.

Fu, E., Zhang, Y., Li, H. et al. (2023). Chitosan reduces damages of strawberry seedlings under high-temperature and high-light stress. *Agronomy* 13 (2): 517. https://doi.org/10.3390/agronomy13020517.

Ganguly, D.R., Crisp, P.A., Eichten, S.R. et al. (2018). Maintenance of pre-existing DNA methylation states through recurring excess-light stress. *Plant, Cell & Environment* 41 (7): 1657–1672. https://doi.org/10.1111/pce.13324.

Gayathri, B., and Anbudurai, P.R. (1997). Oxidative stress injury in tomato plants induced by supplemental UV-B radiation. *Biologia Plantarum* 39: 215–221. https://doi.org/10.1023/A:1000388719570.

Gerganova, M., Popova, A.V., Stanoeva, D. et al. (2016). Tomato plants acclimate better to elevated temperature and high light than to treatment with each factor separately. *Plant Physiology and Biochemistry* 104: 234–241. https://doi.org/10.1016/j.plaphy.2016.03.030.

Gill, S.S., Anjum, N.A., Gill, R. et al. (2015). DNA damage and repair in plants under ultraviolet and ionizing radiations. *The Scientific World Journal* 2015: 250158. https://doi.org/10.1155/2015/250158.

Gollan, P.J. and Aro, E.M. (2020). Photosynthetic signalling during high light stress and recovery: targets and dynamics. *Philosophical Transactions of the Royal Society B* 375 (1801): 20190406. https://doi.org/10.1098/rstb.2019.0406.

Golob, A., Stibilj, V., Kreft, I. et al. (2018). Selenium treatment alters the effects of UV radiation on chemical and production parameters in hybrid buckwheat. *Acta Agriculturae Scandinavica, Section B—Soil & Plant Science* 68 (1): 5–15. https://doi.org/10.1080/09064710.2017.1349172.

Gu, J., Zhou, Z., Li, Z. et al. (2017). Photosynthetic properties and potentials for improvement of photosynthesis in pale green leaf rice under high light conditions. *Frontiers in Plant Science* 8: 235715. https://doi.org/10.3389/fpls.2017.01082.

Guo, H., Li L., Aluru M. et al. (2013). Mechanisms and networks for brassinosteroid regulated gene expression. *Current Opinions on Plant Biology* 16: 545–553. https://doi.org/10.1016/j.pbi.2013.08.002.

Hafeez, M.B., Zahra, N., Zahra, K. et al. (2021). Brassinosteroids: molecular and physiological responses in plant growth and abiotic stresses. *Plant Stress* 2: 1–7. https://doi.org/10.1016/j.stress.2021.100029.

Hamdani, S., Khan, N., Perveen, S. et al. (2019). Changes in the photosynthesis properties and photoprotection capacity in rice (*Oryza sativa*) grown under red, blue, or white light. *Photosynthesis Research* 139: 107–121. https://doi.org/10.1007/s11120-018-0589-6.

Han, C., Wang, L., Lyu, J. et al. (2023). Brassinosteroid signaling and molecular crosstalk with nutrients in plants. *Journal of Genetics and Genomics* 50 (8): 541–553. https://doi.org/10.1016/j.jgg.2023.03.004.

Hasanuzzaman, M., Bhuyan, M.B., Zulfiqar, F. et al. (2020). Reactive oxygen species and antioxidant defense in plants under abiotic stress: revisiting the crucial role of a universal defense regulator. *Antioxidants* 9 (8): 681. https://doi.org/10.3390/antiox9080681.

Hectors, K., Prinsen, E., De Coen, W. et al. (2007). *Arabidopsis thaliana* plants acclimated to low dose rates of ultraviolet B radiation show specific changes in morphology and gene expression in the absence of stress symptoms. *New Phytologist* 175 (2): 255–270. https://doi.org/10.1111/j.1469-8137.2007.02092.x.

Hectors, K., Van Oevelen, S., Geuns, J. et al. (2014). Dynamic changes in plant secondary metabolites during UV acclimation in *Arabidopsis thaliana*. *Physiologia Plantarum* 152 (2): 219–230. https://doi.org/10.1111/ppl.12168.

Hideg, É. and Vass, I. (1996). UV-B induced free radical production in plant leaves and isolated thylakoid membranes. *Plant Science* 115 (2): 251–260. https://doi.org/10.1016/0168-9452(96)04364-6.

Hideg, É., Jansen, M.A., and Strid, Å. (2013). UV-B exposure, ROS, and stress: inseparable companions or loosely linked associates? *Trends in Plant Science* 18 (2): 107–115. https://doi.org/10.1016/j.tplants.2012.09.003.

Huang, X., Ouyang, X., Yang, P. et al. (2013). Conversion from CUL4-based COP1–SPA E3 apparatus to UVR8–COP1–SPA complexes underlies a distinct biochemical function of COP1 under UV-B. *Proceedings of the National Academy of Sciences* 110 (41): 16669–16674. https://doi.org/10.1073/pnas.1316622110.

Ibanez, C., Delker, C., Martinez, C. et al. (2018). Brassinosteroids dominate hormonal regulation of plant thermomorphogenesis via BZR1. *Current Biology CB*. 28: 303–310. https://doi.org/10.1016/j.cub.2017.11.077.

Jiang, K., Guo, H., and Zhai, J. (2023). Interplay of phytohormones and epigenetic regulation: a recipe for plant development and plasticity. *Journal of Integrated Plant Biology* 65 (2): 381–398. https://doi.org/10.1111/jipb.13384.

Jaiswal, D., Pandey, A., Agrawal, M., and Agrawal, S.B et al. (2023). Photosynthetic, biochemical and secondary metabolite changes in a medicinal plant *Chlorophytum borivillianum* (Safed musli) against low and high doses of UV-B radiation. *Photochemistry and Photobiology* 99 (1): 45–56. https://doi.org/10.1111/php.13672.

Jenkins, G.I., (2017). Photomorphogenic responses to ultraviolet-B light. *Plant, Cell & Environment* 40 (11): 2544–2557.76. https://doi.org/10.1111/pce.12934.

Jiroutová, P., Mikulík, J., Novák, O. et al. (2019). Brassinosteroids induce strong, dose-dependent inhibition of etiolated pea seedling growth correlated with ethylene production. *Biomolecules* 9: 849. https://doi.org/10.3390/biom9120849.

Jovanić, B.R., Radenković, B., Despotović-Zrakić, M. et al. (2022). Effect of UV-B radiation on chlorophyll fluorescence, photosynthetic activity and relative chlorophyll content of five different corn hybrids. *Journal of Photochemistry and Photobiology* 10: 100115. https://doi.org/10.1016/j.jpap.2022.100115.

Kalbina, I. and Strid, Å. (2006). The role of NADPH oxidase and MAP kinase phosphatase in UV-B-dependent gene expression in *Arabidopsis. Plant, Cell & Environment* 29 (9): 1783–1793. https://doi.org/10.1111/j.1365-3040.2006.01555.x.

Kataria, S., Jajoo, A., and Guruprasad, K.N. (2014). Impact of increasing ultraviolet-B (UV-B) radiation on photosynthetic processes. *Journal of Photochemistry and Photobiology B: Biology* 137: 55–66. https://doi.org/10.1016/j.jphotobiol.2014.02.004.

Kciuk, M., Marciniak, B., Mojzych, M. et al. (2020). Focus on UV-induced DNA damage and repair—disease relevance and protective strategies. *International Journal of Molecular Sciences* 21 (19): 7264. https://doi.org/10.3390/ijms21197264.

Khan, T.A., Kappachery, S., Karumannil, S. et al. (2023). Brassinosteroid signaling pathways: insights into plant responses under abiotic stress. *International Journal of Molecular Sciences* 24 (24): 17246. https://doi.org/10.3390/ijms242417246.

Khare, S., Singh, A., Amist, N. et al. (2023). Secondary metabolites interference on potential of *Solanum lycopersicum* grown under UV-B stress and its impact on developmental attributes of Capsicum annuum. *Plant Stress* 8: 100167. https://doi.org/10.1016/j.stress.2023.100167.

Kono, M. and Terashima, I. (2014). Long-term and short-term responses of the photosynthetic electron transport to fluctuating light. *Journal of Photochemistry and Photobiology B: Biology* 137: 89–99. https://doi.org/10.1016/j.jphotobiol.2014.02.016.

Krasensky-Wrzaczek, J. and Kangasjärvi, J. (2018). The role of reactive oxygen species in the integration of temperature and light signals. *Journal of Experimental Botany* 69 (14): 3347–3358. https://doi.org/10.1093/jxb/ery074.

Kreslavski, V.D., Strokina, V.V., Khudyakova, A.Y. et al. (2021). Effect of high-intensity light and UV-B on photosynthetic activity and the expression of certain light-responsive genes in *A. thaliana phyA* and *phyB* mutants. *Biochimica et Biophysica Acta (BBA)-Bioenergetics* 1862 (8): 148445. https://doi.org/10.1016/j.bbabio.2021.148445.

Kumar, D. and Ohri, P. (2023). Say “NO” to plant stresses: unravelling the role of nitric oxide under abiotic and biotic stress. *Nitric Oxide* 130: 36–57. https://doi.org/10.1016/j.niox.2022.11.004.

Kumar, D., Manhas, R.K., and Ohri, P. (2023). Deciphering the growth promoting and stress curtailing role of nitric oxide in *Meloidogyne incognita* infested *Solanum lycopersicum* seedlings. *Scientia Horticulturae* 319: 112147. https://doi.org/10.1016/j.scienta.2023.112147.

Kurdziel, M., Filek, M., and Łabanowska, M. (2018). The impact of short-term UV irradiation on grains of sensitive and tolerant cereal genotypes studied by EPR. *Journal of the Science of Food and Agriculture* 98 (7): 2607–2616. https://doi.org/10.1002/jsfa.8753.

Li, C., Gu, L., Gao, L. et al. (2016). Concerted genomic targeting of H3K27 demethylase REF6 and chromatin-remodeling ATPase BRM in *Arabidopsis. Nature Genetics* 48 (6): 687–693. https://doi.org/10.1038/ng.3555.

Li, H., Li, D., Yang, Z. et al. (2020). Flavones produced by mulberry flavone synthase type I constitute a defense line against the ultraviolet-B stress. *Plants* 9 (2): 215. https://doi.org/10.3390/plants9020215.

Li, Z., Ou, Y., Zhang, Z. et al. (2018). Brassinosteroid signaling recruits histone 3 lysine-27 demethylation activity to *FLOWERING LOCUS C* chromatin to inhibit the floral transition in *Arabidopsis. Molecular Plant* 11 (9): 1135–1146. https://doi.org/10.1016/j.molp.2018.06.007.

Liang, T., Mei, S., Shi, C. et al. (2018) UVR8 interacts with BES1 and BIM1 to regulate transcription and photomorphogenesis in *Arabidopsis. Developmental Cell* 44: 512–523 e515. https://doi.org/10.1016/j.devcel.2017.12.028.

Liang, T., Shi, C., Peng, Y. et al. (2020) Brassinosteroid-activated BRI1-EMSSUPPRESSOR 1 inhibits flavonoid biosynthesis and coordinates growth and UV-B stress responses in plants. *Plant Cell* 32: 3224–3239. https://doi.org/10.1105/tpc.20.00048.

Lin, F., Cao, J., Yuan, J. et al. (2021). Integration of light and brassinosteroid signaling during seedling establishment. *International Journal of Molecular Sciences* 22 (23): 12971. https://doi.org/10.3390/ijms222312971.

Lu, D., Zhai, J., and Xi, M. (2022). Regulation of DNA methylation during plant endosperm development. *Frontiers in Genetics*, 13: 760690. https://doi.org/10.3389/fgene.2022.760690.

Lu, F., Cui, X., Zhang, S. et al. (2011). *Arabidopsis* REF6 is a histone H3 lysine 27 demethylase. *Nature Genetic* 43 (7): 715–719. https://doi.org/10.1038/ng.854.

Lu, T., Meng, Z., Zhang, G. et al. (2017). Sub-high temperature and high light intensity induced irreversible inhibition on photosynthesis system of tomato plant (*Solanum lycopersicum* L.). *Frontiers in Plant Science* 8: 365. https://doi.org/10.3389/fpls.2017.00365.

Luo, J., Jiang, J., Sun, S. et al. (2022). Brassinosteroids promote thermotolerance through releasing BIN2-mediated phosphorylation and suppression of HsfA1 transcription factors in *Arabidopsis. Plant Communications* 3 (6): 100419. https://doi.org/10.1016/j.xplc.2022.100419.

Lv, M. and Li, J. (2020). Molecular mechanisms of brassinosteroid-mediated responses to changing environments in *Arabidopsis. International Journal of Molecular Sciences* 21 (8): 2737. https://doi.org/10.3390/ijms21082737.

Lv, M., Su, H.Y., Li, M.L. et al. (2021). Effect of UV-B radiation on growth, flavonoid and podophyllotoxin accumulation and related gene expression in *Sinopodophyllum hexandrum. Plant Biology* 23: 202–209. https://doi.org/10.1111/plb.13226.

Mahdavian, K., Ghorbanli, M., and Kalantari, K. (2008). The effects of ultraviolet radiation on some antioxidant compounds and enzymes in *Capsicum annuum* L. *Turkish Journal of Botany* 32 (2): 129–134.

Mannucci, A., Mariotti, L., Castagna, A. et al. (2020). Hormone profile changes occur in roots and leaves of Micro-Tom tomato plants when exposing the aerial part to low doses of UV-B radiation. *Plant Physiology and Biochemistry* 148: 291–301. https://doi.org/10.1016/j.plaphy.2020.01.030.

Mao, B., Yin, H., Wang, Y. et al. (2017). Combined effects of O3 and UV radiation on secondary metabolites and endogenous hormones of soybean leaves. *PLoS One* 12 (8): e0183147. https://doi.org/10.1371/journal.pone.0183147.

Mariz-Ponte, N., Mendes, R.J., Sario, S. et al. (2021). Physiological, biochemical and molecular assessment of UV-A and UV-B supplementation in *Solanum lycopersicum. Plants* 10 (5): 918. https://doi.org/10.3390/plants10050918.

Martin, L.J. (2008). DNA damage and repair: relevance to mechanisms of neurodegeneration. *Journal of Neuropathology & Experimental Neurology* 67 (5): 377–387. https://doi.org/10.1097/NEN.0b013e31816ff780.

Mathur, S., Bheemanahalli, R., Jumaa, S.H. et al. (2024). Impact of ultraviolet-B radiation on early-season morpho-physiological traits of *indica* and *japonica* rice genotypes. *Frontiers in Plant Science* 15: 1369397. https://doi.org/10.3389/fpls.2024.1369397.

Meijer, D., Meisenburg, M., van Loon, J.J. et al. (2022). Effects of low and high red to far-red light ratio on tomato plant morphology and performance of four arthropod herbivores. *Scientia Horticulturae* 292: 110645. https://doi.org/10.1016/j.scienta.2021.110645.

Mewis, I., Schreiner, M., Nguyen, C.N. et al. (2012). UV-B irradiation changes specifically the secondary metabolite profile in broccoli sprouts: induced signaling overlaps with defense response to biotic stressors. *Plant and Cell Physiology* 53 (9): 1546–1560. https://doi.org/10.1093/pcp/pcs096.

Meyer, P., Van de Poel, B., and De Coninck, B. (2021). UV-B light and its application potential to reduce disease and pest incidence in crops. *Horticulture Research* 8: 194. https://doi.org/10.1038/s41438-021-00629-5.

Mittler, R., Zandalinas, S.I., Fichman, Y. et al. (2022). Reactive oxygen species signalling in plant stress responses. *Nature Reviews Molecular Cell Biology* 23 (10): 663–679. https://doi.org/10.1038/s41580-022-00499-2.

Mu, T. and Han, R. (2024). Magnesium and nitrogen co-doped carbon dots alleviate UV-B radiation damage in wheat seedlings. *Pakistan Journal of Botany* 56 (1): 227–237. https://doi.org/10.30848/PJB2024-1(33).

Müller-Xing, R., Xing, Q., and Goodrich, J. (2014). Footprints of the sun: memory of UV and light stress in plants. *Frontiers in Plant Science* 5: 474. https://doi.org/10.3389/fpls.2014.00474.

Nasibova, A., Khalilov, R., Bayramov, M. et al. (2023). Electron paramagnetic resonance studies of irradiated grape snails (*Helix pomatia*) and investigation of biophysical parameters. *Molecules* 28 (4): 1872. https://doi.org/10.3390/molecules28041872.

Nguyen, N.H. (2020). HY5, an integrator of light and temperature signals in the regulation of anthocyanins biosynthesis in *Arabidopsis. AIMS Molecular Science* 7 (2): 70–81. https://doi.org/10.3934/molsci.2020005.

Nitschke, S., Cortleven, A., and Schmülling, T. (2017). Novel stress in plants by altering the photoperiod. *Trends in Plant Science* 22 (11): 913–916. https://doi.org/10.1016/j.tplants.2017.09.005.

Nitschke, S., Cortleven, A., Iven, T. et al, Feussner, I., Havaux, M., Riefler, M., Schmülling, T. (2016). Circadian stress regimes affect the circadian clock and cause jasmonic acid-dependent cell death in cytokinin deficient Arabidopsis plants. *Plant Cell* 28 (7): 1616–1639. https://doi.org/10.1105/tpc.16.00016.

Nolan, T.M., Vukasinovic, N., Liu, D. et al. (2020). Brassinosteroids: multidimensional regulators of plant growth, development, and stress responses. *Plant Cell* 32 (2): 295–318. https://doi.org/10.1105/tpc.19.00335.

Paik, I. and Huq, E. (2019). Plant photoreceptors: multi-functional sensory proteins and their signaling networks. *Seminars in Cell & Developmental Biology* 92: 114–121. https://doi.org/10.1016/j.semcdb.2019.03.007.

Pandey, A., Kumar, S., Singh, G. et al. (2022). Regulatory role of gamma-aminobutyric acid in cyanobacteria challenged with UV-B: the implication of nitric oxide. *Plant Stress* 5: 100094. https://doi.org/10.1016/j.stress.2022.100094.

Park, C.J. and Seo, Y.S. (2015). Heat shock proteins: a review of the molecular chaperones for plant immunity. *Plant Pathology Journal* 31: 323–333.

Planas-Riverola, A., Gupta, A., Betegon-Putze, I. et al. (2019). Brassinosteroid signaling in plant development and adaptation to stress. *Development* 146 (5): 151894. https://doi.org/10.1242/dev.151894.

Qian, C., Mao, W., Liu, Y. et al. (2016). Dual-source nuclear monomers of UV-B light receptor direct photomorphogenesis in *Arabidopsis. Molecular Plant* 9: 1671–1674. https://doi.org/10.1016/j.molp.2016.10.005.

Qiu, Z., Zhu, L., He, L. et al. (2019). DNA damage and reactive oxygen species cause cell death in the rice *local lesions 1* mutant under high light and high temperature. *New Phytologist* 222 (1): 349–365. https://doi.org/10.1111/nph.15597.

Rai, N., Morales, L.O., and Aphalo, P.J. (2021). Perception of solar UV radiation by plants: photoreceptors and mechanisms. *Plant Physiology* 186 (3): 1382–1396. https://doi.org/10.1093/plphys/kiab162.

Raja, V., Majeed, U., Kang, H. et al. (2017). Abiotic stress: interplay between ROS, hormones and MAPKs. *Environmental and Experimental Botany* 137: 142–157. https://doi.org/10.1016/j.envexpbot.2017.02.010.

Rizzini, L., Favory, J.J., Cloix, C. et al. (2011). Perception of UV-B by the *Arabidopsis* UVR8 protein. *Science* 332 (6025): 103–106. https://doi.org/10.1126/science.1200660.

Robson, T.M., Klem, K., Urban, O. et al. (2015). Re-interpreting plant morphological responses to UV-B radiation. *Plant, Cell & Environment* 38 (5): 856–866. https://doi.org/10.1111/pce.12374.

Roeber, V.M., Bajaj, I., Rohde, M. et al. (2021). Light acts as a stressor and influences abiotic and biotic stress responses in plants. *Plant, Cell & Environment* 44 (3): 645–664. https://doi.org/10.1111/pce.13948.

Roeber, V., Bajaj, I., Rohde, M., et al. (2020). Light as a stressor and its impact on biotic and abiotic stress responses in plants. *Authorea Preprints*. https://doi.10.1111/pce.13948

Ryu, H., Cho, H., Bae, W. et al. (2014). Control of early seedling development by BES1/TPL/HDA19-mediated epigenetic regulation of *ABI3. Nature Communications* 5: 4138. https://doi.org/10.1038/ncomms5138.

Saijo, Y., Sullivan, J.A., Wang, H. et al. (2003). The COP1–SPA1 interaction defines a critical step in phytochrome A-mediated regulation of HY5 activity. *Genes & Development* 17 (21): 2642–2647. https://doi.org/10.1101/gad.1122903.

Santin, M., Becagli, M., Sciampagna, M.C. et al. (2023). Integrative effect of UV-B and some organic amendments on growth, phenolic and flavonoid compounds and antioxidant activity of basil (*Ocimum basilicum* L.) plants. *Horticulturae* 9 (8): 894. https://doi.org/10.3390/horticulturae9080894.

Semenova, N.A., Ivanitskikh, A.S., Uyutova, N.I. et al. (2024). Effect of UV stress on the antioxidant capacity, photosynthetic activity, flavonoid and steviol glycoside accumulation of *Stevia rebaudiana* Bertoni. *Horticulturae* 10 (3): 210. https://doi.org/10.3390/horticulturae10030210.

Shao, M., Liu, W., Zha, L. et al. (2020). Differential effects of high light duration on growth, nutritional quality and oxidative stress of hydroponic lettuce under red and blue LED irradiation. *Scientia Horticulturae* 268: 109366. https://doi.org/10.1016/j.scienta.2020.109366.

Sharlaeva, E.A. and Chirkova, V.Y. (2021). The impact of short-wave UV radiation on peroxidase activity in soft wheat seeds. *IOP Conference Series: Earth and Environmental Science* 670 (1): 012008. https://doi.org/10.1088/1755-1315/670/1/012008.

Sharma, N., Khanna, K., Jasrotia, S. et al. (2023). Metabolites and chemical agents in the plant roots: an overview of their use in plant-parasitic nematode management. *Nematology* 25 (3): 243–257. https://doi.org/10.1163/15685411-bja10220.

Shengxin, C., Chunxia, L., Song, C. et al. (2016). Morphological, photosynthetic, and physiological responses of rapeseed leaf to different combinations of red and blue lights at the rosette stage. *Frontiers in Plant Science* 7: 183164. https://doi.org/10.3389/fpls.2016.01144.

Smith, J.L., Burritt, D.J., and Bannister, P. (2000). Shoot dry weight, chlorophyll and UV-B-absorbing compounds as indicators of a plant's sensitivity to UV-B radiation. *Annals of Botany* 86 (6): 1057–1063. https://doi.org/10.1006/anbo.2000.1270.

Song, L., Chen, W., Yao, Q. et al. (2019). Genome-wide transcriptional profiling for elucidating the effects of brassinosteroids on *Glycine max* during early vegetative development. *Scientific Reports* 9: 16085. https://doi.org/10.1038/s41598-019-52599-3.

Sonoike, K. (1996). Photoinhibition of photosystem I: its physiological significance in the chilling sensitivity of plants. *Plant and Cell Physiology* 37 (3): 239–247. https://doi.org/10.1093/oxfordjournals.pcp.a028938.

Sztatelman, O., Grzyb, J., Gabryś, H. et al. (2015). The effect of UV-B on *Arabidopsis* leaves depends on light conditions after treatment. *BMC Plant Biology* 15: 1–16. https://doi.org/10.1186/s12870-015-0667-2.

Szymańska, R., Ślesak, I., Orzechowska, A. et al. (2017). Physiological and biochemical responses to high light and temperature stress in plants. *Environmental and Experimental Botany* 139: 165–177. https://doi.org/10.1016/j.envexpbot.2017.05.002.

Takahashi, M., Teranishi, M., Ishida, H. et al. (2011). Cyclobutane pyrimidine dimer (CPD) photolyase repairs ultraviolet-B-induced CPDs in rice chloroplast and mitochondrial DNA. *The Plant Journal* 66 (3): 433–442. https://doi.org/10.1111/j.1365-313X.2011.04500.x.

Takahashi, S., and Badger, M.R. (2011). Photoprotection in plants: a new light on photosystem II damage. *Trends in Plant Science* 16 (1): 53–60. https://doi.org/10.1016/j.tplants.2010.10.001.

Takahashi, S., and Murata, N. (2008). How do environmental stresses accelerate photoinhibition? *Trends in Plant Science* 13 (4): 178–182. https://doi.org/10.1016/j.tplants.2008.01.005.

Takahashi, S., Milward, S.E., Yamori, W., Evans, J.R., Hillier, W., and Badger, M.Ret al. (2010). The solar action spectrum of photosystem II damage. *Plant Physiology* 153 (3): 988–993. https://doi.org/10.1104/pp.110.155747.

Tan, Y., Duan, Y., Chi, Q. et al. (2023). The role of reactive oxygen species in plant response to radiation. *International Journal of Molecular Sciences* 24 (4): 3346. https://doi.org/10.3390/ijms24043346.

Tossi, V.E., Burrieza, H.P., Escandón, A.S. et al. (2019). Beyond *Arabidopsis*: differential UV-B response mediated by UVR8 in diverse species. *Frontiers in Plant Science* 10: 459191. https://doi.org/10.3389/fpls.2019.00780.

Tripathi, D.K., Singh, S., Singh, V.P. et al. (2017). Silicon nanoparticles more effectively alleviated UV-B stress than silicon in wheat (*Triticum aestivum*) seedlings. *Plant Physiology and Biochemistry* 110: 70–81. https://doi.org/10.1016/j.plaphy.2016.06.026.

Tuteja, N., Ahmad, P., Panda, B.B. et al. (2009). Genotoxic stress in plants: shedding light on DNA damage, repair and DNA repair helicases. *Mutation Research/Reviews in Mutation Research* 681 (2-3): 134–149. https://doi.org/10.1016/j.mrrev.2008.06.004.

Ulm, R. and Nagy, F. (2005). Signalling and gene regulation in response to ultraviolet light. *Current Opinion in Plant Biology* 8 (5): 477–482. https://doi.org/10.1016/j.pbi.2005.07.004.

Ulm, R., and Jenkins, G.I. (2015). Q&A: How do plants sense and respond to UV-B radiation? *BMC Biology* 13: 1–6. https://doi.org/10.1186/s12915-015-0156-y.

Vanhaelewyn, L., Prinsen, E., Van Der Straeten, D. et al. (2016). Hormone-controlled UV-B responses in plants. *Journal of Experimental Botany* 67 (15): 4469–4482. https://doi.org/10.1093/jxb/erw261.

Van Der Straeten, D., De Coninck, B. et al. (2020). Ultraviolet radiation from a plant perspective: the plant-microorganism context. *Frontiers in Plant Science* 11: 597642. https://doi.org/10.3389/fpls.2020.597642.

Velitchkova, M., Popova, A.V., Faik, A. et al. (2020). Low temperature and high light dependent dynamic photoprotective strategies in *Arabidopsis* thaliana. *Physiologia Plantarum* 170 (1): 93–108. https://doi.org/10.1111/ppl.13111.

Verdaguer, D., Jansen, M.A., Llorens, L. et al. (2017). UV-A radiation effects on higher plants: exploring the known unknown. *Plant Science* 255: 72–81. https://doi.org/10.1016/j.plantsci.2016.11.014.

Vuleta, A., Jovanović, S.M., and Tucić, B. (2016). Adaptive flexibility of enzymatic antioxidants SOD, APX and CAT to high light stress: the clonal perennial monocot *Iris pumila* as a study case. *Plant Physiology and Biochemistry* 100: 166–173. https://doi.org/10.1016/j.plaphy.2016.01.011.

Wang, X., Tan, N.W.K., Chung, F.Y. et al. (2023). Transcriptional regulators of plant adaptation to heat stress. *International Journal of Molecular Sciences* 24 (17): 13297. https://doi.org/10.3390/ijms241713297.

Wei, L., Gu, L., Song, X. et al. (2014). Dicer-like 3 produces transposable element-associated 24-nt siRNAs that control agricultural traits in rice. *Proceedings of the National Academy of Sciences of the United States of America* 111: 3877–3882. https://doi.org/10.1073/pnas.1318131111.

Wu, D., Hu, Q., Yan, Z. et al. (2012). Structural basis of ultraviolet-B perception by UVR8. *Nature* 484 (7393): 214–219. https://doi.org/10.1038/nature10931.

Wu, J., Yan, M., Zhang, D. et al. (2020). Histone demethylases coordinate the antagonistic interaction between abscisic acid and brassinosteroid signaling in *Arabidopsis. Frontiers in Plant Sciences* 11: 596835. https://doi.org/10.3389/fpls.2020.596835.

Wu, K., Liu, J., Liu, Y. et al. (2022). Tissue-specific metabolic response of *Acanthopanax senticosus* (Rupr. et. Maxim.) harms leaves to enhanced UV-B radiation. *Acta Physiologiae Plantarum* 44 (12): 131. https://doi.org/10.1007/s11738-022-03452-0.

Wu, X., Chen, B., and Xiao, J. et al. (2023). Different doses of UV-B radiation affect pigmented potatoes' growth and quality during the whole growth period. *Frontiers in Plant Science* 14: 1101172. https://doi.org/10.3389/fpls.2023.1101172.

Xu, W., Li, M., Li, W. et al. (2024). Effect of H2 treatment under UV-B irradiation on the enrichment of germinated soybean isoflavones and mechanisms based on growth state, antioxidant system and metabolomics. *LWT* 195: 115821. https://doi.org/10.1016/j.lwt.2024.115821.

Yang, L., Gao, N., Yang, Y. et al. (2020). ABA alleviated soybean seedling stress exposed to UV-C radiation. *International Agrophysics* 34 (1): 115–121. https://doi.org/10.31545/intagr/113259.

Yang, Y., Liang, T., Zhang, L. et al. (2018a). UVR8 interacts with WRKY36 to regulate HY5 transcription and hypocotyl elongation in *Arabidopsis. Nature Plants* 4 (2): 98–107. https://doi.org/10.1038/s41477-017-0099-0.

Yang, Y., Yang, X., Jin, W. et al. (2018b). UV RESISTANCE LOCUS 8 from *Chrysanthemum morifolium* Ramat (CmUVR8) plays important roles in UV-B signal transduction and UV-B-induced accumulation of flavonoids. *Frontiers in Plant Science* 9: 339148. https://doi.org/10.3389/fpls.2018.00955.

Yang, Y., Yin, J., Zhu, L. et al. (2024). Genome-wide analysis of the *Liriodendron Chinese Hsf* gene family under abiotic stress and characterization of the *LcHsfA2a* gene. *International Journal of Molecular Sciences* 25 (5): 2733. https://doi.org/10.3390/ijms25052733.

Yin, R., Skvortsova, M.Y., Loubéry, S. et al. (2016). COP1 is required for UV-B-induced nuclear accumulation of the UVR8 photoreceptor. *Proceedings of the National Academy of Sciences of the United States of America* 113: 4415–4422. https://doi.org/10.1073/pnas.1607074113.

Yoshida, H., Shimada, K., Hikosaka, S. et al. (2022). Effect of UV-B irradiation on bioactive compounds of red perilla (*Perilla frutescens* (L.) Britton) cultivated in a plant factory with artificial light. *Horticulturae* 8 (8): 725. https://doi.org/10.3390/horticulturae8080725.

Zha, L. and Liu, W. (2019). Morphological and physiological stress responses of lettuce to different intensities of continuous light. *Frontiers in Plant Science* 10: 472308. https://doi.org/10.3389/fpls.2019.01440.

Zhan, J., Yang, Q., Lin, Z. et al. (2021). Enhanced antioxidant capacity and upregulated transporter genes contribute to the UV-B-induced increase in blinin in *Conyza blinii. Environmental Science and Pollution Research* 28: 13275–13287. https://doi.org/10.1007/s11356-020-11502-8.

Zhang, D., Jing, Y., Jiang, Z. et al. (2014). The chromatin-remodeling factor PICKLE integrates brassinosteroid and gibberellin signaling during skotomorphogenic growth in *Arabidopsis. Plant Cell* 26 (6): 2472–2485. https://doi.org/10.1105/tpc.113.121848.

Zhang, X.R., Chen, Y.H., Guo, Q. S., Wang, W.M., Liu, L., Fan, J., Cao, L.P., and Li, C et al. (2017). Short-term UV-B radiation effects on morphology, physiological traits and accumulation of bioactive compounds in *Prunella vulgaris* L. *Journal of Plant Interactions* 12 (1): 348-–354. https://doi.org/10.1080/17429145.2017.1365179.

Zhang, X.D., Wang, R.P., Zhang, F.J. et al. (2013). Lipid profiling and tolerance to low-temperature stress in *Thellungiella salsuginea* in comparison with *Arabidopsis thaliana. Biologia Plantarum* 57: 149–153. https://doi.org/10.1007/s10535-012-0137-8.

Zheng, J., Ji, F., He, D. et al. (2019). Effect of light intensity on rooting and growth of hydroponic strawberry runner plants in a LED plant factory. *Agronomy* 9 (12): 875. https://doi.org/10.3390/agronomy9120875.

Zhou, X., Lyu, J., Sun, L. et al. (2021). Metabolic programming of *Rhododendron chrysanthum* leaves following exposure to UVB irradiation. *Functional Plant Biology* 48 (11): 1175–1185. https://doi.org/10.1071/FP20386.

Zhu, L., Bu, Q., Xu, X. et al. (2015). CUL4 forms an E3 ligase with COP1 and SPA to promote light-induced degradation of PIF1. *Nature Communications* 6 (1): 7245. https://doi.org/10.1038/ncomms8245.

13

Brassinosteroids: Fascinating Roles on Stomatal Mechanism, Physiological, and Anatomical Responses

Maria Antonia Machado Barbosa[1] and Allan Klynger da Silva Lobato[2]*

[1]*Grupo de Estudos em Fitotecnia, Universidade do Estado de Minas Gerais. Ituiutaba, Minas Gerais, Brazil*
[2]*Núcleo de Pesquisa Vegetal Básica e Aplicada, Universidade Federal Rural da Amazônia. Paragominas, Pará, Brazil*

*Corresponding author: maria.barbosa@uemg.br

13.1 Introduction

The control of the flow of gases between the plant and the atmosphere is governed by a completely mechanical process overseen by the stomata. These structures are on the surface of leaves and stems, composed of two guard cells that control the closing and opening of a pore (Torii 2021; Li et al. 2022). Internal and external stimuli to the leaves can result in changes in the volume of the guard cells that increase or reduce the opening of the stomatal pore through a high or low pressure of leaf turgor (Li et al. 2021). Additionally, the mechanisms related to the opening and closing of these valves are also induced by light and controlled by K^+ channels present in the guard cell (Saito and Uozumi 2019). Such changes affect the CO_2 influx, minimize excessive water losses, and sometimes regulate the internal leaf temperature. Therefore, critical physiological processes for plant life, such as the assimilation of photosynthetic carbon and transpiration, are related to stomatal characteristics, such as the degree of opening (stomatal conductance – g_S), density and shape of these structures (Serna 2014).

Environmental changes – especially those involving hydraulic factors such as air humidity and water availability – strongly influence g_s and, therefore, the survival of plants (Marchin et al. 2022). This is because g_s can modulate a rapid response in plants to environmental fluctuations, such as light intensity (Zhang et al. 2019), water availability (Gleason et al. 2019), changes in vapor pressure deficit (VPD) (Li and Liu 2022), increase in salinity (Karimi et al. 2021), between others. However, stomatal responses to changes in these factors will eventually depend on the type, intensity, duration, and combination of stresses (Aasamaa and Sõber 2011), as well as the genetic material (Faralli et al. 2019; Shekoofa et al. 2020) and hormonal control (Sarwat and Tuteja 2017).

Brassinosteroids (BRs) correspond to a new class of phytohormones involved in plant growth and development. However, various plant functions are regulated by BRs, including

Brassinosteroids and Abiotic Stress: Recent Developments on Physiological and Molecular Aspects,
First Edition. Edited by Anket Sharma, et al.
Published 2026 by John Wiley & Sons Ltd.

physiological and biochemical changes and improved tolerance against multiple types of biotic and abiotic stresses (Sharma et al. 2017; Hafeez et al. 2021). BRs are perceived by family receptor-like kinase, brassinosteroid insensitive1 (BRI1), that interacts with the co-receptor BRI1-associated receptor kinase1 (BAK1) and positively regulates brassinazoleresistant1 (BZR1) and BRI1-EMS suppressor1 (BES1), which are essential positive regulators of BR signaling (Clouse 2011). In these events, several genes related to different physiological functions are expressed through a series of phosphorylation and dephosphorylation (Fang et al. 2019).

The action of BR in the defense of plants against environmental stress includes events related to the development and regulation of stomata. However, the information still needs to be clarified. Notably, this steroid acts on stomatal signaling pathways, altering the degree and duration of stomata opening and the distribution pattern and a number of these structures on the plant surface (Kim et al. 2012; Pereira et al. 2020). Stomata, therefore, play an essential role in the performance of plants under stressful conditions, controlling vital processes such as the rate of CO_2 assimilation and transpiration, as well as their adaptation to different environments (Lawson and Blatt 2014; Xu et al. 2016). Here, we will discuss the role of BR in stomatal regulation mechanisms and physiological benefits in plants under different types of abiotic stresses.

13.2 Role of BRs in Abiotic Stress

The various functions performed by BRs in the plant organism have been described in several studies. BR regulates plant growth and development processes, such as cell elongation, seed germination, xylem differentiation, cell division, and cell wall regeneration (Lin et al. 2013; Wei and Li 2016). However, BRs have recently been described as a promising molecule for increasing productivity under stress conditions (Anwar et al. 2018; Hafeez et al. 2021). This occurs due to the interaction of this steroid with several signaling processes responsible for mitigating the effects of stress on plants. In addition, BRs induce stress tolerance through improvements in biochemical and physiological processes in the plant, such as better CO_2 assimilation, better activation of the antioxidant system, elimination of reactive oxygen species (ROS), and photoprotection, among others (Sharma et al. 2017; Ahammed et al. 2020). Some of the physiological, biochemical, and molecular processes involved in tolerance to abiotic stress induced by BR and structures of two more active BR are shown in Figure 13.1.

Physiological processes such as CO_2 assimilation and perspiration are affected by water stress, heat, cold, salt stress, heavy metals, and other environmental conditions. They can drastically reduce crop growth, development, and productivity (Koyro et al. 2012). Recent studies have shown that the exogenous application of BRs has an effective action in reducing the damage caused by environmental stress, especially with improvements in gas exchange and antioxidant defense system (Lima and Lobato 2017; dos Santos Ribeiro et al. 2019; Cunha et al. 2021).

Exogenous application of BRs in young rice plants (*Oryza sativa*) significantly improved net photosynthesis rate, stomatal conductance, stomatal limitation, and transpiration rate while decreasing intercellular CO_2 concentration and water use efficiency under heat stress

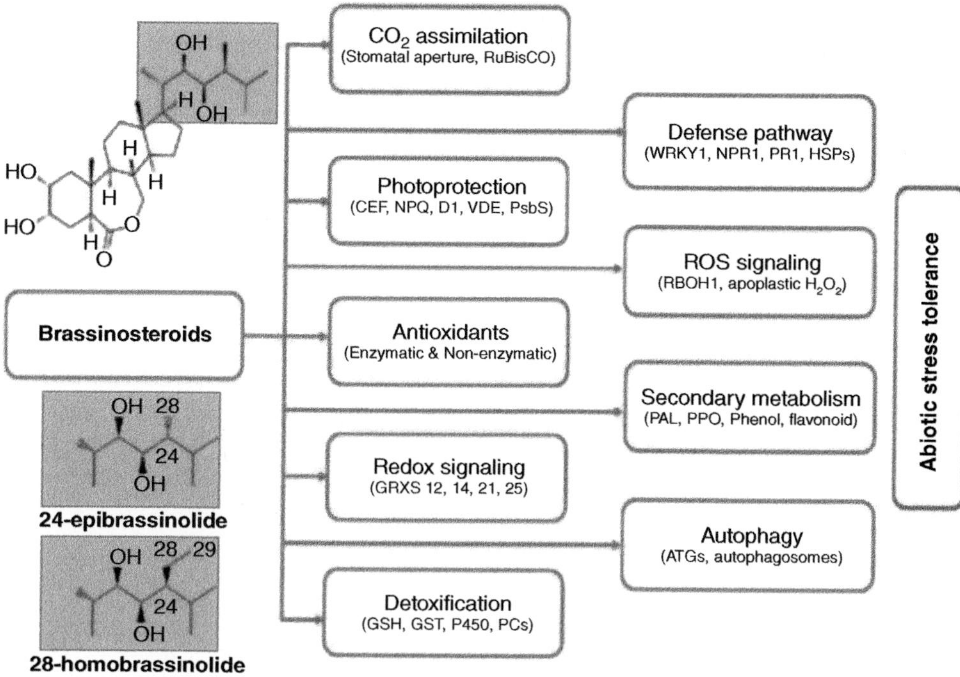

Figure 13.1 BRs promote stress tolerance by positively modulating multiple cellular processes such as CO_2 assimilation, photoprotection, antioxidant defense, redox balance, reactive oxygen species (ROS) scavenging, ROS signaling, defense response, detoxification, secondary metabolism, and autophagy in plants. *Source*: Ahammed et al. (2020)/Springer Nature.

(Sonjaroon et al. 2018). At the same time, there were increases in total soluble sugars due to photosynthetic improvement. In tomato plants (*Solanum lycopersicum* L.), there was sensitivity to chilling stress in *dwf* mutants with a deficiency in the biosynthesis of BRs, whereas, in *DWF* mutants with overexpression of this steroid, there was greater tolerance to stress (Fang et al. 2019).

Characteristics related to stomatal performance, including pore development and stomatal conductance, are involved in critical physiological processes in plants, such as photosynthetic activity and water use efficiency (Terashima et al. 2011; Cunha et al. 2021). Several studies demonstrate that exogenous application of BR at different concentrations positively modulates such characteristics by simply regulating specific proteins that act in stomatal development (Serna 2014). Thus, the beneficial action of BR on stomatal performance suggests a higher concentration of CO_2 absorbed and available at Rubisco carboxylation sites, even under conditions of stress for the plants (Sharma et al. 2017).

Improvements in the photosynthetic activity of plants under stressful conditions and after exogenous BR application are associated with increases in chlorophyll content (Shu et al. 2016; Anwar et al. 2018). For example, stress caused by excess cadmium (Cd) severely reduces the range of photosynthetic pigments in young *Eucalyptus urophylla* plants (Cunha et al. 2021). However, plants sprayed with 24-epibrassinolide (EBR + 0 μM Cd) presented increases of 11, 23, 13, and 45% in levels of chlorophyll *a* (Chl *a*), chlorophyll *b* (Chl *b*), total chlorophyll (Total Chl), and carotenoids (Car), respectively. Similarly, the plants treated with EBR + 450 μM Cd presented increases of 3, 19, 6, and 52%, and those treated with EBR + 900 μM Cd presented increases of 77, 92, 80, and 162%, respectively, when compared to the plants without EBR treatment. Studies suggest that EBR regulates the binding of Chl molecules to membrane proteins required for maintaining thylakoid stability, consequently improving the photosynthetic capacity of plants (Shu et al. 2016).

13.3 Role of BRs in Stomatal Development

Recently, several studies have unraveled stomata development processes in epidermal cells – an event marked by several asymmetric and symmetric divisions associated with cell differentiation (Serna 2014). Stomata gain form and function from a set of specialized cells that occur in young tissues (sprouts, for example) but not in mature tissues (Figure 13.2) (Bergmann and Sack 2007). The process starts from a meristemoid mother cell (MMC) and,

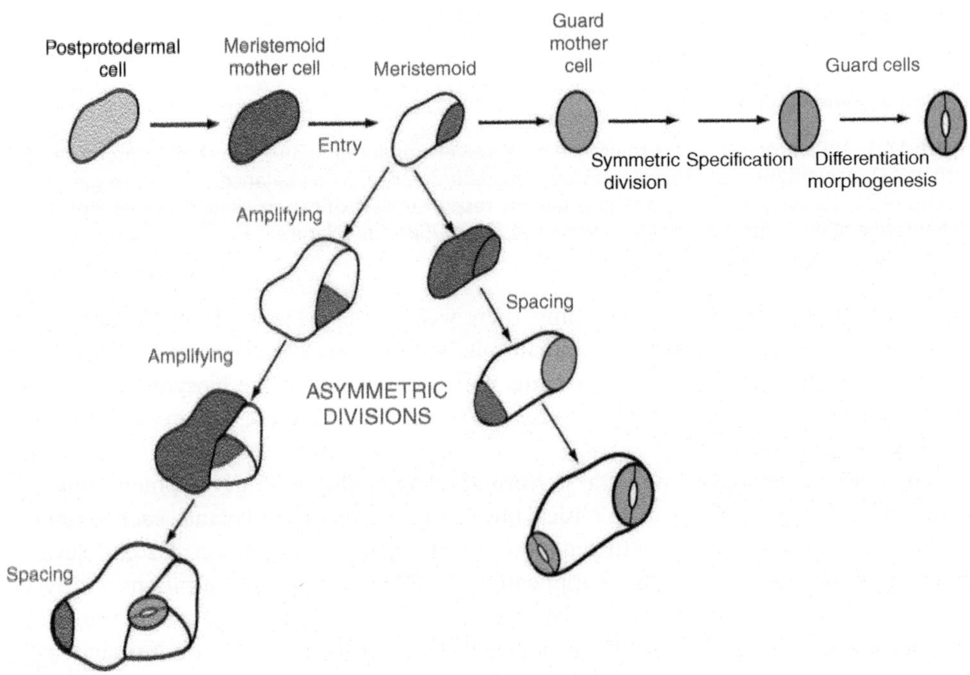

Figure 13.2 Diagram of critical stages and divisions in *Arabidopsis thaliana* stomatal development. Protodermal cells in the epidermis are converted into meristemoid mother cells (MMCs) through an unknown process. MMCs undergo an asymmetric entry division to create a meristemoid. Meristemoids may undergo additional asymmetric amplifying divisions or convert into a guard mother cell (GMC). The GMC will divide a single time, symmetrically, to form the two guard cells. Later, morphogenesis and pore formation create the mature stoma. The division process is reiterative. Cells next to meristemoids, GMCs, and guard cells can become MMCs and undergo spacing divisions to develop new meristemoids. The plane of this division is oriented so that the new meristemoid is placed away from the preexisting stoma or precursor cell. *Source*: Bergmann and Sack (2007)/Annual Reviews.

subsequently, the acquisition of a smaller enclosure (meristemoid, M). The M converts to a guard mother cell (GMC) that divides equally, followed by a coordinated morphogenesis yielding the two guard cells (GCs) of the stoma (Le et al. 2014). A series of asymmetric divisions produce M and a larger sister cell that forms the epidermal floor cells – an important event to ensure the spatial difference between stomata and to favor the chemical process involved in the movement of stomata (Bergmann and Sack 2007).

Studies that have unraveled the signal control that governs stomatal development in *Arabidopsis thaliana* indicate that these processes are well-conserved in virtually all plant species (Chater et al. 2017). However, biotic and abiotic factors also regulate the developmental processes of these critical structures (Wang et al. 2007). For example, the stomatal pattern regulators in *Arabidopsis FOUR LIPS* (*FLP*) and its paralog *MYB88* that encode MYB proteins that establish stomatal patterns, participate in the response to abiotic stress, with the development of interrupted stomata to alleviate tensions caused by saline stress (Xie et al. 2010). Water restriction promotes alterations in stomata development, as verified in plants of *Populus*, in which leaves that developed under irrigated conditions presented higher stomatal indices than leaves that grew under water deficit (Hamanishi et al. 2012). Additionally, stomatal development is negatively regulated by high concentrations of CO_2 (Xu et al. 2016) and influenced in response to pathogen attacks (Wang et al. 2007).

Endogenous factors, such as hormonal action, also play a role in stomatal development, primarily through three hormones: BRs, abscisic acid (ABA), and auxins (Le et al. 2014). The action of ABA on stomatal development involves a limitation at the beginning of stomatal development, with the prolonged formation of meristemoids and GMCs, which are precursors of stomata (Tanaka et al. 2013). A more recent study revealed the importance of ABA metabolism and perception in regulating stomatal development, limiting the formation of new stomata as an essential strategy for plants to conserve water in periods of scarcity (Yang et al. 2022).

BR negatively regulates stomatal development. BR-deficient *Arabidopsis* mutants have excess stomata, while brassinolide (BL) treatment reduces stomatal density (Kim et al. 2012). The signaling pathway that regulates stomatal development includes peptide ligands, a receptor protein (TMM), the ERECTA family (ERf) of receptor-like kinases (ER, ERL1, and ERL2), and a MAPK module (MAPKKK: YODA, MAPKKs: MKK4/5/7/9, and MAPKs: MPK3/6) (Wang et al. 2007). A study carried out by Kim et al. (2012) demonstrates that BR inhibits stomatal development by activating MAPKKK, YODA (YDA; a mitogen-activated protein kinase kinase kinase [MAPKKK]). In addition, Genetic analyses indicate that receptor kinase-mediated BR signaling inhibits development through the GSK3-like kinase BIN2, and BIN2 acts upstream of YODA but downstream of the known ERECTA family.

The hormonal control of ABA and BR in stomatal development includes a signaling pathway well characterized in *Arabidopsis* plants (Figure 13.3) (Serna 2014). BR is detected by a kinase located in the plasma membrane and rich in leucine (BRI1), which in turn activates a family of kinases called BRs signaling kinases (BSKs), with BSU1 domain phosphorylation sequence, which inactivates BIN2, allowing MAP kinase signaling to repress SPCH activity, which negatively regulates stomatal initiation. ABA also promotes stomatal closure, including PP2C sequestration (protein phosphatase 2C), inhibiting its activity and allowing the activation of SnRK2s (sucrose nonfermenting-1-related protein kinases), that finally activate the SLAC1 (slow anion channel-associated1), triggering stomatal closure (Serna 2014).

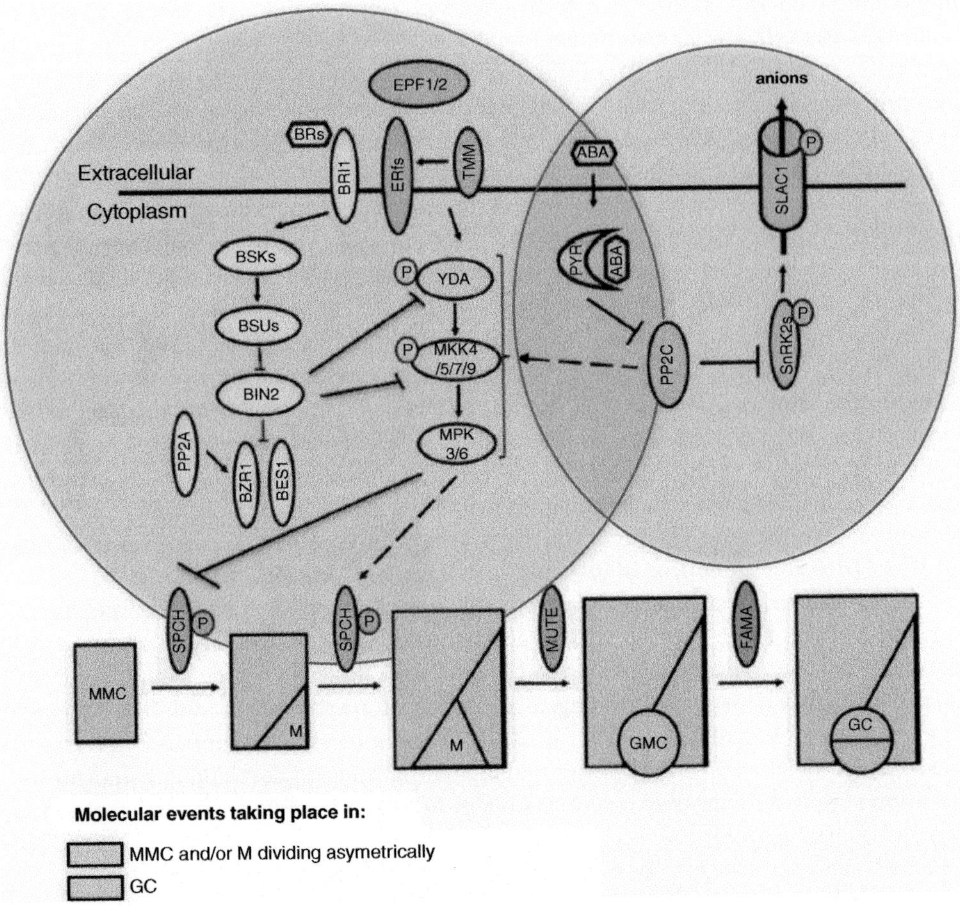

Figure 13.3 Hormonal control of stomatal development in *Arabidopsis* leaves and cotyledons. EPF1, 2, epidermal patterning factor 1 and 2; TMM, too many mouths; ERfs, erecta family receptors; YDA, YODA; MKK4, 5, 7, 9, mitogen-activated-protein kinase kinase 4, 5, 7, and 9; MPK3, 6, mitogen-activated-protein kinase 3 and 6; SPCH, speechless; BRs, brassinosteroids; BRI1, brassinosteroids insensitive 1; BSKs, brassinosteroids signaling kinases; BSUs, BRI1 suppressors; BIN2, brassinosteroids insensitive 2; PP2A, 2C, protein phosphatase 2A and 2C; BZR1, brassinazole resistant 1; BES1,BRI1 EMS suppressor; ABA, abscisic acid; PYR, PIRABACTIN RESISTANCE 1; SnRK2s, sucrose nonfermenting-1-related protein kinases; SLAC1, slow anion channel-associated1. MMC, meristemoid mother cell; M, meristemoid; GMC, guard mother cell; GC, guard cell. Dashed arrows indicate hypothetical steps. *Source*: Serna (2014)/with permission of Elsevier.

13.4 Role of BRs in Stomatal Regulation in Abiotic Stress

As epidermal structures specialized in regulating gas exchange (CO_2 and H_2O) between the plant and the environment, stomata are considered critical pathways for acclimatization or survival of plants against stressful events. Water deficit, for example, is one of the environmental stresses that affect plant growth and development the most and is also one of the most common occurrences (Lima and Lobato 2017). Some studies suggest that drought tolerance responses involve increased BR levels, acting on stomatal closure (Xia et al. 2014), similar to mechanisms mediated by ABA. ABA is involved in many processes of plant drought tolerance, with increases in its levels occurring when there is a reduction in water intake in the plant body. In addition to root elongation in response to drought, ABA controls stomatal closure to prevent excessive water loss to the atmosphere, whose molecular mechanisms are already very well described (Takahashi et al. 2013; Inoue et al. 2017).

The involvement of both ABA and BR hormones in stomatal closure seems to be species-dependent, as observed in a study by Jager et al. (2008), where they did not find a relationship in the increases in the levels of ABA and BR in plants of *Pisum sativum*. Notably, these authors did not observe changes in BR levels under water deficit conditions in apical tissue, internode or leaves in wild-type (WT) plants and *lkb* mutants (BR deficiency mutant) e *lka* (a mutant with BR perception). In contrast, ABA levels were increased in response to stress.

Stomatal regulation involving BR also involves other molecules at the cellular level. For example, studies have shown that this phytohormone promotes stomatal opening through a direct relationship with starch metabolism (Inoue et al. 2017; Li et al. 2020). In combination with hydrogen peroxide levels (H_2O_2), BR induces starch breakdown in guard cells of *Arabidopsis* plants, with consequent opening of the stomatal pore (Li et al. 2020). In WT plants, starch degradation occurs through binding BR to the kinase receptor BRI1, promoting the accumulation of H_2O_2, and increasing the interaction between BZR1 and GBF2. This is an essential mechanism for BAM1 expression that induces starch degradation in guard cells (Figure 13.4). Conversely, the mutant insensitive to BR or with low levels of H_2O_2 showed starch accumulation in the guard cells and reduced stomatal opening.

BRs have been shown to mitigate the adverse effects of water deficit on gas exchange in several crops associated with physiological mechanisms. In *Vigna unguiculata* plants under water deficit, there were reductions in net photosynthetic rate (P_N), stomatal conductance (g_s), and transpiration rate (E) (Lima and Lobato 2017). However, 100 nM EBR

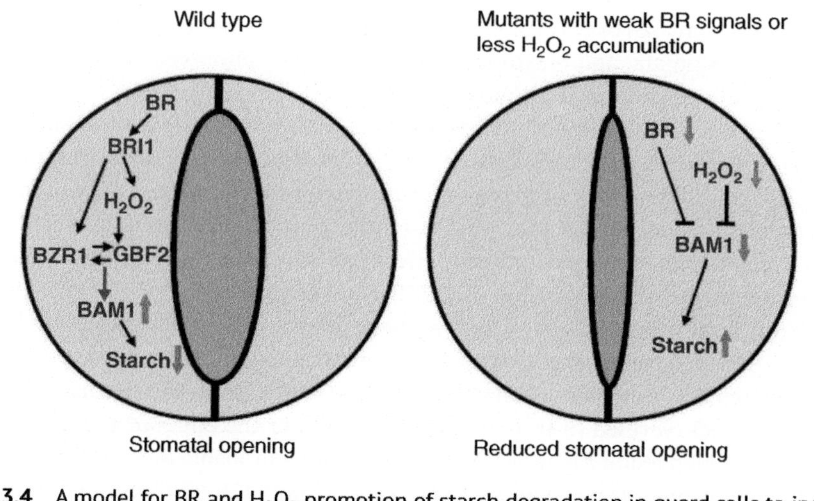

Figure 13.4 A model for BR and H_2O_2 promotion of starch degradation in guard cells to induce stomatal opening. *Source*: Li et al. (2020)/Oxford University Press.

caused increases of 96, 24, and 33%, respectively, in these variables compared with the water deficit + 0 nM EBR treatment (Table 13.1). The C_i levels increased with the water deficit, but the application of 100 nM EBR resulted in a significant decrease of 18% compared with the water deficit treatment without EBR (Table 13.1). Plants subjected to the water deficit had lower WUE and P_N/C_i values, whereas the application of 100 nM EBR caused increases of 49 and 141%, respectively, compared with the water deficit + 0 nM EBR treatment (Table 13.1). EBR promoted increases in gas exchange associated with other benefits, such as the absence of photoinhibition and improvement in the water status of the plant. In addition, EBR increased the activity of RUBISCO, the enzyme responsible for intercellular CO_2 assimilation.

EBR promoted protection against damage caused by flooding in soybean plants, which was explained by the higher density and index of the stomata (Pereira et al. 2020). EBR minimized the adverse effects, with 10 nM EBR generating increases in stomatal density (SD), stomatal functionality (SF), and stomatal index (SI) of 34, 10, and 19% on the adaxial face, respectively, and 10, 13, and 18% on the abaxial side, respectively. Improvements in these stomatal characteristics jointly favored critical physiological processes such as stomatal conductance and liquid photosynthesis, facilitating the capture of CO_2. Positive relationship linked to stomatal density, stomatal conductance, net assimilation, and water use efficiency suggested that this variable and the size of the guard cell suffered changes in response to waterlogging because they are closely associated with the photosynthesis process and the use of water by the plant. EBR induces stomata formation more elliptical, being elliptical shape more efficient and functional if compared with a rounded shape.

EBR maximized the maximal fluorescence yield of the dark-adapted state (F_m), variable fluorescence (F_v) and maximal quantum yield of PSII photochemistry (F_v/F_m) values of plants subjected to waterlogging because EBR minimized photoinhibition and decreased the dissipation of excitation energy in the antennas of photosystem II (Figure 13.5) (Pereira et al. 2020). In addition, EBR increased q_P because this steroid favors the recovery of the D1 protein and minimizes the damage to reaction centers caused by photoinhibition (Yuan et al. 2012). Therefore, EBR demonstrated beneficial action on the photosynthetic machinery, with increases in gas exchange due to direct effects related to the higher density and index of the stomata, and maximized the efficiency of photosystem II and mitigated the deleterious effects promoted by photoinhibition.

Salinity promotes negative interferences on gas exchange and lowers expansion and cell division rates. However, the exogenous application of BRs has been used to mitigate the adverse effects of saline stress, an essential factor compromising agricultural crop growth and development. Sousa et al. (2022) evaluated the application of this steroid in tomato plants. Although stress promoted adverse effects on stomatal characteristics and reductions in gas exchange, the application of 24-EBR at a concentration of 100 nM efficiently mitigated the deleterious effects of salt accumulation. As previously reported, plants treated with EBR and Na+ had increases in SD and SI, suggesting the beneficial effects of EBR on the stomatal mechanism (Chaves et al. 2009). This led to the rise in the stomatal conductance values, verified in this study in plants under the effect of 100 nM EBR and 150 mM NaCl, which led to higher CO_2 absorption and a probable increase in the activity of RuBisCo, an enzyme responsible for CO_2 fixation in the Calvin Cycle, and an ensuing increase in net assimilation (Allel et al. 2018).

Table 13.1 Gas exchange in *Vigna unguiculata* plants splayed with EBR and exposed to water deficit.

Water condition	EBR (nM)	P_N (μmol m^{-2} s^{-1})	E (mmol m^{-2} s^{-1})	g_s (mol m^{-2} s^{-1})	C_i (μmol mol^{-1})	WUE (μmol mmol^{-1})	P_N/C_i (μmol m^{-2} s^{-1} Pa^{-1})
Control	0	15.6 ± 0.5Ca	3.38 ± 0.02Aa	0.37 ± 0.02Aa	247 ± 15Ab	4.62 ± 0.12Ca	0.063 ± 0.005Ba
Control	50	17.0 ± 0.7Ba	3.38 ± 0.05Aa	0.36 ± 0.01Aa	247 ± 20Ab	5.03 ± 0.15Ba	0.065 ± 0.004Ba
Control	100	19.2 ± 0.4Aa	3.46 ± 0.07Aa	0.38 ± 0.02Aa	247 ± 17Aa	5.55 ± 0.16Aa	0.078 ± 0.006Aa
Water deficit	0	5.2 ± 0.2Cb	1.42 ± 0.09Bb	0.09 ± 0.01Bb	303 ± 10Aa	3.88 ± 0.18Cb	0.017 ± 0.001Cb
Water deficit	50	8.0 ± 0.4Bb	1.67 ± 0.09Ab	0.12 ± 0.01Ab	272 ± 15Ba	4.81 ± 0.22Ba	0.029 ± 0.003Bb
Water deficit	100	10.2 ± 0.5Ab	1.76 ± 0.08Ab	0.12 ± 0.01Ab	249 ± 22Ba	5.80 ± 0.31Aa	0.041 ± 0.003Ab

P_N, Net photosynthetic rate; E, Transpiration rate; g_s, Stomatal conductance; C_i, Intercellular CO_2 concentration; WUE, Water-use efficiency; P_N/C_i, Carboxylation instantaneous efficiency. Columns with different uppercase letters between EBR levels (0, 50, and 100 nM EBR under equal water conditions) and lowercase letters between water conditions (control and water deficit under equal EBR concentration) indicate significant differences from the Scott–Knott test (P\0.05). Values described corresponding to means from five repetitions and standard deviations.
Source: Lima and Lobato (2017)/Springer Nature.

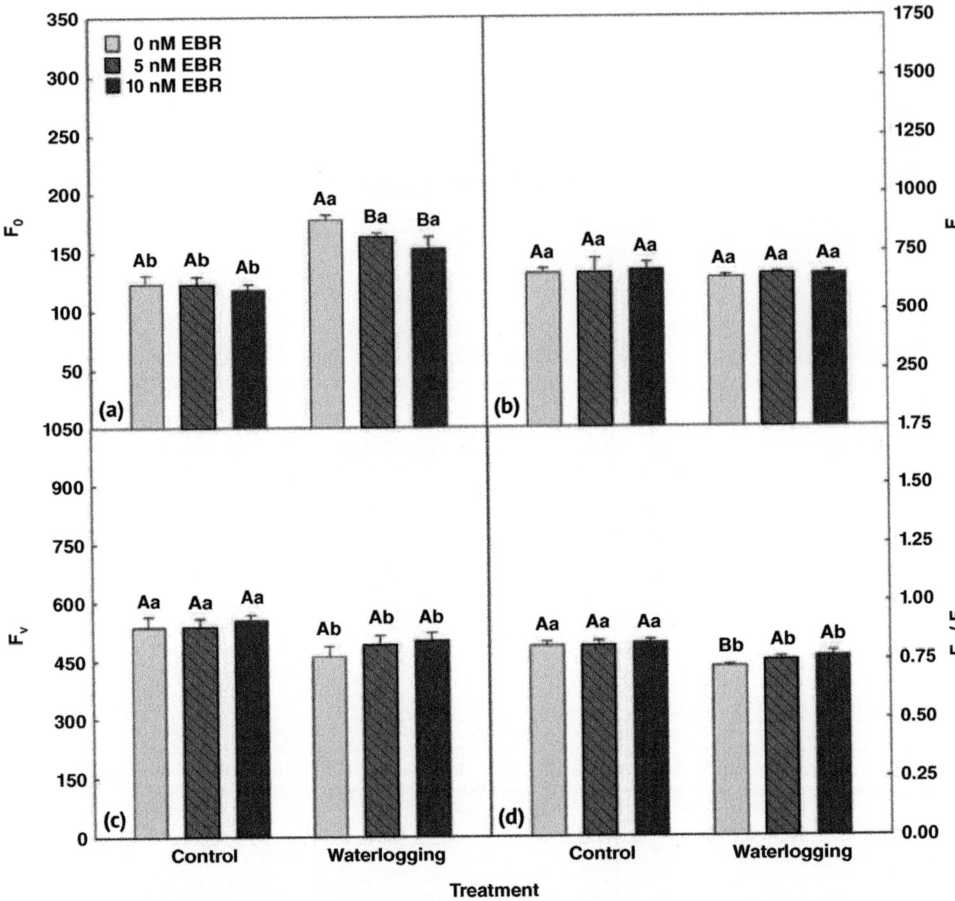

Figure 13.5 Minimal fluorescence yield of the dark-adapted state (F_0) (a), maximal fluorescence yield of the dark-adapted state (F_m) (b), variable fluorescence (F_v) (c), and maximal quantum yield of PSII photochemistry (F_v/F_m) (d) in soybean plants treated with EBR and subjected to waterlogging. Columns with different uppercase letters between EBR levels (0, 5, and 10 nM EBR under equal water conditions) and lowercase letters between water conditions (control and waterlogging under equal EBR concentration) indicate significant differences from the Scott–Knott test ($P < 0.05$). Columns corresponding to means from five repetitions and standard deviations. *Source*: Pereira et al. (2020)/Taylor & Francis Group.

Exogenous 24-EBR application may mitigate the damage caused by magnesium stress (Mg) in higher plants, mainly because it involves cell division and elongation processes. dos Santos et al. (2021) observed that low and high Mg supplies caused adverse effects on stomatal characteristics; however, EBR mitigated these deleterious impacts (Table 13.2). On the adaxial side, plants exposed to low Mg + EBR had increases of 24 and 23% in SD and SI, respectively, and decreases in polar diameter of the stomata (PDS) (10%) and equatorial diameter of the stomata (EDS) (4%). High Mg + EBR resulted in increments of 11 and 13% in SD and SI, respectively, and decreases in PDS (9%) and EDS (5%), respectively, compared with the same treatment without EBR. For the abaxial side, the low Mg and 100 nM EBR

Table 13.2 Stomatal characteristics in soybean plants sprayed with EBR and exposed to different Mg supplies.

EBR	Mg supply	SD (stomata per mm^2)	PDS (μm)	EDS (μm)	SF	SI (%)
Adaxial face						
–	Low	169 ± 8 Bb	12.69 ± 0.3 Ba	20.7 ± 0.5 Aa	0.61 ± 0.03 Aa	5.7 ± 0.3 Ab
–	Control	221 ± 11 Aa	12.0 ± 0.3 Ba	19.8 ± 0.3 Ba	0.61 ± 0.04 Aa	5.8 ± 0.4 Ab
–	High	160 ± 13 Ba	13.2 ± 0.2 Aa	21.3 ± 0.4 Aa	0.62 ± 0.05 Aa	5.4 ± 0.2 Ab
+	Low	209 ± 6 Aa	11.4 ± 0.2 Bb	19.8 ± 0.4 Aa	0.58 ± 0.02 Aa	7.0 ± 0.4 Aa
+	Control	231 ± 24 Aa	11.1 ± 0.3 Bb	18.9 ± 0.3 Bb	0.59 ± 0.03 Aa	7.5 ± 0.6 Aa
+	High	178 ± 6 Ba	12.0 ± 0.2 Ab	20.3 ± 0.4 Ab	0.59 ± 0.03 Aa	6.1 ± 0.3 Ba
Abaxial face						
–	Low	284 ± 17 Bb	11.5 ± 0.4 Aa	19.5 ± 0.7 Aa	0.59 ± 0.03 Aa	8.8 ± 0.4 Ab
–	Control	323 ± 17 Ab	11.0 ± 0.5 Aa	19.2 ± 0.6 Aa	0.58 ± 0.02 Aa	9.2 ± 0.6 Ab
–	High	252 ± 15 Ba	12.0 ± 0.5 Aa	20.5 ± 0.9 Aa	0.59 ± 0.02 Aa	7.8 ± 0.2 Bb
+	Low	368 ± 28 Ba	10.7 ± 0.4 Aa	17.7 ± 0.3 Bb	0.60 ± 0.04 Aa	10.2 ± 0.6 Ba
+	Control	525 ± 70 Aa	9.6 ± 0.3 Bb	16.3 ± 0.2 Cb	0.61 ± 0.04 Aa	11.9 ± 0.5 Aa
+	High	294 ± 7 Ca	11.2 ± 0.6 Aa	18.9 ± 0.5 Ab	0.59 ± 0.03 Aa	9.1 ± 0.4 Ba

SD, Stomatal density; PDS, Polar diameter of the stomata; EDS, Equatorial diameter of the stomata; SF, Stomatal functionality; SI, Stomatal index. Columns with different uppercase letters between Mg supplies (low, control, and high Mg supply under equal EBR level) and lowercase letters between EBR levels (with and without EBR under equal Mg supply) indicate significant differences from the Scott–Knott test ($P < 0.05$). Means ± SD, $n = 5$.
Source: Santos et al. (2021)/Taylor & Francis Group.

spray resulted in increases of 30, 2, and 16% in the variables SD, SF, and SI, respectively, and reductions in PDS (7%) and EDS (9%), when compared with similar treatment without EBR. Under high Mg + EBR, increases in SD (17%) and SI (17%) were observed, and reductions in PDS (7%) and EDS (8%) occurred.

The exogenous application of EBR reduced the deleterious effects related to low and high Mg supply in SD, SI, PDS, and EDS (Figure 13.6) (dos Santos et al. 2021).

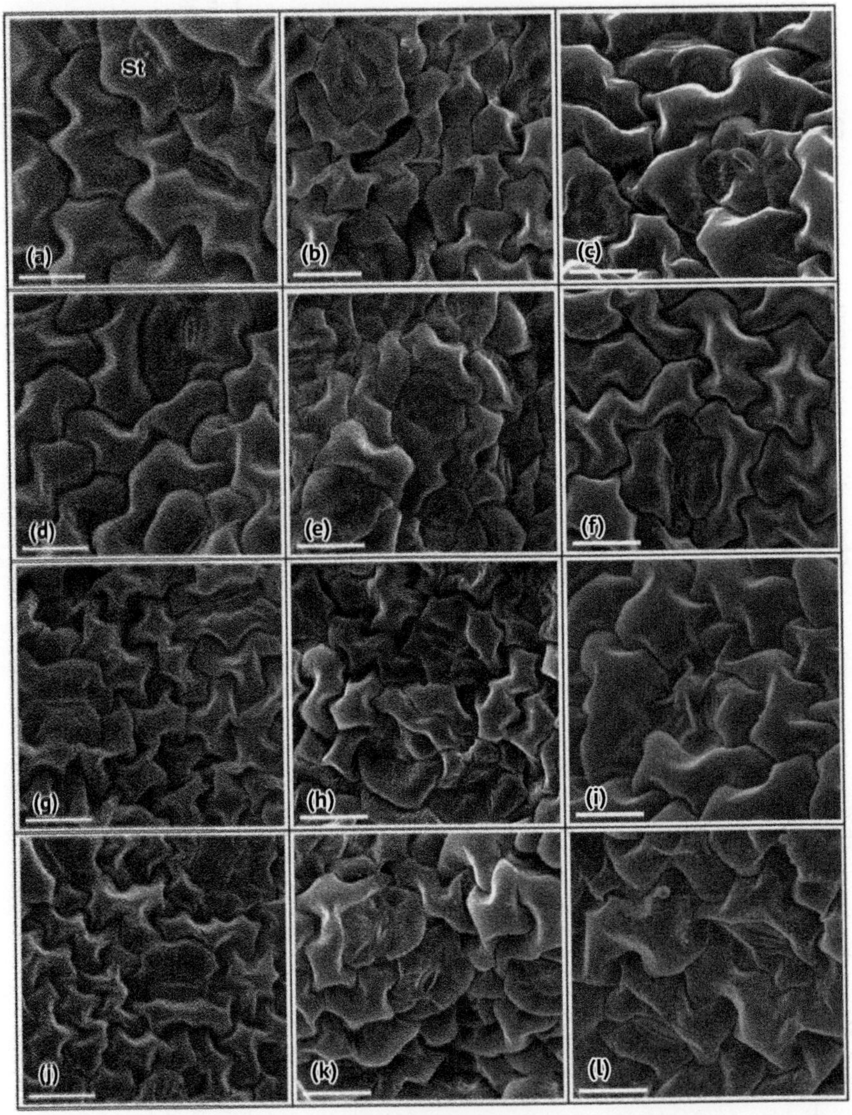

Figure 13.6 Adaxial (a–f) and abaxial leaf surface (g–l) in scanning electron microscopy showing stomata in soybean plants sprayed with EBR and exposed to different Mg supplies. Low Mg without EBR (a and g), low Mg with EBR (d and j), control Mg without EBR (b and h), control Mg with EBR (e and k), high Mg without EBR (c and i) and high Mg with EBR (f and l). Legend: St = Stomata. Bars: 20 μm. *Source*: Santos et al. (2021)/Taylor & Francis.

This steroid stimulates the production of stomata by regulating specific proteins of stomatal development, such as BIN2 kinase, MAP kinase, and SPCH (Serna 2014), and increments in SD suggest a higher concentration of CO_2 that is absorbed and available for the photosynthetic process. Increases in SD regulate transpiration rates, reducing excessive water losses under stress conditions, such as water deficit (Lawson and Blatt 2014). In addition, reductions observed in PDS and EDS reveal the beneficial influence of the EBR on stomatal shape, inducing an elliptical format in which these structures continuously adjust to environmental and intracellular conditions (Drake et al. 2013).

EBR pretreatment promoted improvements in leaf structures in soybean plants (*Glycine max*) in boron (B) stress (Pereira et al. 2022). The low and high B supplies reduced stomatal characteristics, except for PDS and EDS. Plants sprayed with EBR under the low treatment B on the adaxial face presented increases for SD, SF, and SI of 30, 6, and 42%, respectively, compared to the same treatment without EBR. Under the control treatment, the adaxial face had increased for SD and SI by 9 and 18%, respectively; in high B treatment, there were increases of 23, 5, and 28% in SD, SF, and SI, respectively. For PDS and EDS in the abaxial face, EBR + low B induced decreases of 13 and 12%, respectively. Under control treatment combined with EBR suffered declines of 2% (PDS) and 6% (EDS), and under the high B with EBR, 13% in PDS and 12% in EDS were reduced. These results confirm the beneficial action of this steroid on stomatal development through activating proteins involved in the signaling pattern of the stomatal mechanism (Lin et al. 2013). The relationship of these stomatal changes under stress conditions and the benefits of EBR are efficient strategies of plants because the increase in the density and reduction in the size of the stomata contribute to a higher diffusion of CO_2 to the carboxylation sites and additions in water use efficient (Devi and Reddy 2018).

13.5 Conclusion

BR revealed an essential function in the defense of plants against different types of stress, acting on several mechanisms in the development and regulation of stomata. The endogenous processes that govern stomatal development involve the action of BR that downregulates a signaling cascade. Like ABA, this steroid prolongs the formation of stomatal precursor cells and acts in stomatal control under stress conditions. In addition, the exogenous application of BR in plants under different environmental stresses promoted improvements in photosynthetic performance, including stomatal conductance, transpiration, photoinhibition reduction, and stomatal characteristics. Therefore, BR acting in stomatal regulation becomes a vital control pathway in the balance between CO_2 gain and transpiration rate for plants under stress conditions.

References

Aasamaa K. and Sõber A. (2011). Responses of stomatal conductance to simultaneous changes in two environmental factors. *Tree Physiology* 31: 855–864. https://doi.org/10.1093/treephys/tpr078.

Ahammed, G.J., Li, X., Liu, A. et al. (2020). Brassinosteroids in plant tolerance to abiotic stress. *Journal of Plant Growth Regulation* 39: 1451–1464. https://doi.org/10.1007/s00344-020-10098-0

Allel, D., Ben-Amar, A., and Abdelly, C. (2018). Leaf photosynthesis, chlorophyll fluorescence and ion content of barley (*Hordeum vulgare*) in response to salinity. *Journal of Plant Nutrition* 41: 497–508. https://doi.org/10.1080/01904167.2017.1385811

Anwar, A., Liu, Y., Dong, R. et al. (2018). The physiological and molecular mechanism of brassinosteroid in response to stress: A review. *Biological Research* 51: 1–15. https://doi.org/10.1186/s40659-018-0195-2

Bergmann, D.C. and Sack, F.D. (2007). Stomatal development. *Annual Review of Plant Biology* 58: 163–181. https://doi.org/10.1146/annurev.arplant.58.032806.104023

Chater, C.C.C., Caine, R.S., Fleming, A.J. et al. (2017). Origins and evolution of stomatal development. *Plant Physiology* 174: 624–638. https://doi.org/10.1104/pp.17.00183

Chaves, M.M., Flexas, J., and Pinheiro, C. (2009). Photosynthesis under drought and salt stress: Regulation mechanisms from whole plant to cell. *Annals of Botany* 103: 551–560. https://doi.org/10.1093/aob/mcn125

Clouse, S.D. (2011). Brassinosteroid signal transduction: From receptor kinase activation to transcriptional networks regulating plant development. *The Plant Cell* 23: 1219–1230. https://doi.org/10.1105/tpc.111.084475

Cunha, L.F.S., de Oliveira, V.P., do Nascimento, A.W.S. et al. (2021). Leaf application of 24-epibrassinolide mitigates cadmium toxicity in young Eucalyptus urophylla plants by modulating leaf anatomy and gas exchange. *Physiologia Plantarum* 173: 67–87. https://doi.org/10.1111/ppl.13182

Devi, M.J. and Reddy, V.R. (2018). Transpiration response of cotton to vapor pressure deficit and its relationship with stomatal traits. *Frontiers in Plant Science* 871: 1–12. https://doi.org/10.3389/fpls.2018.01572

dos Santos Ribeiro, D.G., da Silva, B.R.S., and da Silva Lobato, A.K. (2019). Brassinosteroids induce tolerance to water deficit in soybean seedlings: contributions linked to root anatomy and antioxidant enzymes. *Acta Physiologiae Plantarum* 41: 1–11. https://doi.org/10.1007/s11738-019-2873-2

dos Santos, L.A., da Silva, B.R.S., and da Silva Lobato, A.K. (2021). Foliar-applied 24-epibrassinolide systemically triggers tolerance to magnesium stress in soybean plants: plausible responses focused on root and leaf structures. *Botany Letters* 168: 558–569. https://doi.org/10.1080/23818107.2021.1911844

Drake, P.L., Froend, R.H., and Franks, P.J. (2013). Smaller, faster stomata: scaling of stomatal size, rate of response, and stomatal conductance. *Journal of Experimental Botany* 64: 495–505. https://doi.org/10.1093/jxb/ers347

Fang, P., Yan, M., Chi, C. et al. (2019). Brassinosteroids act as a positive regulator of photoprotection in response to chilling stress. *Plant Physiology* 180: 2061–2076. https://doi.org/10.1104/pp.19.00088

Faralli, M., Matthews, J., and Lawson, T. (2019). Exploiting natural variation and genetic manipulation of stomatal conductance for crop improvement. *Current Opinion in Plant Biology* 49: 1–7. https://doi.org/10.1016/j.pbi.2019.01.003

Gleason, S.M., Cooper, M., Wiggans, D.R. et al. (2019). Stomatal conductance, xylem water transport, and root traits underpin improved performance under drought and well-watered conditions across a diverse panel of maize inbred lines. *Field Crops Research* 234: 119–128. https://doi.org/10.1016/j.fcr.2019.02.001

Hafeez, M.B., Zahra, N., Zahra, K. et al. (2021). Brassinosteroids: molecular and physiological responses in plant growth and abiotic stresses. *Plant Stress* 2: 100029. https://doi.org/10.1016/j.stress.2021.100029

Hamanishi, E.T., Thomas, B.R., and Campbell, M.M. (2012). Drought induces alterations in the stomatal development program in Populus. *Journal of Experimental Botany* 63: 4959–4971. https://doi.org/10.1093/jxb/err313

Inoue, S.I., Iwashita, N., Takahashi, Y. et al. (2017). Brassinosteroid involvement in *Arabidopsis thaliana* stomatal opening. *Plant and Cell Physiology* 58: 1048–1058. https://doi.org/10.1093/pcp/pcx049

Jager, C.E., Symons, G.M., Ross, J.J. et al. (2008). Do brassinosteroids mediate the water stress response? *Physiologia Plantarum* 133: 417–425. https://doi.org/10.1111/j.1399-3054.2008.01057.x

Karimi, S.M., Freund, M., Wager, B.M. et al. (2021). Under salt stress guard cells rewire ion transport and abscisic acid signaling. *New Phytologist* 231: 1040–1055. https://doi.org/10.1111/nph.17376

Kim, T.-W., Michniewicz, M., Bergmann, D.C. et al. (2012). Brassinosteroid regulates stomatal development by GSK3-mediated inhibition of a MAPK pathway. *Nature* 482: 419–422. https://doi.org/10.1038/nature10794

Koyro, H.-W., Ahmad, P., and Geissler, N. (2012). Abiotic stress responses in plants: an overview. In: *Environmental Adaptations and Stress Tolerance of Plants in the Era of Climate* Change, 1–515. Springer.

Lawson, T. and Blatt, M.R. (2014). Stomatal size, speed, and responsiveness impact on photosynthesis and water use efficiency. *Plant Physiology* 164: 1556–1570. https://doi.org/10.1104/pp.114.237107

Le, J., Liu, X.G., Yang, K.Z. et al. (2014). Auxin transport and activity regulate stomatal patterning and development. *Nature Communications* 5: 1–8. https://doi.org/10.1038/ncomms4090

Li, J.G., Fan, M., Hua, W. et al. (2020). Brassinosteroid and hydrogen peroxide interdependently induce stomatal opening by promoting guard cell starch degradation. *The Plant Cell* 32: 984–999. https://doi.org/10.1105/TPC.19.00587

Li, S. and Liu, F. (2022). Vapour pressure deficit and endogenous ABA level modulate stomatal responses of tomato plants to soil water deficit. *Environmental and Experimental Botany* 199: 104889. https://doi.org/10.1016/j.envexpbot.2022.104889

Li, S., Liu, J., Liu, H. et al. (2021). Role of hydraulic signal and ABA in decrease of leaf stomatal and mesophyll conductance in soil drought-stressed tomato. *Frontiers in Plant Science* 12: 1–12. https://doi.org/10.3389/fpls.2021.653186

Li, S.L., Tan, T.T., Fan, Y.F. et al. (2022). Responses of leaf stomatal and mesophyll conductance to abiotic stress factors. *Journal of Integrative Agriculture* 21: 2787–2804. https://doi.org/10.1016/j.jia.2022.07.036

Lima, J.V. and Lobato, A.K.S. (2017). Brassinosteroids improve photosystem II efficiency, gas exchange, antioxidant enzymes and growth of cowpea plants exposed to water deficit. *Physiology and Molecular Biology of Plants* 23: 59–72. https://doi.org/10.1007/s12298-016-0410-y

Lin, L.L., Wu, C.C., Huang, H.C. et al. (2013). Identification of microRNA 395a in 24-epibrassinolide-regulated root growth of *Arabidopsis thaliana* using microRNA arrays. *International Journal of Molecular Sciences* 14: 14270–14286. https://doi.org/10.3390/ijms140714270

Marchin, R.M., Backes, D., Ossola, A. et al. (2022). Extreme heat increases stomatal conductance and drought-induced mortality risk in vulnerable plant species. *Global Change Biology* 28: 1133–1146. https://doi.org/10.1111/gcb.15976

Pereira, Y.C., da Nascimento, A.N.S., dos Aguiar, B.T.S. et al. (2022). Anatomical modifications modulated by pretreatment with 24-epibrassinolide alleviate boron stress in soybean plants: valuable repercussions on nutrient contents, photosynthesis, and biomass. *Journal of Soil Science and Plant Nutrition* 22 (4): 4533–4550. https://doi.org/10.1007/s42729-022-01053-x

Pereira, Y.C., da Silva, F.R., da Silva, B.R.S. et al. (2020). 24-epibrassinolide induces protection against waterlogging and alleviates impacts on the root structures, photosynthetic machinery and biomass in soybean. *Plant Signaling and Behavior* 15 (11): 1805885. https://doi.org/10.1080/15592324.2020.1805885

Saito, S. and Uozumi, N. (2019). Guard cell membrane anion transport systems and their regulatory components: an elaborate mechanism controlling stress-induced stomatal closure. *Plants* 8 (1): 9. https://doi.org/10.3390/plants8010009

Sarwat, M. and Tuteja, N. (2017). Hormonal signaling to control stomatal movement during drought stress. *Plant Gene* 11: 143–153. https://doi.org/10.1016/j.plgene.2017.07.007

Serna, L. (2014). The role of brassinosteroids and abscisic acid in stomatal development. *Plant Science* 225: 95–101. https://doi.org/10.1016/j.plantsci.2014.05.017

Sharma, I., Kaur, N., and Pati, P.K. (2017). Brassinosteroids: A promising option in deciphering remedial strategies for abiotic stress tolerance in rice. *Frontiers in Plant Science* 8: 1–17. https://doi.org/10.3389/fpls.2017.02151

Shekoofa, A., Safikhan, S., Snider, J.L. et al. (2020). Variation in stomatal conductance responses of cotton cultivars to high vapour pressure deficit under controlled and rainfed environments. *Journal of Agronomy and Crop Science* 207 (2): 332–343. https://doi.org/10.1111/jac.12440

Shu, S., Tang, Y., Yuan, Y. et al. (2016). The role of 24-epibrassinolide in the regulation of photosynthetic characteristics and nitrogen metabolism of tomato seedlings under a combined low temperature and weak light stress. *Plant Physiology and Biochemistry* 107: 344–353. https://doi.org/10.1016/j.plaphy.2016.06.021

Sonjaroon, W., Jutamanee, K., Khamsuk, O. et al. (2018). Impact of brassinosteroid mimic on photosynthesis, carbohydrate content and rice seed set at reproductive stage under heat stress. *Agriculture and Natural Resources* 52: 234–240. https://doi.org/10.1016/j.anres.2018.09.001

Sousa, V.Q., Messias, W.F.S., Pereira, Y.C. et al. (2022). Pretreatment with 24-epibrassinolide synergistically protects root structures and chloroplastic pigments and upregulates antioxidant enzymes and biomass in Na+-stressed tomato plants. *Journal of Plant Growth Regulation* 41: 2869–2885. https://doi.org/10.1007/s00344-021-10481-5

Takahashi, Y., Ebisu, Y., Kinoshita, T. et al. (2013). bHLH transcription factors that facilitate k+ uptake during stomatal opening are repressed by abscisic acid through phosphorylation. *Science Signaling* 6: 1–11. https://doi.org/10.1126/scisignal.2003760

Tanaka, Y., Nose, T., Jikumaru, Y. et al. (2013). ABA inhibits entry into stomatal lineage development in *Arabidopsis* leaves. *The Plant Journal* 74: 448–457. https://doi.org/10.1111/tpj.12136

Terashima, I., Hanba, Y.T., Tholen, D. et al. (2011). Leaf functional anatomy in relation to photosynthesis. *Plant Physiology* 155: 108–116. https://doi.org/10.1104/pp.110.165472

Torii, K.U. (2021). Stomatal development in the context of epidermal tissues. *Annals of Botany* 128: 137–148. https://doi.org/10.1093/aob/mcab052

Wang, H., Ngwenyama, N., Liu, Y. et al. (2007). Stomatal development and patterning are regulated by environmentally responsive mitogen-activated protein kinases in *Arabidopsis*. *The Plant Cell* 19: 63–73. https://doi.org/10.1105/tpc.106.048298

Wei, Z. and Li, J. (2016). Brassinosteroids regulate root growth, development, and symbiosis. *Molecular Plant* 9: 86–100. https://doi.org/10.1016/j.molp.2015.12.003

Xia, X.J., Gao, C.J., Song, L.X. et al. (2014). Role of H_2O_2 dynamics in brassinosteroid-induced stomatal closure and opening in *Solanum lycopersicum*. *Plant, Cell & Environment* 37: 2036–2050. https://doi.org/10.1111/pce.12275

Xie, Z., Li, D., Wang. L. et al. (2010). Role of the stomatal development regulators FLP/MYB88 in abiotic stress responses. *The Plant Journal* 64: 731–739. https://doi.org/10.1111/j.1365-313X.2010.04364.x

Xu, Z., Jiang, Y., Jia, B. et al. (2016). Elevated-CO_2 response of stomata and its dependence on environmental factors. *Frontiers in Plant Science* 7: 1–15. https://doi.org/10.3389/fpls.2016.00657

Yang, X., Gavya, S.L., Zhou, Z. et al. (2022). Abscisic acid regulates stomatal production by imprinting an SnRK2 kinase-mediated phosphocode on the master regulator SPEECHLESS. *Science Advances* 8 (40): eadd2063. https://doi.org/10.1126/sciadv.add2063

Yuan, L., Shu, S., Sun, J. et al. (2012). Effects of 24-epibrassinolide on the photosynthetic characteristics, antioxidant system, and chloroplast ultrastructure in *Cucumis sativus* L. under $Ca(NO_3)_2$ stress. *Photosynthesis Research* 112: 205–214. https://doi.org/10.1007/s11120-012-9774-1

Zhang, Q., Peng, S., and Li, Y. (2019). Increase rate of light-induced stomatal conductance is related to stomatal size in the genus *Oryza*. *Journal of Experimental Botany* 70: 5259–5269. https://doi.org/10.1093/jxb/erz267

14

Regulation of Photosynthesis by Brassinosteroids Under Optimal and Suboptimal Conditions

*Rohit Chhabra[1], Neetu Jagota[1], Kompal Gaur[2], Tavisha Singh[1], and Ashish Sharma[1]**

[1]*Department of Botany and Environment Science, DAV University, Jalandhar, Punjab, India*
[2]*Department of Floriculture and Landscaping, Punjab Agricultural University, Punjab, India*

*Corresponding author: ashish10210@davuniversity.org

14.1 General Introduction

Naturally occurring chemical substances called phytohormones have a significant impact on a variety of plant physiological functions, even at extremely low concentrations. These chemical messengers, integral to plant physiology, facilitate communication and regulation within the plant. In 1937, Went and Thimann showed that they move from one part of the plant to another. Plants produce a diverse range of phytohormones, including ethylene, auxins, gibberellins, cytokinins, abscisic acid, and brassinosteroids (BRs). Additionally, more recently discovered plant growth regulators include salicylic acid, strigolactones, and jasmonic acid. These hormones orchestrate various aspects of plant growth, development, and stress responses, playing crucial roles in optimizing plant health and productivity. All plant species include BRs, a family of naturally occurring polyhydroxylated sterol derivatives. Structurally akin to the steroid hormones found in insects and animals, BRs were first identified by Mitchell and his team at the USDA through the analysis of pollen extracts from over 30 different plant species. The most potent form of BRs, brassinolide (BL), was isolated from *Brassica napus* (*B. napus*) and named "brassins" (Grove et al. 1979), marking a significant discovery in plant growth regulation. BRs are characterized by their complex structure, which includes multiple oxygen moieties located at carbon atoms C2, C6, C22, and C23, with at least one additional oxygen moiety at the C_3 position. According to Symons et al. (2008), BRs play a critical role in modifying the auxin transport system, thereby influencing long-distance signaling within plants. The significance of BRs in controlling many physiological processes and promoting plant growth and development is highlighted by this interaction.

More than 70 different BR compounds have been discovered since BL, the strongest BR, was extracted from *B. napus* pollen grains in the 1970s. These compounds are ubiquitous

Brassinosteroids and Abiotic Stress: Recent Developments on Physiological and Molecular Aspects,
First Edition. Edited by Anket Sharma, et al.
Published 2026 by John Wiley & Sons Ltd.

across the plant kingdom (Bajguz and Tretyn 2003). BRs are present in all plant tissues, including the vascular cambium, pollen, flowers, seeds, roots, shoots, and leaves. Although BRs are essential to plant physiology, they are not considered long-distance signaling molecules due to their limited mobility within the plant. They do not traverse large distances between roots and shoots or between different organs within a shoot. However, as noted by Symons et al. (2008), BRs significantly influence the auxin transport system, which indirectly affects long-distance signaling in plants. The way that BRs alter auxin transport emphasizes how important they are for coordinating several facets of plant growth and development.

Photosynthesis is a crucial process significantly influenced by BRs, which are essential to the general growth of plants. This highly regulated process involves chlorophyll absorbing solar energy, which excites electrons. As a result of the excitation, electrons proceed from water to $NADP^+$, generating reduced power and driving a series of enzymatic reactions that assimilate carbon dioxide (CO_2) to produce carbohydrates while releasing oxygen (Pan et al. 2018). Chloroplasts, the primary sites of photosynthesis, are responsible for both light-dependent and light-independent reactions (Ashraf and Harris 2013). Within the chloroplasts, photosynthesis occurs in the thylakoid membranes, where two reaction centers, photosystem I (PSI) and photosystem II (PSII), operate in succession. PSII is predominantly located in the stacked grana membranes; in contrast, PSI is present in the stroma membranes that are not stacked. PSI absorbs light at a peak wavelength of 700 nm, while PSII absorbs at 680 nm. These PSs facilitate noncyclic electron transport, oxidizing water to produce oxygen and reducing $NADP^+$ to NADPH, with oxygen being produced as the sole byproduct in Earth's atmosphere. Many studies have been done on how BRs affect plants, especially when they are subjected to different abiotic stressors such as drought, salt, temperature fluctuations, and metal toxicity (Ashraf and Harris 2013). These studies demonstrate how BRs can improve photosynthetic efficiency and plant resistance in stressful situations.

For plants to grow and develop, photosynthesis and chlorophyll are essential components. To convert solar energy into chemical energy stored in carbohydrates, the photosynthetic process depends on chlorophyll, the pigment that absorbs light energy. This energy fuels various developmental processes essential for plant health and productivity. Exogenous application of BRs has been shown to significantly improve chlorophyll levels in plants. This boost in chlorophyll content improves the plant's photosynthetic capacity across a range of environmental conditions. One important indicator of photosynthetic activity and chloroplast health is chlorophyll, as its levels directly correlate with the plant's ability to perform photosynthesis effectively. Plants exposed to stressors such as extreme temperatures, drought, salinity, or metal toxicity often experience a reduction in chlorophyll levels, including chlorophyll *a*, chlorophyll *b*, and carotenoids. This reduction adversely impacts photosynthetic efficiency, as chlorophyll is highly sensitive to these external stimuli. However, BRs play a protective role by shielding chlorophyll from such stresses. They enhance chlorophyll concentration and photosynthetic indices, thereby mitigating the harmful effects of stress. BRs also influence the interaction between membrane proteins and chlorophyll molecules by regulating the activity of chlorophyllase, an enzyme involved in chlorophyll degradation. This regulation helps maintain the integrity of the thylakoid membranes within chloroplasts, which is critical for sustaining photosynthetic processes. By preserving chlorophyll and optimizing its function, BRs support the plant's ability to adapt to and recover from abiotic stress, ensuring continued photosynthetic efficiency and overall plant health (Nawaz et al. 2017).

Under water stress, tomato seedlings' photosynthetic activity is regulated and maintained by BRs. BR modifies the defense system and lessens the detrimental impacts of different stressors by regulating the transcription levels of defense genes in cucumber (Li et al. 2013). Furthermore, BR regulates the Rubisco activase (RCA) gene in wheat under drought and heat stress, which is essential to photosynthesis. Furthermore, BR dramatically boosts photosynthesis and antioxidant enzyme activity. Additionally, the CO_2 assimilation and quantum yield of PSII, Rubisco activities, and the expression of the genes encoding the Rubisco large subunit (rbcL) and small subunit (rbcS), which improve photosynthetic capability, were all increased in BR-treated seedlings (Choudhary et al. 2012). These results point to BR's beneficial contribution to photosynthesis (Siddiqui et al. 2019). Additionally, a transcriptome investigation of BR found that it boosts a variety of chlorophylls and genes important for photosynthesis under low-temperature stress. Thus, through regulating several enzymes, including chlorophyllase and Rubisco, and the transcript levels of genes encoded for chlorophyll and photosynthesis under stress, BR encourages the accumulation of chlorophylls and photosynthetic capability (Anwar et al. 2018).

Although the precise mechanisms by which BRs enhance photosynthesis are not fully understood, several hypotheses have been proposed to explain their beneficial effects. One theory suggests that BRs might improve photosynthesis by stimulating or activating the production of the enzymes necessary for the biosynthesis of chlorophyll. Chlorophyll is crucial for capturing light energy and facilitating photosynthesis. By boosting the production of these enzymes, BRs may aid in maintaining the plant's ability to photosynthesize by halting the deterioration of chlorophyll pigments (dos Santos et al. 2021). Another hypothesis posits that BRs could enhance photosynthetic efficiency by affecting stomatal function. Plant leaves have tiny holes called stomata that control gas exchange, including the uptake of CO_2 and the outflow of oxygen. BRs may influence the function of these stomata, potentially increasing internal CO_2 concentrations available for photosynthetic enzymes. This increase in CO_2 availability could improve the efficiency of carbon fixation, a critical step in the photosynthetic process (Li et al. 2013). Normal plant growth and development depends on maintaining an ideal level of phytohormones, such as BRs. BRs play a significant role in regulating their homeostasis through a feedback mechanism. This involves the modulation of BR biosynthesis and inactivation processes to maintain a balanced internal hormone environment. By regulating these processes, BRs ensure that their levels are kept within a range that supports healthy plant development while avoiding potential negative effects associated with excess or deficiency (Shu et al. 2016). Overall, these proposed mechanisms highlight the multifaceted roles of BRs in enhancing photosynthesis and supporting plant health. Continued research into these processes will further elucidate how BRs contribute to plant resilience and productivity.

14.2 Optimal vs. Suboptimal Conditions

Plants must adapt to their constantly changing environment (Steber and Mccourt 2001). A plant has several decisions to make to survive over its existence. Cues from their surroundings influence these choices. "Optimal conditions" in the context of plant biology and physiology refer to environmental parameters that are optimum for a plant's growth and function, such as temperature, availability of water, light intensity, and nutrient levels. These conditions typically allow the plant to photosynthesize efficiently and maximize its growth potential. On the other hand, "sub-optimal conditions" denote environmental stresses or deviations from the optimal range that can negatively impact a plant's growth and photosynthetic efficiency. These circumstances could include biotic stressors like pathogen infections or herbivore damage, as well as abiotic stressors like drought, high or low temperatures, salt, or nutritional deficiencies. Therefore, "optimal" and "sub-optimal" conditions are crucial distinctions in considering how plants respond to their environment and how interventions such as BR treatments can potentially improve plant resilience and productivity (Siddiqui et al. 2018a). Since plants are greatly impacted by sub-optimal growing circumstances, considerable effort is spent on making agricultural inputs available to crops as much as possible. In a nonlimiting environment, genetic features have a vital influence on defining the photosynthetic performance of the species. The idea of optimal conditions for growth is fundamentally connected to species/tree/organ resource-dependent responses (Hayat et al. 2001a). Therefore, to maintain plant functionality and growth in the face of less-than-ideal growth conditions, plants adapt to environmental restrictions through changes in their morphology, anatomy, biochemistry, and physiology. Scientists examine how these plant hormones affect photosynthetic mechanisms and plant reactions to stress in research focusing on how BRs control photosynthesis in both optimal and unfavorable circumstances (Zaharah et al. 2012). For example, by boosting the gene expression involved in photosynthesis, BRs may, under ideal conditions, increase photosynthetic efficiency. By controlling physiological reactions that preserve or enhance photosynthetic activity, they may be able to lessen the harmful consequences of stress in less-than-ideal circumstances (Zhao et al. 2019).

14.3 Impact of BRs on Photosynthetic Activity

The regulation of photosynthesis under both stress and normal circumstances is significantly influenced by BRs. Research has demonstrated that BRs benefit plants experiencing stress from cold, water deficiency, and salt (Clouse and Sasse 1998; Kagale et al. 2007). While the Hill reaction activity initially decreases after HBL treatment, it increases in later developmental stages. The rate of linear electron transport and total photosynthesis are directly impacted by the efficiency of ϕPSII, a crucial measure of the percentage of absorbed photons used in photochemistry. Studies have shown that BRs enhance ϕPSII levels, indicating improved photosynthetic performance (Yu et al. 2004). Chloroplasts, the primary site of photosynthesis, are responsible for both light absorption and carbon fixation. Chlorophyll, the pigment responsible for capturing light, plays a crucial role in converting light energy into adenosine triphosphate (ATP). Estimating chlorophyll content is essential for assessing a plant's photosynthetic rate (Dalio et al. 2011). BRs have been shown to boost photosynthetic capability by raising the concentration of chlorophyll (Alyemeni and Al-Quwaiz 2016). Plants contain two forms of chlorophyll: chlorophyll *a* and chlorophyll *b*. The presence of BRs is associated with increased overall chlorophyll content (Bajguz and Czerpak 1998; Farazi et al. 2015; Gabr et al. 2011; Maity and Bera 2009).

Furthermore, it has been discovered that 24-epibrassinolide (24-EBR) raises the concentrations of carotenoid and total chlorophyll (Asha and Lingakumar 2015). EBL raises the overall amount of chlorophyll in *Cucumis sativus* while lowering the ratio of chlorophyll *a/b* (Yuan et al. 2012).

They have also been found to increase Rubisco activity, a critical enzyme in the Calvin cycle (Xia et al. 2009; Yu et al. 2004). This enhancement in Rubisco and other Calvin cycle enzymes is due to BRs upregulating specific genes encoding these enzymes. As a result, the net photosynthetic rate (PN) and related characteristics improve in the presence of BRs. Improved photosynthetic efficiency is achieved through increased CO_2 assimilation, Rubisco activity, and stomatal conductance (Gruszka 2013).

Bicarbonate and CO_2 can be reversibly converted by the enzyme carbonic anhydrase (CA), which plays a role in maintaining CO_2 levels near Rubisco in C_3 plants (Badger and Price 1994). According to Fariduddin et al. (2006), BRs have been shown to enhance CA activity in *Brassica juncea* and *Vigna radiata* leaves treated with HBL. Research on BR-treated plants and BR-deficient mutants suggests a connection between BRs and genes involved in photosynthesis (Oh et al. 2011; Bai et al. 2012). Researchers have found that *Arabidopsis* mutants with changed BR responses had smaller PSII complexes, larger thylakoids, less CO_2 evolution, and a poorer quantum yield of PSII (Krumova et al. 2013). Plants that receive exogenous BRs can develop stress tolerance or successfully lessen the negative impacts of abiotic stressors (Cutler 1991; Rao et al. 2002). According to Shahbaz et al. (2008), this application modifies the activity of important antioxidant enzymes, particularly in situations involving salt stress. It has been demonstrated that foliar application of 24-EBR and 28-homobrassinolide (28-HBL) increases catalase (CAT) activity in groundnuts (*Arachis hypogaea*). During salt stress, BRs help replenish chlorophyll levels and boost the activity of nitrate reductase, an enzyme critical for nitrogen assimilation and plant development. Although BRs have little impact on cellular structure under normal conditions, they significantly prevent salt stress-induced damage to nuclei and chloroplasts. Furthermore, both salt-tolerant and salt-sensitive wheat cultivars benefit greatly from EBR in terms of growth and photosynthetic ability. Salt-tolerant wheat types are more significantly improved by foliar application of BRs than salt-sensitive ones, according to Shahbaz et al. (2008).

14.4 Mechanism of BR-mediated Regulation of Photosynthesis Under Optimal Conditions

According to Siddiqui et al. (2019), the photosynthetic mechanism controls several photosynthesis-related elements, including the apparatus of the PS, stomatal conductance, stomatal motility, sugar buildup, and Calvin cycle enzymes. BRs can prevent the loss of photosynthetic pigments by encouraging or stimulating the production of enzymes necessary for the formation of chlorophyll, which ensures the general growth of plants. It increases photosynthetic carbon fixation efficiency and gets beyond stomatal limitations. BRs also contribute to increased glucose synthesis by acting at different intensities of light and dark reactions. It accomplishes that by controlling the rate at which ribulose 1,5-bisphosphate (RuBP) regenerates and the rate at which Rubisco carboxylation occurs. An increase in the protein level and profusion of Rubisco mRNA is responsible for the rise in early Rubisco activity and maximum carboxylation rates. According to Xia et al. (2009), EBL increases the rate of photosynthetic electron transport and the production of genes that code for Calvin cycle enzymes that are important in RuBP regeneration. Transgenic plants show enhanced growth and development along with photosynthetic capacity attributable to SLBRI1 overexpression in tomato (*Solanum lycopersicum*) plants and an increase in BR signaling mediated by the BRI1 gene. In comparison to plants of the native type, transgenic plants also have larger leaves, larger blooms, more petals, and a more pronounced style. The rise in net photosynthetic rate and CO_2 assimilation rate in plants overexpressing SLBRI1 suggests that SlBRI1 overexpression-mediated improved BR signaling is linked to the regulation of photosynthetic capacity in tomatoes (Nie et al. 2017).

14.4.1 In Light Reaction

Numerous studies have utilized chlorophyll fluorescence measurement to explore how BRs impact photochemical reactions within the leaf thylakoid membrane. One of the key ways to evaluate electron transport during photosynthesis is by measuring the Hill reaction activity in chloroplasts, which is facilitated by BR treatment (Verma et al. 2011). Interestingly, the Hill reaction activity initially decreased right after homobrassinolide (HBL) treatment but increased at later developmental stages, suggesting a complex regulatory role of BRs in photosynthesis. The PSII quantum efficiency (ϕPSII) is a critical measure of the amount of light that is utilized for photochemical reactions and is absorbed by chlorophyll. As a measure of the total photosynthetic rate, the electron transport rate (ETR) is directly correlated with this efficiency. BRs play a crucial role in regulating these key photochemical reactions (ETR and ϕPSII) in plants, improving PSII efficiency, ETR, and photochemical quenching. The efficiency of PSII (ϕPSII) is particularly significant since it represents the plant's total photosynthetic capability and establishes the pace of linear electron transfer. Studies have indicated that the existence of BRs is frequently linked to a rise in PSII, thereby enhancing plant development and metabolism by improving the formation of assimilatory capabilities necessary for sugar synthesis (Lima and Lobato 2017). Investigating how BRs and genes regulate photosynthesis can be achieved by examining BR-treated plants or mutants with BR deficiencies (Bai et al. 2012). For example, an *Arabidopsis* mutant with a modified BR response showed reduced PSII efficiency, a smaller PSII complex, thylakoid expansion, and a suppression of CO_2 evolution (Krumova et al. 2013). Similarly, Kim et al. (2012) found that downregulated genes linked to photosynthesis in *Arabidopsis* mutants resulted in underdeveloped plants, slowed photosynthetic processes, and disrupted PSII assembly. The primary location of photosynthesis, chloroplasts, comprises two major processes: light absorption and carbon fixation. Chlorophyll, the main pigment responsible for light absorption, is essential for plants to capture and utilize solar energy for photosynthate production. According to Dalio et al. (2011), the leaf greenness index, which correlates linearly with chlorophyll content, can be used to estimate the rate of photosynthetic activity. It has been well documented that BRs enhance chlorophyll levels across various plant species (Farazi et al. 2015; Alyemeni and Al-Quwaiz 2016). Plants contain two forms of chlorophyll: chlorophyll *a* and chlorophyll *b*. The overall chlorophyll concentration is observed to increase in the presence of BRs

(Farazi et al. 2015). According to Asha and Lingakumar (2015), treatment with 24-EBR has been demonstrated to increase the amount of carotenoid and total chlorophyll. According to reports, BRs raise the amount of chlorophyll in a variety of plant species, including *Zea mays* (*Z. mays*), *C. sativus, Phaseolus aureus, V. radiata* (Fariduddin et al. 2006), *B. juncea* (Hayat et al. 2001b), *Cicer arietinum* (Ali et al. 2007), and *Vicia faba* (Piñol and Simón 2009; Alyemeni and Al-Quwaiz 2016). However, in *C. sativus*, the application of EBL results in a decrease in the chl *a/b* ratio, despite increasing the total chlorophyll content in other species (Yuan et al. 2012). BRs not only boost the capacity of plants to absorb light but also trigger the transcription and translation of enzymes required for chlorophyll synthesis. Additionally, following BR treatment, *B. juncea* leaves have been found to exhibit higher magnesium contents along with increased chlorophyll levels (Siddiqui et al. 2018b). This suggests that BRs enhance the photosynthetic machinery by not only increasing chlorophyll content but also optimizing the overall chloroplast function, thus promoting plant growth and resilience under various environmental conditions.

14.4.2 In Dark Reaction

By catalyzing the carboxylation of ribulose bisphosphate (RuBP), Rubisco, the most prevalent protein on Earth, contributes significantly to the Calvin cycle, commonly referred to as the C_3 cycle. RCA controls its activity. It has been demonstrated that the presence of BRs increases Rubisco activity (Portis 1992; Xia et al. 2009). The expression of genes for Rubisco and other Calvin cycle enzymes is upregulated by BRs, which contributes to this increase in activity. As a result, when BRs are present, the net PN and related parameters improve. Since stomata act as gateways for CO_2 entry into cells, enhanced stomatal conductance (Gs) facilitates greater CO_2 uptake and Rubisco activity, thereby boosting photosynthetic efficiency with BR application (Gruszka 2013). Additionally, HBL not only enhances photosynthetic rates and chlorophyll content but also increases the total sugar content in *S. khuzestanica* (Siddiqui et al. 2018b).

CA, which is intimately linked to Rubisco in C_3 plants, catalyzes the reversible interconversion of bicarbonate (HCO^{3-}) and CO_2, increasing the concentration of CO_2 at its active site (Badger and Price 1994). There is an increase in CA activity when BRs are present. Leaves of *B. juncea, Lycopersicon esculentum, C. arietinum*, and *V. radiata* treated with HBL exhibited higher CA activity (Hayat et al. 2001a; Fariduddin et al. 2006; Ali et al. 2006; Alam et al. 2007). The carboxylation efficiency would be impacted by the increased CO_2 availability surrounding Rubisco because CA activity has increased. According to Siddiqui et al. (2018b), increased Calvin cycle activity results in better sugar synthesis, which is subsequently used for metabolism and growth.

Research on plants treated with BRs and BR-deficient mutants demonstrates a connection between BRs and genes related to photosynthesis (Bai et al. 2012). An investigation of a mutant *Arabidopsis* with a changed BR response reveals bigger thylakoids, a smaller PSII complex, limited CO_2 evolution, and a decreased PSII quantum yield (Krumova et al. 2013). A study by Kim et al. (2012) examined the *Arabidopsis* BR insensitive 1 (BRI1) mutation, which revealed that genes related to photosynthesis regulation are downregulated. This led to growth retardation, decreased photosynthetic activity, and disruption of the PSII assembly. Thus, it is evident that BRs boost the expression of many genes involved in the synthesis of these enzymes, raising the concentration of chlorophyll and PSII's photosynthetic efficiency, both of which raise the net photosynthetic rate. The precise method by which BRs regulate photosynthetic activities is still unknown, despite the abundance of data supporting this claim. However, theories explaining the likely mechanism of BR-mediated regulation of photosynthetic processes have been put forth. For example, BRs may overcome stomatal barriers to increase CO_2 entrance into enhanced photosynthetic carbon fixing efficiency due to the leaf's availability for photosynthetic enzymes. Additionally, they might stimulate or activate the enzymes that produce chlorophyll (Holá et al. 2010). Therefore, it is possible to conclude that, in typical circumstances, BRs regulate photosynthesis at different levels. By promoting PSII efficiency and ETR, BRs promote the production of NADP and ATP, which are essential for the Calvin cycle and other activities. Furthermore, it promotes the accumulation of sugars, the activity of the C_3 cycle enzyme, and finally raises the photosynthetic efficiency of plants.

14.5 Influence of BRs on Photosynthetic Performance in Plants Exposed to Environmental Stress

14.5.1 High-temperature Stress

A major environmental component influencing plant growth is temperature. According to Nolan et al. (2020), one of the main abiotic pressures brought on by climate change in recent years has been heat stress. On almost every continent in the world, it negatively affects crop productivity. A plant's physiological metabolism, growth, and development can be severely harmed by high temperatures, claim Zhuang et al. (2020). These damage symptoms include decreased root and shoot plant growth, fruit destruction, senescence, burning of the leaves, and lowered plant productivity. Plants can sustain both direct and indirect damage from heat. According to Zhang et al. (2018), indirect damage includes reduced biological activity of nucleic acid, increased protein degradation, metabolic famine, buildup of harmful compounds including ethanol and acetaldehyde, and inhibited photosynthesis of plants. Direct damage primarily manifests as the denaturation of proteins. According to Sadura and Janeczko (2018), it is notable that the administration of exogenous the negative effects of heat on plants can be mitigated by BR. According to Ahammed et al. (2020), photosynthesis is the physiological function that is most susceptible to heat stress. High temperatures have two effects: they reduce the net photosynthetic rate and hinder the photosynthetic efficiency of PSII and the photochemical activity associated with the PSI (Zhang et al. 2013). It is interesting to note that BR can enhance thermotolerance in plant genotypes that are either heat-sensitive or heat-tolerant. For instance, EBR pretreatment offers significant advantages to both heat-tolerant and heat-sensitive melon ecotypes under heat stress in terms of photosynthetic pigment levels, net CO_2 assimilation rate, stomatal conductance, PSI photochemical activity, and water-use efficiency (Zhang et al. 2013). According to the reports shown in Table 14.1, BRs have specific stimulatory effects on photosynthesis in plants as well as antioxidant capacity, both of which contribute significantly to lessening the negative consequences of heat stress. Few studies have examined detailed mechanisms with genetic evidence, even though many have demonstrated

Table 14.1 Studies documenting effect of BR application on the mitigation of temperature stress in different plant species.

Plant name	Hormone	Stress type	Effects	References
Oryza sativa L.	EBL (0.1 μM)	High temperature	Reduced superoxide concentration, cell membrane damage, and increased expression of the sucrose transporter gene	Chen et al. (2019)
Solanum lycopersicum L.	EBR (0.2 μM)	High temperature	Reduce photosynthesis and boost the action of antioxidant enzymes	Ogweno et al. (2008)
Solanum melongena L.	EBR (0.05–0.2 μM)	High temperature	Boost antioxidant capacity and reduce the build-up of reactive oxygen species (ROS)	Wu et al. (2014)
Triticum aestivum L.	EBR (0.01 μM)	High temperature	Enhanced capacity for photosynthesis, biomass accumulation, growth, and antioxidants	Hussain et al. (2019)
Oryza sativa L.	αDHECD (0.0001 μM)	High temperature	Boost the weight, setting, photosynthesis, and carbohydrate contents of the seeds	Sonjaroon et al. (2018)
Solanum lycopersicum L.	EBR (0, 30, 100, 200, 500, and 1000 Nm)	High temperature	Genes linked to defense and the stress response have elevated transcript levels	Zhou et al. (2014)
Z. mays L.	EBR 0.2 μmol L^{-1}	Cold temperature	The seedlings' SPAD value first grew before declining. Relative conductivity and malondialdehyde (MDA) content dropped	Sun et al. (2019)
Solanum lycopersicum L.	EBR (0.1 μM)	Cold temperature	Improves resistance to freezing by plummeting oxidative imbalances	Xia et al. (2018)

(Continued)

Table 14.1 (Continued)

Plant name	Hormone	Stress type	Effects	References
C. sativus L.	EBR (0.1 μM)	Cold temperature	Enhances the absorption of CO_2 and lessens the photoinhibition of PSII activation of important enzymes	Jiang et al. (2012)
Vitis vinifera L.	EBR (0.1 mg L^{-1})	Cold temperature	Modifies the AsA-GSH cycle component	Chen et al. (2019)
Vitis vinifera L.	EBR (0.3 μM)	Cold temperature	Enhancing the antioxidant capacity reduces the peroxidation of membrane lipids	Xi et al. (2013)
Z. mays L.	EBR (1.0 μM)	Cold temperature	Boost the biomass, height, and concentration of sugar, protein, and chlorophyll in plants	Singh et al. (2012)
Solanum lycopersicum L.	EBR (6 μM)	Cold temperature	Inhibition of the activities of lipoxygenase and phospholipase D	Aghdam and Mohammadkhani (2014)
Mangifera indica L.	EBR (10 μM)	Cold temperature	Raising a group of proteins' concentrations, fatty acids' degree of unsaturation, and fluidity	Li et al. (2012)
Capsicum annuum L.	EBR (15 μM)	Cold temperature	Levels of l-ascorbic acid and chlorophyll as well as the activity of antioxidant enzymes were increased	Wang et al. (2012)

the protective effect of BR against heat stress using exogenous applications (Zhou et al. 2014). It seems that just a small portion of the intricate process by which BRs produce heat tolerance is presently understood. According to research by Zhou et al. (2014), a brief production of hydrogen peroxide (H_2O_2) in the apoplast of tomato plants enables them to tolerate heat stress brought on by BR. NADPH oxidase, encoded by RESPIRATORY BURST OXIDASE HOMOLOG 1 (RBOH1), is necessary for the apoplast's BR-induced H_2O_2 production. Silencing the genes for RBOH1, MITOGEN-ACTIVATED PROTEIN KINASE 2 (MPK2), or MPK1/2 in tomato plants significantly reduces H_2O_2 accumulation and jeopardizes BR-induced heat stress tolerance (Nie et al. 2013). However, when MPK1 is silenced, this effect is absent, suggesting that MPK2 is more important for apoplast H_2O_2 production that is activated by BR and the ensuing heat tolerance. A positive feedback loop involving RBOH1, H_2O_2, and MPK2 may also be responsible for BR-induced tolerance to high temperatures, according to the study.

Along with the antioxidant system, heat-shock proteins (HSPs) are essential for BR-induced thermotolerance, according to Kagale et al. (2007). Dhaubhadel et al. (2002) claim that by protecting different translational machinery components from extended heat stress, EBR treatment enhances the synthesis of HSP. The enhanced production of HSPs in both the det2-1 and dwf4 BR mutants under heat stress contradicts data suggesting that HSP accumulation may not be necessary for BR-induced thermotolerance in *Arabidopsis* (Kagale et al. 2007). Again, on the fifth day, *Arabidopsis* seedlings' resistance to heat or salt stress is not enhanced by overexpressing the BR biosynthetic gene AtDWF4. The investigation of BR mutants in barley that are either BR-deficient (mutations in the HvDWARF or HvCPD) or defective in BR signaling (missense HvBRI1 gene) indicates that all barley mutants are more tolerant of high temperatures than the wild-type (Sadura et al. 2019). According to Nie et al. (2019), these results suggest that endogenously and exogenously applied BRs have quite different physiological effects and methods of action in different plant species. According to Martinez et al. (2018), PHYTOCHROME INTERACTING FACTOR4 (PIF4) levels rise as a result of molecular accumulation of BES1 and BZR1. BZR1's effect on gene transcription via the formation of PIF4-BES1 heterodimers promotes thermogenic growth. However, BRI1 accumulation decreases at high temperatures, affecting BR signaling and leading to an increase in root development (Martins et al. 2017).

14.5.2 Low-temperature Stress

Particularly for thermophilic plants, low temperatures that result in chilling or freezing stress pose a serious challenge to agricultural productivity in many regions of the world (Zhang et al. 2020). Long-term exposure to nonfreezing temperatures below 12 °C will cause "low-temperature damage" in plants, a physiological malfunction that occurs after the crucial period (Seydpour and Sayyari 2015). Low temperatures typically cause impairments in morphology, physiology, biochemical metabolism, membrane fluidity, macromolecule activities, and cell structure of plants. These effects lead to decreased plant growth, mechanical constraints, osmotic potential in the cells, and photoinhibition (Zhuang et al. 2019). In addition, according to Cheng et al. (2014), necrosis of leaves, discoloration, aberrant plant maturation, lipid cell membrane breakdown, inhibited thylakoid electron transport, increased malondialdehyde (MDA) content, and increased relative conductivity can all be caused by sub-optimal temperatures. Reduced activity of the Benson–Calvin Cycle enzyme, a decrease in the rate of CO_2 assimilation, and photoinhibition at PSI and PSII are some of the consequences of cold stress on plant photosynthetic systems. According to Chen et al. (2013), cold stress results in a buildup of reactive oxygen species (ROS), which can harm cell membranes and induce lipid peroxidation. ROS are also caused by heat stress. The amount of MDA that builds up can, in part, reveal the extent of cell membrane damage, and the rate at which it builds up can reveal how well plant free radicals can be scavenged. To avoid excessive ROS accumulation and PS degradation that results in photoinhibition, plants have a variety of ROS scavenging and photoprotective mechanisms (Ahammed et al. 2020). Plants with higher ROS-scavenging potential or genes encoding such enzymes are more resistant to cold stress, claim Xiong et al. (2002). ROS can be produced in large quantities under unfavorable conditions. This can disturb the balance and accumulate to a dangerous level, which will significantly hinder the growth and development of plants. Plants have enzymes like superoxide dismutase (SOD) and peroxidase (POD) that help get rid of free radicals and ROS. When given to KJD6 and KY131 at the rice booting stage, BRs were found to significantly increase SOD and POD activity, increase soluble sugar and soluble protein content, and decrease MDA content. Thus, low-temperature stress was lessened.

When exposed to cold stress, plants employ a vital strategy called photoprotection to avoid photoinhibition. BRs contribute to photoprotection in plants under cold stress. Plants subjected to low temperatures gradually activate BRs and activate BZR1, which increases the transcript levels of apoplastic H_2O_2 production and RBOH1, according to Fang et al. (2019). Conversely, RBOH1 suppression or a mutation in BZR1 eliminates BR-induced photoprotection, aggravating chilling-induced photoinhibition. BR-induced photoprotection under chilling is indirectly linked to the induction of nonphotochemical quenching (NPQ), accumulation of D1 and PSII subunit S (PsbS) proteins, redox signaling, and the PROTON GRADIENT REGULATION5 (PGR5)-dependent cyclic electron flow (CEF). It is important to note that BRs-induced apoplastic H_2O_2 is necessary for these processes (Fang et al. 2019). According to transcriptome analysis, EBR treatment raises the transcript levels of genes involved in photosynthesis and chlorophyll biosynthesis at low temperatures. These genes include those encoding ferredoxin, PSI subunit, light-harvesting chlorophyll protein complexes I and II, PSII oxygen-evolving enhancer protein, and PSI (Zhao et al. 2019). According to Eremina et al. (2017), BRs have been shown to enhance freezing tolerance by activating COLD-RESPONSIVE (COR) genes through both independent and C-REPEAT/DEHYDRATION-RESPONSIVE ELEMENT BINDING FACTOR1 (CBF1)-dependent pathways. Li et al. (2017) claim that during BR-mediated enhanced cold tolerance, BZR1 and BES1 concentrate in their unphosphorylated (active) forms, which in turn triggers CBF1 and CBF2 transcription to create cold tolerance. The transcription factor INDUCER OF CBF EXPRESSION1 (ICE1) is destabilized by BRASSINOSTEROID INSENSITIVE 2 (BIN2) (Ye et al. 2019). Nolan et al. (2020) claim that these findings suggest that BRs can both improve stress tolerance and lessen stress reactivity, which are primarily governed by spatiotemporal regulation. Fruits and other plant-based products have a higher tolerance to cold because of the effects of BRs. It has been shown that BR can improve the quality of fruits and vegetables after harvest by extending their shelf life during low-temperature stress (Aghdam and Mohammadkhani 2014).

14.6 Role of BRs in Mitigation of Heavy Metal Stress-induced Photosynthetic Repercussions

Plants under heavy metal stress experience a number of negative consequences that affect crop quality and yield (Table 14.2). Through the food chain, plants cultivated in heavy metal-contaminated soil frequently collect these metals, posing serious health hazards to humans (Hasan et al. 2019). Because of the possibility of food chain contamination, consuming such tainted plant products is especially dangerous. Numerous approaches have been investigated to lessen these difficulties, such as employing plant growth regulators to modify endogenous hormones and signaling networks. Among several plant species, BRs in particular have demonstrated a promising solution to heavy metal stress (Rajewska et al. 2016; Zhou et al. 2018). By interfering with CO_2 absorption and harming the photosynthetic apparatus, heavy metals harm plants. Research has demonstrated that heavy metals can impede photosynthetic activity by restricting the utilization of energy molecules and decreasing the power, including ATP and NADPH, in the Calvin cycle. According to Rajewska et al. (2016), for example, exposure to cadmium dramatically lowers stomatal conductance, net photosynthetic rate, PSII quantum efficiency, and photochemical quenching coefficients. Moreover, it has been discovered that exogenous 24-EBR reduces cadmium uptake in roots and its subsequent translocation to leaves.

The mesophyll cells of tobacco leaves have been shown to exhibit structural damage, including thylakoid dilatation and weakened cell walls and membranes, when exposed to chromium (Cr) stress (Bukhari et al. 2016). Nevertheless, it has been demonstrated that EBR therapy shields chloroplasts against Cr-induced harm, maintaining the arrangement of grana and thylakoids under stress. As a stress-protector against heavy metals, HBL functions similarly to EBR. HBL treatment can mitigate the negative effects of cadmium on tomato seedling development, photosynthesis, and PSII photochemistry, as Singh and Prasad (2017) showed. Through promoting the activity of detoxifying and antioxidative enzymes, increasing photosynthetic pigment content, and improving photosynthetic efficiency, BRs are known to protect against heavy metal stress. It has been demonstrated that EBR treatment (0.1 μM) increases photosynthetic pigment levels and improves photochemical efficiency, hence boosting tomato plants' resistance to cadmium stress (Ahammed et al. 2013).

Foliar applications of BRs (0.01 μM EBR or HBL) can increase tomato fruit output and quality in soils containing 12 mg kg^{-1} cadmium (Hayat et al. 2012). By increasing the amounts of enzymatic and nonenzymatic antioxidants, HBL treatment also reduces the phytotoxicity of cadmium in *C. arietinum* (Hasan et al. 2008). Furthermore, by boosting antioxidant enzyme activity and preserving redox equilibrium, adding EBR (5 nM) to a half-strength MS medium improves tomato seedling resistance to stress caused by zinc oxide nanoparticles. The use of photosynthetic pigments, the ability to scavenge ROS, and the formation of glutathione and phytochelatins are all significantly improved when exogenous BR treatment is used to increase heavy metal tolerance. These impacts enhance resilience to heavy metal stress and improve carbon usage (Choudhary et al. 2012; Rajewska et al. 2016).

Table 14.2 Studies documenting effect of BR application on the mitigation of heavy metal stress in different plant species (Reports cited: 2015–2024).

S. No.	Plant species	Heavy metal concentration	BR application concentration	Effect	References
1	*Brassica juncea*	150 mg (kg Mn)$^{-1}$	10^{-8} M EBL	Increased production of ROS, enhanced photosynthesis, reduction in electrolyte leakage, and restoration of stomatal opening	Hussain et al. (2019)
2	*Glycine max*	20 μM Zn	100 nM EBL	Improvement of functioning of PS II, replenishment of chloroplast, and mitigation of zinc stress by boost in nutritional content and antioxidant enzyme activity	dos Santos et al. (2020)
3	*Vitis vinifera*	2 GL^{-1} Zn	0.4 mg L^{-1} EBL	Enhancement in productivity due to improved photosynthesis	Tadayon and Moafpourian (2019)
4	*Pisum sativum*	150 mg L^{-1} Cd	10^{-7} M EBL	Increased nutritional uptake and reduction in ROS accumulation	Jan et al. (2018)
5	*Solanum lycospersicum*	3 mg kg^{-1} Cd, 9 mg kg^{-1} Cd	10^{-8} M HBL	Mitigation of toxic effects of Cd by upregulation of photosynthetic and nitrogen assimilation enzymes	Singh and Prasad (2017)
6	*Solanum nigrum*	100 μM Ni	1 μM EBL	Upregulation of superoxide dismutase (SOD) activity for mitigation of Ni stress	Soares et al. (2016)
7	*V. radiata*	200 mg kg^{-1} Zn	10^{-8} M EBL	Upregulation of antioxidant enzyme activity	Mir et al. (2015)
8	*Vitis vinifera*	1 mM Cd^{2+}	0.05 mg L^{-1}, 0.1 mg L^{-1}, and 0.2 mg L^{-1}	Alleviation of oxidative stress caused by Cd toxicity by the upregulation of antioxidant enzyme activities	Li et al. (2022)
9	*Z. mays*	60 mM NaCl, 120 mM NaCl	0.1 and 0.2 μM EBL	Significant enhancement in photosynthetic capacity and antioxidant enzyme activities	Seleiman et al. (2023)
10	*Lactuca sativa*, *Spinacia oleracea*, and *Brassica oleracea*	Heavy metal-infused sewage water	3.5 μM	Restricted translocation of heavy metals Pb and Cd	Balal et al. (2024)

14.7 BRs and Pb Toxicity

According to the research done by Bajguz and Godlewska-Zykiewicz (2004), the use of 20-hydroxyecdysone to *Chlorella vulgaris* cells led to significant improvements in cell proliferation. This treatment not only prevented the loss of vital components such as proteins, sugars, and chlorophyll but also enhanced the production of phytochelatin, a key molecule involved in heavy metal detoxification. The study highlighted the protective and growth-promoting effects of 20-hydroxyecdysone in algae, underscoring its potential for improving cell health and resilience. Further, Bajguz (2002) conducted research which elucidated the role of BRs in enhancing phytochelatin production in *C. vulgaris* cells. The study involved the application of various BRs, including BL, 24-epibrassinolide (24-EpiBL), HBL, castasterone (CS), homo-castasterone (HomoCS), and 24-Epicastasterone (24-EpiCS), in combination with two levels of lead (Pb) exposure. The results indicated that BRs significantly enhanced phytochelatin production in the presence of Pb, demonstrating their role in enhancing metal detoxification and promoting cellular defense mechanisms. Pb toxicity was decreased in radish (*Raphanus sativus* L.) seedlings treated with 24-EpiBL in a related investigation. The activity of several antioxidant enzymes, including as reducing peroxidase (POD, EC 1.11.1.11), CAT, ascorbate peroxidase (APX), SOD, and guaiacol peroxidase (GPX, EC 1.11.1.7), was also enhanced by this treatment. By reducing Pb-induced oxidative stress and efficiently scavenging ROS, 24-EpiBL may improve the plant's resistance to heavy metal stress. These results were expanded upon by Swamy et al. (2014), who investigated the effects of 28-HBL on *Trigonella foenum-graecum* seedlings. The use of 28-HBL led to enhanced plant biomass, faster photosynthesis, higher levels of biomolecules, and a significant decrease in Pb toxicity. With its potential to improve agricultural output and plant health in contaminated environments, this study further supports the significance of BRs in promoting plant growth and stress tolerance under heavy metal stress conditions.

14.8 BRs and Cu Toxicity

Pre-germination treatment with 24-EpiBL in *B. juncea* L. cv. PBR 91 was found to significantly improve biomass production, shoot emergence, and overall plant development under conditions of copper (Cu) toxicity (Sharma and Bhardwaj 2007a). This indicates that 24-EpiBL enhances the plant's resilience to Cu stress from the very beginning of germination, resulting in improved growth and development even when harmful Cu levels are present. According to additional research by Sharma and Bhardwaj (2007b), 24-EpiBL contributes to preventing Cu absorption and buildup in plants. This shows that 24-EpiBL reduces internal Cu concentration and its harmful consequences on plant health by lowering Cu uptake. Fariduddin et al. (2009) demonstrated that applying HBL at concentrations of 10^{-10}, 10^{-8}, and 10^{-6} M to *B. juncea* seedlings resulted in enhanced growth characteristics and improved photosynthetic ability. Additionally, this treatment increased the activity of several antioxidant enzymes that are essential for shielding the plant from oxidative stress, including proline, CAT, POD, and SOD. H_2O_2, a ROS that can harm cellular components, also decreased after HBL was applied. By reducing oxidative damage and boosting the plant's ability to withstand stress, the rise in proline and antioxidant enzyme activity is probably going to offset the negative consequences of Cu toxicity.

14.9 BRs and Cr Toxicity

Applying 24-EpiBL to *Raphanus sativus* L. cv. "Pusa Chetki" seedlings under chromium (Cr VI) stress dramatically increased the levels of numerous metabolites, according to Choudhary et al. (2011). Proline, glycine betaine, ascorbic acid, glutathione (GSH), and phytochelatins were among the substances with elevated amounts. The functions of these metabolites in detoxification, stress response, and oxidative damage defense are well recognized. Important antioxidant enzymes like glutathione reductase (GR), SOD, and APX also showed increased activity following 24-EpiBL therapy. H_2O_2 and MDA, two markers of oxidative stress and cellular damage, were also markedly decreased at the same time. According to this, 24-EpiBL is essential for strengthening the plant's antioxidant defenses and lowering oxidative damage, which helps it better withstand Cr (VI)-induced oxidative stress. 28-HBL was given to *R. sativus* L. cv. "Pusa Chetki" seedlings at a concentration of 10^{-7} M in a related investigation by Sharma et al. (2011). A significant increase in fresh biomass, root and shoot length, and the activity of several antioxidant enzymes was observed following the application of 28-HBL. A stronger resistance against oxidative stress and improved growth are shown in these benefits. Furthermore, MDA levels and lipid peroxidation, two indicators of oxidative stress-induced cellular damage, were decreased with 28-HBL therapy. By increasing the plant's development and boosting its antioxidant defense systems, 28-HBL may help lessen the negative effects of Cr (VI) stress and increase the plant's resistance to damage caused by heavy metals.

14.10 BRs and Nickel Toxicity

Nickel (Ni) toxicity was successfully reduced in a study by Kanwar et al. (2013) when isolated 24-EpiBL was applied to *B. juncea* L. plants that were 30 days old. Antioxidant enzyme activity significantly increased as a result of this treatment, which is important for preventing oxidative stress brought on by exposure to heavy metals. According to the results, 24-EpiBL improves the plant's capacity to control and lessen the negative impacts of Ni, thereby improving overall plant health and resilience. Further research by Kanwar et al. (2012) demonstrated that BRs, including 24-EpiBL, CS, dolicholide, and typhasterol, significantly reduced Ni metal exposure in *B. juncea* L. plants. Increasing the activity of stress-relieving enzymes and lowering the plants' uptake of Ni were the two primary methods used to accomplish this reduction. When the plants were exposed to different concentrations of Ni (0.2, 0.4, and 0.6 mM), the study discovered that applying these BRs successfully reduced the negative effects of Ni, demonstrating their ability to increase plant resistance to metal stress. Sharma et al. (2008) conducted a related investigation in which *B. juncea* L. seedlings were exposed to varying concentrations of Ni (0, 25, 50, and

100 mg dm^{-3}) after being treated with HBL at doses of 0.01–100 nM for 8 hours. The results indicated that HBL treatment led to several beneficial effects, including enhanced protein content, improved antioxidative enzyme activity, and increased growth, as evidenced by the lengthening of shoots and roots. This suggests that HBL not only helps in mitigating the negative impact of Ni but also encourages general plant development and growth in the presence of metal stress. When taken as a whole, these findings highlight how important BRs are for enhancing plant resistance to nickel toxicity. By enhancing antioxidant defenses and reducing metal uptake, BRs contribute to better plant health and growth, making them valuable tools in managing metal stress in crops.

14.11 BRs and Mercury Toxicity

With concentrations of 0.5, 1.0, and 1.5 mM, it has been demonstrated that increasing the activity and protein levels of the enzymes glutathione-S-transferase (GST) and polyphenol oxidase (PPO) reduces the oxidative stress brought on by exposure to mercury (Hg) (Sharma et al. 2014). To decrease oxidative damage, GST and PPO are essential enzymes that detoxify ROS and break down toxic chemicals. The harmful effects of Hg, which is known to induce oxidative stress and damage to plant tissues, can be considerably reduced by increasing the activity of these enzymes. Applying 28-HBL to radish plants successfully lessened the negative effects of Hg in the study by Sharma et al. (2014). In addition to reducing oxidative stress brought on by Hg, this therapy increased the plants' protein content and antioxidant enzyme activity. The results show that by strengthening the plant's antioxidant defenses and general physiological health, 28-HBL improves its capacity to handle Hg toxicity. According to the findings, 24-EpiBL therapy decreased the amount of MDA, a sign of oxidative stress and lipid peroxidation. Proteins, GSH, and other antioxidant enzymes were all elevated concurrently by the therapy. This suggests that by boosting antioxidant capacity and lowering oxidative damage in plants, 24-EpiBL helps reduce Hg poisoning. All things considered, these investigations demonstrate how BRs strengthen plants' antioxidant defense systems, which ultimately help to lessen the negative consequences of Hg exposure. BRs are essential for enhancing plant resistance to heavy metal stress, which promotes improved development and resilience in contaminated settings by raising antioxidant levels and the activity of detoxifying enzymes.

14.12 Conclusion

All plant species contain a family of multihydroxylated sterol derivatives called BRs. By affecting photosynthesis, which is crucial for transforming light energy into chemical energy to support plant development, these substances have a basic impact on plant growth. It is well known that BRs promote the synthesis and activation of enzymes that produce chlorophyll, which is essential for absorbing light energy. Additionally, they control several elements of the photosynthetic apparatus, such as stomatal conductance, photochemical reactions, and Calvin cycle enzymes. By modulating stomatal function, BRs influence how effectively plants can fix carbon during photosynthesis. This adjustment helps optimize the net photosynthetic rate, which is vital for ensuring proper plant growth and development. The regulatory effects of BRs on photosynthesis are crucial for maintaining an efficient photosynthetic rate, which directly impacts plant health and productivity. The constitutive regulation of photosynthetic components by BR continually improves the net photosynthetic rate and, in turn, the growth and development of plants. For that reason, it is crucial to investigate how BRs regulate photosynthesis. Understanding how BRs regulate photosynthetic activities under both optimal and sub-optimal conditions provides valuable insights into plant hormone signaling pathways. This knowledge has significant implications for agricultural practices, particularly in improving crop resilience and productivity. In summary, this chapter sheds light on the regulatory functions of BRs in photosynthesis, offering new perspectives on their role in both ideal and challenging environmental conditions. These insights are pivotal for advancing our understanding of plant hormone signaling and for developing strategies in sustainable agriculture and crop improvement. By further exploring these mechanisms, we can better leverage the potential of BRs to enhance agricultural productivity and resilience.

References

Aghdam, M.S. and Mohammadkhani, N. (2014). Enhancement of chilling stress tolerance of tomato fruit by postharvest brassinolide treatment. *Food and Bioprocess Technology* 7: 909–914.

Ahammed, G.J., Choudhary, S.P., Chen, S. et al. (2013). Role of brassinosteroids in alleviation of phenanthrene–cadmium co-contamination-induced photosynthetic inhibition and oxidative stress in tomato. *Journal of Experimental Botany* 64 (1): 199–213.

Ahammed, G.J., Wang, Y., Mao, Q. et al. (2020). Dopamine alleviates bisphenol A-induced phytotoxicity by enhancing antioxidant and detoxification potential in cucumber. *Environmental Pollution* 259: 113957.

Ali, B., Hayat, S., and Ahmad, A. (2007). 28-Homobrassinolide ameliorates the saline stress in chickpea (*Cicer arietinum* L.). *Environmental and Experimental Botany* 59 (2): 217–223.

Ali, B., Hayat, S., Hasan, S.A. et al. (2006). Effect of root applied 28-homobrassinolide on the performance of *Lycopersicon esculentum*. *Scientia Horticulturae* 110 (3): 267–273.

Alyemeni, M.N. and Al-Quwaiz, S.M. (2016). Effect of 28-homobrassinolide on the performance of sensitive and resistant varieties of *Vigna radiata*. *Saudi Journal of Biological Sciences* 23 (6): 698–705.

Anwar, A., Liu, Y., Dong, R. et al. (2018). The physiological and molecular mechanism of brassinosteroid in response to stress: a review. *Biological Research* 51: 121–132.

Asha, A. and Lingakumar, K. (2015). Effect of 24-epibrassinollide on the morphological and biochemical constitutions *Vigna unguiculata* (L.) seedlings. *Seed* 8 (2): 52–63.

Ashraf, M.H.P.J.C. and Harris, P.J. (2013). Photosynthesis under stressful environments: an overview. *Photosynthetica* 51: 163–190.

Badger, M.R. and Price, G.D. (1994). The role of carbonic anhydrase in photosynthesis. *Annual Review of Plant Biology* 45: 369–392.

Bai, M.Y., Shang, J.X., Oh, E. et al. (2012). Brassinosteroid, gibberellin and phytochrome impinge on a common transcription module in *Arabidopsis*. *Nature Cell Biology* 14: 810–817.

Bajguz, A. (2002). Brassinosteroids and lead as stimulators of phytochelatins synthesis in *Chlorella vulgaris*. *Journal of Plant Physiology* 159 (3): 321–324.

Bajguz, A. and Czerpak, R. (1998). Physiological and biochemical role of brassinosteroids and their structure-activity relationship in the green alga *Chlorella vulgaris* Beijerinck (Chlorophyceae). *Journal of Plant Growth Regulation* 17: 131–139.

Bajguz, A. and Godlewska-Zylkiewicz, B. (2004). Protective role of 20-hydroxyecdysone against lead stress in *Chlorella vulgaris* cultures. *Phytochemistry* 65 (6): 711–720.

Bajguz, A. and Tretyn, A. (2003). The chemical structures and occurrence of brassinosteroids in plants. In: *Brassinosteroids: Bioactivity and Crop Productivity* (ed. S. Hayat and A. Ahmad), 1–44. Dordrecht: Springer.

Balal, R.M., Javed, S.A., Jaffar, M.T. et al. (2024). Heavy metal toxicity induced by sewage water treatment in three different vegetables (lettuce, spinach and cabbage) was alleviated by brassinosteroid and silicon supplementation. *Journal of King Saud University-Science* 36: 103310.

Bukhari, S.A., Wang, R., Wang, W. et al. (2016). Genotype-dependent effect of exogenous 24-epibrassinolide on chromium-induced changes in ultrastructure and physicochemical traits in tobacco seedlings. *Environmental Science and Pollution Research International* 23: 18229–18238.

Chen, M., Zhang, W.H., Lv, Z.W. et al. (2013). Abscisic acid is involved in the response of *Arabidopsis* mutant sad2-1 to ultraviolet-B radiation by enhancing antioxidant enzymes. *South African Journal of Botany* 85: 79–86.

Chen, Z.Y., Wang, Y.T., Pan, X.B. et al. (2019). Amelioration of cold-induced oxidative stress by exogenous 24-epibrassinolide treatment in grapevine seedlings: toward regulating the ascorbate-glutathione cycle. *Scientia Horticulturae* 244: 379–387.

Cheng, S., Yang, Z., Wang, M.J. et al. (2014). Salinity improves chilling resistance in *Suaeda salsa*. *Acta Physiologiae Plantarum* 36 (7): 1823–1830.

Choudhary, S.P., Kanwar, M., Bhardwaj, R. et al. (2011). Epibrassinolide ameliorates Cr (VI) stress via influencing the levels of indole-3-acetic acid, abscisic acid, polyamines and antioxidant system of radish seedlings. *Chemosphere* 84 (5): 592–600.

Choudhary, S.P., Yu, J.Q., Yamaguchi-Shinozaki, K. et al. (2012). Benefits of brassinosteroid crosstalk. *Trends in Plant Science* 17 (10): 594–605.

Clouse, S.D. and Sasse, J.M. (1998). Brassinosteroids: essential regulators of plant growth and development. *Annual Review of Plant Biology* 49 (1): 427–451.

Cutler, G.C. (1991). Brassinosteroids through the looking glass: an appraisal. In: *Brassinosteroids: Chemistry, Bioactivity and Applications* (ed. H.G. Cutler, T. Yokota, and G. Adam), 334–345. ACS Symp Ser 474. Washington, DC: American Chemical Society.

Dalio, R.J.D., Pinheiro, H.P., Sodek, L. et al. (2011). The effect of 24-epibrassinolide and clotrimazole on the adaptation of *Cajanus cajan* (L.) Millsp. to salinity. *Acta Physiologiae Plantarum* 33: 1887–1896.

Dhaubhadel, S., Browning, K.S., Gallie, D.R. et al. (2002). Brassinosteroid functions to protect the translational machinery and heat-shock protein synthesis following thermal stress. *Plant Journal* 29 (6): 681–691.

dos Santos, L.R., da Silva, B.R.S., Pedron, T. et al. (2020). 24-Epibrassinolide improves root anatomy and antioxidant enzymes in soybean plants subjected to zinc stress. *Journal of Soil Science and Plant Nutrition* 20: 105–124.

dos Santos, L.R., Paula, L.D.S., Pereira, Y.C. et al. (2021). Brassinosteroid-mediated amelioration of iron deficiency in soybean plants: beneficial effects on the nutritional status, photosynthetic pigments and chlorophyll fluorescence. *Journal of Plant Growth Regulation* 40: 1803–1823.

Eremina, M., Unterholzner, S.J., Rathnayake, A.I. et al. (2017). Correction for Eremina et al., Brassinosteroids participate in the control of basal and acquired freezing tolerance of plants (vol 113, pg E5982, 2016). *Proceedings of the National Academy of Sciences of the United States of America* 114 (6): E1038–E1039.

Fang, P., Yan, M., Chi, C. et al. (2019). Brassinosteroids act as a positive regulator of photoprotection in response to chilling stress. *Plant Physiology* 180 (4): 2061–2076.

Farazi, E., Afshari, H., and Abadi, H.H. (2015). Effect of different concentrations of brassinosteroid on physiomorphological characteristics of five pistachio genotypes (*Pistacia vera*. L). *Journal of Nuts* 6: 143–153.

Fariduddin, Q., Hayat, S., Ali, B. et al. (2006). Effect of 28-homobrassinolide on the nitrate reductase, carbonic anhydrase activities, and net photosynthetic rate in *Vigna radiata*. *Acta Botanica Croatica* 65: 19–23.

Fariduddin, Q., Yusuf, M., Hayat, S. et al. (2009). Effect of 28-homobrassinolide on antioxidant capacity and photosynthesis in *Brassica juncea* plants exposed to different levels of copper. *Environmental and Experimental Botany* 66: 418–424.

Gabr, M.A., Fathi, M.A., Azza, I.M. et al. (2011). Influences of some chemical substances used to induce early harvest of 'Canino'apricot trees. *Nature and Science* 9 (8): 59–65.

Grove, M.D., Spencer, G.F., Rohwedder, W.K. et al. (1979). Brassinolide, a plant growth-promoting steroid isolated from *Brassica napus* pollen. *Nature* 281 (5728): 216–217.

Gruszka, D. (2013). The brassinosteroid signaling pathway—new key players and interconnections with other signaling networks crucial for plant development and stress tolerance. *International Journal of Molecular Sciences* 14 (5): 8740–8774.

Hasan, M.K., Ahammed, G.J., Sun, S. et al. (2019). Melatonin inhibits cadmium translocation and enhances plant tolerance by regulating sulfur uptake and assimilation in *Solanum lycopersicum* L. *Journal of Agricultural and Food Chemistry* 67 (38): 10563–10576.

Hasan, S.A., Hayat, S., Ali, B. et al. (2008). 28-Homobrassinolide protects chickpea (*Cicer arietinum*) from cadmium toxicity by stimulating antioxidants. *Environmental Pollution* 151 (1): 60–66.

Hayat, S., Ahmad, A., Mobin, M. et al. (2001b). Carbonic anhydrase, photosynthesis, and seed yield in mustard plants treated with phytohormones. *Photosynthetica* 39: 111–114.

Hayat, S., Ahmad, A., Mobin, M. et al. (2001a). Photosynthetic rate, growth, and yield of mustard plants sprayed with 28-homobrassinolide. *Photosynthetica* 38: 469–471.

Hayat, S., Alyemeni, M.N., and Hasan, S.A. (2012). Foliar spray of brassinosteroid enhances yield and quality of *Solanum lycopersicum* under cadmium stress. *Saudi Journal of Biological Sciences* 19 (3): 325–335.

Holá, D., Rothová, O., Kočová, M. et al. (2010). The effect of brassinosteroids on the morphology, development, and yield of field-grown maize. *Plant Growth Regulation* 61: 29–43.

Hussain, M., Khan, T.A., Yusuf, M. et al. (2019). Silicon-mediated role of 24-epibrassinolide in wheat under high-temperature stress. *Environmental Science and Pollution Research* 26: 17163–17172.

Jan, S., Alyemeni, M.N., Wijaya, L. et al. (2018). Interactive effect of 24-epibrassinolide and silicon alleviates cadmium stress via the modulation of antioxidant defense and glyoxalase systems and macronutrient content in *Pisum sativum* L. seedlings. *BMC Plant Biology* 18: 1–18.

Jiang, Y.P., Cheng, F., Zhou, Y.H. et al. (2012). Hydrogen peroxide functions as a secondary messenger for brassinosteroid-induced CO_2 assimilation and carbohydrate metabolism in *Cucumis sativus*. *Journal of Zhejiang University Science B* 13 (10): 811–823.

Kagale, S., Divi, U.K., Krochko, J.E. et al. (2007). Brassinosteroid confers tolerance in *Arabidopsis thaliana* and *Brassica napus* to a range of abiotic stresses. *Planta* 225: 353–364.

Kanwar, M.K., Bhardwaj, R., Arora, P. et al. (2012). Plant steroid hormones produced under Ni stress are involved in the regulation of metal uptake and oxidative stress in *Brassica juncea* L. *Chemosphere* 86 (1): 41–49.

Kanwar, M.K., Bhardwaj, R., Chowdhary, S.P. et al. (2013). Isolation and characterization of 24-Epibrassinolide from *Brassica juncea* L. and its effects on growth, Ni ion uptake, antioxidant defense of Brassica plants and in vitro cytotoxicity. *Acta Physiologiae Plantarum* 35: 1351–1362.

Kim, T.W., Michniewicz, M., Bergmann, D.C. et al. (2012). Brassinosteroid regulates stomatal development by GSK3-mediated inhibition of a MAPK pathway. *Nature* 482: 419–422.

Krumova, S., Zhiponova, M., Dankov, K. et al. (2013). Brassinosteroids regulate the thylakoid membrane architecture and the photosystem II function. *Journal of Photochemistry and Photobiology B: Biology* 126: 97–104.

Li, B., Zhang, C., Cao, B. et al. (2012). Brassinolide enhances cold stress tolerance of fruit by regulating plasma membrane proteins and lipids. *Amino Acids* 43 (6): 2469–2480.

Li, H., Ye, K., Shi, Y. et al. (2017). BZR1 Positively regulates freezing tolerance via CBF-dependent and CBF-independent pathways in *Arabidopsis*. *Molecular Plant* 10 (4): 545–559.

Li, P., Chen, L., Zhou, Y. et al. (2013). Brassinosteroid-induced systemic stress tolerance was associated with increased transcripts of several defence-related genes in the phloem in *Cucumis sativus* L. *PLoS One* 8 (6): e66582.

Li, B.B., Fu, Y.S., Li, X.X. et al. (2022). Brassinosteroids alleviate cadmium phytotoxicity by minimizing oxidative stress in grape seedlings: toward regulating the ascorbate-glutathione cycle. *Scientia Horticulturae* 299: 111002.

Lima, J.V. and Lobato, A.K.S. (2017). Brassinosteroids improve photosystem II efficiency, gas exchange, antioxidant enzymes and growth of cowpea plants exposed to water deficit. *Physiology and Molecular Biology of Plants* 23: 59–72.

Maity, U. and Bera, A.K. (2009). Effect of exogenous application of brassinolide and salicylic acid on certain physiological and biochemical aspects of green gram (*Vigna radiata* L. Wilczek). *Indian Journal of Agricultural Research* 43 (3): 194–199.

Martínez, C., Espinosa-Ruíz, A., de Lucas, M. et al. (2018). PIF 4-induced BR synthesis is critical to diurnal and thermomorphogenic growth. *The EMBO Journal* 37 (23): e99552.

Martins, S., Montiel-Jorda, A., Cayrel, A. et al. (2017). Brassinosteroid signaling-dependent root responses to prolonged elevated ambient temperature. *Nature Communications* 8 (1): 309–314.

Masidur Alam, M., Hayat, S., Ali, B. et al. (2007). Effect of 28-homobrassinolide treatment on nickel toxicity in *Brassica juncea*. *Photosynthetica* 45: 139–142.

Mir, B.A., Khan, T.A., and Fariduddin, Q. (2015). 24-epibrassinolide and spermidine modulate photosynthesis and antioxidant systems in *Vigna radiata* under salt and zinc stress. *International Journal of Advanced Research* 3 (5): 592–608.

Nawaz, F., Naeem, M., Zulfiqar, B. et al. (2017). Understanding brassinosteroid-regulated mechanisms to improve stress tolerance in plants: a critical review. *Environmental Science and Pollution Research* 24: 15959–15975.

Nie, S., Huang, S., Wang, S. et al. (2017). Enhancing brassinosteroid signaling via overexpression of tomato (*Solanum lycopersicum*) SlBRI1 improves major agronomic traits. *Frontier Plant Science* 8: 1386.

Nie, S., Huang, S., Wang, S. et al. (2019). Enhanced brassinosteroid signaling intensity via SlBRI1 over expression negatively regulates drought resistance in a manner opposite of that via exogenous BR application in tomato. *Plant Physiology Biochemistry* 138: 36–47.

Nie, W.F., Wang, M.M., Xia, X.J. et al. (2013). Silencing of tomato RBOH1 and MPK2 abolishes brassinosteroid-induced H_2O_2 generation and stress tolerance. *Plant Cell Environmental* 36 (4): 789–803.

Nolan, T.M., Vukašinović, N., Liu, D. et al. (2020). Brassinosteroids: multidimensional regulators of plant growth, development, and stress responses. *The Plant Cell* 32 (2): 295–318.

Ogweno, J.O., Song, X.S., Shi, K. et al. (2008). Brassinosteroids alleviate heat-induced inhibition of photosynthesis by increasing carboxylation efficiency and enhancing antioxidant systems in *Lycopersicon esculentum*. *Journal of Plant Growth Regulation* 27: 49–57.

Oh, M.H., Sun, J., Oh, D.H. et al. (2011). Enhancing Arabidopsis leaf growth by engineering the BRASSINOSTEROID INSENSITIVE1 receptor kinase. *Plant Physiology* 157 (1): 120–131.

Pan, G., Liu, Y., Ji, L. et al. (2018). Brassinosteroids mediate susceptibility to brown planthopper by integrating with the salicylic acid and jasmonic acid pathways in rice. *Journal of Experimental Botany* 69 (18): 4433–4442.

Piñol, R. and Simón, E. (2009). Effect of 24-epibrassinolide on chlorophyll fluorescence and photosynthetic CO_2 assimilation in *Vicia faba* plants treated with the photosynthesis-inhibiting herbicide terbutryn. *Journal of Plant Growth Regulation* 28: 97–105.

Portis, A.R. Jr. (1992). Regulation of ribulose 1, 5-bisphosphate carboxylase/oxygenase activity. *Annual Review of Plant Biology* 43: 415–437.

Rajewska, I., Talarek, M., and Bajguz, A. (2016). Brassinosteroids and response of plants to heavy metals action. *Frontiers in Plant Science* 7: 629–640.

Rao, S.S.R., Vardhini, B.V., Sujatha, E. et al. (2002). Brassinosteroids–a new class of phytohormones. *Current Science* 82: 1239–1245.

Sadura, I. and Janeczko, A. (2018). Physiological and molecular mechanisms of brassinosteroid-induced tolerance to high and low temperatures in plants. *Biology Plant* 62 (4): 601–616.

Sadura, I., Pociecha, E., Dziurka, M. et al. (2019). Mutations in the HvDWARF, HvCPD and HvBRI1 genes-involved in brassinosteroid biosynthesis/signal ling: altered photosynthetic efficiency, hormonal homeostasis and tolerance to high/low temperatures in barley. *Journal of Plant Growth Regulation* 38 (3): 1062–1081.

Seleiman, M.F., Ahmad, A., Tola, E. et al. (2023). Exogenous application of 24-epibrassinolide confers saline stress and improves photosynthetic capacity, antioxidant defense, mineral uptake, and yield in maize. *Plants* 12 (20): 3559.

Seydpour, F. and Sayyari, M. (2015). Chilling injury in cucumber seedlings amelioration by methyl salicylate. *International Journal of Vegetation Science* 22: 432–441.

Shahbaz, M., Ashraf, M., and Athar, H.U.R. (2008). Does exogenous application of 24-epibrassinolide ameliorate salt induced growth inhibition in wheat (*Triticum aestivum* L.)? *Plant Growth Regulation* 55: 51–64.

Sharma, I., Pati, P.K., and Bhardwaj, R. (2011). Effect of 28-homobrassinolide on antioxidant defence system in *Raphanus sativus* L. under chromium toxicity. *Ecotoxicology* 20: 862–874.

Sharma, N., Hundal, G.S., Sharma, I. et al. (2014). 28-Homobrassinolide alters protein content and activities of glutathione-s-transferase and polyphenol oxidase in *Raphanus sativus* L. plants under heavy metal stress. *Toxicology International* 21 (1): 44–53.

Sharma, P. and Bhardwaj, R. (2007a). Effect of 24-epibrassinolide on seed germination, seedling growth and heavy metal uptake in *Brassica juncea* L. *General and Applied Plant Physiology* 33 (1–2): 59–73.

Sharma, P. and Bhardwaj, R. (2007b). Effects of 24-epibrassinolide on growth and metal uptake in *Brassica juncea* L. under copper metal stress. *Acta Physiologiae Plantarum* 29: 259–263.

Sharma, P., Bhardwaj, R., Arora, N. et al. (2008). Effects of 28-homobrassinolide on nickel uptake, protein content and antioxidative defence system in *Brassica juncea*. *Biologia Plantarum* 52: 767–770.

Shu, S., Tang, Y., Yuan, Y. et al. (2016). The role of 24-epibrassinolide in the regulation of photosynthetic characteristics and nitrogen metabolism of tomato seedlings under a combined low temperature and weak light stress. *Plant Physiology and Biochem*istry 107: 344–353.

Siddiqui, H., Ahmed, K.B.M., and Hayat, S. (2018b). Comparative effect of 28-homobrassinolide and 24-epibrassinolide on the performance of different components influencing the photosynthetic machinery in *Brassica juncea* L. *Plant Physiology and Biochemistry* 129: 198–212.

Siddiqui, H., Hayat, S., and Bajguz, A. (2018a). Regulation of photosynthesis by brassinosteroids in plants. *Acta Physiologiae Plantarum* 40: 1–15.

Siddiqui, H., Sami, F., Faizan, M. et al. (2019). Brassinosteroid mediated regulation of photosynthesis in plants. In: *Brassinosteroids: Plant Growth and Development* (ed. S. Hayat, M. Yusuf, R. Bhardwaj et al.), 185–217. Singapore: Springer.

Singh, I., Kumar, U., Singh, S.K. et al. (2012). Physiological and biochemical effect of 24-epibrassinoslide on cold tolerance in maize seedlings. *Physiology and Molecular Biology of Plants* 18 (3): 229–236.

Singh, S. and Prasad, S.M. (2017). Effects of 28-homobrassinoloid on key physiological attributes of *Solanum lycopersicum* seedlings under cadmium stress: photosynthesis and nitrogen metabolism. *Plant Growth Regulation* 82: 161–173.

Soares, C., de Sousa, A., Pinto, A. et al. (2016). Effect of 24-epibrassinolide on ROS content, antioxidant system, lipid peroxidation and Ni uptake in *Solanum nigrum* L. under Ni stress. *Environmental and Experimental Botany* 122: 115–125.

Sonjaroon, W., Jutamanee, K., Khamsuk, O. et al. (2018). Impact of brassinosteroid mimic on photosynthesis, carbohydrate content and rice seed set at reproductive stage under heat stress. *Agriculture and Natural Resources* 52 (3): 234–240.

Steber, C.M. and Mccourt, P.A. (2001). Role for brassinosteroids in germination in *Arabidopsis*. *Plant Physiology* 125: 763–769.

Sun, Y.J., Wu, Y., Ma, D.Z. et al. (2019). Effects of exogenous brassinosteroid on germination and physiological characteristics of maize seedlings under low temperature stress. *North China Journal of Agronomy* 34: 119–128.

Swamy, K.N., Vardhini, B.V., Ramakrishna, B. et al. (2014). Role of 28-homobrassinolide on growth biochemical parameters of *Trigonella foenu-graecum* L. plants subjected to lead toxicity. *International Journal of Multidisciplinary and Current Research* 2: 317–321.

Symons, G.M., Ross, J.J., Jager, C.E. et al. (2008). Brassinosteroid transport. *Journal of Experimental Botany* 59 (1): 17–24.

Tadayon, M.S. and Moafpourian, G. (2019). Effects of Exogenous epi-brassinolid, zinc and boron foliar nutrition on fruit development and ripening of grape (*Vitis vinifera* L. clv. 'Khalili'). *Scientia Horticulturae* 244: 94–101.

Verma, A., Malik, C.P., and Gupta, V.K. (2011). In vitro effects of brassinosteroids on the growth and antioxidant enzyme activities in groundnut. *ISRN Agronomy* 2012: 1–8.

Wang, Q., Ding, T., Gao, L. et al. (2012). Effect of brassinolide on chilling injury of green bell pepper in storage. *Scientia Horticulturae* 144: 195–200.

Went, F.W. and Thimann, K.V. (1937). *Phytohormones*, 294. New York: Macmillan.

Wu, X., Yao, X., Chen, J. et al. (2014). Brassinosteroids protect the photosynthesis and antioxidant system of eggplant seedlings from high-temperature stress. *Acta Physiologiae Plantarum* 36 (2): 251–261.

Xi, Z., Wang, Z., Fang, Y. et al. (2013). Effects of 24-epibrassinolide on antioxidation defense and osmoregulation systems of young grapevines (*V. vinifera* L.) under chilling stress. Plant Growth Regulation 71: 57–65. https://doi.org/10.1007/s10725-013-9809-4.

Xia, X.J., Fang, P.P., Guo, X. et al. (2018). Brassinosteroid-mediated apoplastic H_2O_2–glutaredoxin 12/14 cascade regulates antioxidant capacity in response to chilling in tomato. *Plant Cell Environment* 41 (5): 1052–1064.

Xia, X.J., Huang, L.F., Zhou, Y.H. et al. (2009). Brassinosteroids promote photosynthesis and growth by enhancing activation of Rubisco and expression of photosynthetic genes in *Cucumis sativus*. *Planta* 230: 1185–1196.

Xiong, L.M., Schumaker, K.S., and Zhu, J.K. (2002). Cell signaling during cold, drought, and salt stress. *Plant Cell* 14: 165–183.

Ye, K., Li, H., Ding, Y. et al. (2019). BRASSINOSTEROID-INSENSITIVE2 negatively regulates the stability of transcription factor ICE1 in response to cold stress in *Arabidopsis*. *Plant Cell* 31 (11): 2682–2696.

Yu, J.Q., Huang, L.F., Hu, W.H. et al. (2004). A role of brassinosteroids in the regulation of photosynthesis in *Cucumis sativus*. *Journal of Experimental Botany* 55: 1135–1143.

Yuan, L., Shu, S., Sun, J. et al. (2012). Effects of 24-epibrassinolide on the photosynthetic characteristics, antioxidant system, and chloroplast ultrastructure in *Cucumis sativus* L. under $Ca(NO_3)_2$ stress. *Photosynthesis Research* 112: 205–214.

Zaharah, S.S., Singh, Z., Symon, G.M. et al. (2012). Role of brassinosteroids, ethylene, abscisic acid, and indole-3-acetic acid in mango fruit ripening. *Journal Plant Growth Regul*ation 31: 363–372.

Zhang, S., Wang, S.J., Lv, J.L. et al. (2018). SUMO E3 ligase SlSIZ1 facilitates heat tolerance in tomato. *Plant Cell Physiology* 59 (1): 58–71.

Zhang, Y.P., Zhu, X.H., Ding, H.D. et al. (2013). Foliar application of 24-epibrassinolide alleviates high-temperature-induced inhibition of photosynthesis in seedlings of two melon cultivars. *Photosynthetica* 51 (3): 341–349.

Zhang, Z., Wu, P., Zhang, W. et al. (2020). Calcium is involved in exogenous NO-induced enhancement of photosynthesis in cucumber (*Cucumis sativus* L.) seedlings under low temperature. *Scientia Horticulturae* 261: 108953.

Zhao, M., Yuan, L., Wang, J. et al. (2019). Transcriptome analysis reveals a positive effect of brassinosteroids on the photosynthetic capacity of wucai under low temperature. *BMC Genomics* 20 (1): 810.

Zhou, J., Wang, J., Li, X. et al. (2014). H_2O_2 mediates the crosstalk of brassinosteroid and abscisic acid in tomato responses to heat and oxidative stresses. *Journal of Experimental Botany* 65 (15): 4371–4383.

Zhou, Y.L., Huo, S.F., Wang, L.T. et al. (2018). Exogenous 24-Epibrassinolide alleviates oxidative damage from copper stress in grape (*Vitis vinifera* L.) cuttings. *Plant Physiology and Biochemistry* 130: 555–565.

Zhuang, K., Kong, F.Y., Zhang, S. et al. (2019). Whirly1 enhances tolerance to chilling stress in tomato via protection of photosystem II and regulation of starch degradation. *New Phytology* 221 (4): 1998–2012.

Zhuang, K.Y., Gao, Y.Y., Liu, Z.B. et al. (2020). WHIRLY1 regulates HSP21.5A expression to promote thermotolerance in tomato. *Plant Cell Physiology* 61 (1): 169–177.

15

Brassinosteroids and Its Crosstalk with Other Plant Hormones Under Stressful Environments

Mohammad Faizan[1], Ghafoor Unnisa[1], Aurosha Fayaz[1], Ira Khan[1], S. Maqbool Ahmed[1], and Shamsul Hayat[2]*

[1]*Department of Botany, School of Sciences, Maulana Azad National Urdu University, Hyderabad, India*
[2]*Department of Botany, Aligarh Muslim University, Aligarh, India*

*Corresponding author: mohammadfaizan@manuu.edu.in

15.1 Introduction

Brassinosteroids (BRs) are a class of polyhydroxylated steroidal phytohormones that play a crucial role in regulating plant growth, development, and stress responses. Since their discovery in the 1970s, BRs have been extensively studied for their involvement in various physiological processes, including cell elongation, division, vascular differentiation, seed germination, flowering, and senescence (Clouse and Sasse 1998; Bajguz and Hayat 2009). Beyond their roles in growth and development, BRs are known to enhance plant tolerance to a wide range of abiotic and biotic stresses, such as drought, salinity, extreme temperatures, heavy metal toxicity, and pathogen attacks, by modulating key metabolic and signaling pathways (Sharma et al. 2015; Jiang et al. 2020). In stressful environments, plants rely on a complex network of hormonal interactions to maintain cellular homeostasis and optimize their responses to adverse conditions. BRs are central components of this network and exhibit extensive crosstalk with other major plant hormones, including abscisic acid (ABA), auxins (IAA), gibberellins (GA), jasmonic acid (JA), salicylic acid (SA), cytokinins (CK), and ethylene (ET). This intricate hormonal interplay allows plants to fine-tune their growth-defense balance, ensuring survival under challenging environmental conditions (Verma et al. 2016; Yu et al. 2018). For instance, BR-ABA crosstalk is critical for stomatal regulation during drought stress, while BR-JA and BR-SA interactions modulate plant defense against pathogens (Wang et al. 2020; Nawaz et al. 2017). Additionally, BRs have been shown to synergize or antagonize the actions of other hormones depending on the specific stress type and developmental stage (Yang et al. 2011; Choudhary et al. 2012).

The interaction between BRs and ABA is particularly important in stress responses, especially under drought conditions. ABA is well known for promoting stomatal closure to reduce water loss during drought stress, whereas BRs have been shown to have an

Brassinosteroids and Abiotic Stress: Recent Developments on Physiological and Molecular Aspects, First Edition. Edited by Anket Sharma, et al.
Published 2026 by John Wiley & Sons Ltd.

antagonistic effect by inducing stomatal opening (Ha et al. 2014). However, under certain stress conditions, BRs and ABA can act synergistically. For example, BR signaling can enhance ABA-induced gene expression, thereby strengthening stress tolerance (Cai et al. 2014; Wang et al. 2020). Crosstalk between BRs and ABA also involves shared signaling components such as SnRK2 kinases, which are central to ABA-mediated stress signaling (Nakashima et al. 2009). BRs and IAA have a well-established synergistic relationship in regulating plant growth and development. Both hormones control processes such as cell elongation, root development, and photomorphogenesis. The crosstalk between these two hormones involves modulation of each other's biosynthesis and signaling pathways. For example, BRs can enhance the expression of auxin-responsive genes, while IAA can upregulate BR biosynthesis genes (Nakamura et al. 2003; Nemhauser et al. 2004a). This synergistic interaction is crucial for coordinating growth responses under stress conditions (Kuppusamy et al. 2009). Moreover, BRs and IAA share common transcriptional regulators, such as BES1 and ARF proteins, which allow coordinated responses to environmental cues (Vert et al. 2008). The crosstalk between BRs and GAs is critical for plant growth and developmental regulation. Both hormones promote cell elongation and division by influencing the expression of common downstream target genes (Bai et al. 2012). BRs have been shown to enhance GA signaling by promoting the degradation of DELLA proteins, which are negative regulators of GA signaling (Gallego-Bartolomé et al. 2012; Li et al. 2018). This interaction ensures a coordinated regulation of growth processes, particularly under environmental stresses that might otherwise restrict development (Che et al. 2016). The BR-GA crosstalk is also involved in seed germination, where both hormones positively regulate the expression of GA biosynthetic genes (Gallego-Bartolomé et al. 2012). BRs also interact with JA and SA pathways, which are primarily involved in plant defense mechanisms. BRs can enhance JA-induced defense responses by upregulating the expression of JA biosynthetic genes and defense-related proteins (Zhang et al. 2009; Wang et al. 2012). Conversely, BRs may suppress certain SA-dependent pathways to prevent excessive defense responses that could compromise growth (Nawaz et al. 2017). The BR-JA-SA crosstalk is thus essential for balancing growth and defense under stress conditions (Yang et al. 2011). For example, BR-activated transcription factors BES1 and BZR1 have been shown to differentially regulate JA and SA signaling components, ensuring appropriate stress responses (Zhang et al. 2009). BR signaling intersects with reactive oxygen species (ROS) pathways, influencing plant stress tolerance. BRs can modulate ROS levels by regulating the activity of antioxidant enzymes and enhancing the expression of genes involved in oxidative stress responses (Koh et al. 2007; Xia et al. 2009). Additionally, BRs have been shown to regulate secondary metabolism pathways like phenylpropanoid biosynthesis, which are crucial for stress adaptation and defense (Sharma et al. 2015). The BR-induced enhancement of antioxidant defense systems reduces oxidative damage under environmental stress, contributing to improved plant resilience (Xia et al. 2009). Understanding the molecular mechanisms underlying BR crosstalk with other plant hormones provides valuable insights into the adaptive strategies of plants under stress. Advances in genomics, transcriptomics, and proteomics have identified key signaling components and transcription factors involved in BR-mediated stress responses (Yu et al. 2018; Wang et al. 2020). This knowledge paves the way for developing stress-resilient crops through targeted genetics and biotechnological interventions, highlighting the potential of BRs in sustainable agriculture.

15.2 Interplay Between BRs and Other Phytohormones and the Underlying Signaling Events

Plant hormones called BRs are essential for controlling growth, development, and reactions to stress. They coordinate intricate signaling networks that regulate plant physiological processes in conjunction with other phytohormones like IAA, GAs, ABA, JA, and melatonin acid (MA). These interactions, which guarantee adaptive reactions to environmental signals, take place at several levels, including hormone production, receptor activation, and transcriptional control.

15.2.1 IAA and BRs

BRs and auxin are essential plant hormones that function together to modulate growth, development, and stress response. Their interactions affect processes including cell elongation, root and shoot development, vascular differentiation, and stress tolerance. Numerous physiological studies indicate that certain auxin-induced responses were synergistically amplified by BR treatments (Clouse and Sasse 1998). According to reports, BRs and auxin work in concert to promote cell growth in the pea (*Pisum sativum*) epicotyl, bean (*Phaseolus vulgaris*) hook bioassays, and azuki bean (*Vigna angularis*) (Yopp et al. 1981). BRs and indole-3-acetic acid (IAA) showed a synergistic effect in promoting the curvature of the leaf lamina in dwarf rice (*Oryza sativa*) seedlings and the angulation of the rice lamina joint (Cao and Chen 1995; Takeno and Pharis 1982). Furthermore, the bending and inclination generated by BR can be suppressed by a high concentration of the IAA transport inhibitor triiodobenzoic acid (Cao and Chen 1995; Takeno and Pharis 1982). ET production in etiolated mung bean hypocotyls is synergistically enhanced by BRs and auxin (Arteca et al. 1983). These physiological data indicate an interaction between BRs and auxin. Recent microarray analyses have further substantiated this connection, demonstrating that numerous early auxin-responsive genes are also modulated by BRs (Goda et al. 2002; Müssig et al. 2002; Yin et al. 2002). Conversely, auxin enhances the expression of the BR-responsive gene TCH4 (Xu et al. 1995).

The interaction between BRs and auxin signaling transpires at various levels, encompassing gene expression, signaling pathways, and protein interactions. BZR1, a crucial transcription factor in BR signaling, interacts with auxin-responsive elements to modulate genes associated with cell expansion. BRs also influence auxin transport by modulating PIN-FORMED (PIN) proteins, which play a crucial role in auxin distribution. This coordination improves organ development and gravitropic responses. BR-auxin interactions at the molecular level are facilitated by the TARGET OF RAPAMYCIN (TOR) pathway, which integrates energy signaling and hormone regulation. The inhibition of the BR negative regulator BIN2 further amplifies auxin signaling, thereby strengthening their collaborative function in plant development. Such interactions facilitate optimal growth and adaptation to environmental challenges. BRs and auxin interact physiologically with hypocotyl elongation or root growth. Previous research suggests that BRs and auxin stimulate cell elongation, and mutants of either route exhibit similar developmental abnormalities, including a severe dwarf phenotype. Using hypocotyl elongation, researchers observed that auxin-responsive mutants' axr1, axr2, axr3, tir1, and arf2 have greatly reduced BR sensitivity suggesting that BR's hypocotyl elongation function depends on auxin signal transduction. BR therapy also increases auxin response in hypocotyl elongation and BR signaling mutant bri1 is insensitive to temperature-induced elongation, possibly via auxin. Overall, the evidence suggests that auxin response requires a functional BR signal transduction system. The root-development essay confirmed that BRs and auxin crosstalk: BR treatment alters the expression of several genes, such as AXR3/IAA17 in the root; BR-deficient mutants repress the transcription levels of PIN4 and PIN7, which encode auxin transporters; and BRs stimulate plant tropisms by regulating root PIN2 localization and auxin polar transport.

Several studies have shown the molecular mechanisms of BR-auxin crosstalk. Many early responsive genes involved in plant growth are shared by BRs and auxin, according to microarray studies. ARF-binding sites in the regulatory domains of BR-responsive genes may link BRs to auxin pathways, according to a recent study. In *Arabidopsis* mutants with altered BRs or auxin signaling, Nemhauser et al. (2004b) showed how these two hormones govern hypocotyl elongation. Recent research showed that the BR signaling component BIN2 can directly interact with the auxin signaling component ARF2, a transcriptional regulator in the Auxin Response Factor family. BIN2 phosphorylation of ARF2 reduces its DNA binding and repression activity. Based on the aforesaid findings, BIN2 directly inactivates the repressor ARF2 to boost auxin-induced gene expression and transcription synergistically. The mechanism of BR-auxin interaction is likely not limited to this sort of crosstalk. Other paths appear to demonstrate their complicated interplay.

15.2.2 GA and BRs

BRs and GA are two major classes of growth-promoting hormones that influence plant height through the regulation of cell elongation. BRs are suggested to be synthesized in the endoplasmic reticulum (ER) and subsequently transported to the apoplast, where they interact with their receptors (Vukašinović and Russinova 2018). There is increasing evidence indicating that crosstalk between BRs and GA signaling pathways may occur in either a synergistic or antagonistic manner. Physiological studies indicate that BRs may partially enhance the germination of GA biosynthetic and GA-insensitive mutants, potentially through the mechanism of embryo expansion. Second, mutants deficient in G-proteins (GCR1, GPA1, and AGB1) exhibit diminished sensitivities to both BRs and GAs during seed germination. Microarray data from *Arabidopsis* and rice indicate that a significant number of genes are coordinately regulated by BRs and GAs. Research on rice indicates that OsGSR1 enhances BR biosynthesis by directly regulating a BR biosynthetic enzyme. Knockdown of OsGSR1 resulted in increased levels of endogenous GAs, implying that OsGSR1 may mediate the two pathways. Research indicates that BRs and GAs exert adverse impacts on the accumulation of mRNAs for the GA-responsive gene GASA1 and the GA-suppressed gene GA5. Furthermore, BRs and GAs modulate the expression of the GA-responsive gene γ-TIP in an antagonistic fashion, with the mRNA levels of γ-TIP exhibiting ectopic accumulation in BR biosynthetic and responsive mutants. Considering that γ-TIP encodes a tonoplast-intrinsic aquaporin, it is plausible to hypothesize that BRs and GAs regulate turgor pressure or solute flow in an antagonistic manner.

The activation of both GA and BR signaling serves as a feedback mechanism that suppresses their respective biosynthesis to refine the hormonal response in vivo, with BZR1 and BES1 playing a role in this process by inhibiting BR biosynthesis (He et al. 2005; Sun et al. 2010; Yu et al. 2011). GA interacts with other phytohormones, such as BR, to govern plant growth and development (Weiss and Ori 2007). Although GA and BR are both growth-promoting hormones, an early study discovered that they work against each other to control the expression of two genes: GA biosynthesis gene GA5 and GA-responsive gene GASA1 (for GA-stimulated transcript in *Arabidopsis*).

A gene involved in GA biosynthesis, GA5. BR stimulates the production of GA5, which encodes a GA 20-oxidase (GA20ox) implicated in GA biosynthesis. In rice (*Oryza sativa*), mutants lacking either GA or BR exhibit a stunted growth phenotype. Tong et al. (2014) and Unterholzner et al. (2015) substantiate the GA synthesis theory by documenting the impact of modified BR status on GA levels. In rice, a mutant exhibiting increased BR levels (m107) possessed fivefold greater concentrations of bioactive GA (GA1) and reduced levels of precursor GAs such as GA20, in comparison to the wild type (Tong et al. 2014). An elevation in GA1 content is anticipated to modify growth, as a prior investigation (Oikawa et al. 2004) demonstrated that a two to three-fold augmentation in GA1 is adequate to stimulate development. Conversely, the alterations in GA content in several rice BR mutants (e.g. d11-2) examined by Tong et al. (2014) were very minimal and of ambiguous statistical significance. Recent studies of the relevant mutations have yielded substantial advancements in the signaling and metabolic mechanisms of both GA and BR.

15.2.3 CKs and BRs

CKs and BRs are two crucial phytohormones that govern multiple facets of plant growth and development, encompassing cell division, elongation, organogenesis, and stress responses. Their interplay is essential for refining plant responses to environmental stimuli, guaranteeing proper growth and adaption. At the molecular level, CKs and BRs have both synergistic and antagonistic interactions. BRs generally facilitate cell growth and elongation, while CKs govern cell division and differentiation. In the dicotyledonous plant *Arabidopsis thaliana*, the total seed count is contingent upon the quantity of siliques and the seed count per silique (SNS). The SNS is influenced by developmental events such as ovule initiation and development, male and female gametophyte development, fertilization, and zygote development (Chaudhury et al. 1997). The quantity of ovules eventually dictates the maximum SNS. Two phytohormone-related mutants, ckx3 ckx5 (Bartrina et al. 2011) and bzr1-1D (He et al. 2002; Huang et al. 2013; Jiang et al. 2013), exhibiting augmented BR and CK signaling, respectively, demonstrate markedly elevated SNS, suggesting that these phytohormones play a role in ovule initiation and SNS regulation. Phytohormones, such as CK and BR, affect the development of floral organs, which is governed by the ABCDE gene model, pertaining to the four-whorl formation of floral organs and ovules. Class D genes regulate ovule identity and the start of ovule primordia, including transcription factors and regulatory proteins, including SEEDSTICK (STK) and SHATTERPROOF 1/2 (SHP1/2).

BZR1 and BES1, crucial transcription factors in BR signaling, engage with type-B response regulators (ARRs) of CK signaling, synchronizing growth responses. They additionally exhibited a direct protein-level interaction between the principal transcription factors associated with these two hormones, BZR1 and ARR1. This relationship influenced the transcription of downstream genes associated with ovule initiation. Their study demonstrates that BR and CK collaboratively regulate ovule initiation and SNS, while identifying candidate genes that could enhance seed quantity and yield, possessing both scientific relevance and practical implications for augmenting crop productivity. According to earlier research, BR and CK positively control plant growth and seed quantity, suggesting that they encourage ovule initiation and SNS increase seed number, which in turn increases seed output (Higuchi et al. 2004; Bartrina et al. 2011; Huang et al. 2013; Jiang et al. 2013).

To elucidate the interaction between BRs and CK in the modulation of light signaling during photo morphogenesis and development, Hwang et al. (2008) conducted a study to investigate the relationship between BRs and photoreceptor activity in regulating hypocotyl elongation responses in *Arabidopsis*, as well as the influence of BRs on light and CK signal transduction pathways. The actions of BRs and CK are independent; nonetheless, they sequentially regulate hypocotyl elongation, which can be attributed to their analogous interaction with light. Consequently, light, BRs, and CKs may independently regulate downstream light-responsive mechanisms, whereas light modulates the effects of BRs and CKs to some degree.

Yuldashev et al. (2012) investigated that by suppressing the transcription levels of the CYTOKININ OXIDASE 3 (CKX3) gene, which is in charge of CK degradation, exogenous application of 24-epibrassinolide (EBL) causes an increase in the accumulation of CKs. Protein phosphatase with Kelch-like domains 1 (PPKL1), recognized as a regulatory element of BR signaling, competes with OsARR21 for association with OsAHP2, thereby decreasing the phosphorelay from OsAHP2 to OsARR21 and diminishing CK responses (Liu et al. 2022). Additionally, BR signaling regulates CK production by modulating essential genes such as IPT (ISOPENTENYL TRANSFERASE), hence affecting CK homeostasis. In root development, BRs and CKs frequently exhibit antagonistic interactions. CKs inhibit root elongation and lateral root development, whereas BRs facilitate these activities by regulating PIN-FORMED (PIN) auxin transporters. Conversely, in shoot growth, the interaction between BR and CK is predominantly cooperative, enhancing meristematic activity and leaf expansion. Their interaction influences stress responses, with BRs augmenting drought and salinity tolerance, whereas CKs govern senescence and nutrient remobilization. The complex interaction between CKs and BRs underscores their function as principal regulators of plant physiology, allowing plants to assimilate various signals for adaptive growth. Additional research will enhance our comprehension of the molecular principles governing their interaction, potentially resulting in increased agricultural output.

15.2.4 ABA and BRs

ABA is a plant stress hormone that plays a significant role in the regulation of growth and the facilitation of adaptation to stressful environments (Yang et al. 2022) when confronted with difficult environmental conditions. There are specific circumstances that cause plants to display a rapid increase in their levels of ABA. The accumulated ABA is detected by its receptors, which are known as Pyrabactin Resistance/Pyrabactin Resistance 1-Like/Regulatory Component of ABA Receptor (PYR/PYL/RCAR). This interaction between PYR/PYL/RCAR and PROTEIN PHOSPHATASE 2C (PP2C) proteins occurs, which ultimately results in the inhibition of the phosphatase activity of PP2Cs (Ma et al. 2009; Park et al. 2009).

The suppression of Sucrose Non fermenting 1–Related Protein Kinases 2 (SnRK2s) by PP2Cs is subsequently alleviated, enabling SnRK2s to phosphorylate and activate ABA–Responsive Element Binding Factors (ABFs), which in turn initiates the transcription of ABA target genes (Fujii et al. 2009). Within the absence of ABA, PP2Cs engage in interactions with SnRK2s and dephosphorylate them, hence limiting the kinase activities of the SnRK2s. According to Fujii et al. (2009), Ma et al. (2009), and Park et al. (2009), this impairment has an effect on the activation of ABA downstream transcription factors and represses ABA signal transmission.

BRs are known to have antagonistic interactions with ABA when observed under normal settings. (Ha et al. 2014) found that the administration of exogenous BL led to a reduction in the levels of ABA by inhibiting the transcription of genes that are involved in the production of ABA. According to Moon et al. (2021), BZR1 is able to bind to the promoter of ABA-DEFICIENT 2 (ABA2), which results in the inhibition of its expression and a reduction in the accumulation of ABA. Contrarily, BIN2 is responsible for phosphorylating and activating SnRK2s, such as SnRK2.2 and SnRK2.3, which ultimately results in an increase in ABA responses (Cai et al. 2014). Additionally, BIN2 phosphorylates and stabilizes the ABA-responsive factor ABA INSENSITIVE 5 (ABI5), which results in an increase in the amount of ABA signaling that occurs (Hu and Yu 2014). Based on the findings of Zhao et al. (2019), BES1 has a negative interaction with ABI5 because it interferes with the transcriptional activity of ABI5 and hinders the output of ABA signaling. According to Yang et al. (2016), BZR1 binds to the promoter of ABI5, which results in a reduction in the amount of ABA signal transduction and a little decrease in the expression of ABI5. According to Ryu et al. (2014), BES1 has the ability to decrease ABA signaling by using its ability to restrict the expression of ABI3, which is a positive upstream regulator of ABI5. PP2Cs, which act as negative regulators of ABA signaling, have the ability to dephosphorylate and inactivate BIN2, which results in an increase in the output of BR signaling (Wang et al. 2018; Jiang et al. 2019). In addition, PP2C phosphatases, such as ABI1, have the ability to dephosphorylate and activate BES1, which in turn has a favorable influence on the BR signaling that occurs in response to heat (Albertos et al. 2022a). According to the findings of the study, the BR and ABA signaling pathways are able to control one another in an antagonistic manner, which helps to maintain a healthy equilibrium between the growth of plants and their reactions to stress.

ABA and BRs are essential components in plant growth and adaptation to stress, and they commonly exhibit antagonistic interactions and interactions with one another. While ABA is largely responsible for regulating stress responses, seed dormancy, and stomatal closure, BRs are responsible for facilitating growth, development, and resilience to stress. The interplay between them makes it easier for plants to maintain a balance between growth and survival in the face of changing environmental conditions than it would be otherwise. Both ABA and BRs are involved in interactions at the molecular level, and these interactions are mediated by the same signaling components. BZR1, a fundamental BR transcription factor, directly represses ABA-responsive genes in order to favor growth, whereas BSK1, a BR-signaling kinase, is responsible for the negative regulation of ABA responses during development. ABA, on the other hand, is responsible for the increased activity of the protein phosphatase ABI1, which is responsible for suppressing BR signaling by dephosphorylating proteins that are responsive to BR stimulus. Additionally, stress responses are included in this interaction. When drought circumstances are present, ABA encourages stomatal closure in order to reduce the amount of water that is lost. BRs, on the other hand, moderate this impact by maintaining stomatal aperture for the purpose of photosynthesis. BRs, when applied to the setting of salt stress, have the ability to prevent oxidative damage by upregulating antioxidant enzymes. On the other hand, ABA is primarily responsible for initiating osmoprotectant pathways. The combination of ABA and BR signaling networks gives plants the ability to accurately change their responses to environmental stimuli, which in turn improves their ability to survive and adapt to their surroundings. By conducting additional research, we will be able to gain a better understanding of these interactions, which will provide us with potential solutions for crop improvement under stressful conditions.

15.2.5 ET and BRs

Both ET and BRs are important phytohormones that regulate the growth, development, and stress levels of plants (Fatma et al. 2022). In contrast to BRs, which improves cell elongation, differentiation, and resilience to environmental stress, ET is responsible for promoting fruit ripening, senescence, and stress signal initiation. When it comes to the coordination of plant responses in a variety of settings, their crosstalk is absolutely necessary (Zheng et al. 2019b; Li et al. 2020b). Different signaling molecules are shared by BRs and ET. Both BZR1 and BES1 interact with ethylene response factors (ERFs) in order to control the expression of genes that are involved in growth and defense (Ju and Chang 2015).

In addition, BR signaling inhibits ACS genes, which allows for the fine-tuning of ET synthesis. In response to drought and pathogen stress, ET has an effect on the activation of BR receptors, which in turn modulates BR responses. The BR-ET crosstalk is a significant factor in the ripening of fruit. The production of ET rises as a result of BRs, which speeds up the maturation process of climacteric fruits. In general, however, BRs offset the growth-inhibitory effects that ET has on vegetative tissues, which assists plants in maintaining a healthy balance between stress and development. Root development is dependent on the interaction between BR and ET. BRs cause cells to extend, whereas ET can either encourage or inhibit root growth, depending on the concentration and the surrounding environment. This delicate equilibrium assures that adaptation will occur as the circumstances of the soil change. The complex link that exists between BRs and ET demonstrates how these compounds combine growth and stress responses in order to encourage the development of plants better. When further research is conducted, the molecular mechanisms that govern this link will be revealed, which will open up new agricultural opportunities. ET is a gaseous hormone that promotes plant tolerance to extreme stress, including drought, flood, and metals (Fatma et al. 2022). It accomplishes this by regulating essential cellular processes such as seed germination, fruit ripening, and senescence. According to Zheng et al. (2019b) and Li et al. (2020b), ET is also involved in the functioning of plant immunity and defensive responses. According to Ju and Chang (2015), ER membrane-localized receptors, such as ETHR1, ETR2, ERS1, ERS2, and EIN4, are responsible for detecting the release of ET and inactivating CTR1, which is a negative effector found in ET signaling. According to Qiao et al. (2012), the destruction of CTR1 causes

EIN2, a positive regulator in ET signaling, to undergo dephosphorylation and cleavage, which results in the release of its C-terminus (EIN2-CEND). According to the findings of Li et al. (2015), EIN2-CEND has the ability to restrict the translation of EBF1 and EBF2 transcripts from the cytoplasm. Factors of transcription that are controlled by ET EBF1 and EBF2 F-box proteins are responsible for targeting EIN3 and EIN3-LIKE 1 (EIL1) for degradation by 26S proteasomes (Guo and Ecker 2003; Potuschak et al. 2003; Gagne et al. 2004). This research was published in three different studies. Having lower levels of EBF1 and EBF2 leads to an increase in EIN3/EIL1 accumulation. As a result of EIN2-CEND's movement from the cytoplasm to the nucleus, EIN3/EIL1 is activated, which in turn promotes the expression of ET-responsive genes (Chao et al. 1997; Solano et al. 1998; Qiao et al. 2012; Wen et al. 2012; Zhang et al. 2017). In the absence of ET, ET receptors establish a connection with CTR1 and activate the kinase activity of this protein. The research conducted by Ju et al. (2012) revealed that CTR1 is responsible for the phosphorylation of EIN2, which results in its degradation and limits the expression of downstream signal transduction and ET response factors. According to Lv et al. (2018), BRs have an effect on ET production that is concentration dependent. According to Wang et al. (2011), and Wei et al. (2015), high levels of BRs have the potential to improve the synthesis of ET by improving the stability and activity of some of the most important enzymes in the ET biosynthetic pathway. These enzymes include ACSs and ACOs. According to Nie et al. (2017), elevating the expression of the BR receptor SlBRI1 in *Solanum lycopersicum* can result in an increase in the transcript abundance of SlACSs and SlACOs, which in turn leads to an increase in the production of ET. The expression of ACS is inhibited in the BR-deficient mutant det2, which results in a decreased buildup of ET (Li and He 2016). According to Lv et al. (2018) and Ma et al. (2021), the exogenous administration of low BR concentrations results in a reduction in ET biosynthesis. As a result of BZR1 being activated by BR, ACS, and ACO transcription is directly inhibited, which results in a decrease in ET synthesis (Lv et al. 2018; Guo et al. 2019). In addition to its influence on the regulation of ET biosynthesis, the BR signaling network also has an impact on ET signal transduction.

According to Wang et al. (2023), BZR1 has the ability to increase the levels of EIN3 protein by inhibiting the transcription of EBF1 and EBF2. As stated by Zhao et al. (2021), EIN3 engages in a reciprocal relationship with BZR1 in order to enhance the latter's transcriptional activity, which ultimately results in positive BR responses. The expression of essential enzymes in the BR biosynthesis pathway is influenced by ET, which in turn has an effect on the BR signaling pathway. According to De Grauwe et al. (2005), the application of exogenous ACC, which is the precursor of ET, leads to an increase in the transcription of the BR biosynthesis gene CPD, which in turn leads to an increase in BR accumulation and signaling. The expression of DWF4 and DWF1 biosynthetic genes is increased by the SUBMERGENCE 1A (SUB1A) ERF found in rice, which results in an increase in the production of BR (Schmitz et al. 2013). The interaction between BRs and ET can either be synergistic or antagonistic, but either way, it achieves a balance between plant growth and stress tolerance.

15.2.6 JA and BRs

Plant hormones JA and BRs regulate growth, development, and stress. JA controls biotic stress defenses including insect herbivory and pathogen attack, while BRs stimulate cell elongation, seed germination, and abiotic stress tolerance. Their connection regulates plant development and defense. Jasmonates (JAs) are a class of lipid-derived phytohormones that regulate various developmental processes, including apical hook formation, root growth, and seedling photo morphogenesis, as well as defense responses such as pathogen resistance and insect defense and mechanical injury (Hu et al. 2023; Zhang et al. 2023). Through signaling components, BRs and JA interact molecularly. By altering JA-responsive genes, BZR1 and BES1 BR signaling transcription factors affect plant immunity and growth trade-offs. JA inhibits BR-responsive genes to prevent cell elongation. BRs regulate JA signaling to prevent growth-compromising defense activation. JA-BR interactions are important in plant stress responses. JA activates MYC2, a transcription factor, to boost defenses against pathogens, while BRs control these responses to maintain growth. JA closes stomata to prevent pathogen invasion, while BRs preserve photosynthetic efficiency. BRs and JAs engage in a mutually antagonistic interaction by repressing each other's biosynthesis and signaling transduction. BR-activated BES1 inhibits the JA signaling pathway through interactions with JA-induced transcription factors MYB34, MYB51, and MYB122 (Liao et al. 2020). OsGSK2, a rice GSK3-like kinase and a crucial suppressor of BR signaling, activates JA responses through the phosphorylation and degradation of OsJAZ4, a significant negative regulator of JA signaling (He et al. 2020).

BRs exhibit a synergistic interaction with JAs in response to viral infection. Following infection by rice stripe virus (RSV), accumulated OsGSK2 dircctly interacts with MYC2, leading to its degradation and a subsequent reduction in the JA-mediated RSV resistance response (Hu et al. 2020). In rice, the exogenous application of MeJA, a derivative of JAs, significantly reduces the transcript levels of BR-related genes such as D11, OsDWARF4, D2, OsCPDs, OsBRI1, and OsBAK1 in an OsCOI1-dependent manner (He et al. 2017). Recent reports indicate that JA-activated MYC2 inhibits the transcriptional activity of BZR1 by obstructing its interaction with partner transcription factors, including PIF (Zhang et al. 2023). Furthermore, JAs inhibit BR signaling by downregulating the expression of the BR biosynthesis gene DWF4, a process that depends on three essential downstream transcription factors: MYC2, MYC3, and MYC4 (Zhang et al. 2023). Following treatment with 1 μmol/l EBL, there is an induction of the JA biosynthesis gene 12-OXOPHYTODIENOATE REDUCTASE 3 (OPR3) and JA-activated genes such as PAP1, PAP2, and GL3, which encode MYB/bHLH transcription factors. This leads to an enhancement of JA signal transduction (Müssig et al. 2000; Peng et al. 2011). The intricate crosstalk between JA and BRs underscores their role in integrating plant growth and stress adaptation. Future research will further unravel the molecular mechanisms governing this interaction, providing insights into improving crop resilience.

15.2.7 MT and BRs

MT and BRs are essential plant signaling molecules that regulate growth, development, and responses to stress. BRs primarily promote cell elongation, differentiation, and tolerance to abiotic stress, whereas MT serves as a crucial regulator of oxidative stress mitigation, seed germination, and defense against biotic stress. Their interactions regulate antioxidant systems, hormonal balance, and transcriptional processes to sustain plant homeostasis.

MT and brassinolide (BL), an important bioactive BR, modulate numerous physiological and biochemical processes in plants. Phytohormones facilitate cell division and elongation, improve photosynthetic efficiency, and elevate the levels of proteins, pigments, and monosaccharides. Furthermore, both MT and BL demonstrate antioxidant properties in response to abiotic stressors, including heavy metal exposure, extreme temperatures, salinity, and drought. MT is synthesized from tryptophan, whereas BL is produced from campestanol (CN) via the CN-dependent pathway or from 22-hydroxy-5α-ergostan-3-one in the CN-independent pathway. BL can also be synthesized through the early C-6 oxidation pathway originating from CN. MT and BRs exhibit shared molecular regulatory mechanisms. BR signaling components, including BZR1 and BES1, regulate MT biosynthesis genes, such as TDC and SNAT. MT regulates BR-responsive transcription factors, thereby controlling cell elongation and stress adaptation. MT enhances BR-mediated antioxidant defenses during abiotic stress by upregulating superoxide dismutase (SOD) and catalase (CAT), thereby effectively reducing oxidative damage. MT–BR crosstalk is essential for plant adaptation under environmental stress. MT enhances BR-mediated stress tolerance in drought and salinity stress situations by modulating osmoprotectant accumulation and ROS scavenging. Furthermore, their synergistic interactions enhance seed germination, prolong senescence, and increase photosynthetic efficiency. This intricate hormonal interaction illustrates the coordination of plant development and defense mechanisms by MT and BRs. Comprehending their molecular mechanisms offers significant insights for enhancing crop resilience and productivity.

Environmental pollution has led to heightened exposure of plants to heavy metal toxicity, notably cadmium (Cd), which impairs growth, diminishes metabolite levels, and triggers oxidative stress. The aquatic plant *Wolffia arrhiza*, characterized by its absence of roots, demonstrates significant biosorption abilities, positioning it as an exemplary model for investigating responses to heavy metal stress. A study conducted by Chmur and Bajguz (2025) examined the impact of Cd exposure on *W. arrhiza*, revealing that Cd stress notably raised MT levels and moderately increased BR content. The findings are consistent with the research conducted, which showed that exogenous BL promotes MT biosynthesis in rice seedlings under both Cd stress and nonstress conditions. Exogenous MT has been demonstrated to upregulate BR biosynthetic genes, while transgenic rice plants lacking MT show decreased levels of bioactive BRs (Hwang and Back 2022). Lee and Back (2019) demonstrated that rice plants lacking MT exhibit markedly reduced BR levels, highlighting the positive regulatory relationship between these hormones. The application of 0.1 μM EBL significantly increased BR content in *Hordeum vulgare* seedlings, highlighting the strong relationship between MT and BRs in stress adaptation (Bajguz, A et al. 2019). This study demonstrates that MT is more effective than BRs in alleviating the negative impacts of Cd stress on growth, photosynthetic pigment levels, protein and sugar content, and the reduction of stress markers in *W. arrhiza*. BRs significantly enhance the synthesis of detoxifying compounds, including glutathione (GSH) and phytochelatins (PCs). MT and BRs both enhance plant resilience; however, MT provides more extensive advantages regarding growth, metabolism, and the reduction of stress markers. The combined functions of these hormones in heavy metal stress tolerance underscore their potential utility in enhancing crop performance under environmental stress conditions.

15.3 Conclusion

BRs play a pivotal role in regulating plant growth, development, and stress responses by interacting with other plant hormones under various stressful environments. The crosstalk between BRs and hormones such as ABA, IAA, GAs, ET, JA, and SA forms a complex signaling network that helps plants adapt to abiotic and biotic stresses. BRs enhance stress tolerance by modulating physiological and molecular pathways, improving antioxidant defense systems, stabilizing cell membranes, and maintaining ion homeostasis. Under drought and salinity stress, BRs interact with ABA to regulate stomatal closure, water balance, and osmotic adjustment. During pathogen attacks, the interaction of BRs with JA and SA enhances defense responses by triggering immune signaling pathways. Crosstalk with ET and IAA influences root and shoot architecture, thereby improving nutrient and water uptake under nutrient-deficient and water-limiting conditions. Furthermore, BR-GA interactions regulate seed germination and plant growth under temperature stress. The coordinated action of BRs and other hormones fine-tunes gene expression and metabolic adjustments to enhance plant resilience. Understanding the intricate hormonal crosstalk involving BRs opens new avenues for improving crop productivity and stress tolerance through genetic and biotechnological approaches. Future research focusing on the molecular mechanisms of BR crosstalk will provide deeper insights into developing stress-resilient crop varieties.

References

Albertos, P., Dündar, G., Schenk, P. et al. (2022a). Transcription factor BES1 interacts with HSFA1 to promote heat stress resistance of plants. *The EMBO Journal* 41: e108664.

Arteca R.N., Tsai D.S., Schlagnhaufer C. et al. (1983). The effect of brassinosteroid on auxin-induced ethylene production by etiolated mung bean segments. *Physiologia Plantarum* 59 (4): 539–544.

Bai, M.Y., Zhang, L.Y., Gampala, S.S. et al. (2012). Functions of OsBZR1 and 14-3-3 proteins in brassinosteroid signaling in rice. *Proceedings of the National Academy of Sciences of the United States of America* 109 (26): 10515–10520. https://doi.org/10.1073/pnas.1201499109.

Bajguz, A. and Hayat, S. (2009). Effects of brassinosteroids on the plant responses to environmental stresses. *Plant Physiology and Biochemistry* 47 (1): 1–8. https://doi.org/10.1016/j.plaphy.2008.10.002.

Bajguz, A., Orczyk, W., Golebiewska, A. et al. (2019). Occurrence of brassinosteroids and influence of 24-epibrassinolide with brassinazole on their content in the leaves and roots of *Hordeum vulgare* L. cv. Golden Promise. *Planta* 249: 123–137.

Bartrina, I., Otto, E., Strnad, M. et al. (2011). Cytokinin regulates the activity of reproductive meristems, flower organ size, ovule formation, and thus seed yield in *Arabidopsis thaliana*. *Plant Cell* 23: 69–80.

Cai, Z., Liu, J., Wang, H. et al. (2014). GSK3-like kinases positively regulate brassinosteroid signaling through phosphorylating BES1/BZR1 in *Arabidopsis*. *Proceedings of the National Academy of Sciences of the United States of America* 111 (16): 9551–9556. https://doi.org/10.1073/pnas.1402907111.

Cao, H. and Chen, S. (1995). Brassinosteroid-induced rice lamina joint inclination and its relation to indole-3-acetic acid and ethylene. *Plant Growth Regulation* 16: 189–196.

Chao, Q., Rothenberg, M., Solano, R. et al. (1997). Activation of the ethylene gas response pathway in *Arabidopsis* by the nuclear protein ETHYLENE-INSENSITIVE3 and related proteins. *Cell* 89: 1133–1144.

Chaudhury, A.M., Ming, L., Miller, C. et al. (1997). Fertilization-independent seed development in *Arabidopsis thaliana*. *Proceedings of the National Academy of Sciences of the United States of America* 94: 4223–4228.

Che, R., Tong, H., Shi, B. et al. (2016). Control of grain size and rice yield by GL2-mediated brassinosteroid responses. *Nature Plants* 2 (2): 15195. https://doi.org/10.1038/nplants.2015.195.

Chmur, M. and Bajguz, A. (2025). Comparative efficacy of melatonin and brassinolide in mitigating the adverse effects of cadmium on *Wolffia arrhiza*. *International Journal of Molecular Sciences* 26 (2): 692.

Choudhary, S.P., Yu, J.Q., Yamaguchi-Shinozaki, K. et al. (2012). Benefits of brassinosteroid crosstalk. *Trends in Plant Science* 17 (10): 594–605. https://doi.org/10.1016/j.tplants.2012.05.012.

Clouse, S.D. and Sasse, J.M. (1998). Brassinosteroids: essential regulators of plant growth and development. *Annual Review of Plant Biology* 49 (1): 427–451. https://doi.org/10.1146/annurev.arplant.49.1.427.

De Grauwe, L., Vandenbussche, F., Tietz, O. et al. (2005). Auxin, ethylene and brassinosteroids: tripartite control of growth in the *Arabidopsis* hypocotyl. *Plant and Cell Physiology* 46: 827–836.

Faizan, M. and Hayat, S. (2024). *Plant Growth Regulators: Resilience for Sustainable Agriculture*. Springer Nature. https://doi.org/10.1007/978-981-97-2918-0.

Fatma, M., Asgher, M., Iqbal, N. et al. (2022). Ethylene signaling under stressful environments: analyzing collaborative knowledge. *Plants* 11 (17): 2211.

Fujii, H., Chinnusamy, V., Rodrigues, A. et al. (2009). In vitro reconstitution of an abscisic acid signalling pathway. *Nature* 462: 660–664.

Gagne, J.M., Smalle, J., Gingerich, D.J. et al. (2004). Arabidopsis EIN3-binding F-box 1 and 2 form ubiquitin-protein ligases that repress ethylene action and promote growth by directing EIN3 degradation. *Proceedings of the National Academy of Sciences of the United States of America* 101: 6803–6808.

Gallego-Bartolomé, J., Minguet, E.G., Marín-de la Rosa, N. et al. (2012). Brassinosteroid signaling involves interactions with DELLA proteins in Arabidopsis. *The Plant Cell* 24 (8): 2977–2994. https://doi.org/10.1105/tpc.112.100072.

Goda, H., Sasaki, E., Akiyama, K. et al. (2008). The at GenExpress hormone and chemical treatment data set: experimental design, data evaluation, model data analysis and data access. *The Plant Journal* 55: 526–542. https://doi.org/10.1111/j.0960-7412.2008.03510.x.

Goda, H., Sawa, S., Asami, T. et al. (2004). Comprehensive comparison of auxin-regulated and brassinosteroid-regulated genes in Arabidopsis. *Plant Physiology* 134: 1555–1573.

Goda, H., Shimada, Y., Asami, T. et al. (2002). Microarray analysis of bassinosteroid-regulated genes in Arabidopsis. *Plant Physiology* 130: 1319–1334.

Guo, H.W. and Ecker, J.R. (2003). Plant responses to ethylene gas are mediated by SCF(EBF1/EBF2)-dependent proteolysis of EIN3 transcription factor. *Cell* 115: 667–677.

Guo, Y.F., Shan, W., Liang, S.M. et al. (2019). MaBZR1/2 act as transcriptional repressors of ethylene biosynthetic genes in banana fruit. *Physiologia Plantarum* 165: 555–568.

Ha, C.V., Leyva-González, M.A., Osakabe, Y. et al. (2014). Positive regulatory role of strigolactone in plant responses to drought and salt stress. *Proceedings of the National Academy of Sciences of the United States of America* 111 (2): 851–856. https://doi.org/10.1073/pnas.1322135111.

He, J.X., Gendron, J.M., Sun, Y. et al. (2005). BZR1 is a transcriptional repressor with dual roles in brassinosteroid homeostasis and growth responses. *Science* 307: 1634–1638.

He, J.X., Gendron, J.M., Yang, Y. et al. (2002). The GSK3- ike kinase BIN2 phosphorylates and destabilizes BZR1, a positive regulator of the Brassinosteroid signaling pathway in *Arabidopsis. Proceedings of the National Academy of Sciences of the United States of America* 99: 10185–10190.

He, Y.Q., Hong, G.J., Zhang, H.H. et al. (2020). The OsGSK2 kinase integrates brassinosteroid and jasmonic acid signaling by interacting with OsJAZ4. *Plant Cell* 32: 2806–2822.

He, Y.Q., Zhang, H.H., Sun, Z.T. et al. (2017). Jasmonic acid-mediated defense suppresses brassinosteroid-mediated susceptibility to Rice black-streaked dwarf virus infection in rice. *The New Phytologist* 214: 388–399.

Higuchi, M., Pischke, M.S., Mähönen, A.P. et al. (2004). In planta functions of the *Arabidopsis* cytokinin receptor family. *Proceedings of the National Academy of Sciences of the United States of America* 101: 8821–8826.

Hu, J.L., Huang, J., Xu, H.S. et al. (2020). Rice stripe virus suppresses jasmonic acid-mediated resistance by hijacking brassinosteroid signaling pathway in rice. *PLoS Pathog* 16: e1008801.

Hu, S., Yu, K.M., Yan, J.B. et al. (2023) Jasmonate perception: ligand-receptor interaction, regulation, and evolution. *Molecular Plant* 16: 23–42.

Hu, Y.R. and Yu, D.Q. (2014) BRASSINOSTEROID INSENSITIVE2 interacts with ABSCISIC ACID INSENSITIVE5 to mediate the antagonism of brassinosteroids to abscisic acid during seed germination in *Arabidopsis. Plant Cell* 26: 4394–4408.

Huang, H.Y., Jiang, W.B., Hu, Y.W. et al. (2013). BR signal influences *Arabidopsis* ovule and seed number through regulating related genes expression by BZR1. *Molecular Plant* 6: 456–469.

Hwang, I., Paudyal, D., and Cheong, H. (2008). Interaction of brassinosteroids and cytokinin in modulating light-mediated signaling in *Arabidopsis. Journal of the Chosun Natural Science* 1 (1): 24–31.

Hwang, O.J. and Back, K. (2022). Molecular regulation of antioxidant melatonin biosynthesis by brassinosteroid acting as an endogenous elicitor of melatonin induction in rice seedlings. *Antioxidants* 11: 918.

Jiang, H., Tang, B.Y., Xie, Z.L. et al. (2019). GSK3-like kinase BIN2 phosphorylates RD26 to potentiate drought signaling in *Arabidopsis. The Plant Journal* 100: 923–937.

Jiang, W.B., Huang, H.Y., Hu, Y.W. et al. (2013). Brassinosteroids regulates seed size and shape in *Arabidopsis. Plant Physiology*. 162: 1965–1977.

Jiang, Y., Yang, B., Harris, N.S. et al. (2020). Comparative transcriptome analysis reveals brassinosteroid signaling pathway in *Arabidopsis* responding to drought stress. *BMC Plant Biology*, 20 (1): 444. https://doi.org/10.1186/s12870-020-02650-7.

Ju, C.L. and Chang, C.R. (2015). Mechanistic insights in ethylene perception and signal transduction. *Plant Physiology* 169: 85–95.

Ju, C.L., Yoon, G.M., Shemansky, J.M. et al. (2012). CTR1 phosphorylates the central regulator EIN2 to control ethylene hormone signaling from the ER membrane to the nucleus in *Arabidopsis*. *Proceedings of the National Academy of Sciences of the United States of America* 109: 19486–19491.

Koh, S., Lee, S.C., Kim, M.K. et al. (2007). T-DNA insertional mutagenesis reveals the role of phospholipase Dα1 in reactive oxygen species-mediated disease resistance in rice. *The Plant Journal* 52 (5): 873–886. https://doi.org/10.1111/j.1365-313X.2007.03271.x.

Kuppusamy, K.T., Chen, A.Y., and Nemhauser, J.L. (2009). Cross-regulation of auxin and brassinosteroid in *Arabidopsis* requires the BR-responsive bHLH transcription factor HBI1. *The Plant Cell* 21 (11): 3369–3383. https://doi.org/10.1105/tpc.109.070235.

Lee, K. and Back, K. (2019). Melatonin-deficient rice plants show a common semidwarf phenotype either dependent or independent of brassinosteroid biosynthesis. *Journal of Pineal Research* 66: e12537.

Li, Q.F. and He, J.X. (2016). BZR1 interacts with HY5 to mediate brassinosteroid- and light-regulated cotyledon opening in *Arabidopsis* in darkness. *Molecular Plant* 9: 113–125.

Li, Q.F., Wang, C., Jiang, L. et al. (2018). Repression of DELLA function by the brassinosteroid signaling effector BZR1 promotes root elongation in *Arabidopsis*. *The Plant Cell* 30 (4): 736–749. https://doi.org/10.1105/tpc.18.00068.

Li, W.Y., Ma, M.D., Feng, Y. et al. (2015). EIN2- directed translational regulation of ethylene signaling in *Arabidopsis*. *Cell* 163: 670–683.

Li, Z., Liu, H.M., Ding, Z.H. et al. (2020b). Low temperature enhances plant immunity via salicylic acid pathway genes that are repressed by ethylene. *Plant Physiology* 182: 626–639.

Liao, K., Peng, Y.J., Yuan, L.B. et al. (2020). Brassinosteroids antagonize jasmonate-activated plant defense responses through BRI1-EMS-SUPPRESSOR1 (BES1). *Plant Physiology* 182: 1066–1082.

Liu, C., Ma, T., Yuan, D.Y. et al. (2022). The OsEIL1-OsERF115-target gene regulatory module controls grain size and weight in rice. *Plant Biotechnology Journal* 20: 1470–1486.

Lv, B.S., Tian, H.Y., Zhang, F. et al. (2018). Brassinosteroids regulate root growth by controlling reactive oxygen species homeostasis and dual effect on ethylene synthesis in *Arabidopsis*. *PLoS Genetics* 14: e1007144.

Ma, X.S., Yuan, Y., Li, C.Q. et al. (2021). Brassinosteroids suppress ethylene-induced fruitlet abscission through LcBZR1/2- mediated transcriptional repression of LcACS1/4 and LcACO2/3 in litchi. *Horticulture Research* 8: 105.

Ma, Y., Szostkiewicz, I., Korte, A. et al. (2009). Regulators of PP2C phosphatase activity function as abscisic acid sensors. *Science* 324: 1064–1068.

Moon, J., Kim, S.Y., Park, C.-H. et al. (2021). BES1 negatively regulates the expression of ACC oxidase 2 to control the endogenous level of ethylene in *Arabidopsis thaliana*. *Plant Signaling & Behavior* 16(2): 1850625.

Müssig, C., Fischer, S., and Altmann, T. (2002). Brassinosteroid-regulated gene expression. *Plant Physiology* 129: 1241–1251.

Müssig, C., Biesgen, C., Lisso, J. et al. (2000). A novel stress-inducible 12-oxophytodienoate reductase from *Arabidopsis thaliana* provides a potential link between brassinosteroid-action and jasmonic-acid synthesis. *Journal of Plant Physiology* 157: 143–152.

Nakamura, M., Katsumata, H., Abe, M. et al. (2003). The role of the *Arabidopsis* BES1 gene in brassinosteroid signaling and leaf development. *Plant Cell Physiology* 44 (9): 904–911. https://doi.org/10.1093/pcp/pcg112.

Nakashima, K., Yamaguchi-Shinozaki, K., and Shinozaki, K. et al. (2009). The transcriptional regulatory network in the drought response and its crosstalk with phytohormone signaling pathways. *Current Opinion in Plant Biology* 12 (3): 315–321. https://doi.org/10.1016/j.pbi.2009.04.001.

Nawaz, F., Naeem, M., Zulfiqar, B. et al. (2017). Understanding brassinosteroid-regulated mechanisms to improve stress tolerance in plants: a critical review. *Environmental and Experimental Botany* 140: 75–89. https://doi.org/10.1016/j.envexpbot.2017.05.003.

Nemhauser, J.L., Mockler, T.C., and Chory, J. (2004a). Interdependency of brassinosteroid and auxin signaling in *Arabidopsis*. *PLoS Biology* 2: 1460–1471. https://doi.org/ 10.1371/journal.pbio.0020258.

Nemhauser, J.L., Hong, F., and Chory, J. (2004b). Different plant hormones regulate similar processes through largely nonoverlapping transcriptional responses. *Cell* 118 (4): 465–474. https://doi.org/10.1016/j.cell.2004.07.024.

Nie, S., Huang, S.H., Wang, S.F. et al. (2017). Enhancing brassinosteroid signaling via overexpression of tomato (*Solanum lycopersicum*) SlBRI1 improves major agronomic traits. *Frontiers in Plant Science* 8: 1386.

Oikawa, T., Koshioka, M., Kojima, K. et al. (2004). A role of OsGA20ox1, encoding an isoform of gibberellin 20-oxidase, for regulation of plant stature in rice. *Plant Molecular Biology* 55: 687–700.

Park, S.Y., Fung, P., Nishimura, N. et al. (2009). Abscisic acid inhibits type 2C protein phosphatases via the PYR/PYL family of START proteins. *Science* 324: 1068–1071.

Peng, Z.H., Han, C.Y., Yuan, L.B. et al. (2011). Brassinosteroid enhances jasmonate-induced anthocyanin accumulation in *Arabidopsis* seedlings. *Journal of Integrative Plant Biology* 53 (8): 632–640.

Potuschak, T., Lechner, E., Parmentier, Y. et al. (2003). EIN3-dependent regulation of plant ethylene hormone signaling by two *Arabidopsis* F box proteins: EBF1 and EBF2. *Cell* 115: 679–689.

Qiao, H., Shen, Z.X., Huang, S.S. et al. (2012). Processing and subcellular trafficking of ER-tethered EIN2 control response to ethylene gas. *Science* 338: 390–393.

Ryu, H., Cho, H., Bae, W. et al. (2014). Control of early seedling development by BES1/TPL/HDA19-mediated epigenetic regulation of ABI3. *Nature Communications* 5: 4138.

Schmitz, A.J., Folsom, J.J., Jikamaru, Y. et al. (2013). SUB1A-mediated submergence tolerance response in rice involves differential regulation of the brassinosteroid pathway. *The New Phytologist* 198: 1060–1070.

Sharma, A., Zheng, B., Liao, P. et al. (2015). Role of brassinosteroids in plant stress responses: a review. *Plant Signaling & Behavior* 10 (1): e1049784. https://doi.org/10.1080/15592324.2015.1049784.

Solano, R., Stepanova, A., Chao, Q.M. et al. (1998). Nuclear events in ethylene signaling: a transcriptional cascade mediated by ETHYLENE-INSENSITIVE3 and ETHYLENE-RESPONSE-FACTOR1. *Genes & Development* 12: 3703–3714.

Sun, Y., Fan, X.Y., Cao, D.M. et al. (2010). Integration of brassinosteroid signal transduction with the transcription network for plant growth regulation in *Arabidopsis*. *Developmental Cell*. 19: 765–777.

Takeno, K. and Pharis, R.P. (1982). Brassinosteroid-induced bending of the leaf lamina of dwarf rice seedlings: an auxin-mediated phenomenon. *Plant and Cell Physiology* 23: 1275–1281.

Tong, H.N., Xiao, Y.H., Liu, D.P. et al. (2014). Brassinosteroid regulates cell elongation by modulating gibberellin metabolism in rice. *The Plant Cell* 26: 4376–4393.

Unterholzner, S.J., Rozhon, W., Papacek, M. et al. (2015). Brassinosteroids are master regulators of gibberellin biosynthesis in Arabidopsis. Plant Cell 27: 2261–2272.

Verma, V., Ravindran, P., and Kumar, P.P. (2016). Plant hormone-mediated regulation of stress responses. *BMC Plant Biology* 16(1): 86. https://doi.org/10.1186/s12870-016-0771-y.

Vert, G., Walcher, C.L., Chory, J. et al. (2008). Integration of auxin and brassinosteroid pathways by auxin response factor 2. *Proceedings of the National Academy of Sciences of the United States of America* 105 (28): 9829–9834. https://doi.org/10.1073/pnas.0803996105.

Vukašinović, N. and Russinova, E. (2018). BRexit: possible brassinosteroid export and transport routes. *Trends in Plant Science* 23: 285–292.

Wang, B.L., Zhang, J.L., Xia, X.Z. et al. (2011). Ameliorative effect of brassinosteroid and ethylene on germination of cucumber seeds in the presence of sodium chloride. *Plant Growth Regulation* 65: 407–413.

Wang, H.J., Tang, J., Liu, J. et al. (2018). Abscisic acid signaling inhibits brassinosteroid signaling through dampening the dephosphorylation of BIN2 by ABI1 and ABI2. *Molecular Plant* 11: 315–325.

Wang, Y., Cao, J., Zhu, G. et al. (2020). Integration of brassinosteroid and ABA signaling pathways in plant growth and stress responses. *Frontiers in Plant Science* 11: 617646. https://doi.org/10.3389/fpls.2020.617646.

Wang, C., Li, T., Meng, S. et al. (2023). Brassinosteroids fine-tune secondary and primary sulfur metabolism through BZR1-mediated transcriptional regulation. *Journal of Integrative Plant Biology* 65: 1153–1169.

Wei, L.J., Deng, X.G., Zhu, T. et al. (2015). Ethylene is involved in brassinosteroids-induced alternative respiratory pathway in cucumber (*Cucumis sativus L.*) seedlings response to abiotic stress. *Frontiers in Plant Science* 6: 982.

Weiss, D. and Ori, N. (2007). Mechanisms of cross talk between gibberellin and other hormones. *Plant Physiology* 144: 1240–1246. https://doi.org/ 10.1104/pp.107.100370.

Wen, X., Zhang, C.L., Ji, Y.S. et al. (2012). Activation of ethylene signaling is mediated by nuclear translocation of the cleaved EIN2 carboxyl terminus. *Cell Research* 22: 1613–1616.

Xia, X.J., Huang, Y.Y., Wang, L. et al. (2009). Brassinosteroids promote stress tolerance through regulation of antioxidant defense in *Arabidopsis. Journal of Experimental Botany* 60 (2): 327–337.

Xu, W., Purugganan, M.M., Polisensky, D.H. et al. (1995). Arabidopsis TCH4, regulated by hormones and the environment, encodes a xyloglucan endotransglycosylase. *The Plant Cell* 7: 1555–1567.

Yang, C.J., Zhang, C., Lu, Y. et al. (2011). The mechanisms of brassinosteroids' action: from signal transduction to plant development. *Molecular Plant* 4 (4): 588–600. https://doi.org/10.1093/mp/ssr018.

Yang, X., Jia, Z.C., Pu, Q. et al. (2022). ABA mediates plant development and abiotic stress via alternative splicing. *International Journal of Molecular Sciences* 23: 3796.

Yang, X.R., Bai, Y., Shang, J.X. et al. (2016). The antagonistic regulation of abscisic acid-inhibited root growth by brassinosteroids is partially mediated via direct suppression of

ABSCISIC ACID INSENSITIVE 5 expression by BRASSINAZOLE RESISTANT 1. *Plant, Cell & Environment* 39: 1994–2003.

Yin, Y., Wang, Z., Mora-Garcia, S. et al. (2002). BES1 accumulates in the nucleus in response to brassinosteroids to regulate gene expression and promote stem elongation. *Cell* 109: 181–191.

Yopp, J.H., Mandava, N.B., and Sasse, J.M. (1981). Brassinolide, a growth-promoting steroidal lactone. I. Activity in selected auxin bioassays. *Physiologia Plantarum* 53: 445–452.

Yu, X., Li, L., Zola, J. et al. (2018). Brassinosteroids modulate root phototropism. *The Plant Cell* 30 (6): 1177–1191. https://doi.org/10.1105/tpc.18.00068.

Yu, X.F., Li, L., Zola, J. et al. (2011). A brassinosteroid transcriptional network revealed by genome-wide identification of BES1 target genes in *Arabidopsis thaliana*. *The Plant Journal* 65: 634–646.

Yuldashev, R., Avalbaev, A., Bezrukova, M. et al. (2012). Cytokinin oxidase is involved in the regulation of cytokinin content by 24-epibrassinolide in wheat seedlings. *Plant Physiology and Biochemistry* 55: 1–6.

Zhang, F., Wang, L.K., Qi, B. et al. (2017). EIN2 mediates direct regulation of histone acetylation in the ethylene response. *Proceedings of the National Academy of Sciences of the United States of America* 114: 10274–10279.

Zhang, J.J., Chen, W.Y., Li, X.P. et al. (2023). Jasmonates regulate apical hook development by repressing brassinosteroid biosynthesis and signaling. *Plant Physiology* 193: 1561–1579.

Zhao, X., Dou, L.R., Gong, Z.Z., et al. (2019). BES1 hinders ABSCISIC ACID INSENSITIVE5 and promotes seed germination in *Arabidopsis*. *New Phytologist* 221: 908–918.

Zheng, X., Xing, J.H., Zhang, K. et al. (2019b). Ethylene response factor ERF11 activates BT4 transcription to regulate immunity to pseudomonas syringae. *Plant Physiology* 180: 1132–1151.

16

Role of Brassinosteroids on Antioxidants Regulation Under Abiotic Stresses

Mohammad Faizan[1], Anjuman Hussain[2], Shamsul Hayat[2], Nayab Ishrat[3], Aurosha Fayaz[1], and S. Maqbool Ahmed[1]*

[1]*Department of Botany, School of Sciences, Maulana Azad National Urdu University, Hyderabad, India*
[2]*Department of Botany, Aligarh Muslim University, Aligarh, India*
[3]*Department of Biotechnology, Era University, Lucknow, India*

*Corresponding author: mohammadfaizan@manuu.edu.in

16.1 Introduction

Through intrinsic defensive systems, plants are better able to survive abiotic challenges (Kanwar et al. 2012). Abiotic stressors are nonliving environmental circumstances that are harmful to plant growth and cause a variety of physiological, biochemical, and growth reactions in plants. According to Shahzad et al. (2018), these reactions include modifications to gene regulation, changes in the synthesis of proteins and metabolites, modifications to hormone signaling, and elevated antioxidant enzyme activity. At low concentrations, naturally occurring chemical molecules called phytohormones have a significant impact on important plant life cycle activities (Rhaman et al. 2020). In response to abiotic stressors, their participation in signal transduction networks improves plant growth and production (Muhammad et al. 2022). Brassinosteroids (BRs) modulate various biological and cellular processes like stem elongation, pollen tube development, leaf morphology, root growth inhibition, fruit maturation, ethylene production, xylem separation, chlorophyll production, and gene expression (Bajguz 2011; Ahammed et al. 2013). Additionally, BRs have been shown to improve plant development and the levels of proline, sucrose, and chlorophyll, as well as alter antioxidant enzyme and nonenzymatic defense mechanisms (Basit et al. 2022a,b; Yusuf et al. 2024). According to earlier research, BRs have a beneficial effect on *Solanum nigrum* L. by increasing the accumulation of sugar and photosynthetic pigment, increasing the efficiency of photosystem II (PSII), and decreasing electrolyte leakage, malondialdehyde levels, and cadmium (Cd) accumulation (Anwar et al. 2018). Furthermore, during chromium (Cr) stress, BRs improve rice biomass and growth by modifying antioxidant enzyme activity and affecting nutrient uptake (Anwar et al. 2018; Basit et al. 2022a). Additionally, when reaction oxygen species (ROS) levels are extremely

Brassinosteroids and Abiotic Stress: Recent Developments on Physiological and Molecular Aspects,
First Edition. Edited by Anket Sharma, et al.
Published 2026 by John Wiley & Sons Ltd.

high, BRs increase enzymatic activity and reduce oxidative damage in *Raphanus sativus* L. (Kapoor et al. 2014). Likewise, it has been demonstrated that exogenous BR treatment boosts rice biomass and general growth (Basit et al. 2022a,b). Abiotic stressors such as metal stress, chilling stress, and high temperatures are also lessened by BRs (Hussain et al. 2019; Nazir et al. 2021). BRs are important for sustained agricultural production because of their critical function in shielding plants from environmental stressors (Yusuf et al. 2024). Thus, there is potential for agricultural enhancement through modifying BR signaling, biosynthesis pathways, or perceptions. The purpose of this chapter is to describe the advantageous contributions of BRs to abiotic stress tolerance by regulating antioxidant defense system of plants.

16.2 Significance in General Plant Growth and Development

Plant growth and development are influenced by BRs, which are steroid hormones (Peres et al. 2019; Lin 2020). Plant growth, cell division, stem cell maintenance, vascular development, elongation of diverse cell types, and floral transition are only a few of the physiological and developmental processes that BRs control (Lv et al. 2018; Wang et al. 2020). Additionally, they have a variety of functions in the following areas: xylem formation (Lee et al. 2019), xylem differentiation, photomorphogenesis and plant reproduction (Ahammed et al. 2020), senescence (Yin et al. 2002), pollen tube growth, seed germination, pollen germination and development (Vogler et al. 2014), stomata patterning (Khan et al. 2013), and hypocotyl elongation (Oh et al. 2014). BRs can trigger excessive growth in hydroponically grown plants (Arteca, J.M. and Arteca, R.N. 2001), control the proliferation of leaf cells (Nakaya et al. 2002), regulate cell cycle progression (González-García et al. 2011), and activate the cell cycle during seed germination (Zadvornova et al. 2005). Additionally, BRs control the formation of stomata and the responses to biotic and abiotic stress (Planas-Riverola et al. 2019; Ackerman-Lavert and Savaldi-Goldstein 2020). Furthermore, according to Hartwig et al. (2011) and Makarevitch et al. (2012), BR is essential for controlling both male and female fertility in crops. According to Zhu et al. (2015), BR contributes to etiolation and encourages the elongation of stigma, plant architecture, thermo-tolerance, proton transport, tiller number, leaf angle, and leaf size. Furthermore, crop yields may be increased by the exogenous application of BR or by changes to their biosynthesis and signaling (Jaiswal et al. 2020). BRs also play a role in controlling a number of plant genes. Chen et al. (2017) discovered that BRs activated the genes WRKY46, WRKY54, and WRKY70, which were proven to be beneficial for BR-regulated plant growth in *Arabidopsis thaliana* (*A. thaliana*). The histone lysine methyltransferase SDG8 has been implicated in BR-regulated gene expression in another investigation. In *A. thaliana*, the knockout mutant sdg8 exhibited a diminished growth phenotype with impaired BR responses (Wang et al. 2014). By transcriptionally altering the genes and pathways that control the development of the seed and ovule, BR controls the growth of seeds in *A. thaliana* and influences their size, weight, and quantity (Jiang et al. 2013). Additionally, under mild nitrogen deprivation, BRs control the root nitrogen foraging response in *A. thaliana*. When there is a slight nitrogen shortage, the elongation of the primary root is caused by a gene called Brassinosteroid Signaling Kinase 3 (BSK3) (Jia et al. 2019).

In rice, BR controls plant architecture and grain yield (Fujioka, S. and Yokota, T. 2003). The height of rice plants is influenced by the gene products of BRD1 and D11, which are involved in the manufacture of BR (Tanabe et al. 2005). A C-6 oxidase required for rice BR biosynthesis is encoded by the OsDwarf2/OsDwarf1 reduction, which decreases the second internode and seed length elongation (Tanabe et al. 2005; Hong et al. 2005). By raising the amount of antioxidant system (AOS) and energy charge during panicle development, BRs in rice moderate the effects of N fertilization on spikelet development and help to increase spikelet growth (Zhang et al. 2019). Male sterility in maize is caused by insufficient BR biosynthesis, which hinders the formation of pollen and anther (Makarevitch et al. 2012). According to reports, BRs are necessary for the fiber start and elongation of cultivated cotton ovules in cotton (Luo et al. 2007). Furthermore, in monocotyledonous wheat (*Triticum aestivum* L.), an exogenous administration of epibrassinolide (EBR) postpones the vegetative to generative transition. According to Janeczko et al. (2015), the transition and heading stage is accelerated by the BR inhibitor brassinazole. According to Wan Lin et al. (2017), the application of BR sped up the maturation of winter rapeseed by 4–8 days.

According to Zhang et al. (2007), priming lucerne seeds with BL significantly enhanced their length, vigor, and fresh and dry shoot and root weights. According to Symons et al. (2006), the application of BR accelerated the ripening of grape berries. Berries treated with EBR were shown to have significantly higher levels of soluble sugars and sucrose synthase (sucrolytic) (Xu et al. 2015). Jujube fruit deterioration was decreased by BR treatment, most likely as a result of its capacity to delay senescence and create disease resistance (Zhu et al. 2010). Furthermore, it has been demonstrated that the BRs in potato tubers stimulate the formation of apical meristems, speeding up the rate of cell division in isolated *Petunia* hybrida protoplasts (Oh 2003). Together with the inductive photoperiod, CS and BL treatments in Pharbitis nil prevent flowering, suggesting that BR cooperates with environmental cues to guarantee the appropriate reproductive transition (Li and He 2020). By adjusting plant development and enhancing plant performance through interactions with plant growth regulators or other plant hormones, BRs can take part in physiological processes in response to stress (Table 16.1) (Trevisan et al. 2020). A number of developmental processes are impacted by the disruption of BR signaling, such as pollen formation (Ye et al. 2010), flowering duration (Domagalska et al. 2010), and seed development (Jiang et al. 2013). BR-deficient plants are dwarfed, have shortened hypocotyls, petioles, and internodes, altered stomatal development, decreased male fertility, downward-curled leaves, and delayed flowering (Gudesblat et al. 2012). Furthermore, because of the decreased lamina inclination, BR-deficient plants also have a compact plant structure. On the other hand, BR shortage might result in reduced germination, incorrect stomatal distribution, and decreased grain size, seed fertility, and tiller number (Zhu et al. 2013). Because of their slower growth, BR-insensitive and -deficient mutants are frequently referred to as late flowering (Zhao and Li 2012). On the other hand, plants that accumulate too many BRs exhibit increased height and elongation of the petioles and hypocotyls (Rozhon et al. 2019).

Table 16.1 Representative studies on BRs in the mitigation of major abiotic stress-impacts by improving AOS in different plant species.

Abiotic stress	Plant species	References
	Brassica rapa	Ahmad Lone et al. (2022)
	Maize	Sun et al. (2022)
	Wheat	Khan et al. (2021)
Drought	*Jatropha curcas* L.	de Oliveira et al. (2021)
	Vitis vinifera L.	Zheng et al. (2022)
	Soybean	Hasan et al. (2020)
	Maize	Rattan et al. (2020)
	Snap bean	Ghoname et al. (2023)
Salinity	*Brassica juncea*	Gupta and Seth (2023)
	Solanum tuberosum L.	Wang et al. (2025)
Heavy metal stress		
Cr	Rice	Basit et al. (2021a)
Zn	*Vigna radiata*	Kumar et al. (2023)
Al	Rice	Basit et al. (2022b)
Pb	Rice	Guedes et al. (2021)
Pb	*Festuca arundinacea*	Zhong et al. (2020)
Mn	*A. thaliana*	Surgun-Acar and Zemheri-Navruz, (2022)
Heat stress	Soybean	Wang et al. (2022)
	Rice	Nazir et al. (2023)
	Brassica juncea	Neha et al. (2022)
	Peanut	Lai et al. (2024)

16.3 Modulation of the Plant Antioxidant Defense System by BRs Under Abiotic Stress

BRs and related compounds have been shown in numerous studies over the years to have a stress-impact-mitigating role in plants exposed to a variety of abiotic stresses, including high temperatures (Janeczko et al. 2011), low temperatures in terms of chilling (Liu et al. 2011; Wang et al. 2014), and freezing (Janeczko et al. 2009). The importance of BRs and related compounds has been reported by various workers in various plants exposed to drought (Anjum et al. 2011; Mahesh et al. 2013), light (Wang et al. 2012; Kurepin et al. 2012), flooding (Liang and Liang 2009), salinity (Abbas et al. 2013), metals/metalloids (Ashraf et al. 2010; Bajguz 2020), herbicides (Sharma et al. 2013a), pesticides, insecticides, and organic pollutants (Ahammed et al. 2012a, 2013a). The following is a critical evaluation of recent studies on the function of BRs (and related substances) in the regulation of both enzymatic and nonenzymatic elements of the antioxidant defense system in abiotically stressed plants.

16.3.1 Heavy Metal Stress

There are various threats to plant survival, but one of the most significant issues in agricultural research is heavy metal stress. Hormones like the BRs that scavenge reactive oxygen species (ROS) and activate antioxidant defense enzymes like glutathione reductase (GR), guaiacol peroxidase (GPOX), glutathione-S-transferase (GST), ascorbate peroxidase (APOX), peroxidase (POD), catalase (CAT), and superoxide dismutase (SOD) are the main factors that mitigate the effects of metal contamination on plants (Bücker-Neto et al. 2017). Exogenous BR treatment has been shown to increase the activity of antioxidant enzymes (CAT, POD, and SOD) (Figure 16.1) and antioxidases involved in the ascorbate-glutathione (AsA-GSH) cycle, hence improving Cd-tolerance in grape seedlings. Furthermore, grape seedlings treated with EBR under Cd stress showed increased expression levels of the genes VvSOD, VvPOD, VvAPX, VvGR, VvDHAR, and VvMDHAR (Li et al. 2022). As a result of 24-epiBL (5 μM)-mediated increases in antioxidant enzyme activity, proline concentration, and subsequent improvements in the membrane stability index (MSI) and relative leaf water content (RLWC), *Phaseolus vulgaris* may have improved Cd-tolerance (Rady 2011). In tomato cultivars K-25 and Sarvodya, the detrimental effect of Cd was found to be nullified as a result of improvements in photosynthesis machinery and antioxidant defense

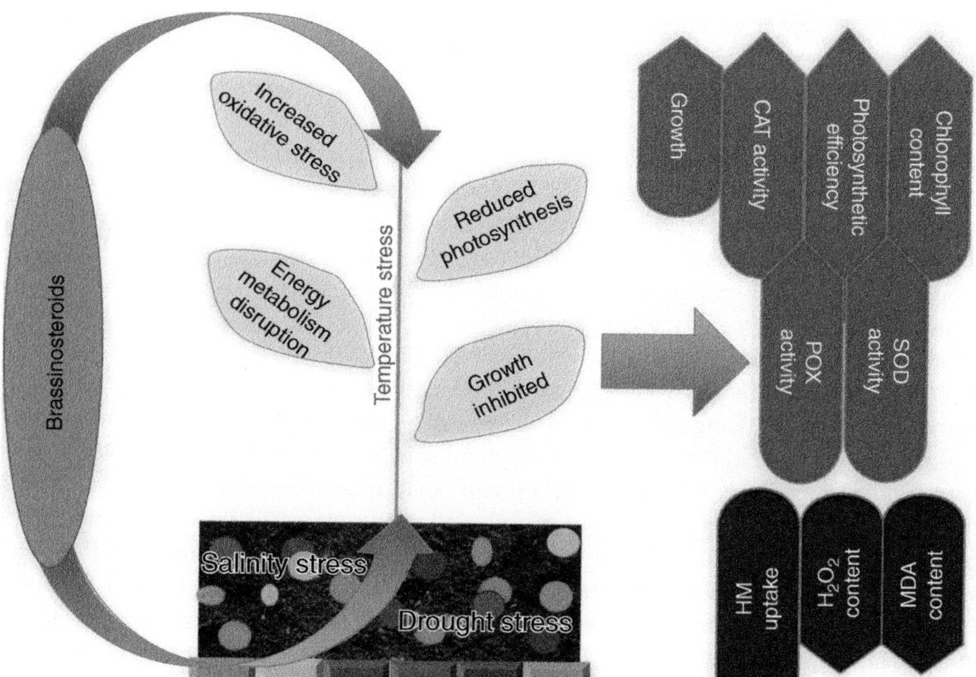

Figure 16.1 BRs mediated fluctuation in antioxidant activity.

systems mediated by 28-homoBL/24-epiBL (10^{-8} M) (Hasan et al. 2020). The application of BRs (10^{-8} M) to stressed Solanum lycopersicum plants treated with Cd (3.0, 6.0, 9.0, and 12 mg kg^{-1}) increased AOS activity and improved fruit yield and quality (Hayat et al. 2012). By primarily increasing the activity of antioxidant enzymes, the application of 24-epiBL alleviated Ni-stress in *Brassica juncea* (Kanwar et al. 2012). These scientists previously observed that via increasing the activity of ROS-metabolizing enzymes (and also by lowering the metal uptake), BRs (24-EpiBL, CS, dolicholide, and typhasterol) significantly mitigated Ni (0.2, 0.4, and 0.6 mM)-stress in *Brassica juncea* plants (Kanwar et al. 2012). EBL seed priming reduces chromium stress in rice cultivars by enhancing antioxidant defense response and ROS metabolism (Basit et al. 2021). It was claimed that five wheat (*Triticum aestivum*) cultivars (PBW-373, UP-2338, DL-LOK-01, DL-373, and HD-2338) were protected against Ni concentrations (50 and 100 μM) by increased CAT, POD, and SOD activity by the spray of 0.01 μM of 28-homoBL (Yusuf et al. 2011). Exogenous EBL treatment enhanced proline and total antioxidant levels in *A. thaliana* while reducing lipid peroxidation during Mn stress. SOD and CAT enzyme activities were found to increase upon exposure to EBL, as were the transcript levels of the genes for copper/zinc superoxide dismutase 1 (CSD1), copper/zinc superoxide dismutase 2 (CSD2), copper/zinc superoxide dismutase 3 (CSD3), Fe superoxide dismutase 2 (FSD2), Fe superoxide dismutase 3 (FSD3), and catalase 2 (CAT2) (Surgun-Acar and Zemheri-Navruz 2022).

16.3.2 Salinity Stress

Numerous reports have documented the modification of several antioxidant defense system components in plants exposed to salt by BRs and related chemicals been extensively reported (Lu and Yang 2013). By promoting the activity of antioxidant enzymes such as SOD, APOX, and CAT, foliar application of Br mitigated the detrimental effects of salt stress in tomatoes and lowered malondialdehyde (MDA) and hydrogen peroxide (H_2O_2) levels (Salehi et al. 2024). The ability of BL treatment to counteract the detrimental effects of salt stress was demonstrated by the rise in CAT, SOD, and GR activities and the decrease in POD and PPO activities of two varieties of sorghum plants ("CSH-5" and "CSH-6") grown in two saline experimental locations in Karikal (Varchikudy and Mallavur) (Vardhini 2011). When epiBL was applied to *Cucumis sativus* seedlings exposed to salinity, the results showed better seedling growth due to elevated SOD, POD, and CAT activities, as well as decreased leaf superoxide anion production rate, H_2O_2, MDA, and cell membrane permeability (Lu and Yang 2013). When exposed to 90 mM NaCl with 0, 0.025, 0.05, 0.10, and 0.20 mg dm^{-3} of epiBL for 10 days, eggplant seedlings showed decreased electrolyte permeability, superoxide production, MDA, and H_2O_2, most likely due to increased activities of SOD, GPX, CAT, and APX enzymes as well as the contents of nonenzymatic antioxidants like AsA and GSH (Ding et al. 2012). 24-epiBL reduced the negative effects of salt stress on two types of pepper (*Capsicum annuum*), possibly via raising the levels of proline, total anthocyanins, and minerals as well as the activities of antioxidant enzymes (Abbas et al. 2013).Treatment with 0.2 μM EBL as exogenous treatment mitigated the salt stress impacts in maize by inducing the activities of antioxidant enzymes like SOD, CAT, and APX (Seleiman et al. 2023). Supplementation of cucumber plants with EBR reduced the damage of salt stress by elevating the activities of antioxidative enzymes and maintaining ion homeostasis (He et al. 2022).

16.3.3 Temperature Stress

It has been widely shown that BRs and related chemicals alter several antioxidant defense system components and contribute to the reduction of the effects in various plants exposed to both high (Xi et al. 2013; Aghdam and Muhammadkhani 2014) temperatures. In tomato seedlings like BR-deficient mutant, extreme dwarf (x), BR-insensitive mutant curl3(-abs) allele (curl3 altered brassinolide sensitivity), and alien overexpressing the dwarf, BR-biosynthesis gene, 35SD) EpiBL-induced tolerance to heat shock (HS) which was claimed to be the outcome of decreased lipid peroxidation and ion leakage due to increased antioxidant enzyme activity (Mazorra et al. 2011). By enhancing the activities of antioxidant enzymes (SOD, CAT, and POD) and antioxidant substances (ascorbic acid and reduced glutathione), foliar spray of EBR mitigated heat stress in soybeans and helped to relieve oxidative damage and maintain ROS homeostasis (Wang et al. 2022). Many plant species produce more antioxidant enzymes when exogenous BR is applied. When BR was applied to heat-stressed rice seedlings, increased SOD and POD production was observed (Cao and Hua 2008). Similarly, when wheat was subjected to heat stress, BR treatment during the pre- and post-flowering stages elevated APX and SOD activities (Kumari and Hemantaranjan 2019). It was observed that exogenous application of EBR in Fenugreek mitigated high-temperature stress by stimulating the biosynthesis processes of enzymatic and nonenzymatic antioxidants (Sheikhi et al. 2023). Seed priming with BR improved the growth and physiological parameters by enhancing antioxidant defense under high-temperature stress in *Brassica juncea* (Neha et al. 2022).

The role of BRs in plants during low-temperature stress is very well documented in the literature. According to Hu et al. (2010), exogenous application of 24-epiBL reduced the reduction of photosynthesis in cucumbers (*Cucumis sativus*) caused by 12/8 °C freezing by lowering ROS formation and accumulation by raising SOD and APX activities. The authors of a different investigation on cucumbers exposed to chilling stress and pretreatment with 24-epiBL (0.3 and 1.0 mmol·L^{-1} chlorpyrifos) showed increases in the activities of APX, GR, CAT, and GPX, which ultimately reduced the chilling-accrued phytotoxicity (Hu et al. 2013). According to Aghdam et al. (2012), applying 3.0 and 6.0 μM BRs to tomato fruits that were kept at 1 °C for 21 days decreased chilling injury, electrolyte leakage, and MDA content; it also increased proline, total phenol content, phenylalanine ammonia-lyase (PAL) activity, and preserved membrane integrity. In their other study, applying 0, 3, and 6 μmol BL to tomato fruits exposed to 1 °C chilling stress can reduce the activities of lipoxygenase (LOX) and phospholipase D (PLD), two key factors that cause chilling injury in tomato fruits (Aghdam and Muhammadkhani 2014). By activating the Calvin cycle enzymes and boosting antioxidant capacity, BRs protected the photosynthetic machinery from cold-induced damage in *Cucumis sativus* plants. This, in turn, reduced photooxidative stress and inhibited plant development during the recovery from chilling injury (Jiang et al. 2013).

16.3.4 Drought Stress

Foliar application of EBL minimized the oxidative stress due to drought by reducing electrolyte leakage, malondialdehyde, and hydrogen peroxide contents whereas it enhanced the activities of antioxidant enzymes, such as CAT, SOD, and POD under drought stress in wheat (Khan et al. 2021). In another study, it was observed that pretreatment of EBR on wheat seedlings reduced the drought stress-induced accumulation of ROS and modulated the activity of antioxidant enzymes (Avalbaev et al. 2024). 24-epibrassinolide, 28-homobrassinolide, and 28-epihombrassinolide ameliorated the negative effects of drought stress in maize seedlings by inducing the increase of antioxidant enzyme activity and proline content and decrease MDA content (Figure 16.1); overexpression of drought tolerance gene *ZmMYB48* was induced (Sun et al. 2022).

16.4 Conclusion

Numerous discussions throughout the world have effectively demonstrated that environmental stressors are the main causes of significant crop output losses worldwide. Plants use a three-step procedure to function as a dependable means of overcoming stress and guaranteeing survival: (i) cellular homeostasis maintenance, (ii) detoxification, and (iii) growth process recovery. It has been observed that the phytohormone BR and several other similar chemicals are excellent, counteracting antidotes to a variety of abiotic stimuli, including heat, cold, drought, salt, heavy metals, etc. BR-induced ROS scavenging, preserving redox equilibrium, enhanced carbon assimilation, photoprotection, and enhancing antioxidant potential (enzymatic and nonenzymatic ROS scavengers) are among the advantageous actions that BRs bring about in reducing the consequences of stress. Other advantageous processes that are initiated by BRs during stress include enhanced autophagy, detoxifying potential, and defense response. Numerous groups' research on this subject has yielded a wealth of knowledge, but the list still has a lot of unchecked items. An increasing number of studies will shed light on the molecular-genetic functions of BRs and other similar substances, which alter the antioxidant defense system and ultimately regulate the harmful stress reactions in plants.

References

Abbas, S., Latif, H. H., and Elsherbiny, E. A. (2013). Effect of 24-epibrassinolide on the physiological and genetic changes on two varieties of pepper under salt stress conditions. *Pakistan Journal of Botany* 45: 1273–1284.

Ackerman-Lavert, M. and Savaldi-Goldstein, S. (2020). Growth models from a brassinosteroid perspective. *Current Opinion in Plant Biology* 53: 90–97.

Aghdam, M. S., Asghari, M., Farmani, B. et al. (2012). Impact of postharvest brassinosteroids treatment on PAL activity in tomato fruit in response to chilling stress. *Scientia Horticulturae* 144: 116–120.

Aghdam, M. S., and Mohammadkhani, N. (2014). Enhancement of chilling stress tolerance of tomato fruit by postharvest brassinolide treatment. *Food and Bioprocess Technology* 7: 909–914.

Ahammed, G.J., Choudhary, S.P., Chen, S. et al. (2013). Role of brassinosteroids in alleviation of phenanthrene–cadmium co-contamination-induced photosynthetic inhibition and oxidative stress in tomato. *Journal of Experimental Botany* 64 (1): 199–213.

Ahammed, G.J., Li, X., Liu, A. et al. (2020). Brassinosteroids in plant tolerance to abiotic stress. *Journal of Plant Growth Regulation* 39 (4): 1451–1464.

Ahmad Lone, W., Majeed, N., Yaqoob, U. et al. (2022). Exogenous brassinosteroid and jasmonic acid improve drought tolerance in Brassica rapa L. genotypes by modulating osmolytes, antioxidants and photosynthetic system. *Plant Cell Reports* 41 (3): 603–617.

Anwar, A., Liu, Y., Dong, R. et al. (2018). The physiological and molecular mechanism of brassinosteroid in response to stress: a review. *Biological Research* 51 (1): 46.

Arteca, J.M. and Arteca, R.N. (2001). Brassinosteroid-induced exaggerated growth in hydroponically grown *Arabidopsis* plants. *Physiologia Plantarum* 112 (1): 104–112.

Avalbaev, A., Fedyaev, V., Lubyanova, A. et al. (2024). 24- Epibrassinolide reduces drought-induced oxidative stress by modulating the antioxidant system and respiration in wheat seedlings. *Plants* 13 (2): 148.

Bajguz, A. (2011). Brassinosteroids–occurence and chemical structures in plants. In: *Brassinosteroids: A Class of Plant Hormone*, 1–27.

Bajguz, A., Chmur, M., and Gruszka, D. (2020). Comprehensive overview of the brassinosteroid biosynthesis pathways: substrates, products, inhibitors, and connections. *Frontiers in Plant Science* 11: 1034.

Basit, F., Bhat, J.A., Dong, Z. et al. (2022a). Chromium toxicity induced oxidative damage in two rice cultivars and its mitigation through external supplementation of brassinosteroids and spermine. *Chemosphere* 302: 134423.

Basit, F., Chen, M., Ahmed, T. et al. (2021). Seed priming with brassinosteroids alleviates chromium stress in rice cultivars via improving ROS metabolism and antioxidant defense response at biochemical and molecular levels. *Antioxidants* 10 (7): 1089.

Basit, F., Liu, J., An, J. et al. (2022b). Seed priming with brassinosteroids alleviates aluminum toxicity in rice via improving antioxidant defense system and suppressing aluminum uptake. *Environmental Science and Pollution Research* 29 (7): 10183–10197.

Bücker-Neto, L., Paiva, A.L.S., Machado, R.D. et al. (2017). Interactions between plant hormones and heavy metals responses. *Genetics and Molecular Biology* 40 (1 suppl 1): 373–386.

Cao, Y.Y. and Hua, Z.H.A.O. (2008). Protective roles of brassinolide on rice seedlings under high-temperature stress. *Rice Science*15 (1): 63–68.

Chen, J., Nolan, T.M., Ye, H. et al. (2017). *Arabidopsis* WRKY46, WRKY54, and WRKY70 transcription factors are involved in brassinosteroid-regulated plant growth and drought responses. *The Plant Cell* 29 (6): 1425–1439.

de Oliveira, D.B., Amaro, C.L., Amaro, L.V. et al. (2021). Importance of brassinosteroids in activating the antioxidant system and growth in '*Jatropha curcas*' L. plants under water deficit. *Australian Journal of Crop Science*15 (3): 369–376.

Dhriti Kapoor, D.K., Amandeep Rattan, A.R., Vandana Gautam, V.G. et al. (2014). 24-Epibrassinolide mediated changes in photosynthetic pigments and antioxidative defence system of radish seedlings under cadmium and mercury stress. *Journal of Stress Physiology & Biochemistry* 10 (3): 110–121.

Ding, H.D., Zhu, X.H., Zhu, Z.W. et al. (2012). Amelioration of salt-induced oxidative stress in eggplant by application of 24-epibrassinolide. *Biologia Plantarum* 56: 767–770.

Domagalska, M.A., Sarnowska, E., Nagy, F. et al. (2010). Genetic analyses of interactions among gibberellin, abscisic acid, and brassinosteroids in the control of flowering time in *Arabidopsis thaliana*. *PloS One* 5 (11): e14012.

Fujioka, S. and Yokota, T. (2003). Biosynthesis and metabolism of brassinosteroids. *Annual Review of Plant Biology* 54 (1): 137–164.

Ghoname, A.A., AbdelMotlb, N.A., Abdel-Al, F.S. et al. (2023). Brassinosteroids or proline can alleviate yield inhibition under salt stress via modulating physio-biochemical activities and antioxidant systems in snap bean. *The Journal of Horticultural Science and Biotechnology* 98 (4): 526–539.

González-García, M.P., Vilarrasa-Blasi, J., Zhiponova, M. et al. (2011). Brassinosteroids control meristem size by promoting cell cycle progression in *Arabidopsis* roots. *Development* 138 (5): 849–859.

Gudesblat, G.E., Schneider-Pizoń, J., Betti, C. et al. (2012). SPEECHLESS integrates brassinosteroid and stomata signalling pathways. *Nature Cell Biology* 14 (5): 548–554.

Guedes, F.R.C.M., Maia, C.F., da Silva, B.R.S. et al. (2021). Exogenous 24-Epibrassinolide stimulates root protection, and leaf antioxidant enzymes in lead stressed rice plants: central roles to minimize Pb content and oxidative stress. *Environmental Pollution* 280: 116992.

Gupta, P. and Seth, C.S. (2023). 24-epibrassinolide regulates functional components of nitric oxide signalling and antioxidant defense pathways to alleviate salinity stress in *Brassica juncea* L. cv. Varuna. *Journal of Plant Growth Regulation* 42 (7): 4207–4222.

Hartwig, T., Chuck, G.S., Fujioka, S. et al. (2011). Brassinosteroid control of sex determination in maize. *Proceedings of the National Academy of Sciences* 108 (49): 19814–19819.

Hasan, M.M., Ali, M.A., Soliman, M.H. et al. (2020). Insights into 28-homobrassinolide (HBR)-mediated redox homeostasis, AsA–GSH cycle, and methylglyoxal detoxification in soybean under drought-induced oxidative stress. *Journal of Plant Interactions* 15 (1): 371–385.

Hayat, S., Alyemeni, M., and Hasan, S. (2012). Foliar spray of brassinosteroid enhances yield and quality of *Solanum lycopersicum* under cadmium stress. *Saudi Journal of Biological Sciences*. 19, 325–335.

He, X., Wan, Z., Jin, N. et al. (2022). Enhancement of cucumber resistance under salt stress by 2, 4-epibrassinolide lactones. *Frontiers in Plant Science* 13: 1023178.

Hong, Z., Ueguchi-Tanaka, M., Fujioka, S. et al. (2005). The rice brassinosteroid-deficient dwarf2 mutant, defective in the rice homolog of *Arabidopsis* DIMINUTO/DWARF1, is rescued by the endogenously accumulated alternative bioactive brassinosteroid, dolichosterone. *The Plant Cell* 17 (8): 2243–2254.

Hussain, M., Khan, T.A., Yusuf, M. et al. (2019). Silicon-mediated role of 24- epibrassinolide in wheat under high-temperature stress. *Environmental Science and Pollution Research* 26: 17163–17172.

Jaiswal, S., Båga, M. and Chibbar, R.N., (2020). Brassinosteroid receptor mutation influences starch granule size distribution in barley grains. *Plant Physiology and Biochemistry* 154: 369–378.

Janeczko, A., Oklestkova, J., Novak, O. et al. (2015). Disturbances in production of progesterone and their implications in plant studies. *Steroids* 96: 153–163.

Jia, Z., Giehl, R.F., Meyer, R.C. et al. (2019). Natural variation of BSK3 tunes brassinosteroid signaling to regulate root foraging under low nitrogen. *Nature Communications* 10 (1): 2378.

Jiang, W.B., Huang, H.Y., Hu, Y.W. et al. (2013). Brassinosteroid regulates seed size and shape in *Arabidopsis*. *Plant Physiology* 162 (4): 1965–1977.

Kanwar, M.K., Bhardwaj, R., Arora, P. et al. (2012). Plant steroid hormones produced under Ni stress are involved in the regulation of metal uptake and oxidative stress in *Brassica juncea* L. *Chemosphere* 86 (1): 41–49.

Khan, I., Awan, S.A., Ikram, R. et al. (2021). Effects of 24-epibrassinolide on plant growth, antioxidants defense system, and endogenous hormones in two wheat varieties under drought stress. *Physiologia Plantarum* 172 (2): 696–706.

Khan, M., Rozhon, W., Bigeard, J. et al. (2013). Brassinosteroid-regulated GSK3/Shaggy-like kinases phosphorylate mitogen-activated protein (MAP) kinase kinases, which control stomata development in *Arabidopsis thaliana*. *Journal of Biological Chemistry* 288 (11): 7519–7527.

Kumar, N., Sharma, V., Kaur, G. et al. (2023). Brassinosteroids as promoters of seedling growth and antioxidant activity under heavy metal zinc stress in mung bean (*Vigna radiata L.*). *Frontiers in Microbiology* 14: 1259103.

Kumari, A. and Hemantaranjan, A. (2019). Mitigating effects of 24-epibrassinolide on heat stress damage by shifting biochemical and antioxidant defense mechanisms in wheat (*Triticum aestivum L.*) at pre-flowering stage and post-flowering stage. *Journal of Pharmacognosy and Phytochemistry* 8 (1): 1157–1161.

Lai, H., Li, X., Chen, Y. et al. (2024). Mitigating heat-induced yield loss in peanut: insights into 24-epibrassinolide-mediated improvement in antioxidant capacity, photosynthesis, and kernel weight. *Field Crops Research* 316: 109521.

Lee, J., Han, S., Lee, H.Y. et al. (2019). Brassinosteroids facilitate xylem differentiation and wood formation in tomato. *Planta* 249: 1391–1403.

Li, B.B., Fu, Y.S., Li, X.X. et al. (2022). Brassinosteroids alleviate cadmium phytotoxicity by minimizing oxidative stress in grape seedlings: Toward regulating the ascorbate-glutathione cycle. *Scientia Horticulturae* 299: 111002.

Li, Z. and He, Y. (2020). Roles of brassinosteroids in plant reproduction. *International Journal of Molecular Sciences* 21 (3): 872.

Lin W., Zhang F., Zhang L. et al. (2017). Brassinosteroids promote seed development and physiological maturity of oilseed rape (*Brassica napus L.*). *Oil Crop Science* 1 (2): 122–130.

Lin, W.H. (2020). Designed manipulation of the brassinosteroid signal to enhance crop yield. *Frontiers in Plant Science* 11: 854.

Luo, M., Xiao, Y., Li, X. et al. (2007). GhDET2, a steroid 5α-reductase, plays an important role in cotton fiber cell initiation and elongation. *The Plant Journal* 51 (3): 419–430.

Lv, B., Tian, H., Zhang, F. et al. (2018). Brassinosteroids regulate root growth by controlling reactive oxygen species homeostasis and dual effect on ethylene synthesis in *Arabidopsis*. *PLoS Genetics* 14 (1): e1007144.

Makarevitch, I., Thompson, A., Muehlbauer, G.J. et al. (2012). Brd1 gene in maize encodes a brassinosteroid C-6 oxidase. *PloS One* 7 (1): e30798.

Mazorra, L. M., Holton, N., Bishop, G. J et al. (2011). Heat shock response in tomato brassinosteroid mutants indicates that thermotolerance is independent of brassinosteroid homeostasis. *Plant Physiology and Biochemistry* 49: 1420–1428.

Muhammad, H.M.D., Abbas, A. and Ahmad, R. (2022). Fascinating role of silicon nanoparticles to mitigate adverse effects of salinity in fruit trees: a mechanistic approach. *Silicon* 14: 8319–8326.

Nakaya, M., Tsukaya, H., Murakami, N. et al. (2002). Brassinosteroids control the proliferation of leaf cells of *Arabidopsis thaliana*. *Plant and Cell Physiology* 43 (2): 239–244.

Nazir, F., Fariduddin, Q., Hussain, A. et al. (2021). Brassinosteroid and hydrogen peroxide improve photosynthetic machinery, stomatal movement, root morphology and cell viability and reduce Cu-triggered oxidative burst in tomato. *Ecotoxicology and Environmental Safety* 207: 111081.

Nazir, F., Jahan, B., Kumari, S. et al. (2023). Brassinosteroid modulates ethylene synthesis and antioxidant metabolism to protect rice (Oryza sativa) against heat stress-induced inhibition of source–sink capacity and photosynthetic and growth attributes. *Journal of Plant Physiology* 289: 154096.

Twinkle, N., Mohapatra, S., Sirhindi, G. et al. (2022). Seed priming with brassinolides improves growth and reinforces antioxidative defenses under normal and heat stress conditions in seedlings of *Brassica juncea*. *Physiologia Plantarum* 174 (6): e13814.

Oh, E., Zhu, J.Y., Bai, M.Y. et al. (2014). Cell elongation is regulated through a central circuit of interacting transcription factors in the *Arabidopsis* hypocotyl. *Elife* 3: e03031.

Oh, M.H. (2003). Brassinosteroids accelerate the rate of cell division in isolated petal protoplasts of *Petunia* hybrida. *Journal of Plant Biotechnology* 5 (1): 63–67.

Peres, A.L.G., Soares, J.S., Tavares, R.G. et al. (2019). Brassinosteroids, the sixth class of phytohormones: a molecular view from the discovery to hormonal interactions in plant development and stress adaptation. *International Journal of Molecular Sciences* 20 (2): 331.

Planas-Riverola, A., Gupta, A., Betegón-Putze, I. et al. (2019). Brassinosteroid signaling in plant development and adaptation to stress. *Development* 146 (5): dev151894.

Rattan, A., Kapoor, D., Kapoor, N. et al. (2020). Brassinosteroids regulate functional components of antioxidative defense system in salt stressed maize seedlings. *Journal of Plant Growth Regulation* 39: 1465–1475.

Rhaman, M.S., Imran, S., Rauf, F. et al. (2020). Seed priming with phytohormones: an effective approach for the mitigation of abiotic stress. *Plants* 10 (1): 37.

Rozhon, W., Akter, S., Fernandez, A. et al. (2019). Inhibitors of brassinosteroid biosynthesis and signal transduction. *Molecules* 24 (23): 4372.

Salehi, H., Miras-Moreno, B., Tolisano, C. et al. (2024). Brassinosteroids alleviate salinity stress in tomato through the modulation of photosynthetic efficiency, antioxidant machinery, and metabolomic profile. *Antioxidant Machinery, and Metabolomic Profile*. http://doi.org/10.2139/ssrn.4752002.

Seleiman, M.F., Ahmad, A., Tola, E. et al. (2023). Exogenous application of 24-epibrassinolide confers saline stress and improves photosynthetic capacity, antioxidant defense, mineral uptake, and yield in maize. *Plants* 12 (20): 3559.

Shahzad, B., Tanveer, M., Che, Z. et al. (2018). Role of 24-epibrassinolide (EBL) in mediating heavy metal and pesticide-induced oxidative stress in plants: a review. *Ecotoxicology and Environmental Safety* 147: 935–944.

Sheikhi, S., Ebrahimi, A., Heidari, P. et al. (2023). Exogenous 24-epibrassinolide ameliorates tolerance to high temperature by adjusting the biosynthesis of pigments, enzymatic, non-enzymatic antioxidants, and diosgenin content in fenugreek. *Scientific Reports* 13 (1): 6661.

Sun, S., Yao, X., Liu, X. et al. (2022). Brassinolide can improve drought tolerance of maize seedlings under drought stress: by inducing the photosynthetic performance, antioxidant capacity and ZmMYB gene expression of maize seedlings. *Journal of Soil Science and Plant Nutrition* 22 (2): 2092–2104.

Surgun-Acar, Y. and Zemheri-Navruz, F. (2022). Exogenous application of 24-epibrassinolide improves manganese tolerance in *Arabidopsis thaliana L.* via the modulation of antioxidant system. *Journal of Plant Growth Regulation* 41 (2): 546–557.

Symons, G.M., Davies, C., Shavrukov, Y. et al. (2006). Grapes on steroids. Brassinosteroids are involved in grape berry ripening. *Plant Physiology* 140 (1): 150–158.

Tanabe, S., Ashikari, M., Fujioka, S. et al. (2005). A novel cytochrome P450 is implicated in brassinosteroid biosynthesis via the characterization of a rice dwarf mutant, dwarf11, with reduced seed length. *The Plant Cell* 17 (3): 776–790.

Trevisan, S., Forestan, C., Brojanigo, S. et al. (2020). Brassinosteroid application affects the growth and gravitropic response of maize by regulating gene expression in the roots, shoots and leaves. *Plant Growth Regulation* 92 (1): 117–130.

Vardhini, B. V. (2011). Studies on the effect of brassinolide on the antioxidative system of two varieties of sorghum grown in saline soils of Karaikal. *Asian and Australasian Journal of Plant Science and Biotechnology* 5: 31–34.

Vogler, F., Schmalzl, C., Englhart, M. et al. (2014). Brassinosteroids promote *Arabidopsis* pollen germination and growth. *Plant Reproduction* 27: 153–167.

Wang, Q., Yu, F. and Xie, Q. (2020). Balancing growth and adaptation to stress: Crosstalk between brassinosteroid and abscisic acid signaling. *Plant, Cell & Environment* 43 (10): 2325–2335.

Wang, W., Xie, Y., Liu, C. et al. (2022). The exogenous application of brassinosteroids confers tolerance to heat stress by increasing antioxidant capacity in soybeans. *Agriculture* 12 (8): 1095.

Wang, X., Chen, J., Xie, Z. et al. (2014). Histone lysine methyltransferase SDG8 is involved in brassinosteroid-regulated gene expression in *Arabidopsis thaliana*. *Molecular Plant* 7 (8): 1303–1315.

Wang, Y., Zhang, R., Wang, X. et al. (2025). 24-epibrassinolide improves potato (Solanum tuberosum L.) tolerance to alkaline salt stress by regulating antioxidant defence and photosynthetic properties. *Journal of Agronomy and Crop Science* 211 (1): e70022.

Xu, F., Xi, Z.M., Zhang, H. et al. (2015). Brassinosteroids are involved in controlling sugar unloading in Vitis vinifera 'Cabernet Sauvignon' berries during véraison. *Plant Physiology and Biochemistry* 94: 197–208.

Ye, Q., Zhu, W., Li, L., et al. (2010). Brassinosteroids control male fertility by regulating the expression of key genes involved in *Arabidopsis* anther and pollen development. *Proceedings of the National Academy of Sciences* 107 (13): 6100–6105.

Yin, Y., Wang, Z.Y., Mora-Garcia, S. et al. (2002). BES1 accumulates in the nucleus in response to brassinosteroids to regulate gene expression and promote stem elongation. *Cell* 109 (2): 181–191.

Yusuf, M., Saeed, T., Almenhali, H.A. et al. (2024). Melatonin improved efficiency of 24-epibrassinolide to counter the collective stress of drought and salt through osmoprotectant and antioxidant system in pea plants. *Scientia Horticulturae* 323: 112453.

Zadvornova, Y.V., Alekseichuk, G.N., Laman, N.A. et al. (2005). Effect of brassinosteroids on activation of the cell cycle during germination of *Brassica oleracea L.* seeds. *Doklady Natsional'noi Akademii Nauk Belarusi* 49: 70–73.

Zhang, S., Hu, J., Zhang, Y. et al. (2007). Seed priming with brassinolide improves lucerne (*Medicago sativa* L.) seed germination and seedling growth in relation to physiological changes under salinity stress. *Australian Journal of Agricultural Research* 58 (8): 811–815.

Zhang, W., Zhu, K., Wang, Z. et al. (2019). Brassinosteroids function in spikelet differentiation and degeneration in rice. *Journal of Integrative Plant Biology* 61 (8): 943–963.

Zhao, B. and Li, J.(2012). Regulation of brassinosteroid biosynthesis and inactivation F. *Journal of Integrative Plant Biology* 54 (10): 746–759.

Zhong, W., Xie, C., Hu, D. et al. (2020). Effect of 24-epibrassinolide on reactive oxygen species and antioxidative defense systems in tall fescue plants under lead stress. *Ecotoxicology and Environmental Safety* 187: 109831.

Zhu, J.Y., Sae-Seaw, J. and Wang, Z.Y. (2013). Brassinosteroid signalling. *Development* 140 (8): 1615–1620.

Zhu, X., Liang, W., Cui, X. et al. (2015). Brassinosteroids promote development of rice pollen grains and seeds by triggering expression of Carbon Starved Anther, a MYB domain protein. *The Plant Journal* 82 (4): 570–581.

Zhu, Z., Zhang, Z., Qin, G. et al. (2010). Effects of brassinosteroids on postharvest disease and senescence of jujube fruit in storage. *Postharvest Biology and Technology* 56 (1): 50–55.

17

Implications of Brassinosteroids Application: A Review

Saurabh Awasthi[1]*, Abhay Chand*[2]*, Om Prakash Narayan*[3*]*, and Sahil Mehta*[4*]

[1]*Department of Botany, University of Delhi, New Delhi, India*
[2]*Department of Life Sciences, Hansraj College, University of Delhi, New Delhi, India*
[3]*Division of Chemical Biology and Medicinal Chemistry, University of Texas at Austin, Austin, TX, USA*
[4]*Department of Botany, Hansraj College, University of Delhi, New Delhi, India*

[*]Corresponding authors: sahilmehtasm21@gmail.com, om.tufts@gmail.com

17.1 Introduction

Ranging from germination to flowering and seed settings, all the processes in the life cycle of a plant have a series of highly regulated events (Yuan et al. 2024; Wu et al. 2025; Pei et al. 2025). Phytohormones or plant growth regulators (PGRs) are crucial compounds for such a controlled pattern of growth and development (Pal et al. 2023). These compounds, along with other factors like edaphic and climatic conditions in the ambience of the plant, bring about miraculous changes in the plant's morphological, anatomical changes (Hu et al. 2016), physiological responses (Weyers and Paterson 2001), and even genetic traits (Ristova et al. 2018). In contemporary times when to ensure global food security and to keep up with the onset of a vegan lifestyle, agriculture suffers great pressure to meet the demands of the world. Climate change, change in soil conditions due to environmental pollution (Mirón et al. 2023; Brinkman et al. 1996), and constantly changing cropping patterns (Liu et al. 2015) due to shrinkage in the cultivable area add on to the existing problems associated with agriculture (Dar and Laxmipathi Gowda 2013; Ali et al. 2024).

Phytohormones have emerged as, none the less than, the knight in shining armor and may just be the much-needed rescue for the agriculture industry (Nadeem et al. 2016; Liao et al. 2025; Hassan et al. 2025). These compounds are being heavily exploited for better crop production yield, sustainable pest tolerant strategy. Of the available plant hormones, brassinosteroids (BRs) are considered to be a class of newly discovered hormones.

Brassinosteroids and Abiotic Stress: Recent Developments on Physiological and Molecular Aspects,
First Edition. Edited by Anket Sharma, et al.
Published 2026 by John Wiley & Sons Ltd.

17.2 Insights into BRs

This novel class of phytohormomes were reported by Mitchell et al. in 1970 in the pollen of *Brassica napus* L. (Rape seed) and were thus termed as "Brassins." Using chromatographic techniques with silica gel, these organic compounds were purified, and upon spectrophotometric analysis, it was concluded that these compounds have a glyceride structure. When applied to the plant internode, they exhibited significant growth, suggesting that these newly discovered compounds were acting as growth elicitors in plants (Li et al. 2024a; Jain et al. 2024).

17.2.1 Occurrence

All the members of the plant kingdom, from algae to land plants, have shown the presence of BRs in them (Chakraborty et al. 2025). In total, 69 BRs have been identified from gymnosperms, angiosperms (including both monocotyledon and dicotyledonous species), *Equisetum arvense* (a vascular cryptogam), and algae such as *Chlorella vulgaris, Hydrodictyon reticulatum*, etc. (Liu and Yang 2025). Certain lower land plants like bryophytes, ferns, lycophytes, and moss have also exhibited the occurrence of BRs in their tissues (Fujioka and Sakurai 1997). In a particular plant, BRs have been found to be occurring ubiquitously in all the parts of the plant, but particularly high concentration of these compounds is reported from seeds, pollen, and fruits. Of all the so far discovered BRs Castasterone (CS), Typhasterol (TY), Brassinolide (BL), 6-deoxocastasterone (6-deoxoCS), 28-norcastasterone (28-norCS), and Teasterone (TE) are the most commonly detected BRs in various plant species although the most commonly used BR is 24-epibrassinolide (EBR), for its commercial availability (Bajguz and Tretyn 2003; Chen et al. 2025).

17.2.2 Biosynthesis of BRs

The synthesis of BRs in the plant cellular machinery begins with campesterol (CR) as their precursor compound (Zhang et al. 2024a; Chakraborty et al. 2025). The pathway is divided into two routes, namely the campestanol (CN)-dependent route and the CN-independent route. CN-dependent route includes the early C-6 oxidation pathway, wherein CN is oxidized to 6-oxocampestanol (6-oxoCN), which subsequently gets converted to CS, and the late C-6 oxidation pathway which gives rise to 6-deoxocastasterone (6-deoxoCS), that is ultimately converted to CS (Hu et al. 2025). Based on the number of carbon units inherited by BRs, they can be distinguished into three types: C_{27}, C_{28}, and C_{29}, all having their unique biosynthetic pathway. The basic structure of these BRs is composed of various sterane-derived alkane skeletons: 5α-cholestane for C_{27} BRs, 5α-ergostane for C_{28} BRs, and 5α-stigmastane for C_{29} BRs. The primary steps involved in the synthesis of BRs are persistent and may proceed either via MVA (Mevalonate) or non-MVA (MEP/ DOXP) pathway. The consecutive steps split up into *cycloartenol-* and *cycloartenol*-dependent pathways for BR synthesis. Though the entire pathways are yet to be understood entirely, the biosynthesis comprises two major stepstones: campesterol formation and hydroxylation at C_{22}, resulting in the formation of 22α-hydroxycampesterol (Figure 17.1).

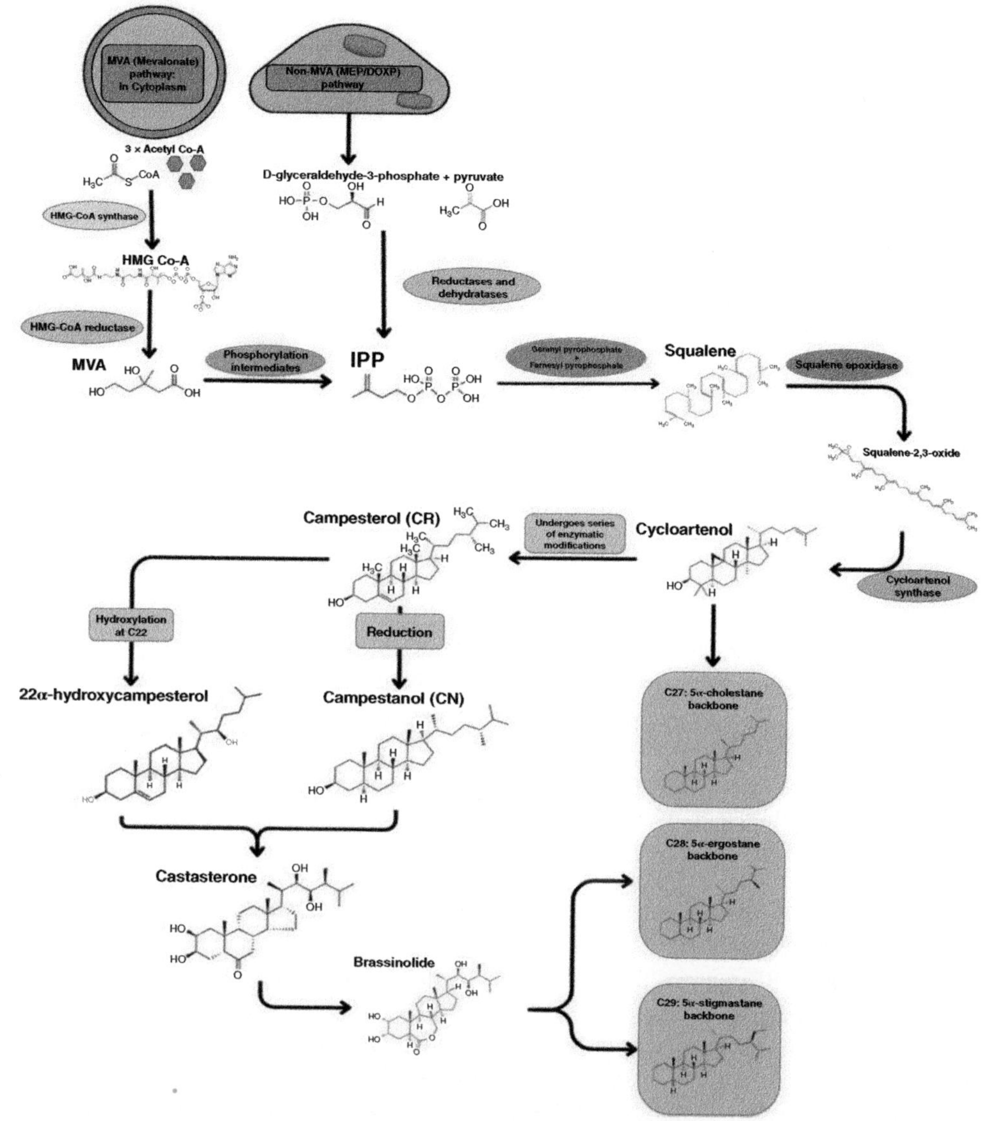

Figure 17.1 Diagrammatic representation of synthesis of BRs.

17.2.3 Initial Steps of BR Biosynthesis

CR synthesis pathway results in the formation of an intermediate blend, known as the Isopentyl pyrophosphate (IPP). The MVA pathway in higher plants and the non-MVA pathway in nether plants are collaterally aligned, which causes the metabolization of CR initially into campestanol (CN) and further into catasterone (CS) (Warang et al. 2025; Li et al. 2024b). In the non-MVA pathway, two biogenic compounds are initially formed: D-glyceraldehyde-3-phosphate and pyruvate. Through the coordinated action of various enzymes, like reductases and dehydratases, these are converted into IPP. During the MVA pathway, an intermediary compound, called the 3-hydroxy-3-methylglutaryl-CoA, is formed due to the coalescing of 3 acetyl Co-A molecules by the action of HMG-CoA synthase. This formed intermediate is then reduced to MVA by the enzyme HMG-CoA reductase. MVA is then utilized for IPP synthesis via phosphorylation intermediates. Biochemical modification of IPP via geranyl pyrophosphate and farnesyl pyrophosphate results in squalene synthesis, which is abraded into squalene-2,3-oxide via squalene epoxidase. Squalene-2,3-oxide, by the action of cycloartenol synthase, is converted to cycloartenol. It is a crucial substrate for multistep reactions in a few pathways of BR synthesis, eventually resulting in the synthesis of C_{27}, C_{28}, and C_{29} BRs (Sedaghathoor et al. 2024).

17.2.4 BR Signaling

Regulation of gene expression by BRs takes place through a complex signaling pathway involving receptor kinases, transcription factors, and phosphorylation/dephosphorylation mechanisms (Han et al. 2022; Ackerman-Lavert et al. 2020). BZR1 (Brassinazole-Resistant 1) and BES1 (BRI1-EMS-Suppressor 1) are the crucial hubs of crosstalk with various transcription factors and histone-modifying enzymes that primarily mediate the regulation of modified gene expression by BRs. This regulation can occur in the absence or presence of BRs.

In the unavailability of BRs, the plasma membrane (PM)-localized receptors BRI1(Brassinosteroid-Insensitive1) and BAK1 (BRI1-Associated Kinase 1) remain inactive, preventing BR signaling. Phosphorylation of BES1 and BZR1 by BIN2 (Brassinosteroid-Insensitive2) makes their form inactive, thus preventing BR-induced gene expression by blocking their entry into the nucleus from the cytoplasm (Figure 17.2). Also, BSS1 binds together BES1 and BZR1, reducing their availability in the nucleus, hence causing suppression of BR-mediated genes. All these events cause repression of BR-inducible genes and activation of BR-repressible genes, leading to a decline in BR-related growth and development responses.

When BRs or its derivatives such as BL are available, they bind to BRI1 which causes conformational changes that promote heterodimerization with four members of Somatic Embryogenesis Receptor Kinases (SERKs) family, resulting in formation of BRI1-SERK, a BR-receptor complex, that initiates signal transduction, that is facilitated by a cytoplasmic protein-mediated phosphorylation/ dephosphorylation telecast. Here, the transduction between the two takes place via BR-signaling Kinases (BSKs). Moreover, homologous proteins: constitutive differential growth1 (CDG1) and CDG-like1 (CDL1) have been recognized as key substrates for BRI1 Kinase. The activation of (BSU1) is facilitated by BSKs, while CDG kinases are involved in its activation. BSU1 is of utmost importance as it inhibits the activity of BIN2, thus functioning as a positive regulator of BR signaling. All these steps together cause activation of BR-inducible genes while inhibiting BR-repressible genes, ensuring efficient proceeding of BR-BR-mediated development processes (Kaushik et al. 2025).

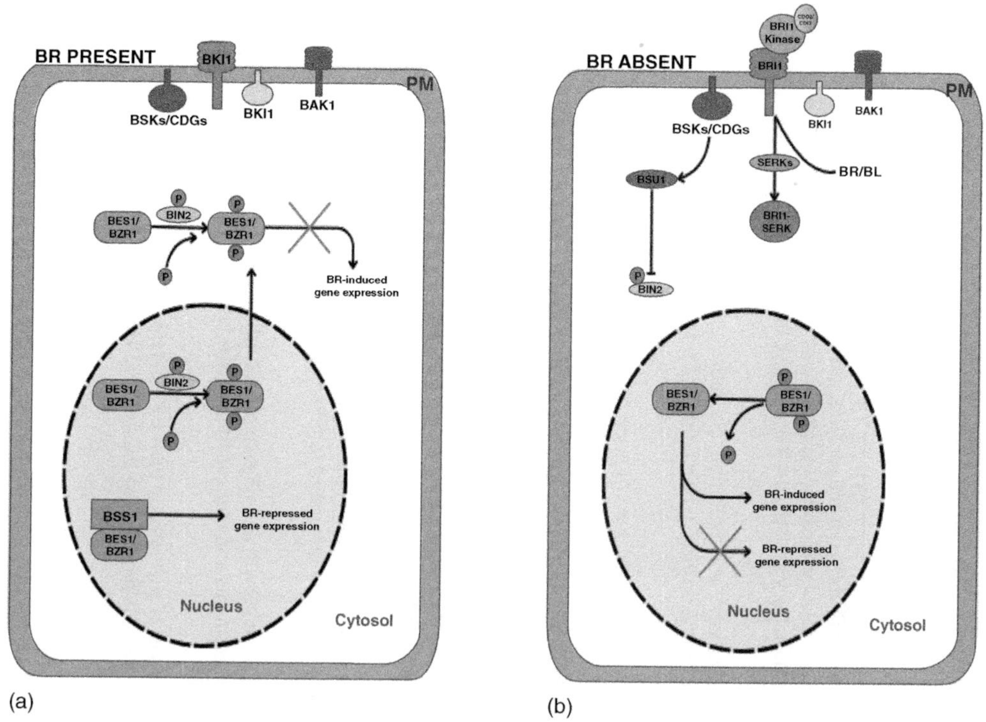

Figure 17.2 Signaling events taking place in cell (a) in the presence of BRs, (b) in the absence of BRs.

17.3 Physiological Functions Performed by BRs

Through signal transduction, transcriptional regulation, and crosstalk with various other phytohormones, signaling molecules BRs regulate plant growth and development both under normal as well as stressed conditions (Miao et al. 2024; Nolan et al. 2020). They bring about the vast majority of normal growth functions like seed germination, root and shoot development, stomata development, vascular differentiation, hypocotyl elongation, skotomorphogenesis (Steber and McCourt 2001; Hussain et al. 2020), etc. BRs have been found crucial for reproductive processes like floral transition in plants. In addition to the normal physiological roles, BRs are also involved in assisting the plant in coping with a plethora of biotic and abiotic stresses. In the subsequent sections, we discuss the various physiological effects that are regulated by BRs. The effects of exogenous application of BRs on various plant species abiotic stresses upon induction is summarized in Table 17.1.

17.3.1 BRs Promote Seed Germination and Elongation

The presence of BRs in the seeds of various plant species has been well demonstrated. The exogenous application of these hormones to the seeds of crops such as rice (Xiong et al. 2022) and barley accelerates their germination process in a fashion similar to Gibberellic

Table 17.1 Effects of exogenous application of BRs on various plant species abiotic stresses upon inducing.

Type of abiotic stress	Plant species	Application of BRs	Mechanism	References
Lead (Pb) (Heavy metal toxicity)	*B. juncea* L. Czern & Coss	EBL concentrations of 0, 10^{-10}, 10^{-8}, and 10^{-6} M through presoaking seeds and on germinating paper	EBL treatment restores normal seedling growth with increased activity of antioxidant enzymes like catalase and peroxidase to overcome heavy metal toxicity	Soares et al. (2020)
Salt stress	*B. rapa* KS101 and KBS3	24-epibrassinolide (EBL) and jasmonic acid (JA)	Increased osmolyte concentration and antioxidant enzyme activity resulting in alleviation of the induced salt stress	Lone et al. (2022)
Water and saline stress	*Fragaria ananassa*	BB16 and EP24 foliar application 0.1 mg/L	Stress tolerance with higher chlorophyll content and increased productivity and dry weight	Furio et al. (2022)
Wounding	*S. tuberosum* L. cv. Atlantic	10 μM 24-epicastasterone	Increased lignin content and suberin polyphenolics inducing the synthesis of caffeic acid, sinapic acid, and cinnamyl alcohol	Han et al. (2022)
Cold stress	*Oryza sativa*	0.1 μmol/L BR solution	QTL mapping reveals that the application of hormones upregulates cold stress-related genes in rice plants treated with combined cold and BR	Guo et al. (2022)
Chilling stress	*S. lycopersicum* L. cultivar 'Roma	Silicon and BR	Increased transcription of antioxidant enzymes along with increased proline and pigment accumulation eventually leading to increment in photosynthetic activity	Bashir et al. (2023)
Drought stress	*S. quitoense* L. cv. *septentrionale*	DI-31 foliar spray 8 mL/L	Increased plant growth and photosynthetic efficiency in the treated plants and reduced lipid peroxidation of membrane	Castañeda-Murillo et al. (2022)
Heat stress	*O. sativa* L.	Ethylene in combination with BR (0.5 mM)	Heat tolerance is alleviated in plants by increased photosynthesis, defense system, and source-sink capacity.	Nazir et al. (2023)

Lead (Pb)	*Pleioblastus pygmaeus*	10 μM JA and 10 μM BR	Reduced ROS stress, higher chelating capacity, phenolics, and antioxidants with pigment concentration	Emamverdian et al. (2024)
Calcium deficiency	*B. rapa* L. ssp. *Pekinensis*	Foliar spray of BL	Increased activities of Ca2+-ATPase and H+-ATPase as well as the total ATP content leading to better transport of calcium ions from roots to shoot	Li et al. (2023)
Nano plastics	*Solanum lycopersicum* cv. *Ailsa Craig*	50 nM BRs added in the medium	Prevents the induction of aquaporin-related genes, thus preventing the accumulation of Polystyrene-Nanoparticles (PS-NPs)	Gao et al. (2023)
Low light intensity	*Lycopersicon esculentum*	0.2 mM 24-EBL	Increased metabolites production along with higher antioxidant activity in the treatment plants	Yousuf et al. (2024)
Salt stress	*Camellia sinensis* "Fuding Dabaicha" (FD) and "Chuancha 2" (CC)	0.2 mg/L BR	Increased photosynthetic activity with higher pigment concentration. With the accumulation of proline and total free amino acid content and upregulation of stress associated genes	Zhang et al. (2024d)
Salt stress	*Malus* "M9T337" and *Malus domestica* "Gala3" and *Arabidopsis*	2,4-Epicastasterone (EBL)	The MdBR6OX2 gene is highly regulated in the salt-treated plants, resulting in elevated levels of the BR-related compounds.	Zhang et al. (2024c)
Salt stress	*Capsicum annuum* L.	2,4-epibrassinolide (EBR)	Elevated sugar, glycine betaine, and proline levels in response to hormonal treatment causing ROS scavenging	Jin et al. (2024)
Drought stress	*Glycine max*, "Tiefeng 8"	Squalene (SQ) (precursor of BRs)	The stress tolerance of the soybean plant increased with the increased expression of the monoxygenase enzyme, which acts on squalene.	Yu et al. (2024)
Low temperature	*Brassica pekinensis* var. *"Huanai"*	0.05 mg/L BL	S-nitrosylation of proteins involved in NO-mediated regulation of plant stress coupled with increased ATP synthesis due to the enhanced efficiency of mitochondrial respiration.	Gao et al. (2025)

acid (GA) (Leubner-Metzger 2003). Moreover, such treatment of the seeds has also been able to overcome inhibition of germination in BR-deficient mutant in *Arabidopsis thaliana* and suppress the dormancy effect of abscisic acid (ABA) (Steber et al. 2001).

17.3.2 BRs in Context to Photosynthesis

Plants, upon receiving exogenous application of BRs, show higher photosynthetic activity as their application triggers higher transcription of photosynthesis-related genes (Xia et al. 2009). Photosynthetic efficiency of such BR-treated plants is maintained even when subjected to certain abiotic stresses like water deficiency (Pereira et al. 2019), low-temperature, calcium nitrate, and weak light, as defense-related genes are upregulated by BR. Seedlings having received treatment with BRs show higher quantum yield in Photosystem II (PSII) (Xia et al. 2009) and RuBisCo activity; moreover, the thylakoid membrane stability (Krumova et al. 2013) and the activity of chlorophyll-degrading enzymes are also maintained by BRs (Siddiqui et al. 2018).

17.3.3 Vascular Tissue Differentiation

Through plant tissue culture studies, it was established that BRs play a significant role in vascular tissue development (Oh et al. 2020; Furuya et al. 2024). Even nanomolar concentrations of BL were enough to bring about xylem differentiation, at a higher rate as compared to that brought about by auxin and cytokinin. This indicates their role in cell differentiation and division. They are involved in secondary cell wall formation and cell death; in vivo BRs are synthesized just before this transition (Furuya et al. 2024).

17.3.4 Reproductive Development

BRs led to delay in the floral transition involving the conversion of shoot apical meristem to inflorescence meristem (Li et al. 2020; Lima and Figueiredo 2024). This repression of flowering is brought by the activation of transcription of the repressor of flowering, FLOWERING LOCUS T (FLT). The transcription factors BZR1 and BES1-INTERACTING MYC-LIKEs (BIMs) bind to the first of FLC, more specifically to the BRRE element located in this region, leading to the expression of FLC. BRs also control flowering in an FLC-indirect way through repression of the autonomous flowering pathway in *Arabidopsis*. The late flowering effect of BRs has been studied through mutants like *bri1* and other synthetic mutants like *det2*, *cpd*, and *dwf*, which exhibit the early flowering phenotype. Other than floral transition BRs also regulate the development of the male floral appendages as indicated by the BR-deficient mutants. In such mutants, stigma elongation does not take place, and anther is unable to release pollen grains. Microspores so developed are ultimately degenerated. Upon activation by the EMS1 and SERK1/2 localized in the tapetal cell membrane, BZR1 and BES1 then initiate transcription of anther development genes like AG, SPOROCYTELESS (SPL), DYSFUNCTIONAL TAPETUM 1 (DYT1), and MALE STERILITY 1 (MS1) as these genes also contain the cis regulatory element BRRE which is the binding site of BES1. Hence, BRs influence male reproductive development (Ye et al. 2010).

17.3.5 Stress Mitigation

In recent decades, various abiotic stresses such as heat, heavy metal, organic pollutants, drought, and salinity stress supplemented by biotic stresses such as pathogen infection and pest infestation have emerged as the biggest threat to global food security and crop yield and production (Kopecká et al. 2023). Climate change has been one of the triggering factors contributing to abiotic stress. In such a scenario, mitigating such stresses and remediating plant health is a crucial affair. Over the years, BRs have emerged as potential agrochemicals and solutions for such problems. In this section of the chapter, we touch upon the role of BRs in alleviating plant stress.

17.3.5.1 BR-induced Thermotolerance in Plants

Heat stress mainly affects the photosynthetic machinery of the plant, resulting in reduced photosynthetic rates and carbon dioxide assimilation, which ultimately leads to decreased crop yield (Zahra et al. 2023). Other than photosynthesis, heat stress results in damage like leaf burn, abscission, aging, and death. BRs can confer tolerance against thermal stress to the plant, as proven by studies involving the exogenous application of BRs. BRs protect the plant from thermal stress by inducing antioxidants production in the plant cell's apoplastic region triggered by NADPH oxidase activity. At the molecular level, BRs induce the transcription of RESPIRATORY BURST OXIDASE HOMOLOG 1 (RBOH1) and MITOGEN ACTIVATES PROTIEN KINASE (MAPK) genes. Under high temperature, BES1 and BZR1 accumulate, which causes higher levels of PHYTOCHROME INTERACTING FACTOR (PIF4), which in turn dimerizes with BES 1, facilitating BZR1 action on gene regulation, resulting in heat tolerance. Upon transcriptomic analysis, it was found that BRs, when supplied exogenously, lead to the upregulation of defense-related genes like WRKY1. APX5. SOD, and those encoding for HEAT SHOCK PROTIENS (HSP).

17.3.5.2 BRs Versus Salinity Stress

High salt concentration in the soil salinity stress is a major issue in areas deemed as arid or semi-arid, having adverse effects on plant health, causing dehydration, osmotic imbalance, and accumulation of ROS (Hasanuzzaman et al. 2021). To cope with such stress, the plant generates and stores certain osmoprotectants like proline, which happens to be involved in maintaining the REDOX harmony and ROS detoxification and maintaining the structural integrity of polypeptides. BRs can increase the proline content in the plant, thereby alleviating the osmotic stress. It was found that plants having been silenced for BR production when supplied with the same hormone exogenously by seed pretreatment or foliar application could withstand the induced salt stress probably on account of an increase in activity of catalase, SUPEROXIDE DISMUTASE, and peroxidase. It is important to mention that BR application in salt stress conditions shows hormonal stress level-dependent biphasic effects. This actually pertains to the fact that depending upon the salt concentration, both excessive and insufficient BR will have an adverse effect on the tolerance of the plants (Mu et al. 2022).

17.3.5.3 Toxic Organic Pollutants and Pesticide Toxicity

In the present scenario, to tackle the burning problem of pest infestation in agriculture leading to lower crop production, the use of synthetic pesticides (for instance,

Imidacloprid, Chlorpyrifos, Phenanthrene, etc.) and similar preparation has emerged as a refuge (Kumar et al. 2025). Although the use of chemical fertilizers and pesticides is unavoidable, yet their unchecked usage poses a major threat to plant physiology and the ecosystem in turn. The application of BRs (like 24-EBL) in such conditions can be a plausible solution to such a problem. As proven in some experimental studies, BRs can provide plants with a tolerance mechanism against such stress through herbicide detoxification by activating the P450 cytochrome-related pathway. Once the plant comes in contact with the pesticide, the P450 monoxygenase and peroxidase activity against the absorbed toxic compound begins. Following this conjugation of glutathione and glucose with the resulting compound takes place by the activity respective transferases. The climax event of this series is the storage of the soluble pesticide molecules in the vacuoles and the apoplasts. Similarly, BRs also enable the plant to withstand and mitigate the adverse effects of organic pollutants such as polycyclic aromatic hydrocarbon (PAHs), benzene, and other persistent organic pollutants (POPs), which cause cytotoxicity, aging, reduced metabolic processes, and cell death, thereby hindering plant growth and development. BRs trigger lipid peroxidation and enhance ROS scavenging by antioxidants production in roots. Many commercially available BRs stimulate the accumulation of many secondary metabolites and flavonoids. It is well understood now that the main function of the BRs is to stimulate the expression of genes encoding Glutathione synthetase (GS), ascorbic acid (ASA), and other such compounds (Hussain et al. 2020).

17.3.5.4 BRs in Mitigating Biotic Stress

Biotic stress involving pathogens like bacteria, fungi, viruses, and actinomycetes, along with pathogens, mainly insects, are a devastating threat to plant health, resulting in loss of crop yield. Upon interaction with a Pathogen or Pest, through either Oral secretions of the insect or pathogen-associated molecular patterns (PAMPs) and their recognition by the plant's pattern recognition receptors, there occurs a triggered immune response in plants, most significantly due to signal transduction (Singh et al. 2024). BRs play an important role in regulating such defense processes. By conferring the ability of higher production of compounds like phytoalexins and upregulating the defense related genes in the plants. They also play a vital role in providing structural defenses to plants like trichrome against herbivory and chitin-chitosan interaction (Sawicka et al. 2024). Other than the direct pathway, BRs are also found to be intertwined with other plant hormones like ethylene and other molecules like H_2O_2 (Hydrogen peroxide), NO (Nitric oxide), H_2S (Hydrogen Sulfide), and sphinolipids help in alleviating biotic stress and conferring tolerance angst such stress (Hu et al. 2021). Although earlier the nature of such immunosresponse wasn't clear but recent studies with important crop plants like *S. lycopersicum* (tomato) show that overexpression of genes responsible for synthesis of BRs enhances the resistance of the plant against the pathogen *Botrytis cinerea* through suppressing Gibberellins synthesis, similar outcomes have been found through application of exogenous BR (Xia et al. 2023). Not only pathogens and pests, pretreatment of seeds with BRs armors the plant against the parasitic plants (Zhang et al. 2024b).

17.4 Advances in BR Research and Their Newly Discovered Applications

So far, we have discussed, in depth, about BRs, their properties, and their regulatory roles in plant systems. From here on in this chapter, we shift our focus to the animal system.

Recent publications, with the majority of them being based on *in silico* studies conducted, show how BRs and some of their analogs have therapeutic roles in the animal system as well, mainly in context to cancer cells. The basis of this application of BRs lies in the fact that they have similarities with human steroids. Upon carrying out competitive binding assay and reporter assay between the two steroids revealed how BRs (namely 24-EBL and 28-homoCS) could manipulate the parthenogenesis (PR) gene expression, suggesting that BRs may be interacting with the Estrogen receptors of the cell. In addition to that, evidence also throws light on the role of the two BRs in the rearrangement of these receptors on the nuclei.

BRs have an impact on the cell cycle as well, as they upregulate the inhibitors of cyclin-dependent kinases (CDKs), retarding the activity of CDK complexes which greatly affects the infinite proliferation capacity of the cancerous cells (those of breast carcinomas). The unavailability of CDKs hinders the activity of Retinoblastoma protein (Rb), causing the cell to get stuck in the G1 phase of the interphase. Consequently, this results in an induced apoptosis in these cells, as proven by various staining procedures like TUNEL (terminal deoxynucleotidyl transferase dUTP nick end labeling, propidium iodide, and acridine orange). Similar evidence was obtained in support of the anticancerous activity of other BR homologs in cases of Prostate, Colon, and Lung cancer (Lorca et al. 2023). Other than anticancerous property, BR homologs have been reported to exhibit antibacterial, antifungal, anticholesteromic, and angiogenesis action (Kaur Kohli et al. 2020). On account of their bioactivity and therapeutic potential, BRs can be exploited for biomedical applications and can pave the path for inventions in the field of health and longevity.

17.5 Concluding Remarks

Considering the discussion above, it is evident that BRs are very potent and useful agrochemicals. Ranging from their effects on normal physiological growth to reproductive development of the plant eventually leading to higher fruit and seed set, with superior traits and including their newly discovered therapeutic roles in medicinal research, these compounds hold great promise for their future applications in diverse fields. Their utilization in agricultural practices can bring about drastic changes in the current scenario and improve the net yield by many folds. Overcoming abiotic and biotic stress and their adverse manifestation in crop health can be achieved through these compounds. Through the maintenance of normal physiological conditions, even in case of unfavorable parameters and increased beneficial traits in case of normal conditions, BRs appear as a promising rescue for the suffering crop yields. Not only are these compounds used while the farming is carried out, but their applications are found even during the post-harvest period (Garrido-Auñón et al. 2024). With the increasing demands for food, there is a requirement to study the effects of their applications in combination with other hormones and compounds in greater depth. Moreover, using nanoparticles for their delivery can enhance their applications by many folds (Feng et al. 2025; Halaji et al. 2024). Nonetheless, risks of biosafety have to be kept in consideration while using such preparations, yet BRs are truly the compounds of utmost importance for the future (Faizan et al. 2024) applications.

References

Ackerman-Lavert, M. and Savaldi-Goldstein, S. (2020). Growth models from a brassinosteroid perspective. *Current Opinion in Plant Biology* 53, 90–97.

Ahmad Lone, W., Majeed, N., Yaqoob, U. et al. (2022). Exogenous brassinosteroid and jasmonic acid improve drought tolerance in *Brassica* rapa L. genotypes by modulating osmolytes, antioxidants, and the photosynthetic system. *Plant Cell Reports* 41 (3): 603–617.

Ali, J., Mukarram, M., Ojo, J. et al. (2024). Harnessing phytohormones: advancing plant growth and defence strategies for sustainable agriculture. *Physiologia Plantarum* 176 (3): e14307.

Bashir, S. and John, R. (2023). Alleviation of chilling stress by supplementation of brassinosteroid and silicon in *Solanum lycopersicum L. Plant and Soil* 486 (1): 165–181.

Bajguz, A. and Tretyn, A. (2003). The chemical characteristic and distribution of brassinosteroids in plants. *Phytochemistry* 62 (7): 1027–1046.

Brinkman, R. and Sombroek, W. G. (1996). The effects of global change on soil conditions about plant growth and food production. *Global Climate Change and Agricultural Production*, 49–63.

Castañeda-Murillo, C.C., Rojas-Ortiz, J.G., Sánchez-Reinoso, A.D. et al. (2022). Foliar brassinosteroid analogue (DI-31) sprays increase drought tolerance by improving plant growth and photosynthetic efficiency in lulo plants. *Heliyon* 8 (2): e08977.

Chakraborty, N., Ganguly, R., Sarkar, A. et al. (2025). Multifunctional role of brassinosteroids in plant growth, development, and defense. *Journal of Plant Growth Regulation* 44: 1–14.

Chen, H., Zhang, S., Chang, J. et al. (2025). Foliar application of 24-epibrassinolide enhances leaf nicotine content under low-temperature conditions during the mature stage of flue-cured tobacco by regulating cold stress tolerance. *BMC Plant Biology* 25 (1): 77.

Dar, W.D. and Laxmipathi Gowda, C.L. (2013). Declining agricultural productivity and global food security. *Journal of Crop Improvement* 27 (2): 242–254.

Emamverdian, A., Khalofah, A., Pehlivan, N. et al. (2024). The exogenous application of jasmonates and brassinosteroids alleviates lead toxicity in bamboo by altering biochemical and physiological attributes. *Environmental Science and Pollution Research* 31 (5): 7008–7026.

Faizan, M., Hussain, A., Ahmed, S.M. et al. (2024). Brassinosteroids and future of crop plants under changing environments. In: *Plant Growth Regulators: Resilience for Sustainable Agriculture*, 1–19. Singapore: Springer Nature Singapore.

Feng, H., Fan, G., Liu, Z. et al. (2025). Nanomediated stimulation: an alternative to brassinolide hormone replacement therapy for plant resistance activation. *Journal of Agricultural and Food Chemistry* 73 (10): 5767–5780.

Fujioka, S. and Sakurai, A. (1997). Brassinosteroids. *Natural Product Reports* 14 (1): 1–10.

Furio, R.N., Salazar, S.M., Mariotti-Martínez, J.A. et al. (2022). Brassinosteroid applications enhance the tolerance to abiotic stresses, production, and quality of strawberry fruits. *Horticulturae* 8 (7): 572.

Furuya, T., Ohashi-Ito, K., and Kondo, Y. (2024). Multiple roles of brassinosteroid signaling in vascular development. *Plant and Cell Physiology* 65 (10): 1601–1607.

Gao, M., Wang, Z., Jia, Z. et al. (2023). Brassinosteroids alleviate nanoplastic toxicity in edible plants by activating antioxidant defense systems and suppressing nanoplastic uptake. *Environment International* 174: 107901

Gao, X., Tie, J., Yang, X. et al. (2025). Brassinosteroid-induced S-nitrosylation of fructose-1, 6-bisphosphate aldolase increased ATP synthesis under low temperatures in mini-Chinese cabbage seedlings. *International Journal of Biological Macromolecules* 308: 142626.

Garrido-Auñón, F., Puente-Moreno, J., García-Pastor, M.E. et al. (2024). Brassinosteroids: an innovative compound family that could affect the growth, ripening, quality, and postharvest storage of fleshy fruits. *Plants* 13 (21): 3082.

Guo, Z., Wang, H., Yao, J. et al. (2022). Quantitative trait loci mapping analysis for cold tolerance under cold stress and brassinosteroid-combined cold treatment at germination and bud burst stages in rice. *Frontiers in Plant Science* 13: 938339.

Halaji, B., Haghighi, M., Kovács, G.P. et al. (2024). The role of brassinosteroids and nano-encapsulated brassinosteroids in capsicum pepper growth and physiological adaptations to high-temperature stress. *Horticulturae* 10 (10): 1062.

Han, C., Wang, L., Lyu, J. et al. (2023). Brassinosteroid signaling and molecular crosstalk with nutrients in plants. *Journal of Genetics and Genomics* 50 (8): 541–553.

Han, Y., Yang, R., Zhang, X. et al. (2022). Brassinosteroid accelerates wound healing of potato tubers by activation of reactive oxygen metabolism and phenylpropanoid metabolism. *Foods* 11 (7): 906.

Hasanuzzaman, M., Raihan, M.R.H., Masud, A.A.C. et al. (2021). Regulation of reactive oxygen species and antioxidant defense in plants under salinity. *International Journal of Molecular Sciences* 22 (17): 9326.

Hassan, M.U., Guoqin, H., Nawaz, M. et al. (2025). The role of brassinosteroids in plant physiological and molecular responses to counter salt stress and ensure food security: a review and future perspectives. *Turkish Journal of Agriculture and Forestry* 49 (1): 1–23.

Hu, D., Wei, L., and Liao, W. (2021). Brassinosteroids in plants: crosstalk with small-molecule compounds. *Biomolecules* 11 (12): 1800.

Hu, Z.L., Wei, H., Sun, L. et al. (2025). Plant steroids on the move: mechanisms of brassinosteroid export. *Trends in Biochemical Sciences.*

Hu, Z., Qi, X., Zhang, M. et al. (2016). Roles of phytohormones in morphological and anatomical responses of plants to flooding stress. In: *Plant Hormones Under Challenging Environmental Factors*, 117–132.

Hussain, M.A., Fahad, S., Sharif, R. et al. (2020). Multifunctional role of brassinosteroid and its analogues in plants. *Plant Growth Regulation* 92: 141–156.

Jain, D., Bisht, S., Parvez, A. et al. (2024). Effective biotic elicitors for augmentation of secondary metabolite production in medicinal plants. *Agriculture* 14 (6): 796.

Jin, Y., Yang, P., Li, J. et al. (2024). Brassinosteroids alleviate salt stress by enhancing sugar and glycine betaine in pepper (*Capsicum annuum L.*). *Plants* 13 (21): 3029.

Kaur Kohli, S., Bhardwaj, A., Bhardwaj, V. et al. (2020). Therapeutic potential of brassinosteroids in biomedical and clinical research. *Biomolecules* 10 (4): 572.

Kaushik, S., Ranjan, A., Sidhu, A. et al. (2025). Phytohormones and metalloids: crosstalk for plant growth and development. In: *Metalloids in Biology*, 177–195. CRC Press.

Kopecká, R., Kameniarová, M., Černý, M. et al. (2023). Abiotic stress in crop production. *International Journal of Molecular Sciences* 24 (7): 6603.

Krumova, S., Zhiponova, M., Dankov, K. et al. (2013). Brassinosteroids regulate the thylakoid membrane architecture and the photosystem II function. *Journal of Photochemistry and Photobiology B: Biology* 126: 97–104.

Kumar, A., Bhople, B.S., and Jeganathan, J. (2025). Balancing food security and environmental health: the dual challenge of pesticide use in India. *Environmental Science and Pollution Research* 32 (9): 4949–4956.

Leubner-Metzger, G. (2003). Brassinosteroids promote seed germination. In *Brassinosteroids: Bioactivity and Crop Productivity*, 119–128. Dordrecht: Springer Netherlands.

Li, L., Wonder, J., Helming, T. et al. (2024a). Brassinosteroid and gibberellin signaling are required for Tomato internode elongation in response to low red: far-red light. *bioRxiv*. https://doi.org/10.1101/2024.02.29.582690

Li, Y., Wu, Y., Tang, Z. et al. (2023). Exogenous brassinosteroid alleviates calcium deficiency-induced tip-burn by regulating calcium transport in Brassica rapa L. ssp. pekinensis. *Ecotoxicology and Environmental Safety* 251: 114534.

Li, Z. and He, Y. (2020). Roles of brassinosteroids in plant reproduction. *International Journal of Molecular Sciences* 21 (3): 872.

Li, Z., Li, Y., Geng, L. et al. (2024b). Genome-wide methylation, transcriptome, and characteristic metabolites reveal the balance between diosgenin and brassinosteroids in *dioscorea zingiberensis. Horticulture Research* 11 (4): uhae056.

Liao, Z., Chen, B., Boubakri, H. et al. (2025). The regulatory role of phytohormones in plant drought tolerance. *Planta* 261 (5): 1–22.

Lima, R.B. and Figueiredo, D.D. (2024). Sex on steroids: how brassinosteroids shape reproductive development in flowering plants. *Plant and Cell Physiology* 65 (10): 1581–1600.

Liu, Y. and Yang, W. (2025). Plant hormones: uneven mitotic division and brassinosteroid activity. *Current Biology* 35 (7): R262–R265.

Liu, Y., Pan, X., and Li, J. (2015). Current agricultural practices threaten future global food production. *Journal of Agricultural and Environmental Ethics* 28: 203–216.

Lorca, M., Cabezas, D., Araque, I. et al. (2023). Cancer and brassinosteroids: mechanisms of action, SAR and future perspectives. *Steroids* 190: 109153.

Miao, R., Li, C., Liu, Z. et al. (2024). The role of endogenous brassinosteroids in the mechanisms regulating plant reactions to various abiotic stresses. *Agronomy* 14 (2): 356.

Mirón, I.J., Linares, C., and Díaz, J. (2023). The influence of climate change on food production and food safety. *Environmental Research* 216: 114674.

Mu, D.W., Feng, N.J., Zheng, D.F. et al. (2022). Physiological mechanism of exogenous brassinolide alleviating salt stress injury in rice seedlings. *Scientific Reports* 12 (1): 20439.

Nadeem, S. M., Ahmad, M., Zahir, Z. A., & Kharal, M. A. (2016). Role of phytohormones in stress tolerance of plants. In *Plant, soil and microbes:* volume 2: *mechanisms and molecular interactions* (pp. 385–421). Cham: Springer International Publishing.

Nazir, F., Jahan, B., Kumari, S. et al. (2023). Brassinosteroid modulates ethylene synthesis and antioxidant metabolism to protect rice (Oryza sativa) against heat stress-induced inhibition of source–sink capacity and photosynthetic and growth attributes. *Journal of Plant Physiology* 289: 154096.

Nolan, T.M., Vukašinović, N., Liu, D. et al. (2020). Brassinosteroids: multidimensional regulators of plant growth, development, and stress responses. *The Plant Cell* 32 (2): 295–318.

Oh, M.H., Honey, S.H., and Tax, F.E. (2020). The control of cell expansion, cell division, and vascular development by brassinosteroids: a historical perspective. *International Journal of Molecular Sciences* 21 (5): 1743.

Pal, P., Ansari, S.A., Jalil, S.U. et al. (2023). Regulatory role of phytohormones in plant growth and development. In: *Plant Hormones in Crop Improvement*, 1–13. Academic Press.

Pei, G., Lyons, H., Li, P. et al. (2025). Transcription regulation by biomolecular condensates. *Nature Reviews Molecular Cell Biology* 26 (3): 213–236

Pereira, Y.C., Rodrigues, W.S., Lima, E.J.A. et al. (2019). Brassinosteroids increase electron transport and photosynthesis in soybean plants under water deficit. *Photosynthetica* 57 (1): 181–191.

Ristova, D., Giovannetti, M., Metesch, K. et al. (2018). Natural genetic variation shapes root system responses to phytohormones in *Arabidopsis. The Plant Journal* 96 (2): 468–481.

Sawicka, B., Barbaś, P., Bienia, B. et al. (2024). Toward enhanced plant immunity: unveiling the role of brassinosteroids in abiotic and biotic stress response.

Sedaghathoor, S., Zare, S.K.A., and Shirinpur-Valadi, A. (2024). Progesterone and steroids In/On. In: *Progesterone-Basic Concepts and Emerging New Applications*, 13. IntechOpen.

Singh, A., Yajnik, K.N., Mogilicherla, K. et al. (2024). Deciphering the role of growth regulators in enhancing plant immunity against herbivory. *Physiologia Plantarum* 176 (6): e14604.

Soares, T.F.S.N., dos Santos Dias, D.C.F., Oliveira, A.M.S. et al. (2020). Exogenous brassinosteroids increase lead stress tolerance in seed germination and seedling growth of *Brassica* juncea L. *Ecotoxicology and Environmental Safety* 193: 110296.

Steber, C.M. and McCourt, P. (2001). A role for brassinosteroids in germination in *Arabidopsis. Plant Physiology* 125 (2): 763–769.

Siddiqui, H., Hayat, S., and Bajguz, A. (2018). Regulation of photosynthesis by brassinosteroids in plants. *Acta Physiologiae Plantarum* 40: 1–15.

Warang, O., Bhattacharjee, P., Debbarma, S. et al. (2025). Brassinosteroids: unveiling their role in fruit ripening and quality. *Applied Fruit Science* 67 (2): 82.

Weyers, J.D. and Paterson, N.W. (2001). Plant hormones and the control of physiological processes. *New Phytologist* 152 (3): 375–407.

Wu, W., Chen, L., Liang, R. et al. (2025). The role of light in regulating plant growth, development and sugar metabolism: a review. *Frontiers in Plant Science* 15: 1507628

Xia, X.J., Huang, L.F., Zhou, Y.H. et al. (2009). Brassinosteroids promote photosynthesis and growth by enhancing activation of Rubisco and expression of photosynthetic genes in Cucumis sativus. *Planta* 230: 1185–1196.

Xia, X., Liu, Y., Zhang, L. et al. (2023). Overexpression of brassinosteroid synthesis gene DWARF promotes resistance to *Botrytis cinerea* by inhibiting gibberellin synthesis in *Solanum lycopersicum L. Plant Stress* 9: 100170.

Xiong, M., Yu, J., Wang, J. et al. (2022). Brassinosteroids regulate rice seed germination through the BZR1-RAmy3D transcriptional module. *Plant Physiology* 189 (1): 402–418.

Ye, Q., Zhu, W., Li, L. et al. (2010). Brassinosteroids control male fertility by regulating the expression of key genes involved in Arabidopsis anther and pollen development. *Proceedings of the National Academy of Sciences* 107 (13): 6100–6105.

Yousuf, W., Bhat, S.A., Bashir, S. et al. (2024). Brassinosteroid improves light stress tolerance in tomato (Lycopersicon esculentum) by regulating redox status, photosynthesis, and photosystem II. *Functional Plant Biology* 51: FP24170.

Yu, T.F., Hou, Z.H., Wang, H.L. et al. (2024). Soybean steroids improve crop abiotic stress tolerance and increase yield. *Plant Biotechnology Journal* 22 (8): 2333–2347.

Yuan, H.Y., Kagale, S., and Ferrie, A.M. (2024). Multifaceted roles of transcription factors during plant embryogenesis. *Frontiers in Plant Science* 14: 1322728.

Zahra, N., Hafeez, M.B., Ghaffar, A. et al. (2023). Plant photosynthesis under heat stress: effects and management. *Environmental and Experimental Botany* 206: 105178.

Zhang, H.Y., Wang, X., Wang, X.N. et al. (2024a). Brassinosteroids biosynthetic gene MdBR6OX2 regulates salt stress tolerance in both apple and *Arabidopsis*. *Plant Physiology and Biochemistry* 212: 108767.

Zhang, N., Ali, S., Huang, Q. et al. (2024b). Seed pretreatment with brassinosteroids stimulates sunflower immunity against parasitic weed (Orobanche cumana) infection. *Physiologia Plantarum* 176 (3): e14324.

Zhang, Y., Dong, G., Zhang, Y. et al. (2024c). BLA1 affects leaf angles by altering brassinosteroid biosynthesis in rice (Oryza sativa L.). *Journal of Agricultural and Food Chemistry* 72 (36): 19629–19643.

Zhang, Z., Ma, X., Tang, D. et al. (2024d). Effects of brassinosteroid on the physiological changes on two varieties of tea plants under salt stress. *International Journal of Molecular Sciences* 25 (24): 13445.

Index

a

aging process 68, 141
angiosperms 2, 36, 63, 140, 191, 268
anthocyanin biosynthesis 111
Altering membrane potential 247
apoplastic signaling molecule 197
apoplastic H_2O_2 production 240, 320
aquaporins 161, 199
autophagy 51, 69, 71, 87, 143, 156, 274, 293, 360

b

biotic stressors 1, 165, 188, 284, 312
biomass accumulation 74, 317
biological contaminants 188
bioactive hormone forms 190
bryophytes 2, 36, 191, 226, 368
brassinazole-resistant 1 174, 198, 199, 370
brassins hormone 190
BR-triggered gene transcription 173
BR-SIGNALING KINASES (BSKs) 173, 370

c

carbon compound synthesis 258
cell proliferation 3, 191, 323
Cell expansion and elongation 247
cellular stem 141
Castanea crenata 226
Catharanthus roseus 164, 226
conjugated forms 190
cellular macromolecules 71
control plant survival 69
climatic extremes 209

d

Decreased membrane damage 213, 247
differentiation of xylem 188, 225
differentiation of cells 188
development of the vascular 141
de novo synthesis 64
drought-induced 26, 55, 249, 254, 306, 362
drought-tolerant 73, 116, 254

e

eggplants 145, 209
endosperm 6, 42, 58, 283
enhanced oxidation 226
environmental elements 74
enhanced proline content 73
Enhanced pigments concentration 247
excessive water retention 188
expression of antioxidant genes 213

f

fluorescent-activated cell sorting 77
flower organogenesis 124
foliar cell plasmalemma 120
foliage 66, 170, 171
foliar treatment 41, 169
fruit formation 210
fruit size 210
fruit softening 149, 150, 155

g

gymnosperms 2, 36, 140, 368
groundnuts 4, 313
glucosinolates content 111
glycolysis 46, 65, 96, 97
glycogen 40, 85, 173, 174
glycosides 91, 93, 105, 271
glycosyl transferase 6
glycoalkaloids 37

h

heat-sensitive rice 74
heat-shock proteins 319
heavy metal ATPases (HMAs) 162
heterodimerization 173, 228, 370
histones 277
histidine 162
histone-modifying enzymes 276, 370
hypocotyl part elongation 141

i

inorganic ions 9, 69, 135, 249
inorganic phosphate 96
influx of ions 9
increased ROS synthesis 123
Iron homeostasis 133
iron deficiency 86, 87, 132, 134, 138, 328
iron-regulated transporters 162
impaired photosynthesis 162
increase lipid peroxidation 162

j

jasmonic acid biosynthesis 109
jasmonate pathway 84, 112
jasmonates 92, 95, 160, 181, 206, 217, 344, 352, 378
japonica 284
juvenile 74

k

kaempferol 93, 271, 272
kelch-like protein phosphatases 43
kinetin 19, 22, 53, 216
kiwifruit 50, 60, 112, 147, 149, 155

l

larger concentration of BRs 191
legumes 31, 109, 140, 201
lipid degradation 75
liverworts 2
limiting agricultural production 120
light energy transformation 9, 69
lycophytes 2, 36, 140, 368

m

metal-contaminated environments 74
metal accumulation in plants 75
metal transporter proteins 161
membrane fluidity 74, 197, 319
metal tolerance protein 1 (MsMTP1) 173
metabolism of osmolytes 41
microarray investigations 68
micronutrients 119, 134

n

nonfreezing temperatures 74, 319
non-photochemical quenching (NPQ) 259
nicotianamine synthase 1 (MsNAS1) 173
nutrient efficiency and resistance 69
non-salinity stress 73
nutrient uptake activity 162

o

osmotic adjustment 3, 50, 123, 214, 346
osmoprotectant compounds 73
osmoprotectants in plants 191
organic extracts 1
oxidative enzyme activity 75
organ structures 69
organic amendments 209, 285

p

pollen tube growth 3, 5, 124, 354
plant kingdoms 35, 139
pteridophytes 2, 36, 191, 226
proton pump 5, 188
phytosterol 43, 56, 164
productivity of agriculture 187
phosphates 161

phosphate transporters 161
phosphorylation cascade reaction 71
photoprotective efficacy 258

q
quenchers 91
quality of fruits 320
quantification 44, 47, 51
quercetin 91, 93, 271, 272

r
radioactive elements 75
Rosaceae family 124
Relative conductivity 317, 319
root development regulators 188
rootknot 84, 112
root structures 245, 249
rhizobium 12
rhizobacteria 176, 232, 239, 253, 254
rhizosphere 16, 69, 159, 161, 241

s
serpentine soils 123
semi-arid regions 123
sodium nitroprusside 15, 26
soil drenching 36
signal transduction regulators 190
signal transduction pathways 51, 70, 174, 224, 273, 340
stem cell maintenance 3, 354
sucrose transporter genes 74
stress circumstances 77, 191, 211
substantial crop plants 175
sulphate transporters 161
sweet cherry cultivars 39

t
tetracyclic structure 37
temperature stressors 191
transcriptional regulation 45, 48, 58, 73, 164, 354, 371
transcription levels 7, 8, 311, 338, 340
transcription regulatory 42
translatomics 77
transporters of hormones 189
thylakoid electron transport 211, 319
thermophilic plants 319

u
ubiquitin 40, 47, 214, 275, 347
uptake of heavy metals 160, 167
UV-induced damage 258, 261, 262, 263, 269
ultraviolet-B (UV-B) 257, 282
ultrastructure analysis 45
under stressful environment 64
UDP-glycosyltransferase 6
utilization of energy 321

v
vascular differentiation 7, 23, 80, 247, 335, 337, 371
vascular development 3, 29, 36, 39, 57, 150, 215, 354, 379, 381
vegetative growth 124, 208, 248
vegetable 39, 85, 243
viability 71, 194, 363

w
water-deficient soybean 41
waterlogging 298, 300, 306
water-stress 84
wheat plants 30, 216, 233, 235
wood formation 28, 39, 363
wood development 39, 47, 50, 57
wild types 67
wild plants 128
wild genotypes 128

x
xylem formation 3, 4, 113, 354
xylem development 141
xylem separation 353
xylem vessels 120, 124
xanthophylls 232, 258
xanthophyll cycle 262, 267, 280
xyloglucan endotransglycosylase 351

y

yellow stripe-like protein 1 (MsYSL1) 173
yellowing 120, 145, 147, 152, 216
yield inhibition 32, 116, 217, 222, 362

z

ZRT–IRT-like proteins 162, 173
zinc-induced facilitator 1 (MsZIF1) 173
zinc toxicity 136, 137, 181, 220
zinc oxide 173, 177, 321

Printed and bound by CPI Group (UK) Ltd, Croydon, CR0 4YY

17/09/2026

14964268-0003